Birds of Rhode Island

Birds of Rhode Island

SEASONAL DISTRIBUTION AND ECOLOGICAL HISTORY

Richard L. Ferren

Edited by Richard R. Veit

Illustrated by Margaret La Farge

Comstock Publishing Associates
an imprint of
Cornell University Press
Ithaca and London

Nuttall Ornithological Club
Cambridge, Massachusetts

Publication of this book was made possible by a generous grant
from the Nuttall Ornithological Club.

First published 2024 by Cornell University Press
Copublished 2024 with the Nuttall Ornithological Club, no. 27
in the series Memoirs of the Nuttall Ornithological Club

Printed in China

Library of Congress Cataloging-in-Publication Data
Names: Ferren, Richard L., 1939– author. | Veit, Richard R., 1957– editor.
| La Farge, Margaret, illustrator.
Title: Birds of Rhode Island : seasonal distribution and ecological history /
Richard L. Ferren, edited by Richard R. Veit, illustrated by Margaret La Farge.
Description: Ithaca [New York] : Comstock Publishing Associates, an imprint
of Cornell University Press, 2024. | Series: Memoirs of the Nuttall Ornithological Club no. 27 | Includes bibliographical references and index.
Identifiers: LCCN 2023024223 | ISBN 9781501771330 (hardcover)
Subjects: LCSH: Birds—Rhode Island. | Birds—Ecology—Rhode Island.
| Ornithology—Rhode Island—History.
Classification: LCC QL684.R4 .F47 2024 | DDC 598.09745—dc23/
eng/20230608
LC record available at https://lccn.loc.gov/2023024223

This book is dedicated to the several generations of birdwatchers who have taken interest in Rhode Island's birdlife and who have worked tirelessly to report their sightings and enhance and preserve what could be called a "Rhode Island ornithological infrastructure." This book is as much a celebration of the efforts of Rhode Island's bird people as it is a study of the life patterns and interactions among the birds themselves.

Contents

Illustrations

Editor's Preface

Richard Ferren worked on this historical ecology of Rhode Island's birds beginning in the late 1960s and continued full tilt until about 2003, at which point Nuttall Ornithological Club editors, P. A. Buckley and A. R. Keith, approached him and asked whether he would be willing to submit it for publication as a Nuttall Ornithological Memoir. As I was on Nuttall's editorial board at the time, and had a fair amount of experience with (and a fondness for) Rhode Island's landscapes and wildlife, especially its birds, I agreed to take on the project. I was impressed with the fine level of detail of documentation of the birds and also the richly detailed history of Rhode Island ornithology dating to the 1600s.

Thus began an extended period of updating the newer records from 2003 to about 2020 and incorporating some recent literature into the text. I was expertly assisted with this task by Robert Emerson, Rachel Farrell, Shai Mitra, and Christopher Raithel.

Roger Tory Peterson, in his forward to *Birds of Massachusetts* (Veit and Petersen 1993), expressed regret that Ludlow Griscom "died before witnessing the full flowering of modern-day birding that he, to a great extent, had stimulated." Griscom could not have imagined birders in the 1990s "armed with not one but several field guides, communicating with each other by CB radios (remember those?), subscribing to at least one birding magazine, and dialing state and national birding hotlines on their car phones." Peterson, let alone Griscom, would be astonished to witness how things have changed since then. The glut of electronic communication has utterly transformed the way people learn about birds and communicate their findings amongst themselves, so much so that one might reasonably ask, "What is a state bird book, and why would I want one?"

I believe this new *Birds of Rhode Island* will help answer these questions. First, and perhaps foremost, bird populations meaningfully change on a timescale much longer than the average human memory. We all remember what we saw last week and last year, but have fairly vague, or at least biased, memories of what things were like 10, 20, or 30 years ago. This book contains quantitative data going back 150 years. Of course, the intensity of effort by birders has increased continuously up to the present, but nevertheless, the changes in status of most species are fairly clear on a broad scale, especially as Rhode Island's landscape has been transformed from agricultural fields to forest and urban sprawl. Second, many birders I know often refer to field observations by birders as "unscientific" or "lacking rigor"; but data are data, and the advantage of data collected by birders is their enormous and consistent geographical scope and temporal intensity. A perusal of Cornell's online reference *Birds of the World* and its predecessor *Birds of North America* will reveal how heavily these scientific publications rely on state bird books. Data collected by birders will only become more important as the world's climate changes: species are shifting ranges now, and they will very likely continue to do so. The way they will accomplish this is by vagrancy, the very process that drives the thrill of birding. Third, the birding community has already organized itself into a continent-wide, standardized sampling scheme with Christmas Bird Counts, Breeding Bird Surveys, and MAPS (Monitoring Avian Productivity and Survivorship) surveys. The data produced by these programs are increasingly important as governments need to make decisions about preserving wild lands and protecting endangered species. The efforts by birders documented within this book have therefore advanced from a consuming pastime to an urgent scientific imperative.

As mentioned above, a perusal of Cornell's *Birds of the World* will show how widely the kinds of data presented in state bird books such as this one are used. Some of the special, perhaps unique, aspects of *Birds of Rhode Island* include a rich, 80-year history of banding and migration watching at Block Island; 70 years of seabird migration quantification at Point Judith by the author and colleagues, especially documenting the fluctuations in numbers of White-winged Scoters and Red-necked Grebes; an exceptionally detailed history of the state's seabird colonies, again by the author and colleagues; multiple surveys of the state's breeding birds; and the exceptional history of Rhode Island ornithology from the 1700s to the present.

Rhode Island has not had a state bird book since 1899, although other New England states, including Maine (2020), New Hampshire (2013), Massachusetts (1993), and Connecticut (1990), have more recent volumes. Thus, *Birds of Rhode Island* by Richard Ferren fills a substantial gap, by bringing to light the very rich history of the state's birds and the people who devoted their lives to studying them.

Richard R. Veit

Acknowledgments

Pulling together what is known about the birds of any state, province, or territory, going back nearly two centuries, is a daunting task, even for a state as small as Rhode Island. Readers can come to appreciate that Rhode Islanders have been studying birds intensively for a long time, and there is much to study. I began pulling together material for this manuscript in 1956, before I entered high school. Now that I have passed the age of 80, after having witnessed vast changes of every sort, events among bird populations remain as dynamic and fascinating as ever.

A substantial number of Rhode Islanders follow events in nature, and dozens have taken part in gathering information on the details of the state's birdlife. Many keep detailed records of birds they see and send their sightings to others who have maintained mechanisms for preserving field notes over many years. Numerous observers have responded to my requests for information about birds encountered in the field. Many cooperatively endured apparently endless questioning about their published and unpublished field records, most notably James Baird, Richard Enser, Richard Bowen, David L. Emerson, Kay Kinsey, Elise Lapham, Christopher Raithel, Dallas Wait, Robert A. Conway, Elizabeth Dickens, Maybelle Davenport, William B. Dean, Geoff Dennis, Helen Lapham, Frances Perry, and Eloise Saunders. These and many others re-examined their records at my request or pursued lost data; some made special field investigations about species that needed clarification.

In this regard, Richard Enser and Christopher Raithel, hired by the Rhode Island Department of Environmental Management to conduct the Rhode Island Natural Heritage Program (part of which evolved into work on the first Atlas of Rhode Island Breeding Birds), became investigators of nongame species within the Rhode Island Division of Fish and Wildlife (RIDFW) and collaborated with me to investigate the distribution of breeding birds within the state. Others in RIDFW, including James E. Myers, Charles C. Allin, and Lori Suprock (who provided innumerable data from division files along with many other courtesies), as well as Enser and Raithel, have collaborated extensively with me to census colonial waterbirds.

Raymond A. Paynter of the Museum of Comparative Zoology, the late Mirabelle Cormack and Marilyn Massaro of Roger Williams Park Museum, John Bull of American

Museum Natural History, Fred C. Sibley of Yale Peabody Museum, George A. Clark Jr. of the University of Connecticut, Tom Webber of Florida Museum of Natural History, and Lloyd Kiff of the Western Foundation for Vertebrate Zoology all arranged for examination of specimens in their collections and often researched their archives. Many birders have aided my efforts to create a "Rhode Island rare birds photo archive" or entrusted me with one-of-a-kind materials. I wish to acknowledge Kenneth P. Able, James Baird, Richard Bowen, Paul A. Buckley, Roland C. Clement, David L. Emerson, Norman Hill, Douglas L. Kraus, J. V. Remsen, Tom Webber, Charles Wood, and Robert E. Woodruff for reviewing parts of the manuscript. In later stages of preparation, Richard R. Veit mounted a massive effort to rearrange chapters and make many edits and suggestions as the Nuttall Ornithological Club editor. Still later, Chris Raithel, Rachel Farrell, Bob Emerson, and Shai Mitra found many important records from recent literature to update the species accounts, shared important information, or served to enhance the project.

A new book on the birds of Rhode Island was a cherished but distant goal of the late Harry S. Hathaway, whose voluminous records would become the backbone of this work. Besides Clement, Bowen, David L. Emerson, and Woodruff, many others have helped in innumerable ways, including the abstraction of field notes in various journals and diaries. Larry Benjamin single-handedly transcribed Rhode Island data from *Records of New England Birds*, and I wish to acknowledge similar work by Robert Emerson, T. Scott Gamwell, Joseph J. (Jay) Mahoney Jr., and Richard Spicer. Chris Raithel and Rick Enser retyped diaries and notes in Harry Hathaway's difficult handwriting. Charles Wood and David L. Emerson were of the greatest aid in ferreting out details of every kind. Al Hawkes provided the personal records and diaries of Harry Hathaway and Elizabeth Dickens from Audubon Society of Rhode Island archives.

In addition to combing the manuscript, a venerable Douglas L. Kraus spent many whole days and spare hours with me over many decades painstakingly reviewing more than 70 years of his own South County field data, the whole of which constitutes the largest and longest piece of individual fieldwork ever assembled on Rhode Island birds. His knowledge, and that of David L. Emerson, about the state's birdlife was indispensable. Practically every observer since the 1940s has had some part in helping this project along its sometimes erratic trajectory. Most notable in the later periods of work were Rachel Farrell, Christopher Raithel, and Robert Emerson. Richard Veit, Allan Keith, and David Larson of the Nuttall Ornithological Club were indispensable in the final stages of preparing the manuscript for publication and with arrangements for its publication as a cooperative venture between Nuttall and Cornell University Press.

Despite my many mentors, associates, and willing collaborators, errors may occur. I wish to take responsibility for any mishaps in this manuscript that has been so thoroughly aided by so many. Readers can appreciate that people have been avidly studying birds in Rhode Island for a long time. This book seeks to summarize knowledge about the state's birds but equally wishes to extol the many Rhode Island ornithologists, birders, and other naturalists who gathered the data that fill its pages.

Richard L. Ferren

Abbreviations

Abbreviations for observers are in Appendix B.

AB	*American Birds*
ABA	American Birding Association
AFN	*Audubon Field Notes*
AMNH	American Museum of Natural History
AOS	American Ornithological Society
AOU	American Ornithologists' Union
ASRI	Audubon Society of Rhode Island
Bay	Narragansett Bay
BBS	Breeding Bird Survey
BIBBS	Block Island Breeding Bird Survey
BISC	Block Island Specimen Collection
BMS	Boston Museum of Science
BOEM	*Bird Observer of Eastern Massachusetts*
BP	before present
BSNH	Boston Society of Natural History
Capt.	Captain
CBC	Christmas Bird Counts Intro; intro, species accounts
cm	centimeter
Col.	Colonel
e.g.	for example
en masse	all together/in a group
et al.	et alia/and others
fide	in good faith
FlMNH	Florida Museum of Natural History
FNRIB	*Field Notes of Rhode Island Birds*
ft	feet
GIS	geographic information system
ha	hectare
in	inch
in litt.	in correspondence
km	kilometers
kph	kilometers per hour
lat, long	latitude, longitude
LRBC	Little Rest Bird Club
Lt.	lieutenant
m	meter
m.obs.	multiple observers
M.V.	Martha's Vineyard
MAPS	Monitoring Avian Productivity and Survivorship
MAS	Massachusetts Audubon Society
MBO	Manomet Bird Observatory
MCZ	Museum of Comparative Zoology

mi	mile
mph	mile per hour
NAB	*North American Birds*
NAO	North Atlantic Oscillation
NBS	Norman Bird Sanctuary
NEBBA	Northeastern Bird Banding Association
NHBC	New Hampshire Bird Club
NOAA	National Oceanic and Atmospheric Administration
NRIORN	*Notes on Rhode Island Ornithology*
NWR	National Wildlife Refuge
PDC	Presidents' Day Counts
pers. comm.	personal communication
Rev.	Reverend
RIARC	Rhode Island Avian Rarities Committees
RIDEM	Rhode Island Department of Environmental Management
RIDFW	Rhode Island Division of Fish and Wildlife
RIFN	Rhode Island Field Notes
RINHS	Rhode Island Natural History Survey
RIOC	Rhode Island Ornithological Club
RNEB	*Records of New England Birds*
RWPM	Roger Williams Park Museum
sic	thus/text quoted verbatim
SP	State Park
TNC	The Nature Conservancy
U.S.	United States
URI	University of Rhode Island
USFWS	U.S. Fish and Wildlife Service
USNM	U.S. National Museum
v.o.	various observers
VDC	Veterans' Day Counts
WPA	Works Progress Administration
yd	yard

Birds of Rhode Island

Introduction

A modern treatment of Rhode Island's birdlife is long overdue. The last account was Howe and Sturtevant's *The Birds of Rhode Island*, published in 1899, during the first widespread wave of interest in nature that swept the Northeast in the late-nineteenth century. This was an era when the mammoth textile industry was in its heyday, when the worldwide millinery industry and local taxidermy enterprises were near their zenith, and when unlimited year-round shooting of birds had only begun to trigger the first faint signs of a protectionist movement that would gradually lead to effective conservation laws and organizations. Howe and Sturtevant wrote their little book at a time when the numbers of many wild bird species had been greatly reduced from direct exploitation and pervasive deforestation, which was leading to a largely agricultural landscape that had barely begun to sustain the heavy suburbanization so evident today. Since Howe and Sturtevant's time, and in decided contrast to popular logic or the experience of birdlife elsewhere in the world, Rhode Island's birdlife has actually increased in numbers and diversity. Many species that were extirpated during the colonial and agricultural eras reappeared phoenix-like in response to protection and reforestation. Other species have been artificially introduced or have spread naturally from elsewhere and have rocketed to unimagined abundance.

Like the avifauna and the landscape, the ways in which we perceive and study birds have also markedly evolved. Today's increasingly ecological and interdisciplinary field ornithology contrasts so greatly with the pursuits of the nineteenth-century cabinet ornithologists and oologists that their nearly forgotten activities are scarcely recognizable as the linear antecedents of modern birding. As elsewhere in the long-settled and well-worked New England region, bird study in Rhode Island has expanded its scope, especially since about 2000, to include ecology, evolution, and conservation, in addition to the detailed documentation of patterns of occurrence and distribution. Members of today's thriving and diverse community, comprising biologists, conservationists, and birders of all kinds, are the direct descendants of a now very unfamiliar past. One aim of this book is to reacquaint today's observers with the peep-shooters, decoy-makers, egg and nest collectors, skin accumulators, taxidermists, and outraged protectionists who built the foundations of Rhode Island ornithology.

As early as the 1940s, Roland C. Clement, Richard Bowen, David Emerson, and other members of the newly formed Rhode Island Ornithological Club (RIOC) felt the need for a new summary of the state's birds and undertook preliminary work with that goal in mind even though circumstances eventually prevented a fuller commitment to such a project. By 1956, Clement was responsible for stimulating the author's interest in developing a new annotated checklist of the birds of Rhode Island, and as the youngest member of RIOC, he unabashedly began work with Robert E. Woodruff on what then seemed to be a relatively modest undertaking. Woodruff worked for several years in abstracting literature and locating references but was ultimately forced to discontinue his efforts because of other necessary work and a move to Martha's Vineyard. After the early 1960s, the author assumed sole responsibility for the project, but it is unlikely that he could have completed the work in its present form without willing cooperation and substantial help from numerous others.

This book describes some of what is known about birds in Rhode Island: when and where they are normally present or absent in the state and in the course of a year, when and why they migrate, and how their numbers vary. It will also outline what is known of each species' history in the state and consider factors that may have contributed to many examples of dramatic changes in status. We can never know the history of our birdlife as well as we know our own history, but a great deal can be inferred from what we know of our continent's landscapes and habitats and of human-caused changes that affect birds. Today, changes in bird populations are occurring as air and seawater absorb combustion gases and heat up because of the use of fossil fuels, while the human population continuously grows and converts wild lands into urban and suburban habitats. The study of avian ecology leads to the affirmation that in nature, little remains the same for very long and that the only unchanging aspect is the prospect for future changes.

History of Rhode Island Ornithology

Tracing the story of birds and bird study in Rhode Island (see Map 1) reveals as much human history and sociology as ornithology. Native American (Niantic) shell heaps at Block Island have yielded bones of Great Auks and Sandhill Cranes among those of many other species still common there. From the 1500s, when the Dutch were trading with the Narragansetts at Dutch Island, to the mid-eighteenth century, there are only casual remarks on birds in Rhode Island. Wild Turkeys, geese, Ospreys, and Passenger Pigeons were mentioned by Roger Williams after his founding of Providence in 1636, but most species lacked documentation. Little was written about birds even up to the time of the Civil War. Nesting Ospreys were described by an anonymous British seaman on Prudence Island in 1748, when the island's vegetation was largely eradicated by numerous large farms (Calder 1935). A more extensive discussion of Rhode Island birdlife, including 79 species, came from Sir Charles Blagden, a British naval officer stationed on a warship in British-occupied Newport during the American Revolution (Howe 1905). Blagden was able to leave his military duties in Newport to visit the fields of lower Middletown and the area of Sachuest Point to comment on such birds as Bobolinks, cormorants, and Least Terns. His specimens of birds and other fauna, shipped to England in barrels of rum, were probably the first ever collected for scientific purposes in Rhode Island and were among the very first from the colonies. More replete with birds is the diary of "Nailer Tom" Hazard, a Quaker inhabitant of Kingston village, who mentions 40 or more species (using period colloquial names) shot near there during the period 1781–1804 (Nichols 1934).

In contrast to Great Britain, where an obsession with natural history surged during the period 1820–1870, the reporting of birds and activities centered around them remained sporadic in New England until the last decades of the nineteenth century. During the 1860s and 1870s, popular interest in field ornithology increased, perhaps stimulated in part by the publication of Charles Darwin's *On the Origin of Species* (1859), a postwar uptick in local attention to shorebird and wildfowl hunting, and the installation of Spencer Fullerton Baird and his immediate successors at the Smithsonian Institution in Washington, DC. These men stimulated a dynamic period in American zoology and western exploration (Baird et al. 1860, 1874, 1884; Bendire 1895). More locally, Massachusetts ornithologists William Brewster and Henry Henshaw formed the Nuttall Ornithological Club in Cambridge during this period, and Brewster conducted extensive surveys of Massachusetts birdlife, circa 1865–1910.

Locally, although Louis Agassiz collected or obtained a few nineteenth-century bird specimens in Rhode Island and Professor John Whipple Potter Jenks (as distinct from Frederick T. Jencks, see below) assembled a period collection of mounted birds at Brown University, the majority of local ornithologists in this early era were not academics but rather sportsmen-naturalists or gentlemen-sportsmen, often with old-family connections, plentiful leisure time, and substantial means to afford the time-consuming hobby of collecting mounts, skins, and eggs. Taxidermy was both an art form and a profitable business during that era. Although some sportsmen mounted their own kills, most relied on professional taxidermy firms in Newport, Providence, and Pawtucket. These establishments not only mounted specimens for local gunners and displayed their work as window dressing, but some firms distributed catalogs with specimens of everything from insects and shells to large mounted mammals. These men and their taxidermists all handled birds directly in the flesh, creating a fortunate by-product in the many records of rare and unusual birds that are substantiated by specimens. Thus, in Providence and Newport, a nucleus of gentlemen hunters (there were no women in this group) became interested in the identification and natural history of the birds they shot, began to accumulate collections of mounted specimens and egg clutches, and acquired an increasing intellectual interest in their quarry.

Among the earliest and most notable of Rhode Island's collectors was S. Newton

Dexter. Starting out as a privileged youth with a gun in the Providence–Cranston area during the 1850s, he became "the first local boy to stuff his own birds," according to longtime Providence taxidermist and field ornithologist F. T. Jencks (in litt. to Harry S. Hathaway). As an adult, Dexter was a bachelor and "something of a dandy" according to Jencks, as well as Rhode Island's quintessential example of the sportsman-naturalist and maker and user of waterfowl and shorebird decoys. He published only a few notes in *Forest and Stream* (Dexter 1879,1887, 1888, 1889) and a single note in *Notes on Rhode Island Ornithology* (Dexter 1900). No personal records survive, but in the period after 1900 he remained well-known through his surviving specimens and from numerous citations of his discoveries in Howe and Sturtevant (1899). Usually alone in the field, he was nevertheless in constant communication with younger field men, who regarded him as the local authority on Rhode Island birds.

Dexter visited and often lived much of the year at his friend Dr. Clarence Tripp Gardner's house east of Round Pond, Sakonnet. A decided gentleman of leisure in that hard-bitten age of immigrants and industrialization, Dexter constantly sought shorebirds and other species with his fowling piece in the general vicinity of Sakonnet Point and lower Little Compton during the last few decades of the nineteenth century. Many of his mounted specimens, which he donated freely to the Franklin Society, are extant (long present at the Roger Williams Park Museum, RWPM), including Long-billed Curlew and Ruff from Sakonnet. (The RWPM is now officially known as City of Providence Museum of Natural History and Planetarium, but it is referred to throughout this work as RWPM.) A well-mounted 1859 Black Guillemot from Mt. Hope Bay also survives, but Dexter's most charismatic specimen is the 1857 Corn Crake from Cranston. While much of his later field experience was in lower Little Compton, 1860–1900, he had a strong familiarity with wildfowl on Narragansett Bay and shot notable birds along the western Rhode Island coast as far afield as Narragansett and Westerly.

Based on 67 citations in Howe and Sturtevant (1899), most of the species Dexter saw or shot were sizable aquatic birds. A few notes refer to Passenger Pigeons and raptors, but almost no notes (and no specimens) are of passerines. Even so, Howe and Sturtevant cite Dexter's records or quote his impressions of Rhode Island birdlife more often than any other field observer of their era. Perhaps half of what is known about Rhode Island's nineteenth-century coastal birdlife was gathered by Dexter. He died of "apoplexy" on 27 July 1901, having been continually in the field for about 50 years. In addition to collecting many rarities, Dexter is better known outside Rhode Island as the carver and co-owner of the only extant shooting-era shorebird decoys known to have emanated from Rhode Island. These were part of a "gunning rig" used by him and Clarence Tripp Gardner of Providence and Sakonnet, some of which fetched US$95,000 in a 1980s auction at Sotheby's. Of course, Dring, the Powels, LeRoy King, Hathaway, and probably many other nineteenth-century gunners possessed homemade shorebird and waterfowl decoys, but where these now reside is a mystery.

Another notable early ornithologist and sportsman with a place in Rhode Island's ornithological history was George H. Mackay of Massachusetts. Mackay (pronounced MACK-ee) visited the Sakonnet islands nearly annually in April or early May, 1869–1890, to shoot scoters and other waterfowl. His accommodations there were at the once-opulent West Island Club, a shooting and fishing lodge used by a tiny membership of the power elite of a gilded age, where visiting U.S. presidents sometimes received artillery salutes from passing warships. The lonely ruins of the West Island Club still stand on West Island, where their granite ramparts now provide nest sites for gulls and cormorants (Ferren and Myers 1998). A detailed record of Mackay's Sakonnet kills and observations may be found in *The Auk* (1891) and in his shooting journal (see Phillips 1929). Some of his specimens found their way into the hands of Massachusetts ornithologist William Brewster, along with other Rhode Island rarities shot by others during the same period. Brewster, however, despite having (through his own fieldwork) advanced the ornithology of Massachusetts (Cambridge and the Concord and Sudbury River valleys) and Maine (Lake Umbagog) to a degree possibly unequaled by any modern New England observer, visited Rhode Island only rarely. He accompanied Mackay on one of his scoter-shooting expeditions to West Island, 16–21 April 1890, which explains Brewster's labels on a few scoter skins in the Museum of Comparative Zoology (MCZ). To our knowledge he visited Providence only once (see Least Bittern account).

Another of the state's noteworthy early bird students and collectors was LeRoy King of Newport, who often boated to the Sakonnet area and Cormorant Rock to shoot shorebirds and other wildfowl at these sites and on the extensive marshes (once the second-best shorebird habitat in the state) that formerly existed behind Second and Third Beaches in Middletown. These marshes were later diked and flooded to create a Newport reservoir called Gardiners Pond. King contributed many records to the turn-of-the-century literature (e.g., King 1900, 1901a, b, 1902a–e, 1903, 1904) and assembled a modest mounted collection, mostly of shorebirds (now displayed at the Norman Bird Sanctuary [NBS]). King was the last survivor among Rhode Island's vintage sportsmen-naturalists and continued to frequent the Newport Reading Room until his death in the late 1950s.

Newport contemporaries of King's included Dr. Edgar A. Mearns and his son, Louis di Zerega Mearns, who for 2 years were stationed at Fort Adams. They are known in Rhode Island mainly from their notes and lists in *Notes on Rhode Island Ornithology* (E. Mearns 1901; L. d. Mearns 1900a–c, 1901, 1902). Edgar Mearns was an army surgeon (as were many of the early ornithologists of his time) and an internationally known taxonomist and collector (Hume 1942). The Mearnses produced an annotated list of the birds of Aquidneck Island and a list of Prudence Island birds. Other contributors to the literature of the times included Lt. Wirt Robinson of Newport and Owen Durfee of Fall River.

Oology, the collection and study of birds' eggs, was popular in nineteenth and early twentieth century Rhode Island, where notable collectors included Walter Angell, Harry A. Cash, William Deardon, Charles E. Doe, Thomas L. and John H. Flanagan,

Harry S. Hathaway (of particular importance in transmitting records of this era to later generations), Rowland G. Hazard II, John Snowdon Howland, Frederick T. Jencks, B. LaFarge, F. E. Newbury, Charles H. Remington, James M. Southwick, James M. Stainton, and others. The most voracious, acquisitive, and indefatigable among them was unquestionably Charles E. Doe, although the Rhode Island portion of his collection is nearly rivaled by the Flanagan collection. Not only a local figure, Doe is nationally recognized today among the most industrious American egg collectors of all time (Lloyd Kiff, Western Foundation for Vertebrate Zoology, pers. comm.). Remarkable in the extreme, however, is the fact that his Rhode Island fieldwork—and even the existence of his collections and decades of diligently handwritten field notebooks—were almost unknown to Rhode Island ornithology until 1992.

Doe was born in Providence in 1864 and active in the field in Rhode Island and nearby Massachusetts between 1879 and 1930, but only a few scraps of information existed in the local literature to suggest that he ever roamed the Rhode Island countryside. In Hathaway's diaries, he appears sparingly, mainly during his relative youth in the 1880s and 1890s. For the last three-quarters of his egging career, Doe rarely interacted with other Rhode Island bird students and contributed nothing directly to appropriate publications of the time. In retirement, and with little in the way of fanfare or farewell, he sold his South Main Street coal business in 1931 and personally trucked his egg and skin collections to the University of Florida in Gainesville. He was still visiting the nests of Ospreys and Bald Eagles in Florida in his late seventies, but as far as is known, never revisited his native Rhode Island nor communicated with any Rhode Island naturalists. None of his finds or fieldwork is directly mentioned in Charles H. Remington and the Commissioners of Birds' *A Checklist of Rhode Island Nesting Birds* (1908), and he was known only indirectly by Howe and Sturtevant (1899). Even Doe's most unique discoveries rarely came to Hathaway's attention (see Hathaway 1913, 1943). He continued curating his collection until his death from tuberculosis in 1946 at nearly 82 years of age.

That Doe collected more than 10,000 egg clutches, 1300 skins and mounts, and 1000 nests and produced 50 volumes of fastidious notes became known to Rhode Islanders only in 1992 with the discovery, intact, of Doe's turn-of-the-twentieth-century specimens and voluminous handwritten notes at the University of Florida, Gainesville. Additional material found its way through Doe's trading to two museums in California: the Santa Barbara Museum of Natural History and the Western Foundation of Vertebrate Zoology. Doe's overall egg collection has recently been rated as twentieth in size and diversity among one-man North American collections (Lloyd Kiff, Western Foundation for Vertebrate Zoology, pers. comm. in 1980s.). The Rhode Island portion, representing one of the state's most sizable troves of specific field data, includes more than 2000 egg clutches and about 500 skins and mounts. The contrast between the sheer bulk of his work and his essential anonymity in Rhode Island constituted the greatest surprise in the author's efforts to grasp the lives and contributions of historical Rhode Island ornithologists.

Far better known than "Charlie" Doe to other collectors and sportsmen were the brothers John H. and Thomas L. Flanagan, a lawyer and jeweler respectively, who both lived on Providence's East Side. John Flanagan was perhaps the more active oologist of the two. Born in 1868, he was a longtime governor-appointed member of the first Rhode Island Bird Commission, the board that appointed deputies to enforce game laws. This organization gradually evolved into the Rhode Island Department of Environmental Management (RIDEM), including the Rhode Island Division of Fish and Wildlife (RIDFW), together with state organizations for conservation officers and environmental police. John Flanagan was chairman of the Bird Commission from 1905 to 1908 (after which he declined reappointment) but had a much longer membership on the commission. He developed a good knowledge of local wildlife conservation problems and enforcement as a lawyer and commissioner and played an important role in formulating Rhode Island's earliest game laws. He was a member of the American Ornithologists' Union (AOU), and, with his brother, had a substantial library of ornithological literature along with their joint egg collection.

Less is now known about jeweler Thomas Flanagan, who survived John and inherited their combined library and egg collection. Both contributed to what is clearly the second-largest collection of Rhode Island bird eggs. Many of the clutches are from west and north of Providence (Johnston, North Providence, Smithfield), which at the turn of the twentieth century were largely unspoiled rural areas containing good early successional landbird habitats, hosting territorial Yellow-breasted Chats among more common species. Their egg records are frequently referenced in *A Checklist of Rhode Island Nesting Birds* (Remington and Commissioners 1908), doubtless because Charles Remington and most of the other authors and contributors happened to be members of the Bird Commission, and, curiously enough, most were also oologists. Contemporary readers will have difficulty rationalizing the commissioners regulating the salvation of parts of the Rhode Island avifauna while conducting a still legal but soon-to-be banned collection of the nests and eggs of local birdlife, including Ospreys. The Flanagan brothers were in the field from the mid-1880s onward. After their deaths, their egg collection vanished from sight for at least 75 years, but in 1994 it came to light intact in the Providence area and was shown by Ferren to a number of RIOC and Audubon Society of Rhode Island (ASRI) and Rhode Island Natural History Survey (RINHS) members.

Acquiring specimens, nests, and eggs of birds was only part of an energetic awakening in the American public to the appreciation of the natural world. During this period, the Franklin Society of Providence (1823 through 1910), the Newport Natural History Society (after 1883), and the Lonsdale Botanical and Field Naturalist Society (ca. 1880–1910) arose from this emerging interest in nature. Quite distinct from the collectors and taxidermists described above, Clarence M. Arnold was an early local example of a bird-watcher. A Woonsocket gentleman who studied birds between 1896 and 1938 around that textile city, Arnold submitted Christmas Bird Counts (CBCs) in the first decades of their existence and kept diaries of arrivals and departures of landbirds near his home. Apart from the egg collecting efforts of Doe, Sprague, and Reynolds, Arnold's diaries represent essentially unique coverage of any part of northern Rhode Island prior to the late 1940s.

In that era, resources for birders were few by modern standards. Nuttall's "Manual" (1832), Chapman's "Handbook" (1895, 1912), Hoffmann's "Guide" (1904), and Chester Reed's illustrated *Bird Guide* (1910) were among the few references to the identification, geographic ranges, habits, and behaviors of birds. Only the most rudimentary optical instruments (2× opera glasses; later 4×) were commonly available until World War I, and German and French prism binoculars were not widespread or affordable until the mid to late 1920s. Zeiss telescopes similar to Griscom's famous 1930s model were never used in Rhode Island, nor did any early Rhode Island birder possess a "pull telescope" like an early model used by Massachusetts's Sam Eliot (Douglas L. Kraus, pers. comm.). Douglas Kraus and his brother Phillip were probably the first Rhode Island birders to use a prismatic telescope, an uncoated 20 × 50 mm Bausch and Lomb (without a tripod) that they shared beginning in the late 1920s.

During the period 1880–1920, all but the very oldest of Rhode Island's vintage naturalists (those who remembered back into the 1870s or earlier) seemed to take for granted the depressed state of many bird populations, and there is little in their writings to suggest any awareness that this scarcity was largely artificial or that improvements in conservation law might ease pressure on everything from ducks and shorebirds to hawks and owls. Haltingly at first, the federal Migratory Bird Treaty (1916) and Migratory Bird Act (1918) limited the acquisition of wild birds for commercial purposes. Although partially ignored at first, both local laws and international treaties governing the taking of birds were increasingly enforced and obeyed after about 1920, allowing bird populations to begin a slow recovery. Many western waterfowl and Arctic shorebirds regained healthy (and sometimes remarkably high) populations by 1940 or earlier, but ground was still being reclaimed by such former game birds of the Atlantic coast as Willets and American Oystercatchers a full century after the beginnings of legislative protection. The return of many once-rare larger species to our area has been as much a response to freedom from shooting as to reforestation or climate change: Great Blue Herons from the north, Turkey Vultures from the south, and Sandhill Cranes—extirpated since colonial times—still expanding into New England as migrants and breeders in the early twenty-first century.

Locally, ASRI supported laws that created the Rhode Island Bird Commission in 1899 (Grant 1899; *Bird-Lore* 2:164). The purpose of the five-member Bird Commission, appointed by Rhode Island's governor, was to relieve the intolerable pressure on larger birds from unlimited and indiscriminate shooting for market, pot, and sport. From 1929 on, as protection became more accepted and effective, the commission began to redefine its role as the state agency for management of bird hunting and propagation, while interest in oology faded away with the gradual resignation of its early commission members and the onset of new federal regulations. Gradually evolving, the commission soon came to control fisheries and the hunting of mammals along with the taking of birds. Needing a more realistic name commensurate with its expanded duties, the Rhode Island Bird Commission became the Department of Fish and Game in 1935, and since the 1950s has been known most of the time as the Division of Fish and Wildlife within the Department of Environmental Management.

Beginning in 1896, the RWPM in Providence took the lead in natural history education and among other activities began to assume importance as a repository for noteworthy specimens of Rhode Island birds. The nucleus was the collection assembled by Charles H. Smith and J. M. Southwick. The collection was soon given to the city by Smith in 1899. Until about 1930, the "Natural History Museum," under the direction of Southwick, was important as a center for nature study. During this period the museum maintained a close contact with the newly formed ASRI, which provided the money for the museum to acquire major portions of its ornithological holdings, such as the collection of Manly Hardy of Brewer, Maine.

Bird study, including lectures, field trips, and bird courses, continued at the museum through the 1920s, but a long decline began at least by 1930 and became particularly acute circa 1940–1990. The decline contrasted with events in Boston, where (with no small contribution by Ludlow Griscom and excellent community financial backing) the old Boston Society of Natural History's museum made a phoenix-like reappearance in the form of the Boston Museum of Science. Roger Williams Park Museum's new curator, Marilyn Massaro, and an enlightened staff reversed the museum's decline beginning in 1990. They brought an influx of lively inspiration, major grants to purchase storage facilities, and a complete revamping of displays and collections, thereby allowing the museum to begin to re-assume part of the state's natural history leadership while continuing its interests in other fields such as ethnology. The bird collection (largely mounted specimens) was entirely conserved and recataloged in the late 1990s.

Following the legal protection of birds and the passage of the old-time collectors, Rhode Island bird study might have faltered but for the work of Harry S. Hathaway, destined to be Rhode Island's leading ornithologist for many decades. Born in Providence in 1869 and living all his life in South Auburn (a part of Cranston), his interest in birds began at the age of 13. In his high school years, he hunted birds with a gun and collected eggs in Roger Williams Park and in the countryside nearby, especially around then-unspoiled Blackmore Pond in Cranston. His later fieldwork centered around favorite haunts in Cranston and areas along the western South Shore. In 1895 he bought a cottage on the dunes at Quonochontaug and began spending his summer vacations there, shooting shorebirds over decoys on the flats and exploring the birdlife of the region at leisure. After retirement in 1924, he spent the warmer months at his cottage at "Quony," returning each fall to his home in South Auburn.

At the publication of Howe and Sturtevant's *The Birds of Rhode Island* in 1899, Hathaway was already 30 years of age and ensconced in a banking career. He was a frequent contributor to *Notes on Rhode Island Ornithology* (Hathaway 1900a–e, g–j; 1901a–e;

1902), and his most important contribution to the published literature was "Notes on the occurrence and nesting of certain birds in Rhode Island" (Hathaway 1913), an important milestone and the first general statement on the state's avifauna since Howe and Sturtevant (1899) and the "Supplement" (Howe 1903). See also an article by Hathaway entitled "American Osprey" (Hathaway 1905).

In many ways Hathaway was the counterpart of his contemporaries: Edward Howe Forbush in Massachusetts, Witmer Stone at Cape May, and Arthur Norton in Maine. Like them, Hathaway collected the records of local Rhode Island gunners and other bird students throughout most of his life and carefully placed the data on cards, the first extensive collection of field data for the state. In 1942, near the end of Hathaway's life, his card file was laboriously typed by Harold N. Gibbs and bound into a large volume entitled *Bird Notes*, of which there are only five known copies. This work also contains invaluable annotations, made by Hathaway late in life, which are lacking in his card file. In addition, *Bird Notes* contains personal accounts about Hathaway's later contemporaries and particulars of their fieldwork. The survival of these records is remarkable because during the hurricane of 21 September 1938, the desk containing the cards and a lifetime of Hathaway's personal notes was swept from the cottage by still-unmatched winds and tidal surge. Instead of floating out to sea, the desk floated with the onshore winds nearly a mile northward across the flooded salt marsh to lodge (with cards and papers intact) on Bills Island in Quonochontaug Pond. Miraculously, the elderly Hathaway was able to find the ruined desk 3 days later and extract his personal notes and the accumulated records of others from its drawers.

Hathaway continued to shoot occasional rarities (sometimes discreetly from the yards of dwellings) with a .410 caliber cane gun. Hathaway's library, as well as his egg, skin, and photographic collections, and personal notes, are now held by ASRI in their Hathaway Library.

Hathaway was the quintessential local ornithologist of his time and had become a local legend well before he died in 1946. Like the tiny remnant of surviving buggy men and ox drivers who in the late 1960s would watch men step onto the moon, Hathaway provided a living bridge between the shotgun era of bird identification and the Peterson era of field identification. His activities were characterized by the careful and methodical approach of a now-extinct generation, particularly an emphasis on a kind of legwork and swamp-trotting now rare among birders. His abilities to find nests and prepare skins have been approached by very few recent observers. With his death and that of Charles E. Doe (both in 1946), along with Walter Angell's death in 1950 and LeRoy King's in the late 1950s, closure came to a long and eventful era of cabinet ornithology and early avifaunal exploration (see Hathaway's autobiographical notes in *Bird-Notes* [1943]; also Clement 1946 and *Bulletin ASRI*, April 1946).

A well-known and much-beloved figure among Rhode Island's early conservationists and bird students was Elizabeth Dickens of Block Island. She was a later contemporary of Harry Hathaway and corresponded for 30 years with him as well as with Arthur

Cleveland Bent, Forbush, and Clement. From 1912 until her death in 1963, she kept daily records of her observations of birds on the isolated southwest moors of Block Island, where she lived all her life on an ancestral and nearly treeless farm near Dickens Point. Her densely packed diaries of bird observations, now owned by ASRI, are certainly among the longest ornithological chronicles in North America. For over half a century, Block Island residents and visitors brought to her a series of avian casualties, including a Magnificent Frigatebird, Northern Lapwing, Ruddy Shelduck, two Bridled Terns, Arctic Tern, and many others. Most of these specimens are on exhibit today at the Block Island School. As the first paid Audubon teacher in America, Elizabeth Dickens was hired in 1914 to teach Block Island schoolchildren to identify and appreciate the many birds of their island, and several generations of island residents remember her contributions, including decades of CBCs. Her more important bird observations appear in Hathaway's *Bird-Notes*. For Dickens's biographical information, see Gleason (1988) and Whitman (1982). Eleven volumes of daily bird observations, written in a fine hand over the course of 51 years, have been carefully examined by the author.

Several other observers of the prewar period are notable. Alice Hall Walter was an indefatigable conservationist of the era from 1906 to 1940 and a major pillar in the early affairs of ASRI. For many years, she also wrote a weekly column in *The Providence Journal* entitled "Birdlife in Rhode Island," while her husband (and Brown University professor) Dr. Herbert E. Walter served as president of ASRI. As the voluntary, long-serving chair of ASRI's Education Committee, Alice Walter arranged for field trips through the 1920s and 1930s to such outlying areas as Roger Williams Park, Fields Point, and elsewhere, often led by such personalities as Albert A. Barden Jr., Phillip and Douglas Kraus, Carlos Wright, and Roland C. Clement. According to Clement (in Weber 1997), she essentially single-handedly kept ASRI from sinking into oblivion during a long period of low membership that spanned the period between the winning of the battles to regulate the unlimited shooting of birds (1916–1918) and the earliest beginnings of the modern birding period after the mid-1930s. Rhode Island may never again know a more tireless amateur conservationist.

Two contemporaries of Alice Hall Walter and Elizabeth Dickens were Mary and Everett Southwick, who came to the Kimball Bird Sanctuary in 1926, 2 years after its acquisition by ASRI. Mary Southwick kept daily records of birds at the sanctuary and environs from 1930 to 1942 and for a few years organized a "Charlestown" Christmas count. After her death in 1942, the incredibly self-sufficient Everett Southwick continued at the sanctuary alone for 33 years until shortly before his death in 1976, a friendly and welcoming figure to several generations of Rhode Island birders. For a number of prewar decades, and through the 1950s, "Kimball" was a standard destination for field-trip groups and a variety of Providence-area birders with sufficient mobility to get that far afield.

The modern era of field ornithology in Rhode Island began during the 1930s. Through the influences of Harry Hathaway's fieldwork, Alice Walter's field trips, the presence of the Southwicks at Kimball, and that of Elizabeth Dickens at Block Island, and with the help of Peterson's new *A Field Guide to the Birds*, bird-watching as we would recognize it began to grow. The emphasis shifted from collections to the sight identification of birds, the discovery of rarities, and the refinement of knowledge about their identification, life histories, periods of occurrence, and threats to their existence. The birding activities of Phillip and Douglas Kraus mark the threshold of this shift, starting as early as 1924. The Kraus brothers led field trips near Providence for ASRI's Education Committee, trips that were instrumental in encouraging both the leaders as well as such followers as Charles Wood and Carlos Wright to get involved in a lifetime of birding. Another of Alice Walter's early trip leaders, Roland C. Clement of Fall River, became a highly insightful and independent birder and took a keen interest in the avifauna of the Tiverton, Little Compton, and Westport areas in 1935. He soon began to correspond with an aging Harry Hathaway and for a time became a central understudy to Hathaway in the pre–World War II era. Except for a hiatus during wartime service in Labrador, Clement took a leading role in encouraging the interests of several younger observers, such as Richard Bowen and David L. Emerson, and was the central figure in encouraging the formation of the Rhode Island Ornithological Club (RIOC) in 1939. Later, Clement would serve as executive director of ASRI for much of the 1950s. Besides Clement, charter or early members of RIOC included Bowen, Emerson, Henry E. Childs Sr., William B. Dean, William H. Drury Jr., Harold N. Gibbs, Douglas L. Kraus, Alvah W. Sanborn, Charles Wood, and Carlos G. Wright. Patterned along the lines of the Nuttall Ornithological Club of Cambridge, Massachusetts, and entirely confined to men until the 1970s, RIOC quickly became a focus for study of the local avifauna. Practically every active male birder in Rhode Island was a member of RIOC until the mid-1980s, and many made noteworthy contributions to knowledge of the local avifauna.

Beginning in the 1970s the work of ornithologists based at academic institutions has been more prominent. Frank Golet, William Eddleman, Peter Paton, and students at the University of Rhode Island (URI) have researched waterbirds, breeding landbirds, and Neotropical migrants. Paton has studied seabirds in offshore waters. Studies of the food habits of landbird migrants at Block Island were undertaken by students under Douglass Morse of Brown University. Jeffrey Parrish showed that ordinarily insectivorous birds such as vireos and warblers shift to frugivory during migratory stops in coastal thickets. Graduate students from other universities have worked on birds in Rhode Island; see especially a banding and life history study of Yellow Warblers in the Great Swamp by Shaibal Mitra (1996, 1999).

Rhode Island Habitats

Rhode Island is in a transition zone where northern and southern avifaunal elements mingle and also is near the eastern limits of some western and midwestern elements. Most of the extremes in these north–south and east–west continua are represented by special habitats inhabited by specialized birds: Carolina Wrens in maritime forest, Hooded and Prothonotary Warblers inland in wooded swamps, and Golden-crowned Kinglets in planted spruce groves. (For scientific names of plants, see Appendix A; for maps of Rhode Island geographic features referenced in this section, see Maps 1–7.)

Some bird species, however, approach their latitudinal limits in habitats that are associated with the opposite end of the north–south continuum. Northern Saw-whet Owls, White-throated Sparrows, Northern Waterthrushes, and Black-throated Green and Canada Warblers (all northern birds) are attracted to southern Rhode Island's Atlantic white cedar groves and sphagnum swamps (Taber 1951). Nashville Warblers and Hermit Thrushes, near their southern limits as breeders, find a local niche in central–southern Rhode Island's pitch pine and scrub-oak habitats. Southern and northern species such as Acadian Flycatchers, Blackburnian Warblers, and Northern Goshawks converge in hemlock ravines. Much of northern and western Rhode Island lies between 200 and 750 ft (61 and 229 m) of elevation (see Map 3), whereas southeastern and coastal sections are largely well below 300 ft (91 m). The acidic tops of most interior hills are frequently dry and rather sterile, with oak and pine forest and huckleberry ground cover, unattractive to most northern breeders except perhaps Hermit Thrush.

Rhode Island lies north of the southern terminus of the Wisconsin glacial lobe, and the present surficial geology of the state is everywhere dominated by poorly sorted post-glacial debris (Quinn 1976). As elsewhere in the Northeast, a potentially more jagged landscape was rendered into flattened, rounded, or fluted landforms, and bedrock from the interior hills to offshore islands was decapitated and polished by rock fragments that were carried in the lower portions of the ice sheet. In all but a few outwash deposits, the soil is littered with rocks. Ubiquitous stone walls and vast deposits of sand and gravel remain as evidence of ice sheets in the Pleistocene that are thought to have exceeded a half-mile to a mile (0.8–1.6 km) in thickness. At the maximum extent of glaciation, sea level dropped over 350 ft (107 m), exposing the continental shelf well to the south of Block Island's present position. Early in its melting process, the ice sheet probably left a narrow vegetated plain available at times to hardy plants and land animals before it was covered by rising seas.

The terminal moraines associated with the Wisconsin glaciations (25,000–18,000 BP) are very conspicuous in Rhode Island. An early glacial advance reached central Long Island and formed the South Fork of that island, including Montauk Point, as well as Block Island, Martha's Vineyard, and Nantucket (Lynch 2019). In addition, this glacial advance left rocky materials on the ocean floor at or near Cox Ledge to the southeast of Block Island. The moraine from a later advance overlies the earlier morainal material in central Long Island and then swings northeastward to form the North Fork of Long Island, including Orient Point. Associated with this advance are the New York islands of Plum, Great Gull, Little Gull, and Fisher's, as well as the till deposits and cobbly flats around Fort Mansfield at the end of Napatree Point at Watch Hill. Eastward, further signs of the same prolonged glacial stagnation are evident in the long "Charlestown" moraine (see Maps 3 and 6), which stretches from Watch Hill to the Wakefield area and then less visibly southward along the Point Judith peninsula and the Elizabeth Islands, Massachusetts (Lynch 2019). Unlike Maine and eastern Massachusetts, Rhode Island has few well-developed drumlins or eskers, and the landscape is covered with a combination of glacial till and outwash deposits left by receding glaciers and local blankets of loess (postglacial wind-blown deposits).

With the gradual rise in sea level to the present, Block Island was soon disconnected from the mainland, and currents and wave action began to create the familiar coastal features we see today, including barrier beaches between more elevated portions of the outer coast. These forces have straightened out an otherwise more contorted coastline and created all the coastal ponds and other features, such as the northwest dunes at Sandy Point, Block Island, and the double tombolo in the Napatree Point–Sandy Point Island area. Alternating tidal currents have also molded sediments along the shores of Narragansett Bay and the Sakonnet Passage into many familiar and characteristic triangular projections known as cuspate forelands, exemplified by Gaspee and Conimicut Points in Warwick, Sandy and McCurry Points in Portsmouth, Sandy Point on the east shore of Prudence Island, and the original Quonset Point (before its alteration by the U.S. Navy).

Some of the state's largest salt marshes and mudflats developed and continue to evolve behind barrier beaches (see Map 6). These features have reached their present form in the brief span of approximately 13,000 years that has elapsed since the beginnings of "final" glacial recession. No one who has seen wheeling flocks of shorebirds or ducks over the salt marshes, or noted tide pools full of herons and egrets, doubts the productivity and value of these fragile ecosystems. Many fishes and invertebrates that later inhabit the offshore marine environment depend on these shallow-water habitats for breeding or for their larval and other immature stages.

Away from the immediate coast, scattered areas of glacial outwash support pitch pine and scrub-oak forests that contain nesting Pine and Prairie Warblers. Repeated forest fires would likely revive these pine barrens at the expense of oak woodland, and almost certainly did with the burning practices of Narragansett, Niantic, and Nipmuck peoples, but bedrock not far from the surface under most Rhode Island sand plains reduces some of the dryness and porosity. Shrublands, both at the coast and in forest clearings inland, are decreasing at a rate of about 1.5% per year (Buffum et al. 2011), and these decreases are likely to persist without active management (Thomson 1958). The coast from Narragansett to Little Compton is less sediment rich than it is westward along the South County coast, and there are fewer salt marshes, so shorebirds, salt-marsh sparrows, and Clapper Rails are less numerous. ("South County" is a widely used local term referring to an area roughly but not exactly equivalent to Washington County.) Instead, the Narragansett Bay and Sakonnet Passage conduct massive tidal flow and feature only a few salt marshes in their upper reaches.

In Newport County, many rocky islands provide nesting habitat for colonial waterbirds. Westward from Point Judith, the dearth of roosting rocks allows for only small numbers of locally feeding cormorants at all seasons, and most cormorants feeding in this area travel to Fisher's Island, the Point Judith breakwater, or Narragansett Bay (Bay) for nocturnal roosting. In contrast, rocky ledges are both abundant in the Bay and along the outer shore of Newport County. Inhabitants of rocky coasts, such as Red-necked Grebes and Purple Sandpipers, are less frequent west of Point Judith, as are such rarities as Barrow's Goldeneyes, King Eiders, Harlequin Ducks, and Black Guillemots. On the other hand, Common Loons, Horned Grebes, and Red-breasted Mergansers (all tolerant of sandy bottoms) are often more numerous to the west.

Within living memory certain bay-mouth bars (e.g., Napatree Point) have moved landward many hundreds of feet. Briggs Marsh was largely a grassy piece of marshland in the nineteenth century but has become a coastal tidal pond. Certain low-lying rocky islands used by nesting colonial birds in the 1920s are now awash at every tide. Rising sea level is now of major concern to species of birds that depend on salt marshes.

Additionally, from before the turn of the twentieth century and continuing into the early 1970s, large segments of the mainland coast and numerous Bay islands were increasingly occupied by the U.S. Navy. With the Navy's precipitous pullout in the early 1970s, titles to many of these properties (or portions of them) were transferred to the state, and many such acquisitions are now administered for the purposes of wildlife conservation, research, or passive recreation.

Black and white oak forests occupy the majority of the interior along with an understory of abundant acid-tolerant plants such as huckleberries and blueberries, sheep-laurel, bracken fern, and clubmosses. Wild sarsaparilla, Canada mayflower, teaberry, spotted wintergreen, pink lady-slipper, cow-wheat, and (more locally) bearberry and trailing arbutus inhabit soils away from coastal winds. Rock polypody ferns crown large boulders and outcrops in the woods, but few Christmas ferns dot the landscape. Scattered stands of dense mountain-laurel are found on the most acidic and rocky terrain. Common plants of early successional stages include little bluestem, sweet fern, whorled loosestrife, hay-scented fern, various goldenrods, northern bayberry, black chokeberry, gray birch, several shadbushes, wild black cherry, staghorn-sumac, and the invasive black locust. Although two species of aspen are often present as early successional species, they are less common than in many other parts of the Northeast. Most wet woodlands and areas with stagnant water are densely vegetated with red maple, sweet pepperbush, highbush blueberry, and cinnamon fern. Some areas with the most sluggish drainage (especially near deposits of till or outwash) have a boreal flavor, with assorted ericaceous plants, pitcher plants, and sphagnum moss. The best developed of these acidic wetlands host stands of Atlantic white cedar. Many small cedar swamps are found throughout the state (usually away from the Narragansett Basin) and are often associated with small bogs and acidic fens that feature a distinctive postglacial flora, including leatherleaf, sweet gale, pitcher plants, sundews, and occasional orchids more characteristic of the Canadian zone. A few groves of black spruce survive in northwestern Rhode Island, not including one that was flooded out by the Scituate Reservoir.

In northern Rhode Island, red oaks increasingly join the ubiquitous black and white oaks, along with scattered groves of eastern hemlocks and chestnut-oaks. There is sufficient black and yellow birch, white ash, American beech, sugar maple, and other flora in a few places (e.g., parts of Foster and Cumberland) to suggest a northern hardwood forest. Elsewhere, sand plains with abundant pitch pine, scrub-oak, scarlet oak, and such flora as shining sumac are present from Kent County southward, while relatively fewer white ash and occasional tulip-trees are interspersed through the more xeric oak–huckleberry woodland. On wetter ground in southern Rhode Island, red maple and Atlantic white cedar swamps are locally accompanied by formidable stands of great laurel, along with groves of American holly. Black tupelo is a common associate of red maple throughout the state, including coastal areas. A few extensive stands of green ash and pin-oak are found in certain areas in swamps and along streams, most notably in the Pawcatuck River drainage and vicinity of the Great Swamp. Seasonal pools, especially those in urban and suburban areas, provide elevated prey abundance for songbirds during both the breeding season and migration (McKinney and Paton 2009) and may increase in conservation importance as urbanization spreads (see Map 2).

The sedimentary and metamorphic rocks of the Narragansett Basin and Bay islands contrast sharply with the older (including pre-Cambrian) plutonic and intrusive rocks to both the east and west, at any distance from Narragansett Bay (Olson et al. 1980). Schists and other metamorphic rocks are prevalent, and beds of coal said to be 20 ft (6 m) thick were once mined in parts of Cranston (Ashley 1915). In the Narragansett Basin, most granite rocks are fragments and erratics carried south by glaciers. Central portions of Aquidneck and Conanicut Islands have among the least acidic and rocky soils in Rhode Island, and this distinction is reflected in the distribution of many birds, especially in the breeding season but at other seasons as well in some cases. Aquidneck Island's once-luxuriant hayfields contained some of the state's best habitat for Red-winged Blackbirds, Bobolinks, Eastern Meadowlarks, Upland Sandpipers, and Savannah Sparrows, but these lush fields did not attract Vesper, Grasshopper, and Henslow's Sparrows. The latter group of species was, in contrast, once common on South County's drier and more acidic till and outwash plains. The remnant upland fields of Aquidneck and Conanicut Islands, now in all stages of succession, still provide rich feeding and wintering habitats at various seasons for Cattle Egrets, Glossy Ibises, Mourning Doves, and Red-tailed Hawks, along with Carolina Wrens, Northern Cardinals, and White-throated Sparrows.

In most coastal areas, mild winter conditions enhance survival of "half-hardy" species, such as Eastern Towhees, White-throated Sparrows, Gray Catbirds, Brown Thrashers, Hermit Thrushes, Ruby-crowned Kinglets, various blackbirds, and certain waterfowl, in contrast to the virtual absence of such moderately hardy species a few miles inland from the mainland coast. "Warm" seawater in late fall continues to prevent or modify low temperatures along the immediate coast; on CBCs, some protected thickets near tidewater consistently yield tropical winterers that have survived into December.

Weather conditions can be harrowing for birdlife even in summer. A persistent series of rainstorms in June 1903 starved most of the Purple Martins and Common Nighthawks in southern New England, and some populations did not recover for decades, if at all. A 6–12-in (15–30-cm) snowstorm on 6 April 1982 caused a temporary exodus of the already well-ensconced American Robins and brought widespread mortality to transient Tree Swallows, Hermit Thrushes, Eastern Phoebes, and Eastern Bluebirds throughout New England. A severe snowstorm on 9 May 1977 stopped spring migration for days, with unknown but probably substantial mortality to many transients, exemplified by such Neotropical migrants as Scarlet Tanagers foraging for food on the ground in suburban backyards.

Along the southern shores of South County and Little Compton, the ocean ameliorates the acidifying effects of granite, both in coastal freshwater ponds and in nearby terrestrial environments. Many ponds contain extensive stands of narrow-leaved cattail and *Phragmites*. These habitats have attracted breeding Pied-billed Grebes; Gadwalls; Green and Blue-winged Teals; Common Gallinules; Least and American Bitterns; King, Clapper, and Virginia Rails; Soras; Marsh Wrens; Willow Flycatchers; and even Ruddy Ducks.

Block Island (see Map 5), 14 mi (23 km) south of the mainland coast, resembles the South County coastal strip. Sirkin (1995a) has carefully defined most of Block Island's surficial landscape as a recessional moraine that once lay to the north of a now-drowned or eroded terminal moraine more similar to the core of the Montauk peninsula. At Block Island, the precolonial biota was strongly modified by colonial woodcutting, agriculture, and pasturage, and very little of the original flora survived this period. In 1524, Giovanni da Verrazzano reported to King Francis I of France that the island was "covered with trees, well peopled" (Livermore 1877). John Winthrop sailed past the island in the 1620s and described the woods on the southern portions of the island to be in "oak sprout, with no good timber in it." The antinomian and Quaker "grandees" of Newport were among the first to respond to Block Island's isolated, wolf-less conditions by stationing some of their livestock there through the mid-1600s. After purchase and colonization by English settlers from Boston in 1661, and for 60 years, "oak, hickory, elm, ash, cedar, and pine were abundant" (Livermore 1877). These tree species were essentially extirpated after 1721 by the settlers' needs for fuel and fencing.

Thereafter, Block Island remained in a state of deforestation through the nineteenth and early twentieth century, with only a few shade and fruit trees near the houses to accent the island's barren fields and stone walls. Post-agricultural succession of goldenrods, raspberries (ca. 1940s), and bayberry (1950s on) was soon supplemented or supplanted after the 1960s by shadbushes, winterberry, blueberries, red maple, wind-dwarfed wild apple, and thickets of white poplar. The oft-heard "treeless" condition of this long-grazed and storm-tossed island gradually succumbed to a brushy and locally reforested appearance less apt to remind visitors of parts of Ireland or Scotland. By the turn of the twenty-first century, several woodland-breeding birds of the mainland had begun to recolonize the island's maturing vegetation.

Compared with the mainland, Hairy Woodpeckers, Brown Creepers, Blue Jays, White-breasted Nuthatches, Northern Flickers, Mourning Doves, and Red-tailed Hawks are relatively much scarcer on Block Island at most times of year. In most cases, these species began increasing on the island starting in the 1950s. Black-capped Chickadees colonized only in about 1970, and Tufted Titmice retain vagrant-only status. Many common breeders of the nearby mainland continue only as migrants on Block Island. On the other hand, a few species are more common or frequent in winter than on the mainland, particularly Yellow-rumped Warblers and Northern Harriers, as well as (after good fall flights) Red-breasted Nuthatches, Snowy Owls, Northern Shrikes, and Northern Saw-whet Owls. Such waterbirds as Great Black-backed Gulls, Black-legged Kittiwakes, Northern Gannets, Red-breasted Mergansers, and Black Guillemots can be more numerous in surrounding waters than near the mainland, while Ring-billed and Laughing Gulls at any season are much less numerous.

Sources of Data

In 1883–1884 and 1887–1888, Dr. William C. Rives (1883–1884), Col. Powel (1883–1884), James M. Southwick (1887–1888), and Charles H. Lawton (1887–1888) published their respective lists in *Proceedings of the Newport Natural History Society*. Most of these were merely bird lists without much annotation, and they contain a number of inaccuracies. They were, however, the first local bird publications of any importance and led the way to more important writings.

In 1876, the newly formed Nuttall Ornithological Club of Cambridge, Massachusetts, began to publish its *Bulletin of the Nuttall Ornithological Club*, which after 1884 became *The Auk*, the journal of the American Ornithologists' Union. *The Auk* and its predecessor contain many of the important early notes and papers on Rhode Island birds, largely by Rhode Island authors, while many more mundane items of interest during the same era (often from gunners) appeared in *Forest and Stream*. Other periodicals that printed material on Rhode Island birds include *The Osprey* and *Ornithologist and Oologist* (published for a time by Frank Blake Webster of Pawtucket), but these journals ceased publication during the shooting era.

Beginning in 1899, Frank Chapman's quarterly *Bird-Lore* recorded the increasing national demand for conservation in the post-shooting era and provided a popular medium of communication at the national level while gradually taking over the task of publishing distributional information and field notes from major journals such as *The Auk*. Many Rhode Island records are preserved here in the "Season Reports" (later becoming the "Regional Reports") from 1917 on. *Bird-Lore* became *Audubon Magazine* in 1940, and the field notes section was published as a supplement during the war years (1939–1946). The supplement became *Audubon Field Notes* after 1947 and was titled *American Birds* between 1971 and 1994. From mid-1994 through 1996 it was published as *National Audubon Society Field Notes* and then simply as *Field Notes* from 1997 to 1999, when, because of financial realities, responsibility for publication was shared with the American Birding Association (ABA). The ABA took over full control of the journal in 1999 and named it *North American Birds*. These successive journals also published breeding bird plot surveys (1937–1992) and winter bird population studies (1948–1992) until control was transferred to Cornell University's Laboratory of Ornithology and the censuses were published in the *Journal of Field Ornithology*.

In the late-nineteenth and early twentieth centuries, taxidermists served as the hub for the activities of Rhode Island's growing number of hunters and collectors. They also performed the function of recording data pertaining to the many rare birds that passed through their hands, particularly by counting the numbers of such species as Northern Goshawks and Snowy Owls during flight years. The most notable of the early Rhode Island taxidermists were Walter Angell, J. M. Critchley, John Hague, Frederick T. Jencks, and James M. Southwick, all of Providence; Harry A. Cash of Pawtucket; and Charles B. Clarke of Newport. Many of these men were inspired field ornithologists in their own right, in contrast to many latter-day taxidermists, and several of them regularly published notes in *The Auk* and elsewhere.

Southwick and Jencks operated their venerable Natural History Store, which included their taxidermy shop, for decades, variously at 258 and 283 Westminster Street in Providence, and for a short time they published a quarterly, *Random Notes on Natural History* (Jencks 1884b–d, 1885, 1886a, b; Southwick and Jencks 1884a–l, 1885a, b, d, 1886a–e). Both a scientific periodical and a catalog, it contains many early bird records of local interest as well as information on other animals and plants. Its fascinating pages are also full of advertisements for native and exotic egg clutches, bird skins, and numerous other biotic curiosities to satisfy the considerable trade in such things that was still legal and characteristic of the times.

The taxidermy shops' role was supplemented in 1896 by the RWPM, under J. M. Southwick's directorship, and later still by ASRI, but in their heyday, the shops served as focal points for early bird study. As a result, a nucleus of interested people gradually came into existence, a local ornithological literature began to appear, and records of birds began to accumulate. After the shooting era the large taxidermy firms fell on hard times. Angell and Cash survived the coming of protection at least to 1934 at 37 North Main Street, and Angell lived until 1950.

Almost prematurely, the increasing interest in birds and the existence of numerous collections led in 1899 to the publication of *The Birds of Rhode Island* by Reginald Heber Howe Jr. and Edward Sturtevant. Howe was a member of an old Bristol family and for years was headmaster of Belmont Hills School in Massachusetts and an early Nuttall Club member (see also Howe and Allen 1901). Sturtevant lived on Paradise Avenue in

Middletown and taught for many years at nearby St. George's School. Both were avid collectors in and about the Newport–Middletown area but rarely went afield elsewhere in the state. They relied instead on an active correspondence with other Rhode Island field men, notably S. Newton Dexter, Harry Hathaway, James M. Southwick, and Frederick T. Jencks, among others. Their book is a concise summary of what was known about Rhode Island birds at that time, although by any recent standard its information is remarkably, if understandably, incomplete. It does, however, benefit from Howe's generally high editorial standards and contains an exhaustive bibliography of Rhode Island's early ornithological literature down to, and including, newspaper pieces. For a greater insight into Howe's relationship to birds and some vignettes into a few Rhode Island birding locations of yesteryear (e.g., Bristol), see Howe's *On the Birds' Highway* (1899).

Howe continued his interest in Rhode Island birds for some years after coauthoring *The Birds of Rhode Island*, and from 1900 to 1902 published a brief but interesting quarterly called *Notes on Rhode Island Ornithology* (see Howe 1900–1902 in References), now one of New England's rarest bird periodicals. Various trips, local lists, and a great amount of the fieldwork of Rhode Island's earliest observers and collectors are preserved here, allowing considerable insight into the ornithological methods and personalities of the times. Howe (1903) also published *A Supplement to the Birds of Rhode Island*, which briefly summarizes Rhode Island records since 1899 but contains few new data points beyond what was published in *Notes on Rhode Island Ornithology*. After about 1919, Howe disappeared from Rhode Island's ornithological landscape, while Sturtevant continued teaching at St. George's School, collecting occasional birds until the beginnings of protection, and recording his own and others' unusual occurrences in a card file and on the pages of his personal copy of *The Birds of Rhode Island*. According to William H. Drury (pers. comm.), Sturtevant was an ardent small sailboat skipper and knew the birds and waters of his native lower Aquidneck Island well. Sturtevant died in 1939. His writings, bird collection, and collection catalog have all been examined by the author, and his specimen collection (at Norman Bird Sanctuary) was conserved with the cooperation of RWPM during the 1980s.

In 1908, on the eve of protection, Charles H. Remington and the Commissioners of Birds published *A Checklist of Rhode Island Nesting Birds* (Remington and Commissioners 1908), in which a single specific instance of the nesting of each of 104 species is cited, with another 15 species given hypothetical breeding status. For confirmed species, the date is given, but location data are confined to the county where the species was noted or collected. Among the commissioners were men such as Edward Armstrong and F. E. Newbury, avid egg collectors whose interests in songbird eggs seem to contrast sharply with the sporting interests of the men who succeeded them as the Bird Commission evolved into the Department of Fish and Game.

A marked hiatus is evident in the Rhode Island ornithological record for the approximate period 1903–1935. Bent's (1919–1958) *Life Histories of North American Birds* contains only a few distributional notes of local interest. Forbush's (1925, 1927, 1929) venerable *Birds of Massachusetts and Other New England States* likewise contains some information on Rhode Island, but the material is usually too general to be useful and lacking in firsthand detail. A serious gap could therefore have developed in what we know about the state's birdlife except for the nearly single-handed efforts of one individual, Harry S. Hathaway, who ultimately contributed more to local ornithological knowledge than any of his contemporaries or predecessors and who came to play the central role in Rhode Island ornithology for more than 50 years.

One of the more tangible contributions of RIOC was the *Checklist of Rhode Island Birds*, spearheaded by David Emerson and repeatedly updated circa 1950–1980. This useful field card presented arrival and departure dates and listed post-1900 records of rarities. Another publication of RIOC, *A Field Checklist of Rhode Island Birds* (Conway 1979, 1992) gives detailed data on periods of occurrence in graphic form, together with habitat information and records of rarities. There is also a guide to birding locations entitled *Bird Walks in Rhode Island* (Fry 1992).

Although RIOC served some of the social functions of the many bird clubs of other states during the era 1940–1970, such organizations were otherwise almost nonexistent in Rhode Island during the twentieth century. The only such club was the Little Rest Bird Club (LRBC) of Kingston, formed in 1945 by Stanley Gairloch and active through a program of field trips and natural history lectures under the longtime leadership of Douglas L. Kraus until it disbanded in the 1970s. Earlier, there was also a more general Rhode Island Field Naturalists' Club in the period circa 1915–1942 or later, which had a decided "scientific" bias and to which some of the early RIOC members belonged. With a tendency for specialization and lack of communication among various kinds of naturalists, this organization disappeared during World War II, but its flavor was independently filled much later by the Rhode Island Natural History Survey (RINHS), formed in 1992 (see below).

The Rhode Island Division of Fish and Wildlife (RIDFW), evolving from the Rhode Island Bird Commission and the Department of Fish and Game, has had varying amounts of interaction with the Rhode Island birding community, the closest (mainly though Roland Clement) when Harold N. Gibbs was administrator in the 1940s, and again after the mid-1970s. As the arm of state government responsible for the fauna, it has been mainly concerned over the years, like similar agencies in other states, with management and investigations of game species of mammals, birds, and fishes. However, management of aquatic nongame birds became part of RIDFW's interests in the 1970s, which represented a long-overdue change when one considers the relatively small and slowly shrinking population of Rhode Island hunters compared with the considerably larger and growing number of voters and state taxpayers who spend time and money interacting with "watchable wildlife." Studies published by RIDFW have been concerned with local populations of Mute Swans, Canada Geese, and Mourning Doves and an Osprey newsletter dealing with the monitoring of breeding numbers and productivity of these birds was issued periodically after the 1970s. After the late 1970s, RIDFW supported a comprehensive censusing effort of the state's colonial nesting birds (Ferren and Myers 1998). Further indication that stewardship of nongame bird populations is further spreading into state government is the support within the Rhode Island Department of Environmental Management (RIDEM) for Richard Enser and Chris Raithel's work in censusing nesting birds on state-owned land as a local contribution of baseline data on breeding densities of landbirds. These data are part of a continent-wide effort to investigate declines in Neotropical migrants. *The Atlas of Breeding Birds in Rhode Island* (Enser 1992) was also accomplished with RIDEM resources, as is the maintenance of a GIS database of rare fauna and flora initiated as the Natural Heritage Program, a program initially funded by The Nature Conservancy and later wholly under state sponsorship. Between 1990 and 2017, Raithel was the nongame coordinator for RIDFW and helped monitor colonial birds while conducting point counts of breeding landbirds, which dealt with federally listed species of birds and other taxa within Rhode Island.

The Audubon Society of Rhode Island has been repeatedly mentioned in this history and undoubtedly has been the organization most central to Rhode Island bird protection and to the support of the birding community over the years while also attending to other environmental needs and accepting gifts of wildlands. Nationally, the Audubon movement began in Massachusetts with the efforts of Minna B. Hall and Harriet Hemenway, thoughtful society women of Boston, whose indignation at the carnage connected with the use of endangered birds' plumes on women's hats (among other abuses such as unlimited hunting) led to the formation of the Massachusetts Audubon Society in 1895. Incorporated on 20 October 1897, ASRI was one of sixteen other Audubon societies in as many states that formed within a 2-year period.

As late as the mid-1950s ASRI's activities were still largely centered around the study and conservation of birds, but its interests were already evolving toward concerns far wider than bird conservation. Beginning with the directorship of Roland C. Clement in 1950, and despite his personal ornithological interests, ASRI began to grow in size, scope, membership, and political influence. It has adapted and evolved with the times and now responds to ever more acute issues of land, water, and air conservation, as well as biodiversity issues. The society has developed a widely diversified and evolving program aimed at better management and conservation of the total environment through education, land protection, and political involvement. The Audubon movement in Rhode Island has taken in stride many newly threatening aspects of human hegemony over the natural world, such as the vanishing ozone layer, acid precipitation, and global warming/climate change, while continuing to work toward protecting the Rhode Island environment.

The Audubon Society of Rhode Island's wide-ranging field-trip programs and annual birding weekend at Block Island have provided the necessary start to many committed naturalists and environmental activists and have led many people into a lifetime of birding and nature study. Public programs at the society's sanctuaries have done much to interest and educate Rhode Islanders about natural history, and the society's school program has brought enlightenment about nature conservation to great numbers of schoolchildren, for example, 20,000 in 1998, and more since. The Birders' Conference, begun in 1992, fostered needed dialogue concerning birds both locally and farther afield. Through tax laws, land bequests have come to be an all-important means whereby the society has preserved important habitats and open space. The society's system of sanctuaries and wildlife refuges grew from a single parcel of 2.9 acres (1 ha) in 1924 to 80 or more parcels totaling 8300 acres (3359 ha) in 1999. A century after its beginnings, the society had several thousand members, a staff of more than a dozen, innumerable land holdings, sanctuaries, programs, an extensive written history (Weber 1997), and an attractive headquarters, research station, and education center. However small by the standards of other Audubon societies and conservation organizations of larger states, the Audubon movement in Rhode Island has increasingly served as a rallying point for many people of diverse interests who are resistant to the primarily economic, extractive, consumptive view of the natural landscape that has been all too pervasive in Rhode Island as elsewhere. The public concern for nature in Rhode Island during the author's birding career since 1952 has multiplied many times.

The Audubon Society of Rhode Island has been directed since the late 1930s by Harold L. Madison (1941–1950), Roland C. Clement (1950–1958), Alfred L. Hawkes (1958–1993), Lee C. Schisler Jr. (1993–2005), and Lawrence J. F. Taft (2005–2022).

For many years ASRI published Rhode Island's bird records monthly, first in its *Bulletin* (1940–1957) and then in *The Narragansett Naturalist* (1958–1967), with the field notes variously under the editorship of Douglas L. Kraus, Harold Madison, William B. Dean, and Severyn S. Dana. From 1969 through 2015, records were published as *Field Notes of Rhode Island Birds* under the successive editorial efforts of Dana, Charles Wood, David L. Emerson, Richard L. Ferren, and Christopher J. Raithel. Many older Rhode Island data also appear in the New England Natural History Museum's *Bulletin of New England Bird Life* (1936–1944), which was later continued intermittently by the Massachusetts Audubon Society as *Records of New England Birds* (1945–1961; 1964–1968). As might be expected, these publications never received full patronage from Rhode Island observers, and editorial standards for Rhode Island records were not as strict as in more local periodicals.

In the early 2000s, Rachel Farrell undertook to put out a daily email newsletter of Rhode Island bird reports for Rhode Island birders, and during the same period one or more electronic mailing lists have served as open forums for online communication about Rhode Island birds. Begun in 2002 and widely adopted by Rhode Island birders about a decade later, Cornell's eBird website invited birders from the whole world to send their observations for permanent storage and analysis. As part of eBird's reviewer framework, which allows some local quality control, Rhode Island eBird reports were reviewed for a time by Rachel Farrell and more recently by Alan Kneidel. This computer-based method of preserving field data has been greeted with mixed reviews by observers, some highly positive and others less so.

Avian rarities committees have become a necessary adjunct to the field-reporting process nationwide, all the more so as methods of reporting and preserving local sight records have become increasingly conducted through global online systems open to everyone. Unknown and sometimes incompetent observers can file reports of unlikely birds, and important reports need to be assessed for accuracy. Most states filter the most notable reports by setting up avian rarities committees and asking observers to send in descriptions and photographs of rare birds for review. Acceptance is through voting by the committee, which consists of birders who are extremely knowledgeable about the state's avifauna; the committee then publishes its decisions and a list of accepted records. Serious birders quickly learn to send in full reports and photographs of rarities under review, often by email. One or more "Rhode Island Avian Rarities Committees" (RIARC) existed before the millennium but never met nor adjudicated any reports. Since 2000, the latest RIARC, conceived by Shai Mitra and formed in 2008, has reviewed reports sent to secretary Doug Wilson, who sends reports to members by computer disks and email. It is a long overdue but effective process that organizes the acceptance of sight records formerly accomplished by individual field notes editors.

For at least a decade, it has been possible to create websites presenting digital photographs of birds. The RIARC website (ri-avianrecords.org) has numerous photographs of recent rarities along with committee transactions. Numerous Rhode Island rarities are now visible on RIBird.org, a site, managed by Paul L'Etoile, that welcomes new photographs of notable Rhode Island rarities. Another website by Geoff Dennis (geoffdennisphotos.com) presents a mixture of rare and common birds, mainly from Little Compton, including an outstanding diversity of extralimital and out-of-season hummingbirds.

Bird-Banding

One of the most important contributions of the past century was that of James Baird, director of the Norman Bird Sanctuary (NBS) in Middletown from 1954 to 1960. While there, Baird instituted a sizable mist-netting effort as part of Operation Recovery (an Atlantic coastal bird-banding project), thereby assembling the first significant data on the fall migration of many landbirds in coastal areas (Baird et al. 1958, 1959; Baird and Nisbet 1959). Together with Ian C. T. Nisbet, Baird also studied diurnal migration patterns among landbirds, cormorants, and other species in coastal Newport County and at Block Island. Of particular importance were Baird's efforts to collect various extralimital forms that had been previously unknown or unsubstantiated in Rhode Island. He spent the fall of 1960 at Block Island banding birds, collecting specimens, and making daily field observations before leaving Rhode Island to join the staff of the Massachusetts Audubon Society. His mist-netting of migrants on Block Island inspired the continued activities to the present day by Merrill Slate, the late Elise Dickerson, and most enduringly, Elise and Helen Lapham. Kim Gaffett now heads the Block Island banding effort. A sizable project near Clay Head coordinated by Brown University's

Jeffrey Parrish combined a major banding effort with studies of the food requirements of frugivorous migrants (Parrish 1997, 2000). Besides the activities of Baird in Middletown and Block Island, and the work of later Block Island banders, the only other long-term, large-scale mist-netting was done at Kingston by Douglas L. Kraus (see above) and later by URI personnel at Kraus's farm, which has evolved into ASRI's Kingston Wildlife Research Station (Osenkowski et al. 2012).

Christmas Bird Counts

From 1901 to 1949, the state's major Christmas Bird Count (CBC) was centered in Providence and incorporated Bay districts south to Warwick and Barrington. A count in eastern Newport County was established by Clement and others as early as 1937 and continued regularly through the 1940s. After 1949, this count reached a more stable status as the Newport–Westport CBC, and it has been characterized by a remarkably uniform area and coverage under the long-term stewardship of David and Robert Emerson. A count at Charlestown was conducted 1929–1941 by Mary Southwick and others. The enduring effort in South County has been the South Kingstown count, including coastal areas from Ninigret Pond to Bonnet Shores, begun by Douglas Kraus and others in 1940 and now managed by Scott Tsagarakis. At Block Island, CBCs were conducted regularly from 1924 through the early 1960s by Elizabeth Dickens and the Block Island schoolchildren and later on a more intensive scale by Susan R. Drennan, Richard Bowen, Chris Raithel, Scott Comings, Shai Mitra, and others. At Block Island, two other island-wide CBC-style counts have been conducted starting in the late 1990s by Mitra, Raithel, and others to determine prewintering bird populations in November (Veterans' Day Counts [VDCs]) and survival of winterers in late February (Presidents' Day Counts [PDCs]). More recently, a Napatree CBC has spanned Westerly and parts of Charlestown west into southeastern Connecticut and parts of Fisher's Island, New York. Northern Rhode Island hosted several counts in the early decades of the 1900s, including Woonsocket. Thereafter, there were no northern Rhode Island CBCs for nearly a century until parts of northern Providence County were included in a mainly Massachusetts count (Uxbridge) south of Worcester. Nationwide maps of CBC data are available in Root (1988).

Pelagic Birds

Prior to 1966, the only extensive observations of pelagic birds came from crossings on the Block Island ferry and from less than a half-dozen summer trips to (largely) inshore waters by local observers. Farther offshore, only the work of Gordon (1955) was available to chart the truly pelagic avifauna. The now-popular trips to Cox Ledge (see Map 5) and waters southeast of Block Island began in 1966 aboard fishing vessels from Montauk Point and, after 1973, from Galilee. Some of these fishing/birding trips would visit the wreck of the *Suffolk*, a sunken vessel about 25 mi (40 km) southeast of Block Island. More recently, whale-watching/fishing expeditions have been available;

many pelagic birds have been seen on these excursions. Peter Paton has studied pelagic birds well offshore in the 2010s. A few birding trips during the 1990s onward have visited the edge of the continental shelf, or "shelf break," including Block Canyon (see Map 5), where the status of a variety of Gulf Stream species such as Black-capped Petrel, Band-rumped Storm-Petrel, Bridled Tern, and others is slowly being elucidated. Most pelagic field observations have been limited to the period between late May and late October, but a few winter trips since the 1990s have also found puffins, murres, Dovekies, and Great Skuas.

Coastal Seawatching

Since birding telescopes became readily available in the 1950s, and as part of a day afield along the coast, some observers have included modest periods of observation from coastal promontories, such as Warren's Point in Little Compton, Point Judith in Narragansett, and from various points on Block Island. In addition to surveying swimming birds, seawatchers often encounter significant flights of seabirds moving along the coast. Most observers have stayed on station for only limited periods, usually in the morning, thereby missing any full accounting of numbers of coastwise migrants, which can differ sharply with the time of day, including early evening. Sizable flights of loons, scoters, gannets, and other species have often been encountered, but the vast numbers of migrants passing coastal points have needed better clarification. After the early 1970s, efforts to count migrants passing Point Judith and Block Island have yielded counts greatly in excess of most previously recorded numbers. Seawatches in spring at Point Judith seem to produce the largest totals because migrants such as loons, two of the scoters, Common Eiders, Brant, and other waterfowl round Montauk Point on Long Island and trend more northeasterly past Point Judith and onward toward Sakonnet and Vineyard Sound. Seawatches in late afternoon and early evening from mid-April to mid-May have also rediscovered a little-known flight of certain waterbirds orienting westward past the mainland coast toward the Connecticut coast and Long Island Sound, in contrast to the preponderant northeastward migration through Block Island and Vineyard and Nantucket Sounds. Best seen at Point Judith, this late-afternoon westward flight can also be seen passing Watch Hill lighthouse and Fort Mansfield at the west end of Napatree Point. Such apparently "backward" spring flights seem to represent preliminary daylight positioning movements to reach western Long Island Sound, after which they may continue overland after dark along a route from southwestern Connecticut and across New York or along the Hudson River to the eastern Great Lakes. Red-necked Grebes, White-winged Scoters, Long-tailed Ducks, some Brant, and occasionally other Canadian and Alaskan breeders, such as Horned Grebes and Bonaparte's Gulls, follow this route.

Breeding Birds

Some of the first efforts to quantify breeding bird status were the surveys of Henry Childs Sr. at the East Providence Reservoir, in which the number of territories of breeding species on standard plots was determined over many decades in the 1940s and 1950s. The overall distribution of breeding birds statewide was addressed by the first Rhode Island breeding bird atlas (Enser 1992). Patterned after similar studies in England, then Massachusetts, and now most states, more than 50 local observers worked to determine the presence of various breeding birds in 165 "blocks" (each a sixth of a U.S. Geological Survey topographic map) covering the entire state. These efforts have gone a long way to clarifying knowledge of the state's breeding birds, which has chronically lagged behind observers' attention to migrants, winterers, and vagrants.

Hurricanes

The following account of hurricanes in Rhode Island is presented in as much detail as it is for three reasons. First, most people interested in birds have a fascination with birds that appear after storms, and this section provides information from the past that can be helpfully contrasted with future storms that are very likely to increase in both frequency and intensity (Walsh et al. 2016). Second, it seems likely that the importance of hurricanes in driving vagrancy in birds is overstated, and, instead, birds appearing after storms are probably dispersing anyway and aided in their progress by winds associated with the storm. This is evident because of the wide disparity in species composition that follows storms (as is evident in the species accounts). Third, vagrancy is a way that organisms in general may be able to escape "climate extinction," so the continued evaluation of what birds do and do not appear after storms may inform the likelihood of any one species colonizing new habitats (Veit and Manne 2015, Zawadzki et al. 2019).

Owing at least partly to coastal topography, Rhode Island lies perpendicular to the tracks of northward-moving hurricanes, and there is a history of coastal devastation and transport of tropical seabirds to the state. Most notably, the "Great New England Hurricane" of 21 September 1938 made a direct hit on Long Island and Narragansett Bay and caused tremendous damage. On 25 August 1635, a year after the arrival of Rhode Island's first settler, William Blackstone, a tremendous "gale" pushed a 20-ft (6-m) storm surge up to the head of Narragansett Bay and obliged some Native Americans to climb into trees to save themselves. Other major storms occurred in 1638, 1641, 1675, 1683, 1723, 1727, 1743, 1749, 1761, 1770, and 1778. The "Great September Gale" in 1815 created new outlets to many coastal ponds and devastated shipping and shore structures at the head of the Bay with high winds and a tidal surge of 17 ft (5 m). Other hurricanes or strong tropical storms passed through southern New England in 1821, 1841, 1856, 1866, 1869, 1877, 1878, 1879, 1896, and 1904. The "Portland Gale" in November 1898 washed several acres of Rose Island's northern peninsula into the Bay.

The 1938 hurricane toppled almost all the white pines in southern New England through the combined effects of wind and rain. This depletion likely slowed the (re-)colonization of the state by Red-breasted Nuthatches and Blackburnian and Yellow-rumped Warblers. By contrast, some coastal species have benefited positively from historic storms. Piping Plovers and Least Terns, nesting mainly on sandy and gravelly beaches along the outer coast, reached historically high population levels in the years immediately following the two historic hurricanes (1938 and 1954). Beaches at Napatree Point, Quonochontaug Pond, and Ninigret Pond (see Map 6) have been overwashed and even rechanneled by waves from the south (and, during Hurricane Bob, overwashed from the north at Napatree Point and Sandy Point Island, at least). Such erosion and redistribution of sediments benefits the plovers and terns because it creates nesting habitat. American Oystercatchers of Sandy Point Island benefited for a decade from hurricane overwashes at Sandy Point Island that occurred in 1985 and again in 1991.

A chronology of hurricanes and other significant storms striking Rhode Island since 1888 is as follows:

1888. A violent coastal storm, 23–27 November, described as a "hurricane" on the basis of wind speeds (Ludlum 1976), deposited hundreds of Killdeers in coastal areas during an era when these birds were otherwise considered to be excessively rare. Even with a century of additional storms and a vast increase in the species, only one other storm (December 1973) has ever brought as substantial a flight of Killdeers to Rhode Island, suggesting that factors other than the storms may have been important in these influxes.

1901. A gale that swept Cape Cod on 6 October produced a Leach's Storm-Petrel on Potter Avenue in Providence.

1924. The first tropical hurricane to produce rarities in Rhode Island occurred 26 August. Numerous Chimney Swifts, Black Terns, and two Black Skimmers appeared at Block Island, and two skimmers at Quonochontaug were associated with the storm.

1927. "A gale of near hurricane force" 19–20 June was associated with the appearance of a Brown Pelican on 21 June.

1928. A devastating hurricane in the Caribbean later passed west of New England on 19 September, amounting to a heavy northeaster in Rhode Island. A Brown Noddy (especially unusual, even during hurricanes, in New England) and numerous Black Terns were recorded at Block Island.

1932. An exceptionally violent Atlantic storm, 17–19 November, caused the most notable wreck of Dovekies ever locally recorded, depositing birds on land from the Northeast to Florida (Murphy and Vogt 1933). The occurrence of a Magnificent Frigatebird at Block Island on the 16th (when winds had a southerly component) and a Northern Lapwing at the same location on the 20th (when winds had shifted to the northeast) may also be at least partly related to this storm. Veit and Manne (2015) have shown how the southward irruptions of Dovekies and Thick-billed Murres to New England are related to climatic conditions in the North Atlantic, presumably through a scarcity of their food farther north, so storms such as this one may be a secondary influence on wrecks of Dovekies.

1936. A minor hurricane on 27 August brought few vagrants to Rhode Island, but Clement saw a Snowy Egret at Briggs Marsh 3 days later and observed hundreds of terns migrating west along the shore. Another storm on the morning of 19 September, described as a full gale with 70 mph (113 kph) winds, managed to blow several Red-necked Phalaropes northward along the Sakonnet Passage to Boyd's Marsh, Portsmouth, and three Black Skimmers were seen at Briggs Marsh the next day.

1938. Without question, this was the most formidable hurricane in recent New England history, causing $100 million in damage and 262 human deaths in Rhode Island. It crossed Long Island on 21 September with a forward speed of 50 mph (80 kph) and shortly thereafter moved northward through eastern Connecticut and Rhode Island with sustained winds in the easterly sector of 95 mph (153 kph) and gusts to 180 mph (290 kph). Its unprecedented winds and coincidence with a high tide were the cause of the now-infamous devastation, both through wind action and from storm surge flooding, of downtown Providence and other Bay and shore areas (Nichols and Marston 1939). Observers were unable to get to coastal locations because of the unparalleled destruction to trees and property; there were also legal restrictions on travel in shore areas. A Leach's Storm-Petrel at Point Judith a week after the storm may have been displaced by this storm, but otherwise no vagrants were found. An arm of Napatree Point, now called Sandy Point Island, was amputated during the storm and has since shown no sign of becoming reconnected; instead, it has been gradually migrating northward and eastward toward the Connecticut shore.

1944. Another violent hurricane on 14 September, although less severe than that of 1938 and not accompanied by massive flooding, passed northeastward directly across Narragansett Bay after skirting the south Atlantic coast. The next afternoon, Lt. Richard Bowen happened to be commanding a patrol boat in lower Narragansett Bay and observed 12 Sooty Terns, 7 Leach's Storm-Petrels, Rhode Island's first Gull-billed Tern, and a possible Sandwich Tern. Elsewhere, Rhode Island's first Royal Tern and several Caspian Terns were observed. In ensuing weeks, coastal ponds and Narragansett Bay were choked with an estimated 1000 Black Skimmers, and some of these remained into late October.

1950. An offshore hurricane on 11 September caused extremely high tides but no reports of unusual birds. A violent southeast storm with winds to 100 mph (161 kph) on 25 November brought a sizable flight of large alcids (probably Razorbills) and a Black Guillemot past Point Judith during the storm, and Rhode Island's only inland observation of a Northern Gannet was made the next day in Johnston.

1954. Hurricane Carol struck Rhode Island to the west of Narragansett Bay on

31 August, and after the 1938 hurricane, proved to be the second-most destructive storm in New England history to date. Striking about noon at high tide, Carol had sustained winds of 90 mph (145 kph) and gusts to 115 mph (185 kph). Flooding of 13 ft (4 m) above normal tides inundated 3500 cars in Providence; 19 people were killed while 3800 houses and 2000 boats were destroyed. Despite major difficulties in getting to shore areas, observers were able to record numerous Caspian Terns, 8 or more Royal Terns, 2 Gull-billed Terns, more than 50 Forster's Terns, 1 Leach's Storm-Petrel in the Sakonnet Passage off Tiverton, 1 Wilson's Plover, and 3 American Oystercatchers. The storm produced only about half the Black Skimmers reported during the 1944 hurricane. Hard on the heels of Carol came Hurricane Edna on 11 September, which also had passed close along the Atlantic coast. Two oystercatchers at Napatree and a Tricolored Heron in Tiverton on 12 September may have been associated with the storm. Hurricane Hazel, the remnants of which passed through on 16 October, had little known effect on birdlife in Rhode Island.

1955. Hurricane Diane, which followed an east–west trajectory just off the Rhode Island coast 17–19 August, produced the heaviest regional rainfall (to 20 in; 51 cm) and interior flooding known in southern New England history ($170 million damage in Rhode Island). A sprinkling of rarities resulted from the cumulative effect of this storm, mainly west of New England.

1960. Hurricane Donna, after devastating parts of Florida, proceeded to skirt the immediate Atlantic coast before passing through eastern Connecticut and Rhode Island on the afternoon of 12 September. This storm was less violent than those of 1938, 1944, or 1954, and the modest destruction to trees and property ($5 million) allowed observers unobstructed coverage of coastal areas of Little Compton, Newport, South County, and Block Island on the following day. The storm produced 60 Royal Terns, 2 Sandwich Terns, nearly 30 Sooty Terns, 1 Bridled Tern, 3 Brown Noddies, at least 3 Gull-billed Terns, 5 American Oystercatchers, several Caspian and Forster's Terns, numerous Black Terns (many pushed inland), and 6 Brown Pelicans. Numerous Wilson's Storm-Petrels were seen flying overland at Block Island and even in East Providence at the height of the storm, and a Leach's Storm-Petrel was recovered from a garage in Newport. In contrast to some earlier storms, only 30 Black Skimmers were reported.

1961. Hurricane Esther on 21 September destroyed shore property but produced no rarities.

1971. Tropical Storm Doria on 28 August brought 1 Royal Tern and 22 Forster's Terns on the day of the storm, and about 200 Black Terns were present during the ensuing week.

1972. Tropical Storm Carrie of 3 September did not produce any southern rarities, but during the storm a remarkable 47 Hudsonian Godwits flew in over Trustom Pond without alighting.

1976. Hurricane Belle passed across western Long Island in the early morning hours of 10 August after an entirely pelagic journey from near the Bahamas. Few rarities reached New England, and a survey of Rhode Island coastal areas produced only 2 Gull-billed Terns, a half-dozen Forster's Terns, perhaps 2 Royal Terns, about 20 Black Skimmers, and a few Wilson's Storm-Petrels at Quonochontaug. A single Sooty Tern at Cox Ledge the day after the storm was unique in New England.

1979. Hurricane David of 6 September devastated parts of the Caribbean and then slowly moved along the east coast of Florida and into Georgia. As a tropical depression, its center passed rapidly northeastward through Pennsylvania and northern New York. David then created 30–40 mph (48–64 kph) southeast winds on the Rhode Island coast for most of the morning of 6 September; in the afternoon, the winds slowly shifted to the southwest at similar velocities. Observers at Point Judith and elsewhere saw relatively little (e.g., one Sandwich Tern) until the wind changed to the southwest. Thereafter, particularly in the late afternoon and evening, an unexpected and ultimately unparalleled procession of rarities deluged the coast. A minimum of 62 Sooty Terns, at least 5 Bridled Terns, and 1 Brown Noddy passed Point Judith heading northeastward in the very late afternoon hours, with additional Sooties audibly passing the point after dark. Other Sooties were found dead at Napatree Point and seen exhausted inland at Kingston, while 8 oystercatchers appeared at Napatree. More Sandwich Terns were seen the next day. Five Royal, 16 Black, and 1 Gull-billed Tern and 3 Black Skimmers were also seen. A flock of 68 Red and 6 Red-necked Phalaropes at Galilee, and a Red-necked Phalarope and 7 Black Terns inland at Richmond, were seen the day of the storm; 8 oystercatchers at Napatree the day after was then a large number. A final total of 7 Sandwich Terns was a maximum count for the state at the time.

1985. On 27 September, Hurricane Gloria came directly over water from the south-southwest, skirting the area from Cape Hatteras to New Jersey about 50 mi (80 km) offshore, and struck land along a path centered on the middle sections of Long Island and the Connecticut River valley. Only a few jaegers were seen while the wind remained southeast, but after winds veered to the south and increased to more than 90 mph (145 kph), a gannet appeared in the Harbor of Refuge, with sandpipers and an oystercatcher in the parking lots and cormorants landing among summer cottages. A substantial number of storm-petrels were swept in over the beaches and north into the interior, and two *Pterodroma* petrels were closely observed over a parking lot west of Point Judith. Later, as winds turned southwest, an easterly flight of more than 30 jaegers passed by and over Point Judith. Early the next morning, 1 White-tailed Tropicbird closely passed the Point Judith lighthouse, 1 Wilson's Storm-Petrel was found resting in Galilee Harbor, and 13 Forster's Terns were found inland at Wordens Pond. The next day, 16 Sandwich Terns (more than in all previous state history), 25 Royal Terns, 6 Caspian Terns, 93 Forster's Terns, 69 American Oystercatchers, 97 Black Skimmers, and several Piping Plovers were observed at various sites. Three Tricolored Herons the day after Gloria may have been displaced northward by the storm. Many of the terns and oystercatchers were noted actively moving west along the shore in the days following the storm, and a large westward movement of Forster's Terns was seen on a crossing of the Block Island ferry on 30 September. Gloria ripped gaping holes in the few spruce plantations that represented the only habitat in the state for breeding Golden-crowned Kinglets.

1991. Hurricane Bob (19–20 August), after forming off Florida and passing very close to Cape Hatteras, passed directly over Block Island and north along the east side of Narragansett Bay about 2 p.m. on 19 August. Easterly winds at Block Island reached 125 mph (201 kph). A Black-capped Petrel and Wilson's Storm-Petrels were seen over Wakefield as the hurricane's eye passed up the east side of Narragansett Bay. Examination of the mainland coast after the storm produced a Leach's Storm-Petrel, a Brown Pelican, a Red Phalarope, and an excellent assortment of rare terns: 5 Gull-billed, 2 Arctic, 7 Caspian, more than 30 Royal, approximately 15 Sandwich, 125 Forster's, 2 Bridled, 1 Sooty, about 30 Black Terns, as well as nearly 300 Black Skimmers. Only one oystercatcher was seen. Two Black-necked Stilts and an American Avocet that appeared immediately after the storm were thought to have been brought northward from the Carolinas by the storm. More than during Gloria, barrier beaches west to Napatree were substantially overwashed (at Napatree primarily from the north), thereby renewing the habitat of beach nesters. A Wilson's Storm-Petrel was picked up after the storm at URI.

1993. A major storm 13–14 March deposited record snows from the Carolinas northward. An adult-plumaged Purple Gallinule that was found in a snowbank in Wakefield may have been related to this storm.

1996. Hurricanes Edouard (2 September), Fran (6–7 September), and Josephine (8–9 October) produced no tropical species in Rhode Island.

2007. A spring storm ("Patriot's Day Storm") 16–17 April brought, somewhat unexpectedly, a variety of seabirds to the Rhode Island coast.

2011. Hurricane Irene of 28–30 August produced 1 Brown Pelican, 250 Red-necked Phalaropes, 5 Sooty and 16 Bridled Terns, 9 *Onychoprion* spp., 190 Black Terns, 13 Royal Terns, 10 Sandwich Terns, 5 Black Skimmers, and 20 Parasitic and 6 Long-tailed Jaegers.

2012. Hurricane Sandy of 29 October–1 November brought a Leach's Storm-Petrel to East Providence, a male Magnificent Frigatebird to the south coast, and a Red Phalarope to Goosewing Beach.

Migration

All but a few of the approximately 311 bird species that regularly occur in Rhode Island migrate to some extent, even supposedly "resident" species such as Blue Jays, American Crows, White-breasted Nuthatches, Black-capped Chickadees, and American Goldfinches (Newton 2008). Banding records show that the supposedly sedentary (but highly mobile) House Finches that now breed in New England have frequently appeared in states farther south during nonbreeding seasons, as have supposedly nonmigratory Barn Owls. Flocks of "residents" such as Northern Cardinals, Tufted Titmice, House Sparrows, Red-bellied Woodpeckers, and White-breasted Nuthatches often move with other migrants past the north end of Block Island or the tip of Point Judith, and some of them actually migrate or disperse hundreds of miles (Newton 2008). Parrish (2000) has shown how many passerines that are insectivorous during breeding switch to a diet of fruits (especially berries) at Block Island during fall migration. During fall, berries provide a reliable source of food, compared with insects, despite their lower nutritional value.

Rhode Island is either within the wintering range or on the migration route of many migratory breeders of the high Arctic and subarctic (many shorebirds, jaegers, American Pipits, Horned Larks) or close enough to be reached by winter incursions of irruptive northern birds (several owls and finches, Northern Shrikes). Rhode Island's coastal location aids the regular appearance of many southern breeders that commonly wander northeastward along the coast in late summer and fall (certain herons and terns, Laughing Gulls, Black Skimmers, Blue Grosbeaks, Summer Tanagers, southerly breeding warblers).

Populations of western and midwestern species periodically grow and shrink, and during periods of population growth, multiple individuals (mostly young birds) appear in Rhode Island. This occurs most often along the immediate coast, which presumably serves as barrier to their continued dispersal (e.g., Lark and Clay-colored Sparrows, Dickcissels, Western Kingbirds, Western Tanagers, Yellow-headed Blackbirds [Baird and Nisbet 1960; Veit 2000, 2008; Zawadzki et al. 2019, 2021; Acosta et al. 2021; Veit et al. 2021]).

Some oceanic species breed in the Southern Hemisphere and migrate to our area during our summer. South Polar Skuas, Sooty and Great Shearwaters, and Wilson's Storm-Petrels move north and south on the same general schedule as North American migrants, but their breeding and wintering periods are reversed. They winter at our latitude from May through October. Similarly, most first-year Common Terns stay in the tropics until their second summer, although a few nonbreeders (and rarely flocks) visit breeding regions in their first summer (Nisbet 2002).

Waterbirds

Sizable spring flights of White-winged Scoters, Red-necked Grebes, and Long-tailed Ducks pass westward or southwestward through Block Island Sound in April and May. The largest flights since about 1950 have occurred in the late afternoon hours during clear, calm conditions. The westward flights along the Rhode Island coast are likely an efficient path from wintering grounds in eastern New England (e.g., Monomoy and Nantucket Shoals) toward breeding grounds in western Canada and Alaska by way of western Long Island Sound and the Great Lakes. Breeding in western Canada and wintering on Rhode Island's salt waters, a great many common seafowl wintering on the southern New England coast probably reach our area by an unnoticed overland flight from the vicinity of the Great Lakes.

Superimposed on the roughly north–south axis of many shorebird routes in our latitude is an easterly shift in fall and a westerly shift in spring for Marbled Godwits, American Golden-Plovers, Lesser Yellowlegs, Long-billed Dowitchers, and Buff-breasted, Baird's, Western, and Stilt Sandpipers, whose spring and fall routes are separated by as much as 2000 mi (3219 km). Spring coastal storms often result in increased shorebird numbers, unusually early appearances by certain species, or occasional great flights along the outer coast. Considering that shorebirds have the longest migrations of any group of birds, it is not surprising that some are moving almost throughout the year. Starting with the appearance of Killdeers in February and Greater Yellowlegs, Piping Plovers, Pectoral Sandpipers, and American Oystercatchers in March and April, a steady procession passes through, with a peak in those species destined for Arctic breeding grounds in late May. Late individuals and small flocks of Red Knots, Semipalmated Sandpipers, White-rumped Sandpipers, and others continue to pass northward through mid-June.

In fall, the majority of species migrate in July and August, much earlier than most landbirds. In the last week of June, barely a week after the last spring shorebirds have departed northward, a scattering of the earliest southbound shorebirds often appears on the coast, particularly Short-billed Dowitchers, Least and Semipalmated Sandpipers, and Lesser Yellowlegs. These include wandering 1-year-olds as well as failed breeders. Adults of many shorebirds, including especially Semipalmated Plovers, dowitchers, and "peeps," appear in peak numbers in late July. Juveniles of these species reach peak abundance in late August and early September.

As human populations, and their consequent use of intertidal habitats, increase, it is inevitable that migratory shorebirds will face increased disturbance (Koch and Paton 2014). The early history of shorebird hunting is perceptively described by Mackay's (1929) *Shooting Journal*, which also served as a strong impetus for subsequent conservation efforts. There has certainly been an increasing presence of picnickers and beachgoers, often with unleashed dogs, at Charlestown Breachway and Napatree Point; Koch and Paton (2014) and Mayo et al. (2015) show how certain human activities can affect the foraging efforts of shorebirds on these tidal flats.

Hawks

Large numbers of hawks migrate along the outer coast. This movement can be observed near and just to the north of Sakonnet Point and Watch Hill. Accipiters, Northern Harriers, Ospreys, and falcons appear in moderate numbers at Block Island, but buteos and vultures are rare there. Unlike the other diurnal raptors, Peregrine Falcons and Merlins pass in highest numbers across the outermost islands of southern New England. These two species, particularly Peregrines, are therefore much more numerous at Block Island and can regularly be seen leaving the island at Dickens Point and migrating toward Montauk Point, Long Island, even when the latter is invisible to humans. Kestrels appear to pass infrequently over inshore waters and are never abundant at Block Island, but hundreds are sometimes observed migrating over the mainland. Broad-winged Hawks, which are abundant in late September in the interior of New England, have only rarely passed through Rhode Island in numbers. Counts of raptors at hawk watches have proven useful for monitoring population trends (Bednarz et al. 1990; Bildstein et al. 2008).

Passerines and Other Landbirds

Diurnal as well as nocturnal passerine migrants in May (as with spring hawks and some others) tend to travel northeastward through the interior of New England rather than through coastal regions, and the bulk of this movement usually remains northwest of a line running from New York City through coastal New Hampshire (Nisbet and Drury 1967). Howe and Sturtevant were aware of the coast-shunning nature of the spring landbird migration as early as 1899 and stated that "the spring migration along the coast seems to turn in somewhere on the Connecticut coast, cutting across through Providence, to the vicinity of Boston . . . and records of arrival at these three places follow in order. The birds which breed in southern Rhode Island seem to work down as offshoots from this main migration, for arrival records for southern Rhode Island are invariably later than for Providence and vicinity" (Howe and Sturtevant 1899).

Nocturnal migrants passing northeastward over the greater Providence metropolitan area tend to drop into Swan Point Cemetery and gather at earliest light in its wooded, undeveloped northern sections. Here, in perhaps Rhode Island's most heavily birded location, concentrations of Canadian-zone transients may be seen in numbers almost never recorded elsewhere in Rhode Island. Records of southern vagrants such as Summer Tanagers and Prothonotary Warblers are also much scarcer there when compared with coastal records. Even so, the few southern landbirds that find their way into the interior of Rhode Island are most often reported from Swan Point, and this mecca for warbler enthusiasts has produced 37 of the 38 warbler species that are known for the state.

Despite its southeasterly location, and consequent low volume of migrant landbirds, Block Island is often inundated by large numbers of migrants in early spring. Perhaps the largest historical flight of April migrants may have resulted at least in part with the passage of a large low-pressure system during the period 23–25 April 1983. During and after this storm, southern New England was deluged with tropical winterers that more usually arrive 2 or 3 weeks later. Reports in Rhode Island included at least 30 Indigo Buntings, 13 Rose-breasted Grosbeaks, 10 Northern Waterthrushes, 7 Eastern Kingbirds, 8 Veeries, 4 Gray-cheeked Thrushes, 4 Baltimore Orioles, 4 Common Yellowthroats, 1 Summer and 2 Scarlet Tanagers, individuals of Black-throated Green Warbler, Tennessee Warbler, and Ovenbird, along with numerous Gray Catbirds and other species. Most birds noted were reported near the coast, some exhausted on the beaches.

The advent of radar studies (Williams et al. 1977, 1978; Williams and Williams 1979) has shown that small passerines regularly make nocturnal departures from New England and eastern Canada toward the southeast in fall, particularly after the passage of cold fronts (but see Murray 1989; DeLuca et al. 2015).

Range Expansions

Several bird species that once nested in southern New England are still in the process of recovery from nineteenth-century hunting, most notably the eastern race of Willet, American Oystercatchers, and (perhaps still) Black Skimmers. All were unknown in Rhode Island, even as vagrants, until well after the turn of the twentieth century, but all three have now returned to breed in southern New England. Breeding Willets remained extirpated from most salt marshes between Nova Scotia and Virginia for most of the twentieth century, but the southern population at last expanded into our area and began to breed in Rhode Island in 1987. American Oystercatchers required nearly 40 years from their first known return as vagrants (1938) to the beginnings of regular nesting (1976). By the mid-1980s, they were furiously dividing up available scraps of breeding habitat on the smaller bay islands. Even Black Skimmers have attempted to nest here in 1985 and 1991 but have not become established. Soon after oystercatchers, cormorants began nesting in Rhode Island in 1981 and have increased exponentially since then. Great Blue Herons' southward movement through the increasingly wooded interior of central New England (first Rhode Island nest in Burrillville, 1984) is reminiscent of a northward push two decades earlier by Turkey Vultures through the same hills. Both are birds of the interior that have paralleled the coastal species in their advances northward and southward. These and many others are reoccupying habitats that in some cases have been open to them for centuries and where meaningful human disturbance or exploitation has now been negligible for many decades.

Oil Spills

One consequential oil spill occurred off Block Island in January 1930, when numerous Razorbills were killed and a single oiled Common Murre and an Atlantic Puffin drifted ashore. Larger and more recent spills have occurred in nearby Massachusetts waters (e.g., the sinking of the *Argo Merchant* on Nantucket Shoals in December 1976), but major spills from tanker collisions or groundings have mercifully been relatively few along the southern New England coast so far. Two tankers (S.S. *Gulfoil* and M.V. *S.E. Graham*) collided and were set ablaze in the lower East Passage between Fort Adams and The Dumplings on 7 August 1958 and then drifted together close to Rose Island. Eighteen human lives were lost and 22,000 gallons (83,279 liters) of gasoline were burned, blackening the Rose Island lighthouse on the southwest corner of the island and forcing evacuation of the lightkeepers.

In 1960 the tanker *P.W. Thirtle* went aground off Jamestown and dumped several hundred thousand gallons of heavy oil (bunker C) into Narragansett Bay. Thousands of birds are said to have died, but little documentation of the actual kill was made or reported. Miraculously, when the M.V. *World Prodigy* (another tanker) grounded on Brenton Reef in late June 1989, spilling nearly 300,000 gallons (over 1 million liters) of home heating oil, better documentation indicated that few avian casualties occurred. Oil from this spill entered Narragansett Bay and was reported as far west as Moonstone Beach, likely affecting submarine aquatic life in ways that were not adequately studied.

Another oil spill followed a fire on the tugboat *Scandia* northeast of Block Island on 19 January 1996. The tug and its barge, the now-infamous *North Cape*, drifted onto rocky Nebraska Shoal off Moonstone Beach in a severe southerly storm. Over the course of 3 days, 828,000 gallons (over 3 million liters) of #2 home heating oil (a fraction of its four million gallons [15 million liters]) entered the surrounding seawater. More than 12,000 lobsters, large windrows of surf clams, and innumerable other invertebrates washed up, along with at least 303 birds (including eiders, loons, gulls, grebes, Razorbills, and one Common Murre). Despite hastily erected barriers, some of the oil entered South County's salt ponds, particularly Ninigret Pond, Trustom Pond, and Cards Pond, as well as Point Judith Pond at Galilee/Jerusalem. Mammoth slicks were even seen south of Block Island, where seabirds may have been affected. A week or two later most of the oil had apparently evaporated or biodegraded, and little permanent damage was recorded. Some reports, however, stressed the continuing presence of hydrocarbons in sediments off Moonstone Beach and in nearby estuaries.

Introduced Species

Several native species of birds have been affected by the accidental introduction of plant diseases. Warbling Vireos and Least Flycatchers formerly bred regularly in suburban elm trees in the outskirts of Providence, Cranston, East Providence, and Pawtucket. With the decimation of the planted rows of American elms by Dutch elm disease during the 1950s and early 1960s, these birds gradually disappeared. To a lesser extent, Baltimore Orioles have declined in the same habitats. They too were once more common in the immediate vicinity of human dwellings in Providence and its surrounding towns-turned-suburbs, as well as in the hearts of nearly all outlying towns and along rural roads. In these places, the once-beautiful groves of American elms provided a canopy for nesting, often directly over city streets.

Earlier, a similar but much more dramatic decline of Yellow-throated Vireos apparently resulted from the American chestnut blight during and after 1917, when most of the American chestnuts of the Northeast were killed by an accidentally introduced Asian fungus (*Cryphonectria parasitica*) that appeared in New York City during 1904. These vireos formerly bred regularly in the chestnuts and other large suburban shade trees in Cranston, East Providence, Woonsocket, and Kingston. The American chestnut became extinct as a large forest tree in our area after 1925 (sprouts still come up from the roots), and as a result, this vireo disappeared. They are now confined to rural western Rhode Island.

Even though outbreaks of Spongy Moths (*Lymantria dispar*; formerly known as Gypsy Moth) do not normally kill the deciduous trees that they defoliate, the impact of these moths on small white pines and older eastern hemlocks during heavy infestations can be serious. Other introduced insects have affected, or seem destined to affect, native as well as introduced trees in ways that change bird populations. The Red Pine Scale (*Matsucoccus matsumurae*) has gradually been killing large tracts of the Works Progress Administration (WPA) era red pine groves in western Rhode Island, beginning in the late 1980s (Burlingame, Scituate Reservoir). The death of the red pines has removed an important habitat for Pine Warblers but has temporarily created nest sites for Red-breasted Nuthatches. Earlier, an introduced scale insect causing beech bark disease and two fungi killed many of the American beech trees in southern New England during the 1950s and 1960s, while the introduced Pear Thrip (*Taeniothrips inconsequens*) has been defoliating many of the sugar maples in northern New England.

The Woolly Adelgid (*Adelges tsugae*) defoliates and kills eastern hemlocks; it arrived in Virginia in the early 1930s and has gradually moved northward through New Jersey and New York. It arrived in southwestern New England in 1985, and by 1988, most of southern Connecticut's hemlocks were dead (Orwig and Foster 1998). The newly resident Acadian Flycatcher, the only southern bird species partial to hemlocks, may also be affected. Progressively lesser effects could befall evergreen breeders such as Blackburnian Warblers, Black-throated Green Warblers, and Blue-headed Vireos, which (like Yellow-rumped Warblers) also breed regularly in white pines.

Climate Change

Earth's rapidly changing climate has had profound effects on bird populations in Rhode Island as it has elsewhere (Terborgh 1989; Pearce-Higgins and Green 2014). See Griscom (1949) for avian effects of "the year without a summer" in 1832. Wet or cold periods during 1875, 1903–1904, 1907, 1934, 1942–1943, and 1948 as well as unusually warm periods during 1876, 1892–1894, 1932, 1936–1942, and 1943–1947, may have similarly affected populations of New England birds.

The advance of many southern birds into the Northeast is shared with such newly arrived mammals as gray foxes (*Urocyon cinereoargenteus*) and opossums (*Didelphis virginiana*) and with many invertebrates. Bird species that were once rare or unknown in Rhode Island either as breeders or visitors, whose breeding ranges lie primarily to the southwest of our area, and whose increases cannot be immediately connected to human influence include Fulvous Whistling-Ducks; King Rails; Clapper Rails; Acadian Flycatchers; Northern Rough-winged Swallows; Blue-gray Gnatcatchers; Orchard Orioles; Worm-eating, Pine, Prothonotary, and Hooded Warblers; Louisiana Waterthrushes; and Blue Grosbeaks. Even so, increases by Louisiana Waterthrushes; Worm-eating, Pine, and Hooded Warblers; and the more midwestern Cerulean Warblers and Acadian Flycatchers may be largely attributed to twentieth-century reforestation and therefore caused by a reduction in anthropogenic disturbances to woodlands. Fish Crows have also increased. Certainly, increases of Northern Cardinals, Tufted Titmice, and Red-bellied Woodpeckers must be at least partly influenced by the increase of winter bird feeders. Nesting Purple Martins irregularly increased nearly parallel with increasing suburbanization and the resulting increase in available nest sites. Southern herons have multiplied in the middle Atlantic coastal region since about 1910 and have expanded northward to create new breeding populations in southern New England and in Maine and the Maritime Provinces. Two other southern species, Barn Owl and Carolina Wren, have increased erratically during this century, at least partly owing to changing climate.

On the other hand, there are several southern landbirds that have significantly declined as breeders in the Northeast, particularly Yellow-breasted Chats and Red-headed Woodpeckers. The latter were common in western New England a century ago but have declined in part from usurpation of their nesting holes by the aggressive introduced European Starling, as well as highway mortality.

Species whose ranges lie to the west or northwest of Rhode Island have also undergone historical increases as breeders and transients here. Brown-headed Cowbirds were apparently absent in the East before the mid-1700s and appeared only after agricultural corridors through the Appalachians linked established eastern agricultural lands with the prairies (Mayfield 1977). Likewise, "Prairie" Horned Larks reached New England from the Midwest around the beginning of the twentieth century and have been breeding here so long that their former novelty as a breeding bird has been nearly forgotten (Forbush 1927). Ring-billed Gulls have prodigiously increased here as nonbreeders, although their former rarity and present burgeoning numbers have unquestionably resulted from human activities (protection on breeding islands, unlimited food at all seasons). Gadwalls, ducks of western marshes, have colonized Atlantic seaboard marshes and are now breeding on rocky offshore islands.

No parallel diminution of the northern breeding element has occurred in Rhode Island as has been suggested for eastern Massachusetts (Griscom 1949; Griscom and Snyder 1955), and for Cape Cod (Hill 1965). Instead, many such species have prospered as breeders. Increases among relatively northern breeders involve Great Blue Herons; Green-winged Teals; Long-eared and probably Northern Saw-whet Owls; Hermit Thrushes; Veeries; Black-throated Blue, Blackburnian, and Canada Warblers; Northern Goshawks; Red-breasted Nuthatches; Golden-crowned Kinglets; Winter Wrens; Yellow-rumped Warblers; Dark-eyed Juncos; and White-throated Sparrows. Most breeders seem to be responding to the increasing maturity of Rhode Island's woodlands, particularly natural coniferous woodlands, or the appearance of artificial habitats such as conifer plantations. Hooded Mergansers now breed because of the availability of nest boxes. The increases of Northern Waterthrushes and Canada Warblers may stem from a greater availability of sphagnum swamps within regenerating woodlands.

That North American wintering may lead to breeding on this continent has been demonstrated by Black-headed Gulls, Little Gulls, and Manx Shearwaters. Northern Wheatears, vagrants to New England, have similarly invaded Alaska as well as Greenland and the Canadian Arctic from both eastern and western portions of the Old World. Increasing records of vagrant Little Egrets may yet lead to North American nesting, as they have been breeding on Barbados and Antigua from 1994 on (Buckley et al. 2009).

Merlins are also northern breeders that have remained in coastal regions in winter where they were not recorded earlier. Their gradual increase in winter in the last half century seems to have been enhanced by the transformation of many coastal fields to brushland (including multiflora-rose) and by the consequent increase in prey species such as White-throated and Song Sparrows, cardinals, and a wide variety of other landbirds.

Especially noticeable (sometimes only by banders) are the invasions of Canadian species such as Snowy and Northern Saw-whet Owls, Rough-legged Hawks, Northern Shrikes, Northern Goshawks, many northern finches, Red-breasted Nuthatches, Bohemian Waxwings, both chickadees, and the three-toed woodpeckers. It should be emphasized that incursions or increases of northern species may not be entirely propelled by deteriorating conditions of food or weather on the breeding grounds but may result from surpluses there. Increased food resulting from multiple-year infestations of Spruce Budworm (*Choristoneura fumiferana*) in Canada can cause increased breeding productivity in such species as Bay-breasted and Cape May Warblers, the three-toed woodpeckers, and Evening Grosbeaks.

In inshore waters, changing baitfish conditions play a primary role in attracting (or sometimes unseasonally retaining) numbers of Northern Gannets, Cory's Shearwaters,

Black-legged Kittiwakes, Razorbills, Laughing Gulls, Common Terns, Red-breasted Mergansers, loons, grebes, and other marine birds, thereby creating constant variations in numbers ranging from notable scarcities to historic invasions. Large counts of Red-breasted Mergansers and Common Terns at Block Island; Northern Gannets in Block Island Sound; Common Loons along the South Shore; unusual concentrations of kittiwakes, jaegers, Razorbills, and Cory's Shearwaters near the mainland; and other such phenomena all appear to be influenced by occasionally sharp variations in populations of sand lance (*Ammodytes* spp.), herring (*Clupea* spp.), or other baitfish. While unusual weather conditions such as major storms have frequently created concentrations of certain marine birds, changes in abundance of food resources may exceed unusual weather in producing high counts as well as scarcities. See Kenney (1995) for a discussion of the "biomass flip" between herring and sand lance populations; see Veit and Petersen (1993) for discussion of the effect of sand lance abundance on marine bird populations in Massachusetts.

Species Accounts

DUCKS, GEESE, AND SWANS (ANATIDAE)

Black-bellied Whistling-Duck (*Dendrocygna autumnalis*)

Status

Vagrant. One record.

Nonbreeding

A Black-bellied Whistling-Duck was found at Apponaug Cove 8 August 2021 and seen and photographed by numerous observers (eBird; see Photo 1). While this was the first record for Rhode Island, Black-bellied Whistling-Ducks are rapidly expanding their range, and this expansion has generated records of vagrants north to Manitoba, Québec, New Brunswick, and Nova Scotia (eBird).

Fulvous Whistling-Duck (*Dendrocygna bicolor*)

Status

Vagrant. Occurrence highly irregular, sometimes in flocks, with major irruptions reaching Rhode Island in 1961–1963 and 1975.

Nonbreeding

The first record was of a flock of 30, a remarkably large number for the Northeast, which was found at Quicksand Pond on 23 December 1961 by a hunter, who reportedly shot 2 that day and 2 more the next day. One of these, obtained by Brian Harrington, is now #7071 in the skeleton collection at the MCZ. A flock of 8 was seen by Henry E. Childs Sr. at Briggs Marsh 14 September 1963. More than a decade later, 3 were seen at Trustom Pond 10 July 1975 by W. Quist and C. Allin, with 1 still present 20–23 July 1975 (L. Fish, DLK, KKi, FPe). An irruption to the Northeast occurred during 1975, with 5 records in Massachusetts during the period 30 May–18 November of the same year (Veit and Petersen 1993). A single bird was seen at Trustom Pond 9 November 1996 (J. Gass). While Rhode Island records seem to have declined since the incursions of the early 1960s and mid-1970s, vagrant flocks continue to visit the northeastern states (e.g., near Stonington, Connecticut, in 1987 and 3 at Jamaica Bay NWR, Long Island, New York, 28 May 2006).

Snow Goose (*Anser caerulescens*)

Status

Uncommon transient and winter resident.

Nonbreeding

Fall. Large flocks of Snow Geese, probably of the northeastern-most subspecies *atlanticus* ("Greater Snow Goose"), were known to occur during colonial times in New England, but by the turn of the twentieth century, they had been almost eradicated by overhunting. The only surviving report is of a flock of 250, possibly *atlanticus*, that landed in the ocean off Quonochontaug on 15 April 1916 after "a hard northeast blow" (fide HSH).

With decades of complete protection on their wintering grounds, Snow Geese have made a profound recovery since the turn of the twentieth century to a current total population of ~7 million birds, representing a 14- to 25-fold increase since 1900 (Mowbray et al. 2020). Their rarity at the turn of the twentieth century is suggested by the first Rhode Island records: a single bird at Block Island on 21 October 1932 (EDi), and 7 seen and 2 shot there 27 November 1933 (EDi). The first large flock seen was 120 (1 collected) at Block Island on 16 November 1936 (EDi). On 15 November 1945, however, 1000 alighted in the Quonochontaug marshes (HSH), and 7 were seen the next day at Block Island (EDi). This was an exceptional number at the time.

Fall maxima. 7000 flying over Block Island 4 October 1998 (ASRI): this count was unprecedented for Rhode Island and coincided with a similarly large flight of 10,000 birds over Long Island the same weekend (Schiff and Wollin 1999); 2000 flying past Napatree Point 7 November 1987 (LAH; fide HCa); 400, North Kingstown, and 84, South County, 12 November 1979 (Jarman); 300–400, Pachet Brook Reservoir, 20 October–4 November 1995 (JO, DLE et al.). There has been an increasing tendency for numbers to appear earlier in the fall, parallel with growth in the population, and such early flights included 4 reports in 1986 totaling 620, with 495 over Napatree Point on 2 October (CRa) and 118 at Block Island on 4 October (RFa et al.). Flocks of 100 over Barrington on 6 October 1989 (JCB) and 110 over Wallum Lake 14 October 1989 (SMi) were large for inland locations.

Winter. There have been records of a few individuals most winters in coastal areas since the 1930s. Counts are usually of fewer than 10 birds.

Winter maxima. Most birds in recent years appear to be *atlanticus*: 160, Point Judith, 6 December 1998 (STs); 99, Newport–Westport CBC, 3 January 2009 (REm et al.); 98, Newport–Westport CBC total, 17 December 2005 (REm); 80, Wood River Junction, 9 December 2013 (J. Beuth; photo); 65 (including 2 "Blues") migrating over Burlingame State Park, 10 December 2005 (STs); 16, Middletown, 29 December 2012 (B. Sherman).

Spring. The earliest report of spring migrants is of 5 flying north over North Kingstown on 21 February 1971 (DWa); another appeared with migrant Canada Geese at Peckham's Marsh on 1 March 1958 (SSD et al.). Peak numbers appear 20 March–20 April.

Spring maxima. One of the largest spring flights involved 3 flocks totaling over 600 birds over Jerusalem, Jamestown, and Riverside on 31 March 1992 (DLK, ChP, DFi). Another flock of 200 passed over the East Providence Reservoir 31 March 1951 (FPe, EWe). Most flocks are smaller than 30 birds in spring, when the migration route appears to be even more persistently through the interior than in fall. Smaller numbers are seen at inland locations.

Extreme dates. 30 September 2000, Charlestown Breachway (CRa), and 30 September 2008, Matunuck (1 bird; P. Budlong), to 26 April 2015, Seapowet (1 juvenile; REm).

Subspecies

Anser c. caerulescens, with forms "Blue" Goose and "Lesser" Snow Goose, also occurs in Rhode Island. These 2 freely interbreeding forms nest largely northwest of Hudson Bay (including Alaska and Siberia) and winter mainly in California, eastern Texas, and Louisiana. Birds of this race in either color morph were formerly rare but have been increasing in Rhode Island, as illustrated by reports of blue-morph birds in 9 of 10 years since 2005 and by the significant proportion of blue-morph birds present in the huge flight of 4 October 1998.

Older records of "Blue Geese" include a young male shot at Westerly, 16 March 1894 (E.R. Lewis; recently examined at RWPM); 1 shot on Dyer Island, 9 November 1912, by Sinclair Tucker (formerly BMS #12241 but since presumed lost), and 3 adults seen and 1 caught alive off Sakonnet, 15 February 1940 (Foote, fide HNG; photo in *Providence Journal,* 29 August 1940). The only record between 1940 and 1970 is of 2 at Easton Pond on 14 February 1962 (REW).

Maxima. 13, Watson Reservoir, 4 November 1995 (DLE); 5, Gardiners Pond, 28 November 1999 (SSM).

Ross's Goose (*Anser rossii*)

Status

Vagrant. About 10 records, 1974 on.

Nonbreeding

A white-morph individual photographed at Block Island 12 October 1974 (PAB) was judged to be a hybrid Ross's × Snow Goose. A blue-morph individual seen at Fresh Pond, Quonochontaug, and examined at 100 ft (30 m) on 7 October 1992 (MJM) seemed to be a Ross's Goose. A similar bird, possibly the same individual, was examined by many observers at Brigantine NWR in New Jersey a few days later. All others have been white-morph individuals: at Tiverton in October 1998 (RBo); Trustom Pond, 4–8 December 2001 (JSJ; photo); Bristol, 8–12 March 2011 (J. Koger, DFi; photos); Slocum, 23 November–1 December 2013 (STs, PLE, DFi; photos); Boom

Bridge Pond, Westerly, 27 December 2017 (DPe; photos); Trustom Pond, 27 January 2018 (DPe; photos).

Graylag Goose (*Anser anser*)

Status

Vagrant. One record.

Nonbreeding

A Graylag Goose found at Watchemocket Cove 16 December 2016 (JSJ, CPe et al.) was seen and photographed by many; it remained until at least 5 February 2017 (eBird; see Photo 2). While Graylag Geese and their "barnyard" derivatives are commonly kept and bred in captivity, the substantially growing population in Iceland (Kolbeinsson 2017), combined with the increasing number of records since about 2007 in northeastern North America, suggests, as for other Palearctic geese, that these birds are wild vagrants. The Watchemocket bird appeared to be healthy and "trim" and behaved like a wild bird.

Greater White-fronted Goose (*Anser albifrons*)

Status

Rare migrant, mainly in late fall. Since 2005, occurs annually, usually one to two birds per year.

Nonbreeding

An adult with an orange bill was found by Eloise Saunders at Trustom Pond NWR 22 October–10 December 1972 (m.obs.). Another bird that remained at Trustom 13–20 December 1981 (RTL, DLK, RAC et al.) was described as an immature with a pinkish bill (DLK). In the winter of 1985–1986 (during an "unprecedented" invasion of these geese to southern New England, when up to 7 were present from Connecticut to Massachusetts), another adult with a "yellow" bill was seen at Cross Mills Pond on 13 February 1986 (DFi, DLK, JGe, RGe). A third bird, also apparently *flavirostris*, remained at Matunuck from 27 November 1988 to at least 26 December (RAC, DLK). This individual, observed at close range, was described as having a bill color that, depending on the lighting, varied from pinkish or orange to yellowish green (DLK). This bird was reportedly accompanied by 2 others that were shot in November by hunters. There is a fifth report of a bird seen at Moonstone on 5 November 1990 (WBo). Another adult with orange bill and legs was present on Newport Neck 9 February–14 March 1995 (BG, JMu, RS, DLE, RLF et al.). One was seen in the Little Compton area 5 November 1995–6 January 1996 (JO, RBo). An adult at the Richmond turf farms 25 January 1999 was critically identified as *flavirostris* (SSM).

Maxima. Five, Weekapaug, 2 March 2013 (DFi, STa); 4, Sunset Farm, Narragansett, 26–31 January 2008 (m.obs.).

Extreme dates. 4 October 2006, Tiverton (1; RFa, RBo), to 4 April 2004, Napatree (2 migrating; CRa), and 4 May 1999, Warren Reservoir (1; RBo).

Subspecies

Both North American-breeding *A. a. gambelli* and Greenland-breeding *A. a. flavirostris* occur in the eastern United States. Most Rhode Island records seem to be of *flavirostris*, which have orange rather than pink bills.

Pink-footed Goose (*Anser brachyrhynchus*)

Status

Vagrant. Three records.

Nonbreeding

Two Pink-footed Geese were present at Fort Adams in Newport 6 January–19 February 2007 (R. Weaver et al.; photos: see Photo 3). A second bird was found at Slocum turf fields 11–16 November 2016 (Steve Brenner, m.obs.). One was in Westerly 21 December 2017 (DPe). These first records for Rhode Island are consistent with an expanding European population and an increasing number of records in northeastern North America (Mitchell and Hearn 2004).

Brant (*Branta bernicla*)

Status

Fairly common migrant and winter resident.

Historical

Sir Charles Blagden, among the occupying British forces in Newport during the American Revolution, stated that in 1776–1780, "in spring, with a southerly wind, prodigious flocks pass close to Sachawest [*sic*] Point" (Blagden 1903; Howe 1905). Later, Brant were severely persecuted during the shooting era (ca. 1860–1918) and had become scarce in Rhode Island by the turn of the twentieth century. With protection, they increased until the 1931 blight that devastated beds of eelgrass, their principal food. By 1935, Hathaway noted that Brant had become rare. With return of the eelgrass and a partial switch in food habits (to include sea lettuce), Brant recovered during the 1940s and have been numerous since then. During the shooting era Brant did not winter in Rhode Island. Since 1970 there has been a growing tendency for flocks to remain all winter (especially on Narragansett Bay and in the Sakonnet Passage) and for migrant flocks to linger into late May (e.g., at Napatree Point).

Considering the numbers of late-spring migrants and the increase in wintering birds, overall numbers in Rhode Island seem to be as great or greater in recent years than at any period since the Civil War.

Nonbreeding

Spring. Field data show 2 migration peaks, 1 from mid-March through late April and a smaller peak (especially in recent decades, and usually for resting flocks) in mid-May or later.

Spring maxima. 1500, Warwick, 13 May 2012 (v.o.); 1300 flying east in 1 hour off Narragansett 23 March 1958 (CWo, DLK); 1268, upper Narragansett Bay, 4 April 2004 (RFa); 1000 at Barrington, 22 March 1989 (JCB); 906 past Point Judith, 3 April 1993 (RLF); 900, Napatree Point, 30 March 1988 (CRa); 850+ in upper Narragansett Bay, 14 March 1995 (RLF). Although there was little suggestion of any late-spring flight in the first half of the twentieth century, there are now many records of sizable flocks in May, especially at Napatree Point; e.g., 460, 11 May 2001 (CRa).

Pelagic. Three, Cox Ledge, 19 March 2005 (GW).

Fall. A "great flight" occurred along the outer coast 26 October 1985. Flocks totaling 5096 birds were seen flying westward past Point Judith from the area of Martha's Vineyard throughout the morning hours (RLF). Small flocks of migrants are regularly seen through most of November, but migration usually ends in early December. Hard-weather movements involving small numbers have been reported from Block Island later in December. Rare before mid-September.

Fall maxima. See above; also, 1535, Barrington area, 30 November 2003 (RFa); 550 past Point Judith, 24 October 1981 (RLF); 300 there on 29 October 1984 (ESa); 300, Barrington, 30 November 1989 (JCB).

Winter. Some of the first recent reports of large flocks in late winter included 450 at Matunuck, 5 February 1978 (DWa), and 600 at Barrington, 20 February 1984 (JCB). Since 1980, wintering groups have finally taken up residence most years in the upper Sakonnet Passage through December (161 off Portsmouth 22 December 1985; RLF), and flocks have now been reported in numerous other locations, often in upper Narragansett Bay and less frequently in Mount Hope Bay. Particularly encouraging is the regular presence of flocks in the Barrington–Warwick–Bristol area; e.g., 1400, Brushneck Cove, Warwick, 17 January 2006 (DFW); 995, Bullock's Cove, 2 February 2003 (RFa); 896, Barrington, 9 January 1993 (CBC); 800, Bristol, 31 January 1990 (DFi, DLE); 600, Barrington, 27 February 1986 (JCB); 400, Barrington, throughout February 1989 (JCB).

Summer. There are records of <5 birds in most years between mid-June and mid-September as far north as Mount Hope Bay. There are 5 records of 1–3 birds each at Napatree Point since 2005 (CRa).

Summer maxima. 30, Sandy Point Island, 31 July 1971 (RLF).

Inland. Brant are rare inland during migration. One was reported shot in central Exeter in April 1894 (J. M. Southwick); 1 was seen at Diamond Hill Reservoir, Cumberland, 18 November 1987 (DMK), and another was there 16 October 2004 (MLy). In fall, they have only recently been recorded as diurnal migrants crossing over or emerging from the Rhode Island interior, although several questionable reports may be correct.

Subspecies

The subspecies that regularly occurs in the East is the "Pale-breasted" Brant (*B. b. hrota*), breeding largely in the eastern Canadian Arctic. Individuals described as "Black" Brant (*B. b. nigricans*), the race normally wintering on the Pacific coast, have been reported on 2 occasions. Sturtevant knew of 1 that had been shot by the market gunner R. L. Dring 28 November 1907 in the brook at Third Beach, Middletown, and mounted by C. B. Clarke of Newport. The whereabouts of this specimen are unknown. Another was found dead on the beach at Middletown 24 December 1932 by J. J. Lynch, as reported to Hathaway by RCC.

Barnacle Goose (*Branta leucopsis*)

Status

Vagrant. About 12 records.

Nonbreeding

The first Rhode Island record was of a bird shot in October 1910 at Block Island by Lovell Dickens. The wings were saved, and 1 in the Hathaway collection was examined by Clement and the author. The other wing was sent to RWPM; neither wing appears to be extant. The wings were examined by Bent but the record was omitted by him (Bent 1923), suggesting suspicion about its authenticity. Another was shot in North Scituate on 5 October 1925 by M. Salisbury (extant at BMS, #17491). Yet another was reportedly seen in Rhode Island in December 1925 (Forbush, "Items of Interest," no. 57).

There were no records for many years until 11 November 1978, when 1 was shot in fields north of Trustom Pond NWR. A photograph by DVT in the possession of C. Allin revealed no evidence of captivity. One was present at Trustom Pond NWR 2–15 July 1983 (DVT, JHW). One was seen at West Kingston 4 March–7 April 1989 and judged to be a wild bird by its extremely wary behavior (DLK, AFr, GTa, ATa, SHa, et al.). Others have been seen at Newport, 17 February 1990 (H. W. Harrington; photo, *AB* 44:236); Maschaug Pond, 27 December 1988 (CRa); and Greenville, 18 March 1991 (v.o.). Singles were seen at Newport–Westport CBC, 15 December 2001 (Westport, Massachusetts; RBo et al.); Lawton Valley Reservoir, 16 January 2002 (S. Wheelock); Portsmouth, 13 January 2007 (DFi, STa), and continuing in that area until 28 January; and Shartner Farms, Exeter–North Kingstown, 6 November 2008 (R. Miller). One was at Middletown, 9–19 January 2014 (P. Champlin, m.obs.); 1 at Wallum Lake, 6 April 2014 (DPe; photo); 1 remained at St. Mary's Pond January through March 2017 (Michelle St. Sauveur et al.; photos); 1, Slocum, 29–30 January 2019 (CPe); 1, Moonstone, 17–23 March 2019 (D. Mozzonni et al.), and probably the same bird 24 March 2019 at Beavertail (Bill Thompson).

Elsewhere in the Northeast, particularly in Atlantic Canada, bona fide stragglers have appeared (e.g., a bird banded in Norway was shot in Newfoundland, 1981), as have flocks of undoubtedly legitimate vagrants.

Cackling Goose (*Branta hutchinsii*)

Status

Rare winter resident.

Nonbreeding

Records have increased substantially since the taxonomic separation of this species from the Canada Goose in 2003 (Banks et al. 2003) and in conjunction with an ongoing massive increase in abundance of all geese in the Canada/Cackling complex. At present, small numbers of Cackling Geese of the eastern Arctic-breeding subspecies *hutchinsii* ("Richardson's Cackling Goose") occur every winter. In addition, there is 1 well-described report of a Cackling Goose of the subspecies *B. h. minima*, which breeds in western Alaska and ordinarily winters in the Pacific Northwest: Usquepaugh, December 1998 (SSM, PAB et al.).

Maxima. Ten, Newport, 14 March 2011 (DFi, STa); 9, Fort Getty, 9 March 2012 (T Au); 3, Newport–Westport CBC, 17 December 2011.

Extreme dates. 19 November 2015, Newport (3; R. Weaver), to 25 March 2007, Matunuck (CNu).

Canada Goose (*Branta canadensis*)

Status

Abundant migrant and winter resident; common breeder. Has greatly increased since the 1970s with the establishment of a large nonmigratory population and increased overwintering of migratory flocks.

Historical

Canada Geese might have bred as far south as Rhode Island historically, based on the observed readiness with which feral birds have recently spread throughout southern New England. Some evidence for this supposition can be found in the writings of Gosnold and Audubon for eastern Massachusetts (Griscom and Snyder 1955), and an anonymous sailor described the abundance of "wild geese" in the vicinity of Prudence Island in 1748 (Calder 1935). By 1900, Canada Geese had been eradicated from all but northernmost New England as breeders. Rives (1883–1884) stated that "they hardly ever stop [in the Newport area] but are known to have done so on a few occasions." They were especially rare in winter, and the first winter flock of 100 appeared at Middletown on 10 December 1912 (ESt), when the unlimited shooting era was over but waterfowl were still hunted in spring. In later years, with stricter hunting seasons and bag limits, transient numbers recovered slowly, and small flocks of presumably wild Canada Geese were regular in winter by the 1940s.

Rives (1883–1884) stated that Canada Geese were frequently "domesticated" or kept captive in the Newport area. Many of these captive birds were probably live decoys or birds descended from such decoys. The use of live decoys was banned in the 1930s, and these birds were often allowed to go wild. A few fugitive or semicaptive individuals or pairs were always present in Rhode Island's coastal regions during the 1930s and 1940s.

Breeding

The earliest Rhode Island breeding by Canada Geese was in close association with human habitations. Nesting began to pick up momentum and assume some regularity around 1960 and was recorded near Briggs Marsh in 1957, Potowomut Pond in the early 1960s, Cards Pond in 1961, and Ocean Drive, Newport, through the 1960s. The gradual spread of these birds was augmented by the RIDFW during the period 1963–1967, when 120 adults were liberated in statewide locations. During and after this period, others probably dispersed into Rhode Island from releases in neighboring states. They bred in Wakefield in 1964, Wickford in 1965, on Block Island in 1967 or before, and at Quonochontaug in 1968. By the mid-1970s, geese were widely if not universally distributed in coastal areas and had begun to nest in the interior. At the same time, unprecedented flocks of nonbreeding or molting birds began to be encountered in the breeding season: e.g., 117, Scituate Reservoir, 21 June 1973 (DWa); 100, Jamestown, throughout June 1976 (RLF); 150, Trustom Pond, 15 June 1982 (DVT); 144, Sachem Pond, 19 June 2000 (CRa, REn).

Canada Geese often choose offshore islands for breeding, particularly in or near Narragansett Bay, where they nest in close proximity to gulls, cormorants, and other marine birds. Breeding has been particularly concentrated on the islands off Sakonnet and Newport, on Hope and Rose Islands (16 nests on Rose, 1981; 22 in 1984, CCA), on the Dumpling Islands off Jamestown, on Sandy Point Island, and elsewhere. Others now breed in the vicinity of many of the brackish and freshwater ponds in coastal areas as well as on many inland ponds. The species did not occupy Rhode Island's inland ponds, especially in western sections, with the same rapidity noted farther north in the interior of New England. While the island-breeding coastal population seems to have leveled off in the 1980s, increases still continued through the same period in the Rhode Island interior.

Nonbreeding

Spring. Migrant flocks appear in late February and March, reaching peak numbers in late March.

Spring maxima. 3500, Slocum, 11 March 2009 (SSM); 2400, Tiverton–Little Compton, 15 March 1995 (RBo, DLE, RLF); 1000, Briggs Marsh, 14 February 1949 (DLE).

Fall. A few early transients occasionally appear in late summer (20, Block Island, 25 August 1942; EDi) or in early September, but peak movement is in mid to late October or later. Some flocks are still observed migrating south in December and January after severe winter weather.

Winter. Canada Geese increased from rare winter residents in the first quarter of the twentieth century to common winter residents during 1940–1965. Since the mid-1970s, however, a much steeper increase has occurred. Large resting aggregations are increasingly found during the fall and winter months at the Trustom Pond NWR and at St. Mary's Pond in Middletown, with daytime dispersal to fields throughout coastal areas and Narragansett Bay, where geese had seldom fed in the past.

Away from the coast, Canada Geese were rare in winter until the 1980s or so, but flocks now winter on a regular basis.

Winter maxima, coastal. 11,700, Newport–Westport CBC, 3 January 2009 (REm); 10,249, Newport–Westport CBC, 17 December 2005 (REm); 7537, South Kingstown CBC, 21 December 2013 (STs).

Winter maxima, inland. 1200, East Providence Reservoir, 18 January 1986 (DLE); 700, Barrington, 18 January 1990 (DLE); 450, East Providence, 10 January 1992 (v.o.).

Mute Swan (*Cygnus olor*)

Status

Abundant and conspicuous resident on coastal ponds, increasing inland.

Breeding

Native to Asia, Mute Swans were introduced on the Hudson River above New York City in 1910 and at 2 locations on outer Long Island in 1912 (Bull 1964). They remained uncommon and local along the south shore of Long Island until about 1926, after which they began to spread into our area. While some early Rhode Island records of this species may have resulted from captive birds released elsewhere, our local population appears to have been entirely derived from an expansion and spread of the Long Island population.

A swan seen at Trustom Pond on 7 September 1910 (Hathaway 1913), and the same or another seen flying over the Point Judith marshes "in the first week of September" of the same year, are almost certainly the present species, since the date would be very early for Tundra Swan. Otherwise, Mute Swans were first recorded at Block Island, where a flock of 8 appeared on 17 June 1923 (EDi). Thereafter, small flocks were recorded at Block Island on 4 additional occasions in summer during the period 1923–1929. They were first reported on the mainland 11 July 1930, when a flock of 8 was seen at Trustom Pond (MSo). They were observed annually after 1937, and a flock of 40 was noted at Block Island as early as 1941.

Numbers increased dramatically in the 1940s, particularly at Briggs Marsh. Nesting was confirmed there in 1948 (RCC) and 1949 (DLE), with probable breeding as early as 1947 according to local residents. The species nested at Easton Pond in 1950 and at both Almy and Easton Ponds in 1951. It first nested at Charlestown in 1953, Quonochontaug and Narragansett in 1954, Westerly in 1954 or 1955, and Block Island in 1957. From these early beginnings, the species has spread to virtually all of the brackish and freshwater coastal ponds in the state. As many as 35 pairs were located statewide by 1962, and saturation began to occur at the best locations between 1959 and 1965.

In the 1970s, Mute Swans began to breed or attempted to breed at a minimum of 14 inland locations (fide JEM, CCA), including White Brook Pond, Wordens Pond, and the Great Swamp impoundment (all in 1970); Pachet Brook Reservoir in Little Compton and Frenchtown in East Greenwich (1972); East Providence Reservoir and Pawtucket Pond on the Ten Mile River (1974, 1976); a pond in central Exeter (1976); Cranston (1977); Lonsdale Marsh (1979); and Barneys Pond in Lincoln and Echo Lake in Barrington (1982). Additional locations were noted in the 1980s atlas period at Cranston (3 locations), Watchemoket Cove, and the East Providence Reservoir (3 pairs). By the mid-1990s there were 7 nests at the East Providence Reservoir and resident pairs along the Seekonk and Blackstone Rivers north to Manville (CCA). Early spring marks the beginnings of nesting activity in this species, which is one of the earliest breeders in the state. Surviving winterers return to the best breeding habitat in March or earlier, and incubation is underway in April.

Nonbreeding

The highest nonbreeding count for the period 1946–1950 was 21 birds; for 1951–1955, 41 birds; for 1956–1960, 150 birds. Statewide aerial counts (exclusive of Block Island) totaled 416 in 1965 and 560 in 1971. Continual increases noted in the vicinity of Providence are an important sign of further increase in the statewide population despite control measures (see below).

Maxima. 514, Briggs Marsh, 7 August 1990 (CCA); 437, Napatree CBC, 18 December 2005; 422, Conimicut Point, 30 December 2010 (DFW); 347, Occupessatuxet Cove, 14 December 2014 (DFW); 302, East Providence Reservoir, 25 September 2005 (MLy); 215, Watchemoket, 24 November 1995 (DLE).

Population control measures. A control program was begun by RIDFW in 1970, the first such program in the Atlantic states. Between 1972 and 1978, approximately 500 fully grown swans were removed from the population and 300 eggs were sterilized in a process called "egg-shaking"; during the period 1970–1974, the number of known nests dropped from 60 to 25. After 1979, control measures were limited to continued egg sterilization; 2121 eggs were shaken during the period 1979–1986 (about 80% of Rhode Island nests), and 500 eggs were shaken in 1988 alone. In the 1980s, even with an undoubted influx from uncontrolled populations in neighboring states, these efforts had nearly stabilized the local population, despite its high proportion of breeding-age adults. Allin et al. (1987) discussed the population expansion of this species, the ongoing control program in Rhode Island, and presented a model of probable statewide numbers had no controls been implemented. Despite annual egg shaking at the majority of the state's nests, however, continued population increase has been noted. David Emerson remarked in 1989 that the species was still spreading inland rapidly during the last few years, and that total numbers had nearly doubled since 1984. From a statewide population of 560 in 1971, the population had in fact nearly tripled to 1400 in the 13 years through 1993.

Trumpeter Swan (*Cygnus buccinator*)

Status

Vagrant. Two records.

Nonbreeding

Erroneously attributed to Rhode Island by Southwick (1888). This species may have occurred along the New England coast in colonial times, when it bred as far east as Indiana and James Bay. One was found on the Seekonk River 11 May 2014 (B. Shamgochian; photos) and remained at least to 21 May; it was seen and photographed by many. Another, an apparent immature, appeared on a small pond at Clay Head, Block Island, 8–10 April 2018 (Hatfield, KGa, CPe et al.; photos) There are now many records of Trumpeter Swans in the Northeast, most likely deriving from stocking programs in Ontario (Carroll and Swift 2000).

Tundra Swan (*Cygnus columbianus*)

Status

Rare migrant and winter visitor. Reported every year since 2000.

Historical

This native swan became very rare during the shooting era. One of two shot over Quonochontaug Pond 12 November 1879 was the first collected in Rhode Island in more than 30 years, according to Newton Dexter. Others appeared in 1911, 1917, and 1928. They gradually increased thereafter and are now reported almost annually.

Nonbreeding

Tundra Swans have most often been seen at Block Island or on the coastal ponds but have been recorded at several inland locations including Exeter, Wordens and Watchaug Ponds, Tiogue Lake (now annual in small numbers), Scituate Reservoir, Blackstone River, Edgewood, Barrington, Providence River, and Pawtucket.

Most migrants are seen in November and early December. Only very rarely do individuals or flocks spend longer periods in Rhode Island during winter; e.g., 2 that spent most of the winter of 1968–1969 on a pond at Cross Mills (DLK et al.); 2 on Tiogue Lake all February 1991 (JO et al.).

Maxima. 14, Bristol, 10–15 December 1993 (MHC); 11, Watson Reservoir, 14–30 November 1993 (DLE, REm, RBo et al.); 10 (including 2 shot), Block Island, 8 March 1932 (EDi); 10, Block Island CBC, 22 December 2016 (m.obs.); 8, Block Island (after a northeast storm), 5–27 March 1960 (MSl).

Extreme dates. 22 October 1997, 2 each in Warren and Touisset (GBi), to 11 May 1963, Bonnet Shores (CWo, DLK).

Common Shelduck (*Tadorna tadorna*)

Status

Vagrant. Three to four records.

Nonbreeding

Two adults were closely observed at Jamestown on 2 August–4 September 1960 (MDa et al.). A male appeared 9 October 1980 at Easton Pond (SDz, TTr). Another male was seen repeatedly in Barrington, Tiverton, and in the Seekonk River in April 1981 (v.o.), and in Hundred Acre Cove on 26 April 1982 (JCB). A flock of 7 was seen at Quicksand Pond 6 September 1993 (DLE, REm).

While many North American records of Common Shelducks have been dismissed, without supporting evidence, as being of escaped cage birds, recent occurrences in North America are more suggestive of wild origin, as are the dates of occurrence listed above. Common Shelducks have dramatically increased as breeders in Iceland since about 1990 (Kolbeinsson 2017; Tierney et al. 2020). There is growing quantitative evidence for the link between population growth, vagrancy, and colonization by birds (Zawadzki et al. 2019, 2021; Veit et al. 2021).

Ruddy Shelduck (*Tadorna ferruginea*)

Status

Vagrant. Three to four records.

Nonbreeding

An adult first seen at Sachem Pond, Block Island, 1 October 1951 (E. Lee) was later shot by an unidentified gunner on 8 October. The specimen was salvaged by E. Dickens and is in the Block Island Collection, where a recent examination by the author has shown 2 previously overlooked signs of confinement: (1) the breast has a large area that was apparently worn extensively from lying on the ground, with most of the breast feathers replaced shortly before the date of collection; and (2) several rectrices were apparently broken before the bird was collected. In addition, it is known that the bird allowed a close approach by the hunter who shot it. Another was seen flying over Ninigret Pond 1 June 1978, during a period when several were reported on Long Island (RLF). One was at Trustom Pond, 31 July–31 August 1987 (v.o.), and again 20 November 1987 (v.o.). Two were at Hope Valley 29 February 1992 (JZy); and 2 were at the East Providence Reservoir 22–24 March 1992 (BCG, WG). The occurrence of Ruddy Shelducks throughout the western Palearctic has been examined by Vinicombe and Harrop (1999), who concluded that "periodic invasions [of wild birds] do occur," as was especially true during a major irruption during 1994.

Wood Duck (*Aix sponsa*)

Status

Fairly common but somewhat local summer resident and fairly common transient; rare in winter.

Historical

As the most desired waterbird among gunners of the shooting era, Wood Ducks declined and were uncommon migrants and rare summer residents during the period 1870–1920. They were recorded breeding only at Pascoag in 1900 and in Providence County in 1903. They were still relatively rare summer residents in the mid-1930s, and the 1938 hurricane removed most of the remaining natural nesting cavities.

Breeding

Numbers increased again by the late 1940s, after which the statewide population surged dramatically with the widespread provision of nest boxes by RIDFW starting in 1950. There were 762 and 1000 available boxes in 1969 and 1977, respectively; 315 boxes held eggs in 1969 (fide CCA). They now breed widely near suitable freshwater ponds almost throughout the state, even in such urbanized areas as Providence, Pawtucket, and Central Falls. They are particularly common in northeastern Rhode Island but breed fairly commonly throughout western Rhode Island south to Westerly and Matunuck and occasionally Newport. They are still uncommon on Conanicut and Aquidneck Islands and at Little Compton. They were first confirmed breeding at Block Island in 1981, despite the availability of nesting boxes there for much longer.

Nonbreeding

Spring. The earliest probable spring transients were 1 at Cranston on 5 February 1972 (FPe), and 1 at Wickford, 6 February 1987 (v.o.)). Notable are occurrences on the late-winter counts on Block Island, following absences on the CBC: 2 on 20 February 2006 (SSM) and 2 on 1 March 2003 (PJL). Wood Ducks generally arrive in early March (e.g., 5 at the Great Swamp 7 March 1998; SSM), but sometimes not until late March, as during the unusually cool season of 1941. Spring aggregations are usually small, infrequently exceeding a dozen.

Spring maxima. 50, Pawcatuck River, 13 March 1992 (WRE et al.); 50, Lonsdale, 15 April 1992 (REn). Later in spring, they become quiet during the period of incubation and are seldom seen in numbers (25 at Lonsdale, 7 June 1974; RLF).

Fall. Wood Ducks are common fall transients on most freshwater ponds and streams throughout the state, including Block Island. While they have only rarely been reported resting on salt water, they are not infrequently seen migrating off coastal promontories with flocks of scoters. Occasional individuals are reported from waters south of Block Island and Cox Ledge.

Postbreeding concentrations are seen from August onward, increasing during early fall with the arrival of migrants. Large numbers have been seen at the East Providence Reservoir, where flocks that spend the day on the lower Ten Mile River pass over the reservoir at dusk to reach their night feeding areas along the upper Ten Mile River in Pawtucket. Counts here formerly exceeded most other areas.

Fall maxima. All from Great Swamp impoundment unless stated otherwise: 400, 22 October 1976 (JEM); 210, 15 October 1956 (JSu); 200, 17 September 1957 (JSu); 200+ in early October 1966 (Larson); 164, Lonsdale 12 November 1993 (REn); 61, 30 September 2014 (CRa); 20, Block Island, 28–30 September 1990 (RFa).

Winter. Most birds are gone from the state by late November, but there are many records of lingerers into December. A flock of 22 was seen at East Greenwich 11 December 1984 (fide KKi). They have been recorded at the East Providence Reservoir

10 January 1965 and 17 January 1953 (HEC, REW, RLF). Three were in the Saugatucket River 17 January 1988 (REn). A male was present at Lonsdale until 11 February 1992 (REn). They were said to have wintered in Johnston in 1951–1952 and at Hope (a pair) in 1970–1971. There are about 17 reports in January and February since 2005. They can winter where water does not freeze along rivers or springs and sometimes in salt pond breachways.

Blue-winged Teal (*Spatula discors*)

Status

Fairly common migrant, especially in early fall, and rare and sporadic summer resident. Very rare in winter.

Historical

Blue-winged Teals bred at Sakonnet, probably at Long Pond, in 1888 or 1889 (Drowne) and in 1890 (N. Dexter). A pair with young at Quonochontaug was reported to Hathaway in 1902. Their status in subsequent decades is unknown because little field work was done after 1900, particularly in Newport County. In 1932, Bowen found a nest in the Seapowet area of Tiverton, and Clement found the state's largest aggregation at Peckham's Marsh, with up to five nesting pairs in 1934 and for a few years thereafter. Other nests were found at Briggs Marsh (1938 on), Middletown (1937 on), Carolina (1939); they were suspected of breeding in Portsmouth, Matunuck, and Weekapaug.

Breeding

The number of nesting pairs in Newport County remained relatively stable through the 1950s but then declined. Little suggestion of breeding was seen until 1984, when a brood was seen at Long Pond (REn, CRa); 2 broods were seen there 3 August 1985 (RBo). Elsewhere, pairs have been noted occasionally since 1960 in the Matunuck and Galilee areas, and a brood was seen at Block Island in 1970 (ELa). Most breeding has been at the coast, but 1 was seen at Lonsdale Marsh 22 May 1983 (CRa), and 2 were there 1–20 May 1988 (REn).

Summer maxima. 34, Little Compton, 26 July 1941 (RCC, WHD). More recently, an agitated female was watched at Long Pond, Little Compton, 4 July 2010 (REm), and pairs were seen at the Great Swamp 5 May 2011 and at the Charlestown Breachway 29 June 2012 (TAu).

Nonbreeding

There has been a decline since about 2000.

Spring. The earliest records in spring are for 1 at Perryville 26 February 1983 (RAC), and a pair in Tiverton 1 March 1940 (Taber). They are fairly common as migrants in spring (up to 30 per day per locality).

Fall. One of the earliest migrant waterfowl species, Blue-winged Teals usually increase in mid-August and peak in late August and September.

Fall maxima. 400, Little Compton, 27 August 1976 (DLE, REm); 350, Briggs Marsh, 3 September 1964 (Cobb); 330, Middletown, 5 October 1941 (WHD); 50, Briggs Marsh, 3 October 2015 (REm). A few birds are often present to about 10 October, and there are reports of small flocks through late October and into November (20 at Napatree Point 4 November 1989; CRa). Blue-winged Teals are rare in winter; there have been 10 records on the Newport–Westport CBC since 1949, and none since 2005.

Blue-winged Teals have been declining as fall transients for at least several decades. A count of 115 in Little Compton 8 September 1985 was considered to be good by late-1990s standards, and subsequent maxima have rarely exceeded 50.

Extreme dates. One to 2 birds have been recorded at Block Island on the Presidents Day Count, 2008–2011 (CRa et al.). Other extreme dates include 26 February 1983, Perryville (RAC); 9 January 1993, Roger Williams Park (CBC); and 29–30 January 1992, Lonsdale (REn).

Cinnamon Teal (*Spatula cyanoptera*)

Status

Vagrant. One record.

Nonbreeding

A male Cinnamon Teal was seen and photographed by Mike Tucker at Tiogue Lake 25 November 2003. While a local breeder claimed loss of a Cinnamon Teal at about the same time, no conclusive link with the Tiogue Lake bird could be established. Given the recent increase in records of Cinnamon Teals in the Northeast, it seems appropriate to list this record.

Northern Shoveler (*Spatula clypeata*)

Status

Uncommon but regular migrant and uncommon but increasing winter resident.

Historical

Gunners from Newport and Point Judith felt that Northern Shovelers were fairly regular up to perhaps 1875, but that they then decreased to very rare (seven records between 1894 and 1913). They were not recorded again until 1931. Reports gradually increased thereafter, and Northern Shovelers were fairly regular on freshwater ponds in fall but very scarce at other times between the 1940s and 1960s. More recently, they have further increased and have appeared more frequently in winter and spring.

Nonbreeding

Spring. Many spring records are likely of birds that wintered in, or close to, Rhode Island. The only spring records during the historical period were in 1894 and again in 1911, with a gradual increase thereafter. There were still fewer than 60 spring records up to the mid-1980s, including many known winterers. The main period of occurrence is from February through early May. A drake appeared on the Seekonk River 10 May 1992 (DLE), and another drake was seen in Barrington 11 May 1999 (RFa, DFi). Two males were seen at Galilee 25 May 1957 (DLK). Spring counts normally range from 1 to 4 individuals.

Spring maxima. 12, Sisson Pond, 17 April 1999 (REm); 7, Galilee, 19 April 1964 (CWo).

Fall. The earliest record in fall is 1 at Middletown 25 August 1959 (JBa). There are scattered records in September and early October, but most are seen between early October and early December, with a concentration of reports in the first half of November.

Fall maxima. 25, Sisson Pond, 19 December 1998 (REn et al.); 17, Sisson Pond, 13 November 1949 (CWo); 16, St. Mary's and Sisson Ponds, 26 October 2001 (JSJ). Shovelers used to be commonest at Easton Pond in Middletown, where cattail beds once provided the best fall habitat in the state for western pond ducks. More recently, White Brook and Sisson Ponds in Middletown have had the largest numbers.

Winter. There are many records of individuals or small flocks that have lingered into early winter, but these birds rarely remain into mid-January on the coastal ponds. The latest records for Easton Pond are for 4 on 10 January 1954 (RBo) and 9 on 17 January 1975 (DLK, KKi, FPe). Elsewhere, however, there are numerous midwinter records, including 2 males, Roger Williams Park, 17 February 1952 (RBo). The most frequent reports of overwintering come from the pond downstream from the fish hatchery on White Brook, Carolina. The water rarely freezes there, and hatchery waste provides food for the ducks. Records there include 1 from 9 February into April 1964 (DLK); 2 from 6 February to 12 April 1975 (v.o.); 3 from 10 January to 21 March 1976 (v.o.); and 1 to 4 birds from 2 January to 16 April 1977 (v.o.).

Winter maxima. 32, Newport–Westport CBC, 19 December 2015 (REm); 19, St. Mary's Pond, 6 February 2016 (CPe).

Extreme dates. 1 August 1997, Seapowet (1; Suzanne Vetromile); 18 August 2008, Briggs Marsh (1 female; REm). There is increasing overwintering; 1 spent 26–28 February 2011 at URI East Farm in a small pond with a fountain/aerator that kept the water open (S. Milton, m.obs.).

Gadwall (*Mareca strepera*)

Status

Uncommon and local summer resident and fairly common migrant. Substantial increase since the 1940s.

Historical

Gadwalls are one of the "western" ducks whose range has expanded most dramatically during the last century, partly owing to natural dispersal and colonization and partly assisted by introduction programs. There were only four records of Gadwalls prior to 1934, as follows: one shot in Newport 26 February 1882 (F. T. Jencks); one shot at Carolina 4 May 1892 (Howe and Sturtevant 1899); one shot at Point Judith 1 December 1902 (BMS #10024); one shot near Point Judith 11 November 1909 and obtained by Hathaway. The first modern Rhode Island records were two at Portsmouth 6 October 1934 (J. Nute, fide RCC); one at Middletown 7 April 1937 (WHD); one at Briggs Marsh 18 April 1937 (RCC); two males at Watchaug Pond 4 November 1941 (DLK); and two at Portsmouth 15 November 1941 (Williston). After the mid-1940s, Gadwalls appeared annually in Rhode Island. Gadwalls were usually found among flocks of American Wigeons, and counts before about 1958 never exceeded eight per day and were almost all during fall. As with Northern Shovelers, most Rhode Island Gadwalls were seen at Easton Pond, and birds normally disappeared by early January.

Bellrose (1976) indicated that some were breeding on the Atlantic coast at Bombay Hook, Delaware, as early as 1939, as well as at Pea Island, North Carolina, in 1940. By the mid-1940s, they were breeding more widely on the mid-Atlantic coast, with more than 40 pairs at a single Long Island location in 1950 (Bull 1964). As early as 1954, juvenile Gadwalls from Manitoba were liberated in Concord, Massachusetts, including 27 in 1957 and at least 77 in 1965. They began nesting at Barn Island, Connecticut, in the late 1950s.

Breeding

The first Rhode Island breeding records were of a pair with a single, large juvenile at Trustom Pond 11 July 1959 (DLK) and a pair with 6 flying young at Maschaug Pond 24 August 1960 (DLK). On 25 June 1965, 3 females were flushed from 3 nests on Sandy Point Island (RLF, REW). Nests are often but not always on islands that are more or less free from terrestrial predators. Breeding sites include

Sandy Point Island. Regularly 1–2 nests until the 1980s: 1 nest with 11 eggs, 11 June 1997 (RLF) and again 8 June 1999 (RLF).

Napatree Point. Suspected in the 1980s.

Maschaug Pond. Three broods, 1979 (CCA).

Trustom Pond. Regularly, 3 broods in 1978 (CCA).

Perryville. Adult and 6 young, 12 July 1976 (DLK); pair with young there 29 May 1978 (CWo).

Sakonnet islands. First nest in 1976; 3 nests in 1977 (RLF, JEM); a pair and brood there 25 April 1983 (RBo); female and nest with 11 eggs 27 May 1998 and again 23 May 2002 (RLF, CRa); also female and nest 3 June 2003 (RLF,CRa).

Long Pond, Sakonnet. Female and 4 young 13 August 1988 (DLE).

Block Island. Broods in the Sachem Pond area, 1980, 1981 (CCA, RBo).

Few nesting locations remain in Rhode Island. Gadwalls may still nest at Briggs Marsh and Trustom Pond but have not been seen at former nesting locations (mostly islands in Narragansett Bay and elsewhere) for many years.

Nonbreeding

Parallel with the increase in breeding, a substantial increase in numbers of migrants seen has occurred since about 1960.

Spring maxima. 150, Cards Pond, 21 March 1999 (STs).

Fall. The first sizable flock was 25 at Westerly on 16 November 1966 (ESa).

Fall maxima. 198, Echo Lake in Barrington, 30 November 2003 (RFa); 160, Echo Lake, 29 November 1987 (REn); 115, Long Pond, 30 October 1983 (v.o.).

Winter. After 1970, a concentration of Gadwalls persisted during the colder months on the ponds along Ocean Drive, Newport, particularly at Almy Pond. This Ocean Drive flock grew from 34 in 1971 to 275 in the 1980s.

Winter maxima. 293, Providence CBC, 9 January 1993; 196, South Kingstown CBC, 23 December 2006 (STs et al.); 177, Newport–Westport CBC, 17 December 2016 (REm et al.); 154 Barrington 16 January 2005 (RFa).

Eurasian Wigeon (*Mareca penelope*)

Status

Rare but regular migrant and winter resident. They are nearly always found with American Wigeons.

Historical

The first record of Eurasian Wigeon for New England was a bird shot on 20 September 1889 at Middletown by an unsuspecting Edward Sturtevant and recorded as an American Wigeon (Howe and Sturtevant 1899). This specimen was subsequently sent to William Brewster and redetermined to be a Eurasian Wigeon (Brewster 1909; specimen formerly BMS #16216, now apparently discarded). After this, Eurasian Wigeons were unrecorded, but were likely overlooked, until 4 November

1934, when two were seen at Gardiners Pond, Middletown (Lynch 1935). Frequency of observations increased during the 1940s and became almost annual in the 1950s. Since then, Eurasian Wigeons have been recorded annually in fall or winter, with occasional birds remaining into early spring.

Nonbreeding

Fall. Many of the earliest fall records are from the East Providence Reservoir: 1–12 September 1953 (DLE, RLF); 9 September 1966 (BRi, RRi); and 12 September 1977 at Barrington (BSp, CWo). Eurasian Wigeons have been most commonly seen between mid-October and mid-December. They have been much less numerous since the 1970s, as is also true of American Wigeons.

Fall maxima. 17 drakes statewide in November 1941; 6, East Providence Reservoir, 16 November 1952 (RCC, RLF); 6, East Providence Reservoir, 3 October 1953 (DGa et al.).

Winter. There are many winter records of groups of 1 to 4 birds that often survive the winter.

Spring. Many spring records are of overwintering birds, but in addition, migrants appear at sites where none were present during the previous winter. Two were seen at Briggs Marsh 23 April 1938 (DLK), and 1 remained at Carolina 16–25 April 1982 (DLK et al.). The latest spring dates are for 1 in a pond south of Watchemoket Cove 1 May 1995 (DFi) and another at Pawtucket 2–5 May 1999 (DVa, RFa). A drake was reported at Newport, 25 June 1993 (C. Manning; *AB* 47:5). There are 4 records for Block Island: female, 27–29 September 1996 (ASRI); male, 22 December 2005 (CRa); male, 20 February 2006 (CRa); female, 15 February 2010 (PDC).

Inland. The only inland record is for 1 at the Cumberland Reservoir, 23 October 1984 (REn).

Extreme dates. 1 September 1953, East Providence Reservoir, to 5 May 1999 and 2003, Pawtucket; also 25 June 1993, Newport.

American Wigeon (*Mareca americana*)

Status

Common to abundant fall migrant and winter resident; common in spring. Very rare in summer.

Historical

Formerly a rare or uncommon transient 1880–1913, although reported as common in the fall of 1882 (Jencks). Like many other "western" ducks, wigeon gradually increased during and after the "duck depression" in the 1920s and reached present-day abundance before 1930. American Wigeons are prone to shifting between ponds as food supplies change. It is possible that numbers in Rhode Island are related to the breaching of freshwater ponds on Martha's Vineyard (Griscom and Snyder 1955; Veit and Petersen 1993); as the coastal ponds fill with salt water, wild celery and other plants are killed, and the ducks move elsewhere to feed. Fall counts in Rhode Island continued to increase to a peak during the period 1948–1957, after which a general decline took place through the 1960s, 1970s, and beyond, especially at outer coastal locations. In the 1980s, Watchemoket Cove began to attract more wigeon than any other place in New England, and this state of affairs continues.

Nonbreeding

Wigeons have sharply declined since the mid-1950s. This reduction is particularly acute on the coastal ponds where Mute Swans have become a nuisance, but it is also noticeable at locations such as the East Providence Reservoir, where swans were not present in numbers until about 1985. Wigeon counts at the East Providence Reservoir, which before 1956 exceeded 3000 birds, plummeted to fewer than 500 for most of the next 2 decades, rising to 1100–1800 on only a few occasions during the period 1974–1985. Bellrose (1976) noted a southward shift in Atlantic coast wintering numbers from the Chesapeake area to the Carolinas during the 1960s, timing that fits well with above-mentioned reductions in the population in Rhode Island.

Fall. These ducks are most common in late fall and winter. The earliest fall arrival dates are 7 August 1984 at the East Providence Reservoir (DLE) and 13 August 1988 at Trustom Pond. There are numerous other records in late August, usually for less than 10 birds (but 34 on 29 August 1980 at the East Providence Reservoir; RLF). First arrivals normally appear from 10 September onward, and the population builds up to a well-defined peak 15 October–15 November at historically favorable locations such as the East Providence Reservoir, Easton Pond, Briggs Marsh (formerly), Trustom Pond, and other freshwater locations.

Fall maxima. 3000+ regularly in fall at the East Providence Reservoir during the early 1950s; 1290, Watchemoket Cove, 1 January 1992 (DLE et al.); 1200, Watchemoket Cove, 18 November 1993 (DLE); 1200, Watchemoket Cove, 28 December 1991 (DLE); 1100, Barrington, 31 January 1953 (RBo). Maximum counts are in the hundreds during the last 20 years. Numbers decline by early December on freshwater ponds, but in some years, many spend the winter in tidal estuaries near the head of Narragansett Bay (Barrington River and, more recently, Watchemoket Cove) in flocks of several hundred or more.

Winter. The wintering population varies with the severity of the weather.

Winter maxima. 400, Quicksand Pond, 17 January 2016 (REm).

Spring. There is little evidence of northward migration through Rhode Island in spring, but flocks overwintering in the region sometimes concentrate to create a minor peak in mid to late March.

Spring maxima. 857, Newport County, 23 March 1941 (RBo); 400, Watchemoket Cove, 29 March 1988 (DLE). There are less than a dozen May records, mainly of individuals; the latest include 1 at Block Island 12–14 May 1999 (CRa, REn), 1 on the Seekonk River 14–20 May 1992 (DLE), another at Jerusalem 22 May 1990 (REn et al.), and 2 at Briggs 23 May 1954 (Cobb).

Summer maxima. Six, Trustom Pond, 14 June 1981 (DVT, JHW). American Wigeons bred in 1972 on Penikese Island in Buzzards Bay, Massachusetts, and in the early 1980s on Monomoy (Veit and Petersen 1993). The presence of 3 males and a female at Galilee 19–20 June 1972 (KKi) suggests that nesting may have been attempted, especially in light of the Massachusetts records.

Extreme dates. Less than 10 records of summering; otherwise, 7 August 1984, East Providence Reservoir, to 23 May 1954, Briggs Marsh.

Mallard (*Anas platyrhynchos*)

Status

Common to abundant migrant and common resident; has greatly increased since the 1960s.

Breeding

Prior to 1900, Mallards were restricted as breeders to the western and northwestern part of North America. They have gradually extended their range eastward, and this expansion has been partly influenced by the deliberate introduction of birds into areas where they did not previously breed. The eastward expansion was not entirely caused by stocking, however, and it is impossible to tell to what extent they would have spread east without assistance. Rhode Island Division of Fish and Wildlife liberated 4440 captive birds in Rhode Island between 1955–1966, with the largest releases being 1034 in 1957, 978 in 1958, and 1992 in 1961. Such release programs proved generally ineffective at amplifying populations (Drilling et al. 2020), however, and they were halted in Rhode Island after 1966.

Regardless of their exact origins, semi-tame Mallards sharply increased in abundance in Rhode Island during the early 1960s. They are now ubiquitous at all seasons and nest in many of the same habitats as American Black Ducks, including most freshwater, brackish, and saltwater ponds and on most of the vegetated offshore islands (260 in Cumberland 23 August 1986; DMK). Wintering numbers have continued to increase from maximum of 100 per day in the early 1960s.

Maxima. 1695, South Kingstown CBC, 22 December 2007; 1683, Newport–Westport CBC, 17 December 2005; 699, Providence area, 1 January 1992 (CBC); 325, Diamond Hill Reservoir, 9 September 1991 (LS); 312 in Sachuest area, 16 December 1984 (RBu).

American Black Duck (*Anas rubripes*)

Status

Very common fall migrant and winter resident, less numerous in spring. Uncommon breeder.

Breeding

A long-term decline in American Black Ducks seems to have reached a nadir in the mid-1990s, and the population has been slowly increasing since then (Longcore et al. 2020). The decline seems to have been checked by changed hunting regulations, but American Black Ducks may also be affected by acid rain depleting their food supply in northeastern wetlands. Though Mallards may displace American Black Ducks from some Rhode Island nesting locations, hybridization between the 2 species appears to have had a minor, if any, impact (Longcore et al. 2020). They breed or have bred commonly in the vicinity of nearly all of the less-disturbed ponds and marshes of the state, both saltwater and freshwater, as well as on many islands in Narragansett Bay. Aggregations of 100 or more were once commonly seen in late summer on local ponds, these being local adults with full-grown young; recent counts are lower but perhaps increasing.

Nonbreeding

Spring. Spring concentrations are much lower than in fall and much reduced after mid-March.

Fall. Numbers increase gradually during fall to maxima between 15 November and mid-December, when 4500–10,000 have been counted statewide by RIDFW.

Fall and winter maxima. 3354, South County census, 28 December 1974 (v.o.); 2000, Watchaug Pond, 11–18 December 1948 (DLK); 1500, East Providence Reservoir, 10 December 1950 (BRi); 1250, Galilee, 21 January 1965 (KKi). Most of the larger counts are now decades old, and recent counts, even in winter, are greatly reduced. Counts of 700 at Trustom Pond, 27 November 1988 (SHa); 723, Sachuest, 16 December 1984 (RBu); and 850, Napatree Point, 11 January 2015 (CRa); are among the larger recent single-site counts. Recent maxima on CBCs: 1817, Napatree CBC, 27 December 2009; 1425, South Kingstown CBC, 22 December 2007.

Northern Pintail (*Anas acuta*)

Status

Fairly common fall and uncommon spring migrant; uncommon winter resident. Rare in summer.

Historical

In the early 1900s, Northern Pintails were scarce in Rhode Island, but they increased gradually to peak abundance in the 1940s and early 1950s, then declined. They occurred most commonly on freshwater ponds along the coast; many of these ponds are now overrun by Mute Swans, which may have degraded the habitat for pintails and other waterfowl.

Nonbreeding

Northern Pintails are now fairly common migrants in small numbers on ponds and marshes along the coast, but they are uncommon in the interior.

Fall. The earliest record in fall is 2 at Long Pond 30 July 1976 (REm), another appeared at Briggs Marsh 3 August 1952 (DLE), and 1 at was Quicksand Pond 7 August 1993 (DLE). There are more than a dozen other August records of from 1–5 birds, with a maximum of 8 flying past Napatree Point 12 August 1966 (DLK). Reports increase gradually in September (30 at Briggs, 5 September 1957; R. West), and pintails sometimes reach substantial numbers in late September (80 at Briggs on 21 September 1952; S. Cobb). The usual peak is in late October and early November.

Fall maxima. 130, Newport County, 1 November 1950 (P. Brownell).

Winter. Numbers taper off gradually in early winter but in most seasons, some remain into mid to late December, often concentrating in a few most suitable sites. A few remain most years on the freshwater coastal ponds until freeze-up in January or later; greater numbers remain in mild seasons.

There is usually a period in winter when this and many other species of freshwater waterfowl are forced to move out of Rhode Island, but Northern Pintails are some of the first to return in February with the first thaw.

Winter maxima. 239, Newport–Westport CBC, 16 December 2000 (DLE et al.); 168, Newport–Westport CBC, 17 December 2005; 125, Portsmouth to Middletown, 10 February 1952 (RBo); 100, White Brook, Carolina, 21 December 1991 (GLB et al.); 64, Little Compton, 17 January 2016 (REm); 30, White Brook fish hatchery, 1 February 2010 (B. Glemboski); 19, Narrow River, 21 February 2014 (J. Beuth).

Spring. Peak abundance in spring occurs late March to early April. As with other ducks, pintails remain for a much shorter period in spring than in fall, but counts can be larger.

Spring maxima. 200, Little Compton, 5 May 1941 (NPH). Rare after late April.

Summer. One at Adamsville 4 July 1986 (RBo), and 1 at Quicksand Pond 4 July 1993 (DLE). There have been no records in June or July since 2005.

"Eurasian" Green-winged Teal (*Anas (crecca) crecca*)

Status

Vagrant. Now annual in spring, rarer in fall. Although considered conspecific with *A. c. carolinensis* by AOS, "Eurasian" Green-winged Teals, as well as hybrids with *A. c. crecca*, can be distinguished in the field, and we therefore list records of "Eurasian" Green-winged Teals separately.

Nonbreeding

The first "Eurasian" Green-winged Teal recorded in Rhode Island was seen 30 March 1946 at Trustom Pond (DLK). Another was seen at nearby Cards Pond, Matunuck, 29 March 1948 (DLK). There were several records in the Seapowet area during the early 1950s: a pair 11–13 April 1950 (RBo, Hendershee); 2 males 18 March–1 April 1951 (RBo, DLE, HMP); a pair 30 March 1952 (CWo); and 1 male 14 March–21 April 1953 (CWo, RBo, NPH). Other records include 1 at Easton Pond 27 December 1959 and for a week thereafter (A. Craig, JBa); a male at St. Mary's Pond 24 April 1961 (KKi); and 1 at Little Compton 14 March 1965 (B. Harrington). Several other records come from Perryville: 26 March–4 April 1974 (RAC, DWa, DLK); 2 on 2–16 February 1975 (ESa, DLK et al.); 23–29 February 1976 (ESa, RBu, SSD); and 25–27 February 1977 (DWa, JC). One was at Cards Pond 30 April 1989 (DLK). Individual males were at White Brook 24 March 1987 (DFi, DLK); 12 March 1988 (R. Noss); 9–31 January 1993 (JMu, RS, RBo, BCG, WG); 7 February–9 March 1993 (BCG, WG, MJM); 18 January 1994 (RBo); and 5 February 1994 (RS, JMu), plus 7 other records through 1999. The only other report of 2 males together in Rhode Island is also from White Brook on 7 March 1998 (SSM). A bird described as a "Eurasian race intergrade" was present at White Brook 1–6 April 1992 (DFi, WG, BCG, RBo et al.). All reports since 2005 are singles: Moonstone, Mud Pond, 26 December 2007 (B. Heck); Quonochontaug, 5–8 April 2008 (STs); Weekapaug, 15–19 February 2011 (JSJ et al.); Trustom Pond NWR, 4 March 2012 (R. Askins); Portsmouth, 7 April–7 May 2013 (T. Auer); Weekapaug, 23–24 March 2013 (DFi, STa); Matunuck, 11 February 2014 (JMu).

"American" Green-winged Teal (*Anas (crecca) carolinensis*)

Status

Common to abundant spring and fall migrant. Uncommon winter resident and rare breeder.

Historical

These teals were much less common during and after the shooting era than in the modern period, having substantially increased since about 1935. This increase tapered off after the 1940s, and some decline seems to have occurred in transient numbers since then. Green-winged Teals continued to increase as summer lingerers at least into the late 1960s and now breed irregularly.

Breeding

The most consistent location for nesting in Rhode Island has been south of the Escape Road, Galilee, where in the early to mid-1950s a vast amount of dredging spoil was dumped on the salt marsh. This caused the obliteration of the formerly excellent stands of *Spartina alterniflora* and the appearance of a vast sea of *Phragmites*. Soon after these changes, several small cattail ponds were found in the *Phragmites*, and these seem to have been where the teals nested. According to Kraus, teals may have begun breeding in 1955, based on the observation of a pair there 24 June 1955. Another pair was observed 13 May 1956. On 12 July 1962, Kraus found a pair with 6 young. They continued to breed and apparently increased: 1963, 1 brood seen; 1964, 7 adults and 2 broods; 1965, 1 brood; 1966, female showing broken-wing display; 1968, 6 males seen on 27 June; 1972, 5 adults and 9 young; 1973, 4 adults and 6 young; 1974, 2 adults seen on 6 June. Since then, however, they have not been reported in the breeding season at this location.

Elsewhere, most records of suspected breeding are based on adults seen in summer: Easton Pond, 5 birds seen there 20 July 1955 (JBa); Boyd's Marsh in Portsmouth, adults seen periodically in June during the late 1950s and early 1960s; a pair in a beaver flowage along Trestle Trail at Summit on 20 June 1986 (RLF); a female seen 5 June 1986 in Crow Brook, Coventry (CCA); and an agitated pair in the Great Swamp 10 May 1983 (CRa). A male was present at Hope Island 1 June 1978 (RLF), and the first Rhode Island nest was discovered there in an upland section of the island (CCA); the incubating female and 9 eggs were shown to the author and J. E. Myers 29 May 1980. A female was flushed from dense cover in the same section of Hope Island in May 1981 (RWa). The only location for breeding inland is at the Lonsdale marshes, where 2 pairs were seen 7 June 1974 (RLF), and 1 and 4 birds were seen on 22 May 1983 and 21 May 1984, respectively (CRa, REn). A nest with 9 eggs was discovered about 20 ft (6 m) from the Blackstone River at Lonsdale Marsh, 7–13 May 1987 (REn). More recently, 3 were seen at Long Pond, Little Compton, 25 May 2015 (REm); and 2 were at Adamsville Pond, 29 May and 17 June 2011

(REm). Two were at Sachuest NWR 25 June 2007 (RTL), and 1 was at Seapowet 26 June 2010 (REm).

Nonbreeding
Spring. The first arrivals appear by late February, and peak numbers occur in April. Migrants are uncommon after 5 May (21 at Napatree 6 May 1999; CRa), and birds seen during the second week in May are usually at known or suspected breeding locations.

Spring maxima. 140, Lonsdale, 15 April 1992 (REn); 80, Little Compton, 5 April 1951 (NPH); 65, Lonsdale, 3 April 1991 (REn); 60, Trustom Pond, 4 April 1979 (DLK).

Fall. Birds regularly appear at various coastal locations in July and August: 27 at Little Compton 14 August 1995 (REm); 5, Briggs Marsh, 30 July 2016 (REm); and 6, Briggs Marsh, 3 August 2002 (REm). Migration increases in early September, with peak numbers reached in late October and early November.

Fall maxima. 400 in Portsmouth 5 November 1999 (DLE); 300, Sisson Pond, 22 October 1999 (DLE); 250 at Great Swamp 2 October 1995 (CRa); 225 in Newport 16 November 1978 (KKi, FPe); 180 at Sisson Pond 8 November 1995 (DLE); and 220 at Briggs Marsh the same day (CCA). Maximum counts since 2005 have been less than 100 per day per locality.

Inland maxima. Counts are lower; e.g., 54 at Lonsdale 16 October 1993 (REn), and 74 there 13 November 1995 (REn).

Winter. Numbers decrease by early December, and Green-winged Teals are rare inland after the first week in December. While Blue-winged Teals have declined in fall in recent decades, Green-winged Teals have remained more stable as transients, and numbers have increased in winter. Although they were rare during the colder months in the first third of the twentieth century, they are now annual winter residents in a few locations near the coast. Superior wintering sites include Green End Pond (up to 30–40 birds through the winters of 1948–1949 and 1949–1950; up to 25–30 in the mid-1970s); at the mill pond in Perryville (up to 40 in the 1970s); and at the fish hatchery on White Brook in Carolina (35 on 14 January 1978 [RAC] and also 35 on 18 January 1992 [RS]). At Trustom Pond, 33 were seen 20 January 1974 (DLK et al.). Elsewhere in coastal areas, they are uncommon in winter, and they are rare inland from December through February; e.g., 2 at Lonsdale 29–30 January 1992 (REn).

Canvasback (*Aythya valisineria*)

Status

Formerly common to abundant migrant and fairly common winter resident around Providence but has decreased over past 20 years. Very rare in summer.

Historical

At the turn of the twentieth century, this excellent "table" bird was a very rare migrant and did not occur in Rhode Island during winter. Before 1934, when a flock of 20 was seen, the largest number ever recorded in the state was 6. Larger flocks began to appear after 1939. They continued to increase here during the modern period and were found frequently in sizable flocks in the larger estuaries as well as in freshwater and saltwater ponds near Narragansett Bay and along the outer coast during the late twentieth century. They are rare on freshwater ponds.

Nonbreeding

There was a substantial increase in the number of Canvasbacks wintering in southeastern New England beginning in the 1940s. By the mid-1970s, a few thousand birds regularly wintered in Rhode Island, Connecticut, and Massachusetts, and up until about 2000, numbers on the Newport–Westport CBC were often substantial; e.g., 693 on 20 December 1986 and 732 on 21 December 1996. The entire wintering population wandered between the Providence area; the Thames Estuary near New London, Connecticut; and the Taunton–Assabet River area in Massachusetts (Veit and Petersen 1993). Thus, to a certain extent, variability in numbers in Rhode Island was due to the local shifting of this wintering population. That being said, however, Canvasbacks have been much less numerous since about 2000, with most counts being less than 20 birds (R. Farrell, pers. comm.); the maximum count on the Newport–Westport CBC has not exceeded 4 since 2005.

Spring. Spring flocks used to gather in Hundred Acre Cove, beginning in mid-February and sometimes as early as mid-January, depending on ice. They shun the freshwater locations that they often frequent in late fall and early winter. Spring flocks were bigger than those of fall or winter until the decline beginning in about 2000.

Spring maxima. 1500, Hundred Acre Cove, 29 February 1985 (BSp); 850 along the South Shore, 25 February 1976 (DLK, FPe); 800, Barrington, 12 February 1967 (WBD); 800 there, 18 February 1970 (CWo). Lingering individuals include 1 at Trustom Pond 28 May 1966 (DLK) and another there 4 June 1972 (NCl).

Summer. Summering birds have been noted at Barrington 15–31 July 1982 (JCB) and Little Compton 11 August–1 September 1973 (REm, DLE).

Fall. The earliest fall records are for 12 at Trustom Pond and 2 at Ninigret 30 September 1985 (LAH); peak numbers arrive in November. Unlike many other ducks, fall counts have generally been lower than those in spring, rarely exceeding 400–500 until the 1990s. Exceptional at the time were 1360 at Wordens Pond 11 November 1965 (DLK).

Fall maxima. 2000 at Brickyard Pond and 1000 at the East Providence Reservoir, all on 8 December 1984 (DLE, REn); 2000+, Brickyard Pond, 11 December 1974 (BSp, JCB); 1800, Barrington, 29 October 1976 (HWi); 1100 at the East Providence Reservoir, 25 November 1984 (DLE); 1000, Seekonk River, 26 November 1977 (CWo), 16 December 1987 (DLE), and 17 January 1988 (fide RLF).

Winter maxima. No single location had more than 350 birds until the mid-1970s; then 2500 in Watchemoket Cove, 25 January 1980 (CWo, FPe); 2100 in upper Narragansett Bay, 7 February 1982 (DLE); 750, Warwick, 15 February 1995 (DLE); 458, Seekonk River, 8 February 2003 (MLy); 124, East Providence Reservoir, 28 January 2006 (RFa).

Inland. Away from the coast, Canvasbacks are rare on fresh water. There is only one modest count for the Scituate Reservoir, 88 on 23 March 1982 (REn), although a few occasionally appear there or on nearby Moswansicut Pond in spring (RBu). There are now 5 records of flocks of up to 200 birds on Tiogue Lake, Coventry, all since 1982. There are only a handful of records for northern Rhode Island, most notably 17 on the Blackstone River at Lonsdale, 5 April 1978 (REn) and 12 at Diamond Hill Reservoir, 30 October 1988 (DMK).

Extreme dates. 30 September 1985, Ninigret, to 4 June 1972, Trustom Pond.

Redhead (*Aythya americana*)

Status

Uncommon and irregular visitor in small numbers during fall and winter; occasionally more numerous.

Historical

Considered by Howe and Sturtevant (1899) and Hathaway (1943) as rare, Redheads have never occurred at Rhode Island coastal ponds in numbers comparable to those on Martha's Vineyard or Long Island. However, they were apparently numerous prior to 1887 at Block Island in "the salt pond" (probably Great Salt Pond), which was then more restricted in tidal flow (Lawton 1887–1888). They were also said to be numerous on Ninigret Pond about 1913. In the twentieth century, they were most numerous in fall between 1951 and 1956, and in the 1970s were again relatively common in winter at Sachem Pond, Block Island. Otherwise, they have remained relatively scarce. No flocks were reported that exceeded 34 birds during the 1980s, and most of the scattered counts were of 1 to 5 individuals.

Nonbreeding

On the mainland, Redheads are most regularly seen on Trustom Pond, Potter Pond, Gardiners Pond, and Briggs Marsh and less frequently among rafts of scaup on Narragansett Bay or the Sakonnet Passage. It is exceptional to find Redheads in the interior; there are a small number of records for the East Providence Reservoir but few elsewhere (1 male, Coventry, 26 March 2004; MT).

Fall and winter. The earliest fall occurrences are 1 at Trustom Pond 26 September 1977 (DLK), 3 at Sakonnet 27 September 1980 (RBo), and another at Trustom 28–30 September 1976 (KKi). The largest numbers occur from late October into late December, with a peak in late November.

Fall maxima. 300 at Gardiners Pond throughout November 1955 and again in late October 1956 (JBa).

Winter maxima. 251, Block Island CBC, 20 December 1975; 120, Middletown, 21 December 1952 (v.o.); 50 on the South County CBC (mostly Trustom), 27 December 1986; and 12, Trustom Pond, 6 February 2016 (CPe).

Spring. Scattered spring migrants or wanderers often appear in mid to late February and small flocks may occasionally be present through mid-April.

Spring maxima. Flocks of as many as 50 have been reported in spring. The latest records are for 2 at Middletown 21 April 1953 (RCC, R. T. Peterson, J. Fisher); 1 at Trustom Pond 22 April 1976 (RLF); 1 at Tunipus Pond 4 May 1989 (RBo, HMP); and 2 at Trustom 7 May 1980 (DRo, STs).

Extreme dates. 26 September 1977 to 7 May 1980, both at Trustom Pond.

Ring-necked Duck (*Aythya collaris*)

Status

Fairly common to common migrant and uncommon to fairly common winter resident.

Historical

Before the twentieth century, Ring-necked Ducks were of the greatest rarity in New England, a fact discussed by Forbush (1925) and Griscom and Snyder (1955). There is only one definite Rhode Island record prior to 1933, a bird shot at Easton Pond, Newport, on 11 November 1871. Another was reportedly shot by Col. Powel before 1899 and sent to the Academy of Natural Sciences in Philadelphia, but an inquiry by Hathaway was unable to confirm that any such specimen had been received. This species dramatically extended its breeding range into the Northeast during the 1920s and 1930s (Griscom 1939; Mendall 1958). On 22 February 1933, one was found on the Blackstone River in Lonsdale by Alfred E. Eynon (1935), and this record was soon followed by other observations. Migrants appeared regularly by the late 1930s, and there was a gradual increase in numbers during the next two decades. They are now most numerous on the Scituate Reservoir, Belleville Pond, Pachet Brook Reservoir, Watchaug Pond, and Great Swamp impoundment, and sometimes on other reservoirs in northern Rhode Island. On shallow coastal ponds, they are less frequent and never occur in substantial numbers.

Nonbreeding

Substantial flocks of spring migrants appear at the Great Swamp impoundment, and this heavily forested area provides what looks like plausible nesting habitat. Given nesting records from Massachusetts and Connecticut, it seems possible they could nest in Rhode Island.

Spring. The peak of spring migration occurs from early March to mid-April. Numbers of migrants are generally higher, but more irregular, in fall than spring.

Spring maxima. 355, Great Swamp, 4 March 2007 (STs); 350, Great Swamp, 26 March–14 April 1988 (CCA); 210, Great Swamp, 29 March 1991 (CRa); 200, West Kingston, 16 March 1993 (MJM); and 54, Lincoln Woods, 28 February 1990 (REn). Smaller numbers occur to the end of April, and there are more than a dozen records for the first week of May, the latest of these being a pair in Middletown on 22 May 1998 (SSM) and a female in the Great Swamp 28 May 2001 (SSM); also, 1 pair, Great Swamp, 18 May 2002 (CNu). The male of this pair continued through 4 June 2002 (RLF, SSM); perhaps the same bird was reported from Trustom Pond 31 July 2002 (CNu).

Fall. Peak counts occur mid-November to mid-December.

Fall maxima. 750, Pachet Brook Reservoir, 18 November 1993 (DLE, JO); 700, Pachet Brook Reservoir, 11 November 1995 (DLE); 401, Glocester, 22 November 2001 (MLy); 400, Chapman Pond, 28 October 2011 (CRa); 350, Scituate Reservoir, 26 November 1962 (SSD).

Winter. Abundance depends on ice conditions.

Winter maxima. 150 at Watchaug Pond 12 January 1992 (DLK).

Extreme dates. 17 August 2008, Long Pond, Little Compton (REm), to 27 May 2015, Great Swamp (CRa), and 4 June 2002, Great Swamp.

Tufted Duck (*Aythya fuligula*)

Status

Vagrant. Now annual during winter.

Nonbreeding

The first Tufted Ducks for Rhode Island were adult males seen in Barrington 20 January 1991, observed by Chester Hobbs, and in the Saugatuck River, Wakefield, during the same month (fide CRa). The first bird remained in the Barrington–Warren area through February to 10 March, allowing ample study and close photographs by numerous observers (JMe, DMe, JCB et al.). A female was also reported from the Barrington River 4 February 1991 (HWi, DFi, JCB et al.). Another drake appeared on Trustom Pond 12 March–6 April 1992 (SHa, WG et al.). A drake was again at Trustom Pond 27 March–6 April 1993 (MJM, RBo et al.). Other records include 1 female, Trustom, 14 January 1995 (DFi); 4 (2 pairs), Tunipus Pond, 20 February–22 March 1995 (RLF, J. Osborn); 1, Tunipus Pond, 22 October and 27 November 1995 (DLE, J. Osborn); 1, Sabin Point, 1 March 1996 (DFi). A male was at East Providence Reservoir 16 April 2001 (JSJ, RFa) and another at India Point 15 February–9 March 2003 (MLy et al.). One female was at Wakefield/Saugatucket 28 December 2002–10 April 2003 (PAB, STs et al.) and perhaps the same bird 22 November 2003–28 February 2004 (m.obs.). A female was at Turner Reservoir 8 November 2004 (MLy), plus 5 other records in 2003–2005. One male was at the East Providence Reservoir 21 January 2008 (RFa), and perhaps the same bird was there 30 December 2008–15 January 2009 (E. Slattery). Subsequent records are 1 male, Apponaug Cove, 11 January 2009 to 10 February 2010 (JSJ, MLy); 1 male, East Providence Reservoir, 3 December 2009 to 12 March 2010 (JSu, m.obs.); 1 male, Conimicut Point, 15 January 2011 (DFW); 1, Seekonk River, 27 January 2014 to February 2015 (MLy, m.obs.); 1 female, Seekonk River, to at least 19 February 2015 (DFi, STa); a pair, Seekonk River, 30 January–22 March 2017 (DFi, STa); 1, East Providence Reservoir, 28 January 2020 (v.o.; photo); 1, Easton Pond, 28 January–13 March 2020 (v.o.); 1, Trustom Pond, 27 November 2020 (v.o.). While Tufted Ducks are commonly kept in captivity, the simultaneous increase in records from North America and Iceland, along with increasing populations in Europe and Iceland, strongly suggests that most North American occurrences are of wild birds.

Greater Scaup (*Aythya marila*)

Status

Common migrant and abundant winter resident in sheltered bays. Common on saltwater, brackish, and freshwater ponds near salt water but uncommon to rare inland. Regular on some of Block Island's ponds and estuaries but often absent there in midwinter. Rare anywhere in summer.

Historical

Scaup of either species were found in much smaller numbers during winter at the turn of the twentieth century. Although there is a record of 5000 off Gaspee Point 5 March 1905 (Sheldon), the great aggregations of modern times have been almost wholly a phenomenon of the post-protection era beginning about 1920. After about 1930, wintering populations of Greater Scaup on Narragansett Bay rose to exceed nearly all other avian spectacles in Rhode Island. Counts regularly exceeded 20,000 in upper Narragansett Bay alone, and wintering numbers statewide sometimes exceeded 40,000. Sizable aggregations were a regular feature in the lower Providence River until the early 1970s, after which much smaller flocks have been the rule, and larger numbers, while occasional, are no longer encountered with any regularity.

Nonbreeding

A substantial downward trend in continent-wide Greater Scaup abundance since the 1950s is evident in winter survey data (Kessel et al. 2020) and is apparent in data from Massachusetts (Veit and Petersen 1993) and New York (Levine 1998). Other factors possibly affecting the trend in scaup populations are gillnet mortality on the southeastern U.S. coast (D. Forsell, USFWS, retired, pers. comm.) and the new food source provided by zebra mussels (*Dreissena polymorpha*), imported from Europe in the mid-1980s and now abundant in the Great Lakes (Wormington and Leach 1992).

On 1 January 1992, only 1000 were present in upper Narragansett Bay, and no counts over 10,000 were reported during the years 1975–1992. In 1992, 12,000 were noted in upper Narragansett Bay 14 December 1992 (RBo). The largest concentrations have usually been off the Warwick shore from Edgewood to Conimicut Point, or off Sabin Point in Riverside; lesser concentrations of a few thousand or fewer have been regularly found elsewhere in Narragansett Bay, Mount Hope Bay, and the Sakonnet Passage. Flocks of a few hundred are the usual rule on the outer coastal ponds (maximum count at Trustom Pond, 1200 on 18 April 1993; CHa), but numbers sometimes alight on freshwater ponds near Narragansett Bay (4000 and 6000 on St. Mary's and Brickyard Ponds, 4–24 November 1995; DLE).

Fall and winter. The earliest arrivals of scaups (spp.) in fall are 2 on 9 September 1960 in upper Narragansett Bay (CBl); the first flocks of any size arrive in mid-September, and peak numbers occur from mid-October to December.

Fall maxima. The all-time maximum count for the state is an aerial survey of 40,000 in upper Narragansett Bay 6 November 1973 (RIDFW). Other high counts are 32,000 in upper Narragansett Bay 23 November 1974 (RIDFW), and 25,000, Barrington, 14 October 1945.

Winter maxima. 30,000 off Warwick 5 December 1945 (BD); 25,000, Warwick, 15 Feb 1995 (DLE); 10,700, East Providence to Barrington, 22 January 2005 (RFa).

Spring. After early March, a return of migrants or concentration of the wintering population may temporarily swell numbers to nearly equal those of midwinter.

Spring maxima. 20,000, Conimicut Point, 28 March 1971 (DWa). There are scattered reports to 30 May. Summer lingerers are occasionally seen; there are about 15 records for the months June through August for 1–2 birds.

Summer maxima. Four, Providence River off East Providence, 9 June 1977 (RLF); 3, Quicksand Pond, 13 July 2000 (RBo, REm); 3, Watchemoket, 30 July 1996 (DLE).

Lesser Scaup (*Aythya affinis*)

Status

Common migrant and fairly common winter resident. Very rare in summer. More numerous on freshwater and coastal ponds than Greater Scaup.

Nonbreeding

As in Massachusetts and Connecticut, the recent tendency is for Greaters to dominate on saltwater bays, and Lessers to dominate on freshwater lakes.

Summer. An adult male was seen at Tunipus Pond on 23 June 1985 (RBo).

Maxima. 1400, Gardiners Pond, 7 November 1989 (REn); 680, Newport County, 9 November 1982 (RBo, HMP); 164, Newport–Westport CBC, 18 December 2010; 100, East Providence Reservoir, 26 March 2006 (MLy); 100, Trustom Pond, 6 January 2007 (STs).

Extreme dates. (Aside from summer record above) 13 October 2007, Trustom Pond (STs), to 7 May 2012, Napatree (2; CRa).

King Eider (*Somateria spectabilis*)

Status

Uncommon winter visitor; very rare in summer and at any time inland.

King Eiders have occurred in greater numbers off the shores of Rhode Island; neighboring Acoaxet and Westport, Massachusetts; and Montauk, Long Island; than elsewhere on the Massachusetts or Maine coasts.

Historical

There are many pre-1900 records of birds seen or shot about the shores of Middletown, Newport, and Sakonnet. One was shot at Wickford, and two were collected (27 November 1879; 1 January 1884) off Nayatt Point, Barrington, but in the modern period they have not been recorded in Narragansett Bay north of Bonnet Shores. They have always been most numerous off Middletown, Little Compton, and in certain rocky areas along the South County shore, particularly Matunuck. They were most frequent and numerous during the late 1930s and early 1940s and have decreased since.

Nonbreeding

Fall. Most arrive in November, but there are a number of earlier records, including flocks: 11, Point Judith, 28 October 1995 (STs, DLK); 6, Little Compton, 12 October 1970 (RBo); 1 shot at Cormorant Rock 15 September 1902 (Howe and King); and 1, Block Island, 25–27 September 1976 (MDa, DLE, REm, STs).

Winter maxima. Eight, Wheeler Beach, 1 February 1998 (JMu); 4, South Kingston CBC, 26 December 1995 (DLK et al.); 4, Napatree, 4 March 2009 (CRa).

Spring maxima. A remarkable concentration occurred in 1941: Drury and Bowen saw 45 birds off Sachuest Point on 23 March, and this number grew to 78 birds on 13 April (WHD), including 20 adult males, probably the largest count ever for New England. Also, 4, Brenton Point, 1–6 April 2009 (B. Harris). There are a number of records in April, including a small series collected by Mackay at Sakonnet, 16–22 April in the years 1877–1890. There are 2 specimens in the RWPM collection for 29 April 1895 and 30 April 1892 labeled "Narragansett Bay." There are also several May records including single immature males at Napatree Point, 16 May–18 June 2003 (m.obs.); Quonochontaug, 17 May 2003 (REm, DFi, CNu et al.); and Green Hill, 14 May 2005 (STs, CPe et al.).

Summer. Summering birds include 1 at Point Judith, 2 June 1976 (STs); adult male in the mouth of Narragansett Bay just south of Dutch Island, 6 June 1990 (JEM, RLF); 1 off Middletown, 8 June 1969 (Larsen); 2 males, Little Compton, 26 June 1988 (JCB, DLE); 1, Little Compton, 2 July 1941 (DLE); 1, Napatree Point, 10 July 1968 (DLK, KKi); 1 male off Ninigret Breachway, 13 July 1991 (PAB, DRo); 3 (including 2 males), Sakonnet, 26 June–31 August 1969 (DLE); and an oiled bird off Quicksand Pond, 1–16 July 1986 (RBo). Also, 1, Little Compton, 21–31 July 1997 and again 30 August 1997 (DLE).

Inland. An immature male appeared at the East Providence Reservoir 8 November 1970 (RLF).

Common Eider (*Somateria mollissima*)

Status

Common migrant; fairly common and increasing winter resident. Breeds in small but increasing numbers beginning in the late 1990s.

Historical

Common Eiders were said to have been common winter residents before 1900, when there were reports of "thousands." Then, with severe exploitation in eastern Canada, they sharply declined and were rare by 1900 (an exceptional flock of 100 was seen and 1 shot at West Island 5 February 1897 by Sturtevant). For the first half of the twentieth century, they were uncommon and were almost never seen in sizable flocks.

Few flocks in the 1950s exceeded 200 birds. Coincident with their recolonization of the Maine coast, Common Eiders increased slowly in Rhode Island after 1920 but have increased more sharply since 1950. The most dramatic increase, including a number of nesting records, began in the early 1990s. Common Eiders are rare but increasing in Narragansett Bay and the Sakonnet Passage, and there are vagrants to inland localities.

Nonbreeding

In Rhode Island, Common Eiders winter in inshore waters "with relatively fine sediments and a high probability of a hard bottom" (Beuth et al. 2017).

Pre-1990s maxima. 2600, Point Judith, 10 March 1978 (DLK); 950 from Narragansett Pier to Point Judith, 16 March 1974 (DLK, KKi, FPe, CWo); 700 off Bonnet Shores, 12 February to 6 March 1972 (DLK); 400 off Sakonnet, 3 March 1935 (WBD); 515, Little Compton, 18 March 1951 (DLE).

Post-1990s maxima. 16,000 off the Cliff Walk in Newport, 23 January 1996 (DLE); 11,000 between Beavertail and Little Compton in 1993 or 1994 (CPe et al., RBo); 10,800 between Watch Hill and Quicksand Pond, 12–15 March 1995 (RLF); 7236, South Kingstown CBC, 26 December 2005; 6932, Newport–Westport CBC, 18 December 1999 (DLE et al.).

Early migrants arrive in late September, and flocks arrive in mid-October. The bulk of birds arrive after late November, with the largest concentrations in late winter or early spring. Records of smaller numbers extend through April into early May.

Breeding and Other Summer Records

The first birds summered in 1934, but numbers of summerers did not really increase until the 1970s. Eight remained through the summers of 1970 and 1971 at Sakonnet Point (v.o.), and 30 were off Warren's Point 30 July 1973 (REm). An unprecedented 180 remained off Briggs Beach through May 1978 (DLE, REm), and 171 were still present as late as 4 June 1978 (RLF, JEM), with 50 on 13 September 1978 (CWo, REm). Numerous records of smaller summering flocks have accumulated in recent decades; some of these flocks have contained adult drakes. Such flocks have been most regular along the rocky outer coast of Little Compton, particularly in the neighborhood of the Sakonnet islands, and have increased parallel with the recolonization by breeding birds of the Maine, New Hampshire, and Massachusetts coasts.

With the tremendous increase of wintering flocks, an unprecedented sprinkling of eiders has begun to appear in unusual places. An immature was at the Dumpling Islands 2 June 1994 (JEM, RLF), and a subadult male was resting on the north tip of Dyer Island off Melville, 26–30 May 1995 (JEM, DLE, RLF et al.). An adult male was on the south beach of Dyer Island 1 June 1996 (RLF, JEM et al.).

The entire coast of Massachusetts was colonized from the north by the early 2000s. In 1998, a female was seen with a brood of half-grown young in the vicinity of Sakonnet Point (fide PAB), indicating the first Rhode Island nesting record. On

17 June 1999, 2 females leading 10 downy young were seen along the beach at Sakonnet Point; they remained nearby for the rest of the month and another brood was seen in 2000 (F. Norton, fide DLE). Two females and 6 young were along the shore of Little Compton 11 June 2012 (GDe; photo), and 4 females and 13 young were there 9 June 2014 (GDe; photo). A female and 2 young were photographed southeast of Quicksand Pond in the same year (GDe).

Inland. The only suggestion that this species might pass over the interior of Rhode Island during migration, as it sometimes does in spring through the interior of Maine, is 4 birds reportedly shot in Scituate, on unstated date (in Howe and Sturtevant 1899).

Subspecies

A bird labeled "King Eider" in the collection of Edward Sturtevant, collected at Newport 16 December 1898, was redetermined by Bowen and Baird as *S. m. borealis*, "Northern" Eider, but a check of the collection in 1973 indicated that the specimen had disappeared. On 12 October 1960, a bird observed by Kraus at very close range off Matunuck was identified as *borealis*, and another *borealis* was seen 15 January 2015 at Jamestown (SSM).

Harlequin Duck (*Histrionicus histrionicus*)

Status

Fairly common winter resident along rocky shores, especially Sachuest and Beavertail Points. Very rare in summer.

Historical

There are many old records of Harlequin Ducks being seen or collected off the shores of Middletown, Newport, and Sakonnet. Early maxima include a flock of 12, from which 2 were shot, in December 1904 at Cormorant Rock, Middletown (C. B. Clarke), and a flock of 5 from which 4 were shot at the same location 9 February 1911 (fide Sturtevant). One was shot from a flock of five at Block Island 14 January 1917 (EDi); the specimen is in the Hathaway collection. One was seen at Block Island in 1919, but none were then reported until 1937.

Nonbreeding

Harlequin Ducks were reported sporadically during the 1950s but increased to occur annually during the 1960s. Most were seen near Whale Rock off Narragansett, but since the 1970s they have been more numerous near the "Island Rocks" and along the adjacent shore of Sachuest Point in Middletown. They have, however, now been recorded from a wide range of other locations along the outer coast. Favorite locations since the 1990s have included rocks off Briggs Beach, rocks off Long Pond near Sakonnet and along the shore just south of Sakonnet village, and at various places along the Newport shore

FIGURE 1. Harlequin Ducks at Sachuest Point, © Margaret La Farge.

(Land's End and Brenton's Rock). Harlequin Ducks are only rarely observed passing coastal promontories but instead are found in small flocks that may remain for weeks, especially along rocky shores.

Summer. A drake summered at Napatree Point, 31 July–9 October 1965 (Dewire, Bissell, Kraus et al.). An immature male was in the Sakonnet Passage at Fogland Point, Tiverton, 11 July–23 August 1990 (DLE et al.), unusual so far inland. A male was at Sakonnet Point 28 July 2007 (REm).

Maxima. Caron and Paton (2007) show logistic or near logistic growth to a maximum of about 150 birds in Rhode Island 1976–2004.150, Sachuest Point, 10 February 1996 (Connie Bruyiere); 144, Newport–Westport CBC, 19 December 1998 (DLE et al.); 125, Sachuest Point, several times in the early 1990s; 124, Sachuest Point, 18 December 2004 (CBC). West of Narragansett Bay, the maximum is a flock of 6, including 4 drakes, which ranged between Bonnet Shores and Scarborough Beach, 9 January–5 April 1972; the flock increased to 8 birds on the last date (m.obs.).

Extreme dates. 12 September 2009, Napatree (CRa), to 26 April 2010, Sachuest Point (3; B. Murphy).

Labrador Duck (*Camptorhynchus labradorius*)

Status

Extinct. Presumably occurred in Rhode Island, given many specimens from adjacent parts of New York and Massachusetts. These northern-breeding ducks were fairly well known in Massachusetts in the nineteenth century, and 15 specimens are still extant from Long Island (Bull 1964). Although there are no specimens from Rhode Island, they must have occurred here.

Surf Scoter (*Melanitta perspicillata*)

Status

Abundant migrant and common winter resident. Regular in the Sakonnet Passage and lower Narragansett Bay. Uncommon lingerer in summer and rare inland at any season.

Nonbreeding

Surf Scoters are the most numerous of the 3 scoters during migration in Rhode Island, and second commonest after White-winged in winter. Flocks of Surf Scoters feed most commonly off the rocky shores of Newport County, less commonly off the South Shore, and least often in upper Narragansett Bay and Mount Hope Bay.

Fall. Probable transients occur as early as 1 September, and numbers are observed migrating off the outer shore from late September through November.

Fall maxima. 8500+ past Point Judith, 28 October 1979 (RLF); 3519 past Point Judith, 27 October 1985 (RLF); 3401, Block Island VDC, 10 November 2011 (SSM et al.); 2700 past South County, 4 October 1941 (DLK); 300 in the Sakonnet Passage, 8 October 1939 (RBo); 267 past Point Judith in 1 hour, 27 September 1941 (DLK).

Inland. Four, Diamond Hill Reservoir, Cumberland, 16 October 2010 (MLy).

Winter. By December, most of the migration has ceased, but birds move back and forth along the outer coast in both directions throughout the winter. It is unusual to find more than 200 at any one place on one day in winter.

Winter maxima. 3025, Block Island (CBC), 18 December 2008; 750, Warren's Point, 15 January 1956 (RBo); 500, Napatree Point, 8 February 2012 (CRa).

Spring. Spring migration begins in March and peaks in late April, when the 2 smaller scoters are more abundant, or more concentrated, than at any other time of year. There is a spectacular migration through Block Island Sound at that time.

Spring maxima. 23,000 past Point Judith, 27 April 1985 (RLF); 12,000 past Point Judith, 24 April 1982 (RLF); 9700 past Point Judith, 30 April 1995 (RLF). Unlike White-winged Scoters, which often migrate westward in Rhode Island during spring, nearly all of the flocks of Surf and Black Scoters pass eastward.

Summer. There are many records of small flocks in midsummer.

Summer maxima. 60, Little Compton, 13 August 2000 (REm); 33, Seapowet, 21 July 1951 (RBo); 30, Prudence Island, 15 June 1946 (WBD); 17, Tiverton, 17 June 1951 (DLE).

Inland. There are 5 inland records of Surf Scoters: 4, Diamond Hill Reservoir, 16 October 2010 (MLy); 2, Watchaug Pond, 22 November 1941 (DLK); 2, Wordens Pond, 8 December 1945 (DLK); 1, East Providence Reservoir, 8 May 1953 (RLF, HEC, REW); another at the same location 12 November 1983 (DLE).

White-winged Scoter (*Melanitta deglandi*)

Status

Very common migrant and common winter resident. More frequent in upper Narragansett Bay and Mount Hope Bay than other scoters and appears with some regularity on the coastal ponds. Commonest of the 3 scoters in summer; rare inland at any season.

Historical

A large-scale westward movement of White-winged Scoters along the Rhode Island coast in May was well known to numerous nineteenth-century sportsmen and writers, including Mackay (1891), Howe and Sturtevant (1899), Col. Powel (1899), Dexter (1900), and Forbush (1916). This flight took place about 10–20 May according to Col. Powel, but Mackay stated that it might start "May 7, which is unusually early, the customary time being from the 12th to the 15th, and the latest the 25th." Dexter placed the flight on or about 17 May. Mackay reported seeing as many as 10,000 flying west past the Sakonnet islands in a single day in late May, and all early writers agreed that the flight was composed largely of "adult" birds. Howe and Sturtevant rated this westward migration as the most pronounced and interesting movement of any waterbird through Rhode Island waters, and even in this early period these writers suggested that this movement was an attempt to enter Long Island Sound for passage overland toward the Great Lakes. Mackay (see Phillips 1929) described scoter shooting at Sakonnet before the turn of the twentieth century. A long line of up to 20 boats would assemble off The Clumps to intercept the migrant scoters during late April, with substantial bags of up to 100 per day per person.

Nonbreeding

Fall. Numbers begin to build in September (1000 on Narragansett Bay 10 September 1938; HNG), but peak numbers generally pass by in October. They are commonly seen migrating along the outer coast in flocks of hundreds from mid-September on, with a peak during October.

Fall maxima. 15,000 past Point Judith, 12 October 1956 (DLK); 4057, Block Island, 11 November 2010 (VDC); 4000 past Quonochontaug, 24 September 1933 (HSH). Flocks of 1500 and 1000 were seen on Mount Hope Bay 18 October 1944 (RBo) and 27 October 1987 (CPe), remarkable numbers for that location.

Winter. White-winged Scoters are the commonest scoter in Rhode Island during winter. Single-day counts are usually less than 500 and rarely exceed 1000.

Winter maxima. 2040, Block Island, 18 December 2008 (CBC); 1742, Block Island, 23 February 2015 (PDC).

Spring. Coastwise movements in spring are prolonged from early March through early June with a peak in late April and early May.

FIGURE 2. White-winged Scoters and Red-necked Grebes flying past Point Judith, © Margaret La Farge.

While movement of small flocks past Point Judith is most often to the east in April and May, especially early in the season when conditions are frequently poor for an overland flight, there is often an excess of birds (and larger flocks) moving west from late April on. This pattern is most pronounced when clear, calm conditions prevail, and the amount of such westward migration frequently increases sharply during the last few hours of daylight. Some of these flocks cut westward to the north of the Point Judith lighthouse, suggesting an interest in heading overland to the northwest.

Whatever their direction of movement, spring counts of White-winged Scoters seldom approach totals for the other 2 species, especially Surf Scoters.

Spring maxima. 5000 at Point Judith 25 April 1981 (RAC); 4900 past Point Judith, 14 May 1995 (RLF); 3000 past Jamestown (presumably east) 18 April 1954 (MDa); 2000 at Narragansett Pier 5 May 1951 (DLK); 820 flying east at Middletown 30 May 1937 (WHD). Some movement is evident in early June: 200 off Quonochontaug, 2 June 1967 (DLK); 23 migrating northeast past Tiverton, 11 June 1972 (RLF); 24 migrating north over the interior of Tiverton in broad daylight, 13 June 1940 (RCC).

Summer. White-winged Scoters are the commonest scoter in summer, usually in small flocks, both along the outer coast and (more often than the other scoters) in Narragansett Bay. A flock has often been present all summer near Dyer Island; e.g., 60 off Prudence Island, 15 June 1946 (WBD). Other summer counts include 35 at Sakonnet, 18 July 1950 (DLE), and 100 in the Sakonnet River off Portsmouth, 13 August 1966 (DLE).

Inland. Three, Wordens Pond, 9 October 1949 (DLK); 2, Scituate Reservoir, 31 October 1981 (RBu); 4, East Providence Reservoir, 8 May 1953 (RLF, HEC, REW); 3 migrating over Hope Valley, 24 May 2011 (CRa).

Black Scoter (*Melanitta americana*)

Status

Very common migrant and fairly common winter resident. Uncommon in summer; rare inland.

Nonbreeding

Using satellite tracking, Loring et al. (2014) have shown concentrations of Black Scoters during winter near Block Island and along the southeast coast of Rhode Island, as well as off Cape Cod, the Massachusetts islands, and eastern Long Island.

Spring. The largest migratory movements and the highest counts for the year are in spring, when Black and Surf Scoters move eastward through Block Island Sound in large mixed flocks. Migration begins in March or earlier, and peaks in mid to late April.

Spring maxima. 8000+ and 3200+ past Point Judith, 24 and 25 April 1982 (RLF); 3000+ past Point Judith, 3 April 1982 (RLF); 2300 at Charlestown Beach, 21 April 1989 (RBo, DLK); 1400 off Briggs Marsh 30 April 1988 (DLE); 1400+ past Point Judith, 16 May 1982 (RLF). Scattered flocks have been seen moving east through late May, and some migration continues into June; e.g., 70 resting off Quonochontaug, 5 June 1978 (RLF).

Summer. A few spend the summer in small flocks along the outer coast.

Summer maxima. 54, Little Compton, 16 July 1941 (DLE).

Fall. Migrants appear during mid-August (50 and 100, Green Hill, 12 and 20 August 1982; DVT), and migration is well under way in September, earlier than for other scoters. Unlike other scoters, Black Scoters do not occur in large numbers along the outer coast in fall. Perhaps some are passed off as Surf Scoters, but counts have seldom exceeded 150 per day.

Fall maxima. 11,857, Block Island VDC, 10 November 2011 (SSM); 11,005, Block Island VDC, 7 November 2013 (SSM et al.); 800, Tiverton 21 October 1954 (Hendershee); 627 past Point Judith 27 October 1985 (RLF); 600, Sachuest Point, 8 November 1989 (GS); 450, Point Judith, and 140, Napatree Point, 22 October 1988 (RBo; CRa); 300, Moonstone, 9 November 1974 (FAP).

Winter. Black Scoters are usually the least common and the most variable in abundance of the 3 scoters, sometimes missed entirely on Christmas Bird Counts.

Winter maxima. 5500 off Watch Hill, 25 March 2015 (CRa); 5196, Block Island CBC, 18 December 2008; 3201, South Kingstown CBC, 26 December 2011; 800, Little Compton, 29 January 1994 (DLE); 700, Moonstone, 17 February 1971 (FPe, KKi).

Inland. Two, Watchaug Pond, and 1, Wordens Pond, 4 April 1951 (DLK); 1 female, Sneech Pond, Cumberland, 1 November 2008 (MLy); 10, Woonsocket Reservoir, Smithfield, 1 November 2008 (MLy). There are several records, mainly in fall, for flocks of 10–15 birds.

Long-tailed Duck (*Clangula hyemalis*)

Status

Fairly common migrant and winter resident on coastal waters. Rare but regular in summer.

Historical

These sea ducks have declined in Rhode Island waters since the turn of the twentieth century. Writing of the Newport area, Rives (1883–1884) described Long-tailed

Ducks as "among the commonest and best known [ducks]. . . . These ducks are shot at the Dumplings from row boats at sunrise when they fly up the Bay, often in immense flocks, and also at sunset when they return." Lawton (1887) called it the most common sea duck, while Howe and Sturtevant (1899) ranked it third after two of the scoters in abundance. Sturtevant saw over 100 and shot one off Middletown 25 November 1897. There were observations of "great numbers" off Newport in December 1900, and Mearns (1902) called it "abundant." Hathaway (1943), on the other hand, called it "common" but gave no counts of midwinter flocks greater than nine birds. While there are a great many late fall and midwinter counts of small numbers from recent years, almost no counts of feeding flocks have exceeded 25 birds. Davis (1997) and White et al. (2009) detail the massive increase of Long-tailed Ducks in Massachusetts in the early 1970s

Nonbreeding

Fall. The earliest fall migrant was 1 on Wordens Pond, 15 October 1950 (DLK). A flock of 20 was recorded 17 October 1970 off Napatree Point (ESa). The largest numbers of Long-tailed Ducks are seen from coastal promontories in late fall and early winter.

Fall maxima. 300 past Point Judith on 27 October 1985 (RLF); 100 along the South Shore 31 January 1971 (DLK).

Winter. Long-tailed Ducks are uncommon in midwinter; migrants appear in February and March.

Winter maxima. 200, Block Island, 9 March 2015 (M. Resch); 120, Watch Hill, 18 December 1995 (DWa).

Spring. Long-tailed Ducks often move west along the outer coast during March and April, as on 20 April 1985, when 8 birds were recorded moving east, but 169 were moving west among flocks of White-winged Scoters, mostly in the last 2 hours of daylight (RLF). A spring maximum of 733 was moving west through Block Island Sound 18 April 2004 (RLF). Numerous flocks totaling 575 were again seen heading west past Point Judith under clearing skies in the early evening of 25 April 1987 (RLF); flocks totaling 115 accompanied scoters westward past Point Judith 27 April 1997 (RLF). Another indication of northwestward migration in spring is the occasional appearance of flocks in Narragansett Bay during late May: 150 were off the south end of Prudence Island 14 May 1949 (RBo); 80 were between Prudence and Patience Islands 18 May 1941 (RBo, RCC); and 200 rested off the south end of Prudence Island 19 May 1940 (RBo). Migrants have been seen in Narragansett Bay into early June.

Summer. In recent decades oiled or crippled Long-tailed Ducks have summered with some regularity along the outer coast, especially at Napatree Point, where 1 or more (up to 3) have been seen every few years.

Long-tailed Ducks are uncommon in the Sakonnet Passage and lower Narragansett Bay and are now only occasional in upper Narragansett Bay. While wintering birds were formerly regular in Greenwich Bay in the 1940s (RCC), and they have been recorded occasionally off Riverside, Edgewood, and in Mount Hope Bay, there are few recent reports from these areas. One was seen at India Point, Providence, 4 February 1976 (RKl), and another at Watchemoket Cove 15 March 1984 (SDz). Small numbers occasionally alight on inland lakes and reservoirs in October–November and March–April; e.g., 6, Greenwich Cove, 31 January 2005 (GDe); 1 male, 27 November 2010, Diamond Hill Reservoir, Cumberland (MLy); 1, Saylesville Reservoir, 18 March 2001 (MLy).

Extreme dates. 15 October 1950, Wordens Pond (DLK), and 25 October 2006, Upper Narragansett Bay (8; GDe), to 21 May 2012, Allens Harbor (RLF, CRa), and 1 June 2008, Green Hill (3; RLF).

Bufflehead (*Bucephala albeola*)

Status

Common transient and winter resident. Rare anywhere in summer. Uncommon inland.

Nonbreeding

Buffleheads were described as "uncommon fall migrants" by Howe and Sturtevant (1899). Howe (1899a) did not list them among birds known from Bristol, where they are now common. Numbers gradually increased during the twentieth century and now average about 20% of those of Common Goldeneyes. In general, Buffleheads inhabit the same or slightly more protected waters than the latter.

Fall. Migrants arrive in early October (e.g., 14, Gorton's Pond, Warwick, 2 October 1977; GRa), but most appear after early November. Late fall and winter counts are usually less than 100 birds.

Fall and winter maxima. 1199, Newport–Westport CBC, 17 December 1994 (DLE et al.); 1000, Ninigret Pond, 7 February 2004 (STs, CPe).

Spring. There is sometimes an increase in March and early April before most of the wintering population departs northward.

Spring maxima. 465, Ninigret Pond, 14 March 1989 (LAH); 400, Ninigret Pond, 14 March 1991 (RSt).

Summer. There are several records for June–August; e.g., 1, Cards Pond, 23 July 1984 (CRa); 1, Pawtucket, 28 July 1902 (in RWPM); and 1, Warren, 26 August 1939 (RBo).

Inland. Buffleheads are uncommon in the Rhode Island interior at any season. Fall records run from 14 October to 29 December.

Inland maxima. 55, Diamond Hill Reservoir, 1 November 2008 (MLy); 25, Larkin Pond, South Kingstown, 8 December 2014 (CRa).

Extreme dates. 2 October 1977, Warwick, to 12 May 2007, Sachuest (RTL).

Common Goldeneye (*Bucephala clangula*)

Status

Very common migrant and winter resident. Less numerous and regular on fresh water. Rare in summer.

Nonbreeding

Common Goldeneyes commonly appear on sizable freshwater ponds and reservoirs as long as these remain unfrozen. Counts have reached 50 at the East Providence Reservoir and 70 at the Scituate Reservoir. Several hundred have been seen on Watchaug Pond, which is nearer to the ocean.

Fall. The earliest record of probable transients is for 10 off Bullocks Point on 28 September 1958 (CBl). More certain arrival dates are from mid-October on, and the wintering population builds up slowly through November, but large numbers are usually not reported until late November or early December.

Fall maxima. 2500, Bullocks Point, 8–11 November 1957 (CBl). Relatively few are seen passing coastal promontories during daylight hours either in spring or fall.

Winter. Common Goldeneyes are ubiquitous winter residents and have not fluctuated much in abundance.

Winter maxima. 1800, upper Narragansett Bay, 5 January 1941 (RCC, HNG by boat); 633, Newport–Westport CBC, 18 December 2013.

Spring. Numbers of Common Goldeneyes increase during March and early April, likely reflecting arrival of migrants from farther south.

Spring maxima. 1260, Tiverton, 9 March 1952 (DLE). Most of the wintering birds leave in or before early April, and numbers steadily decline and disappear in the first week of May. Three were at Block Island 12–14 May 1999 (CRa, REn). The latest records of probable transients include 3 at Tiverton 21 May 1950 (DLE), 3 at Warren 30 May 1950 (RBo), and 2 on the Seekonk River 4 June 1988 (REn).

Summer. There are about 30 records of individuals summering on coastal waters.

Barrow's Goldeneye (*Bucephala islandica*)

Status

Rare but regular winter visitor on salt water; very rare inland.

Nonbreeding

The first record of Barrow's Goldeneye for Rhode Island was 1 collected at Providence on 2 January 1872; the specimen was acquired by Brewster (MCZ #291356). They then remained unreported but doubtless overlooked until 1935. They now occur annually, with 4–5 individuals per year in most years (Hasbrouck 1944). They are rare at Block Island, where the first record was a drake seen 20 December 1988, 19 December 1989, and 18 December 1990 (CBCs). Immature males constitute the majority of reports. Females, usually readily identifiable, are occasionally seen alone. Barrow's Goldeneyes rarely appear before early or mid-December. The bulk of reports are for late winter and early spring. Inland records include 2, Cumberland Reservoir, 13 December 2013 (BCG); 1 male, Woonasquatucket Reservoir, 2 December 2007 (MLy); 1 male, East Providence Reservoir, 3 December 1950 (HEC). There are also records from coastal freshwater ponds such as Gardiners Pond, Watchaug Pond (8–14 March 1964 [DLK, CWo]; 15 March 1967 [DLK]; 15 March 1969 [DLK]), Cross Mills Pond (29 March 1969; LRBC), and Pachet Brook Reservoir (18 November 1993; JO).

Extreme dates. 16 November 2010, near Fox Island (v.o.), to 14 April 1987, Napatree Point (CRa), and 20 April 1996, Sabin Point (DFi).

Hybrids. An apparent Common Goldeneye × Barrow's Goldeneye hybrid was seen at Warren, 26 January 2000 (Paul Lehman, RFa). A possible Common Goldeneye × Barrow's Goldeneye hybrid was photographed at Jamestown 3 January 2016 (ChP).

Smew (*Mergellus albellus*)

Status

Vagrant. One to three records.

Nonbreeding

An adult male Smew was discovered at Easton Pond, Middletown, on 3 January 1976 by Willoughby and Bushnell, seen by hundreds of observers from all over the country through April, and photographed in color (*AB* 30:691). There were unconfirmed reports of the bird being present in mid-December 1975. Consorting with Common Mergansers, the bird flew well and was excessively wary, suggesting a wild origin. Another or the same bird appeared at Quonochontaug Pond 6–27 January 1978 (E. Saunders, R. Lougee et al.). A report in the *Providence Journal* of yet another from northern Rhode Island about 18 February 1983 was never confirmed, and another

FIGURE 3. Smew, © Margaret La Farge.

bird was reported 9 September 1984 in South Kingstown (SDz). Vagrant Smews first appeared in Iceland in 1974 and increased slightly to 2–3 birds per year during the period 2004–2010 (Kolbeinsson 2017).

Hooded Merganser (*Lophodytes cucullatus*)

Status

Common fall and fairly common spring migrant; fairly common winter resident. Uncommon but increasing as breeder.

Breeding

The Wood Duck nesting box program, begun by RIDFW in 1950, provided a wealth of nest sites for Hooded Mergansers as well. A female was observed among duck boxes at the East Providence Reservoir 1 May 1955 (RLF), and a female and eggs were discovered in a box at Dolly Cole Marsh in Foster 18 April 1956 (Zurlinden, Clement). In 1958, 3 females were banded at their nests, 2 in Foster and 1 in Richmond, and a fourth brood was seen. One of the females banded was brooding both her own young and several Wood Duck chicks.

Mathew C. Perry, then of RIDFW, examined Wood Duck boxes and found a substantial number of these occupied by Hooded Mergansers: a female with 10 eggs 11 May 1967 at Longridge Marsh, Harkney Hill Road, Coventry; 2 eggs at the same location in 1968; abandoned eggs found in 1968 at Pulaski Marsh, Glocester, and in the Buck Hill Marsh, Burrillville. Other instances of breeding include a female and 9 eggs photographed at the Dolly Cole Marsh 24 May 1970 (CCA) and another nest near Carbuncle Pond, Greene, in 1969 or 1970 (Follet, fide CCA). A female with 8 young was seen in Pulaski State Forest, Burrillville, on 2 June 1985, and a female with young was noted at the same location in June 1989 (REn). A female with 11 young was seen at Summit 22 June 1993 (RLC). Away from the acid uplands of western Rhode Island, a female was seen standing on top of a box in a pond on White Brook, Carolina, 21 May 1961 (DLK), and another female was under a box near Nonquit Pond, Tiverton, 5 August 1962 (RLF, M. Gardler). Several other observations are also suggestive of breeding: 2 at Weekapaug 16 May 1978 (ESa); at Durfee Hill 24 May 1992 (JMu, RS); a female in North Smithfield 1 June 1991 (LS); 2 at Portsmouth 22 August 1959 (SSD); and 4 at Westerly 23 August 1962 (ESa).

Nonbreeding

Fall. Migrants are seen in small numbers in October, but most records fall between early November and late December, with a peak in late November. They have been numerous at Scituate Reservoir in recent years.

Fall maxima. Total of 384 on all northern Rhode Island ponds, 22–25 November 2001 (MLy); 150, Watchaug Pond, 21 November 2012 (CRa); 146, Cumberland, 21 November 2004 (MLy).

Winter. Hooded Mergansers are sensitive to cold spells and freezing over of swamps and marshes, so abundance can diminish quickly during cold winters.

Winter maxima. 448, Napatree CBC, 23 December 2007; 378, Ninigret NWR, 14 February 2010 (RDe); 357, South Kingstown CBC, 27 December 2008 (STs et al.).

Spring. Migrants begin to appear again in mid-February, and there is a peak in mid to late March.

Spring maxima. 115, Trustom Pond, 7 March 1993 (JMu, RS).

Common Merganser (*Mergus merganser*)

Status

Common migrant and winter resident on fresh water. Rare in summer.

Nonbreeding

Common Mergansers are birds of lakes and reservoirs. They move to salt water during severely cold weather; they return inland at the first trace of open water. They are not found in any numbers on Rhode Island's rather small rivers.

They have greatly increased since about 1930, probably at least partly because of the construction or enlargement of several reservoirs. Since 1940 they have been most numerous at the East Providence Reservoir, Wordens and Watchaug Ponds, and on coastal ponds of Middletown, Newport, and South County. They are sometimes moderately numerous at Block Island's Sachem Pond.

Fall and winter. The earliest arrivals are 1, Watchaug Pond, 1 October 1939 (DLK), and 4, Wordens Pond, 12 October 1957 (DLK). Usual arrival is in mid-November or later.

Fall maxima. 1000, Watchaug, 4 December 1938 (DLK).

Winter maxima. 539, East Providence Reservoir, 21 January 1951 (HEC); 528, Newport–Westport CBC, 19 December 2009; 500, Aquidneck Island, 9 January 1982 (RBo); 353, Diamond Hill Reservoir, 17 November 2002 (MLy); 163, St. Mary's Pond, 16 December 2006 (SRi et al.). Numbers at Block Island are lower than in other coastal areas.

Spring. Numbers diminish through March and April (to 50 per day in late April), and a few small flocks and individuals occasionally remain through May.

Summer. One crippled bird remained on East Providence Reservoir through the summers 1951–1955 (DLE, RLF et al.). Two flightless birds were closely observed at Ninigret Pond 1 June 1978 (RLF, JEM). A crippled bird was seen off Napatree Point

FIGURE 4. Common Mergansers on Scituate Reservoir, © Margaret La Farge.

and Weekapaug 27 June–5 July 1976 (MPe). Some observations during summer on the Blackstone and Seekonk Rivers are suggestive of breeding: 2 were present at Swan Point 1–19 June 1990 (DFi), and 4 were there 29 May 1991 (DFi). Two were present on the Blackstone River in June and July 1991 (LS). Five (a flock of 2 adults and 3 immatures) were present on the Seekonk River 13–20 September 1991 (JO, DLE), prompting a remark by Emerson in the *Field Notes* that this flock "looked suspiciously like a family group [and] raises the question as to whether the species could have bred in the state this summer." Again, 3 were noted on the Seekonk 24 May 1992 (DLE et al.), 1 was there 22 June 1994 (RLF, JEM), and 1, 3 July 2005 (MLy).

Red-breasted Merganser (*Mergus serrator*)

Status

Very common migrant and abundant winter resident on salt water. Rare inland. Uncommon but regular in summer; 2 nesting records.

Breeding

There are 2 records of breeding by Red-breasted Mergansers in Rhode Island. First, Elizabeth Dickens's diaries mention a female and 6 young seen at Block Island 26 August 1938. It is not known if Dickens saw these birds herself or if (more likely) they were reported to her by others. Second, 2 adults were seen on Quicksand Pond 6 July 1997, accompanied by a half-grown, flightless juvenile (REm). The closed state of the pond's breachway at the time would seem to preclude the birds having moved there from any nesting in immediately adjacent Massachusetts. On several occasions, Red-breasted Mergansers have bred on Long Island (Bull 1964; Andrle and Carroll 1988) and in Massachusetts (Veit and Petersen 1993). Red-breasted Mergansers occasionally oversummer on the coastlines of southeastern New England, and breeding attempts and the occasional success seem inevitable. A pair was present at Sakonnet Point for most of July 2006 (REm, CRa).

Nonbreeding

Peak abundance occurred during the late 1970s and early 1980s, both in Rhode Island and Massachusetts, very likely due to the high abundance of sand lance (*Ammodytes* spp.) at that time.

Fall. This saltwater merganser has long been one of the most abundant ducks in Rhode Island during winter, even though they are greatly outnumbered by scoters during migration. Sir Charles Blagden described the "Sheldrake" as "very plenty" around Newport in 1776–1780 (Howe 1905). The wintering population gradually increases in November and early December, reaching peak numbers between late December and mid-March. Numbers have declined since the peak in the 1970s and 1980s.

Winter maxima. 11,506 on the Block Island CBC in 1982; 10,000+ off southern Block Island, 29 December 1986 (CRa, REn, RBo et al.); 3410 between Green Hill and Narragansett, 28 December 1985 (CBC); 1500 at Point Judith, 9 December 1937 (WBD).

Summer. This species is one of the most commonly reported northern ducks as a nonbreeding summer resident. Individuals or small flocks are reported annually, but larger numbers are rare.

Summer maxima. 20 in the Sakonnet Passage 5 July 1937 (RCC); 20, Dyer Island, 13 August 1950 (RBo).

Inland. While merely uncommon on larger lakes elsewhere in New England, this merganser is a rather rare migrant on lakes and reservoirs in Rhode Island. All such reports have thus far been from the East Providence Reservoir, where a few individuals (and, more rarely, small flocks) have been reported.

Inland maxima. 25, 29 November 1976 (ILe); 25, 14 December 1978 (ILe). The only midwinter records are 1, 18–25 January 1986 (DLE), and 20, 7 February 1953 (Romaine). Two were noted on the Blackstone River 19 February 1992 (LS). There are several records of migrant individuals at the East Providence Reservoir in April, and the latest state record inland in spring is 3 there on 23 May 1955 (RLF).

Masked Duck (*Nomonyx dominicus*)

Status

Vagrant. One record.

Nonbreeding

One was seen 3–5 June 1927 at Island Park (Horton); the identification was said to have been confirmed by Forbush from a written description. Johnsgard and Hagemeyer (1969) list about 7 other records from the northeastern United States during summer. Masked Ducks are rare in captivity, and were especially so in 1927, so the Rhode Island record seems likely to have been of a wild bird.

Ruddy Duck (*Oxyura jamaicensis*)

Status

Common migrant and fairly common winter resident. Former breeder, now rare in summer.

Historical

The "dipper" was said by Sir Charles Blagden to be common in the vicinity of Newport in 1776–1780 (Howe 1905). Unlike many other freshwater ducks, it was said to have remained relatively common during most of the shooting era, perhaps because of its customary refusal to fly in daylight hours unless pursued by boat. Nevertheless, Hathaway noted a considerable decrease between about 1910 and 1935 and considered it to be an uncommon migrant. This impression was not shared by Sturtevant, who saw 300 (14 October 1913) and 100 (27 October 1916) on the then-new Gardiners Pond in Middletown. Since then, it has been common and sometimes abundant on certain favored ponds during fall, early winter, and early spring.

Ruddy Ducks apparently bred before the turn of the twentieth century at Long Pond, Sakonnet. One was shot at this location in June 1884, the specimen formerly in the URI collection. Two others were killed there in July 1889 and mounted by Newton Dexter. One of these may be the same as a specimen of an adult male in the RWPM collection labeled merely "July 6, 1890." Newton Dexter also noted two broods of young at Long Pond in 1895, but the young disappeared when half-grown (Howe and Sturtevant 1899). While there are no recent reports of breeding in the state, individuals and small groups have occasionally been seen since 1937 during June and July at Briggs Marsh, Gardiners Pond, Foster Cove, and Easton Pond. In particular, potential breeding birds were present at the last location in 1937, 1938, 1952, and 1971, but most of the great cattail beds have inexplicably disappeared at this location since the early 1970s, and breeding of Ruddy Ducks or other species such as American Coots, Common Gallinules, or Marsh Wrens has now become unlikely there.

Nonbreeding

Fall and winter. Records increase in mid-September, and general arrival begins about 1 October. Peak counts are made from 20 October–30 November or sometimes later. Sizable aggregations of these freshwater ducks may remain into mid-December in mild seasons.

Fall maxima. 5000, Watson Reservoir, 22 October 1999 (DLE); 3100, Pachet Brook Reservoir, 4 November 1995 (DLE); 650 in Middletown, 28 October 1956 (JBa); 600 in the South Shore ponds, 31 October 1948 (DLK).

Winter maxima. 1835, Newport–Westport CBC 17 December 1995 (DLE et al.); 1348, Trustom Pond NWR, 14 December 2006 (JSJ).

Spring. Even when they are frozen out of the state in midwinter, numbers usually return for a short period to favored ponds with the first open water in late February and early March.

Spring maxima. 550, Middletown, 3 April 1996 (DLE); 525, Trustom Pond, 15 March 1998 (STs); 500, East Providence Reservoir, 28 February 1960 (DGa). A few have remained irregularly into June in recent years (2–3 at Trustom Pond, 5–12 June 1993; JMu, RS).

Maxima, late spring and early summer. 19, St Mary's Pond, 23 July 2002 (John Magill); 18 at Easton Pond from 1 May into July 1952 (RCC, DLE).

NEW WORLD QUAIL (ODONTOPHORIDAE)

Northern Bobwhite (*Colinus virginianus*)

Status

Extirpated resident. No breeding records since about 1990, and all populations only sustained by stocking since 2000.

Historical

Wright (1941) summarized the early distributional and numerical changes of bobwhites in Rhode Island. Their status in precolonial times is conjectural, but they may have been plentiful about Native American agricultural fields. Early colonization brought a crude kind of agriculture that was particularly suitable for bobwhites. After 1880, agricultural land began to decline, and bobwhites seem to have steadily decreased since that time, though factors other than habitat change may have been influential in this decline (Phillips 1915).

Hunting pressure, especially market hunting, during the years 1890–1904 apparently had a strongly negative impact, despite the widespread availability of good habitat. There are unofficial reports of daily bags ranging from 25 to 100 birds. Despite a closed hunting season in 1904, prohibition of market hunting in 1910, limitation of daily bags in 1917, and the liberation of propagated birds, bobwhites continued to decline through the first quarter of the twentieth century. Severe snows in the winter of 1919–1920 are said to have nearly caused their extermination in Rhode Island, and this calamity was followed by heavy snows the following winter and a large incursion of Northern Goshawks in 1926–1927. Wright (1941) stated that the decade 1920–1930 was the most difficult for bobwhites in Rhode Island up to that time. Continued low populations in the mid-1930s are indicated by the fact that RCC then knew of only four localities where they could still be found in Tiverton and Little Compton (previously two of the prime areas in the state), and bobwhites nearly disappeared in the Westerly area during the same period (ESa).

There is anecdotal information pointing to a much larger population in upstate areas and in the vicinity of Providence (Scituate, Rumford, and elsewhere) during the years 1880–1930 than at any time since. They were common in the Chepachet area in 1888–1890 (Sprague). By 1940 there had been some increase, particularly in southern parts of the state, doubtless mostly as a result of the liberation of captive birds. Wright (1941) gave a map showing bobwhite distribution throughout the state, with concentrations along the shore as well as in Cranston, Lincoln, and other upstate areas. Many other sources indicate a generally high population level during the 1940s, which CBC data suggest continued to the 1970s. Since then, bobwhites have declined sharply.

Breeding

Introductions by the Commissioners of Birds and others began in 1904. During the peak of the state release program in the 1940s, 1200 birds were released annually. In recent years, the liberation of quail has been discontinued by the state, but intermittent releases by local hunting clubs have continued. Aside from an old Rhode Island specimen without data in the RWPM, the only surviving specimens of the original native population are a small series (now in USNM) collected by HSH (Aldrich 1946). The last locality where "wild" birds were found was near the former Ninigret Airfield in Charlestown, and the last CBC record for that area was 1998.

Bobwhite survival on islands has been poor. There was formerly a native population on Aquidneck Island that was reportedly extirpated by hunting during the nineteenth century. A large number of birds from the Carolinas were liberated by Col. Powel and others about 1897; they were reported to be quite abundant in Newport and Middletown. They may have been decimated by the cold spring rains of 1903 and 1904, although some individuals from this or another introduction were present in 1916 (ESt). Juanita Gee remembers that bobwhites were common in Middletown in the late 1920s and early 1930s, but there are virtually no records on the island after 1938, and they were then largely absent on Aquidneck for decades, despite extensive agricultural habitat. A few introductions on the island in the mid-1950s were apparently unsuccessful; others seen at 3 scattered localities from Sachuest to Portsmouth in 1982–1984 derive from additional introductions. On Block Island, birds of (allegedly) Mississippi stock were liberated in 1928 but disappeared after about 5 years. Birds were reintroduced there in 1943, again in 1962, and again about 1973 (100 in 1974–1975; ASRI), but each time they disappeared after a few years (EDi, HNG, MSl, RLF). Bobwhites have been introduced in recent years to Prudence Island. They were present in 1970 (RSt), and there were 2 in 1990 (REn, CRa), but there is no record of their presence on Conanicut Island. On nearby Dutch Island, however, 7 were reported 7 August 1991 (WG), and 2 were shot there in 1993 (fide JEM).

PHEASANTS AND ALLIES (PHASIANIDAE)

Wild Turkey (*Meleagris gallopavo*)

Status

Common resident.

Historical

After the species was gradually extirpated from New England by about 1850 and eventually from New York as well, turkeys native to southern Pennsylvania were reintroduced (via New York and Vermont) into parts of their original range in southern New England during the 1980s and 1990s.

Wild Turkeys occurred commonly throughout southern New England (as far east as Bar Harbor, Maine, in earliest colonial times), but specific references to the original population in Rhode Island may be limited to Roger Williams's *A Key into the Language of America*. Here, Williams said, "[T]he turkey is called 'neyhom.' The Indians lay nets on shore, and catch many fowls upon the plains, and feeding under oaks upon acorns, such as geese, turkey." In Rhode Island, turkeys were probably exterminated from most areas before 1750. Mention of turkeys at Kingston in Tom Hazard's diary (Nichols 1934) for 1790 may not refer to wild birds. The last turkeys in Connecticut and Massachusetts were killed in about 1813 and 1850, respectively.

Breeding

During the late 1950s, 50–60 pen-reared birds were released by RIDFW among 3 state management areas: Great Swamp; Buck Hill, Burrillville; and in Glocester. A year after the releases, a bird was reportedly being fed during the winter by a Scituate resident but survival of most of these birds was short. In 1967, about 25 birds were released at Block Island (and not by RIDFW, fide CCA). There was a minimum of 22 still present in 1972 (MSl), and broods were noted in 1974, 1979, and 1981. Broods of Wild Turkeys were reported in 1973 in Burrillville, and the species began to be seen in western Exeter and West Greenwich after 1975. These birds probably derived from private releases, since Massachusetts Fish and Wildlife was just beginning to transplant turkeys in the Berkshire Hills in 1972, and Connecticut Department of Environmental Protection began to release birds in eastern Connecticut only in 1978. Spurred by success with Pennsylvania turkeys that had been gradually transplanted across New York into southern Vermont and later into western Massachusetts, RIDFW captured 29 birds (8 toms, 21 hens) at Rupert, Vermont, on 7 March 1980 and released them the same day near Liberty in Exeter (Chadwick 1982). Some of these birds were noted with broods both in 1980 and 1981, and an estimated 120–150 were present in South County at the end of the summer of 1981. In the first 2 years, they had reportedly been seen east to Kingston, south to Perryville, and west nearly to Hope Valley, a maximum of 10 mi (16 km) from their point of introduction at Liberty.

Turkeys now occur throughout mainland sections of the state, including areas east of Narragansett Bay, and are also seen frequently in urban areas such as Swan Point Cemetery in Providence. Birds also occur on Aquidneck Island. There was a flock there derived from pen-reared stock, but birds could probably also reach the island naturally. Feral Wild Turkeys still occur on Block Island—there may have been additional releases by residents after 1993, but RIDFW did not put them there.

Maxima. 142, South Kingstown CBC, 26 December 2011; 127, South Kingstown CBC, 22 December 2012; 104, Napatree CBC, 26 December 2010; 80, North Kingstown, 15 January 2016 (JMa). At least 1 bird was calling during November 2014 on Block Island (CRa, PP).

Ruffed Grouse (*Bonasa umbellus*)

Status

Rare and declining resident.

Historical

Ruffed Grouse were not common during the nineteenth century when grazing and "poor cutting" dominated many areas of northern and western Rhode Island and when market gunning and snaring were practiced. Illegal "partridge hedges"—brush fences with snared openings—were reportedly in regular use in Rhode Island before the turn of the twentieth century and as recently as about 1930. Regrowth of woodlands throughout mainland portions of the state and relaxed hunting pressure allowed Ruffed Grouse to become common in many areas after the 1920s. A population crash occurred a few years before 1900; the young birds were said to have died from a disease known as "the pips." They were more common by 1900, when three market gunners at Rockville are recorded as having made a shipment of 34 birds. They again declined throughout the state until 1917, when an increase was noted. They were once again common in 1930. The year 1983 may have marked a recent high in the population, when the species was present on territory at two places in Warren, two were seen near Simmons Pond in Little Compton, a brood was observed in Wakefield, and individuals were seen at two locations in late May and June south of the South County moraine (Trustom Pond NWR, Ninigret NWR; DVT).

Breeding

Ruffed Grouse prefer early successional habitats and have declined because of the recent lack of disturbance that would create new patches of high-quality habitat (Blomberg et al. 2012). There has been a decrease since the 1990s, with only a single pair of birds known in 2016 (RFa, CRa). Before then, Ruffed Grouse were found in the western and northern parts of Rhode Island, the range extending east to the edges of the Providence metropolitan area; south to the edges of the wooded morainal hills of South County; and east to Wakefield, Saunderstown, and Greenwich Bay. They are still occasionally found along the fringes of suburban areas in Warwick, Cranston, Providence (Neutaconkanut Hill), North Providence (all winter, 1981–1982), Pawtucket, East Providence, and Barrington. Vagrants appeared on the east side of Narragansett Bay in 1950, 1961, and 1963. They used to be found on territory in modest numbers in the heavily wooded parts of central and eastern Tiverton, southward into eastern Little Compton. One was seen west of Briggs Marsh 18 December 1993 (GLB). There are perhaps 3 unsatisfactory reports from Aquidneck Island and no records of any kind for other islands in Narragansett Bay.

The effect of hunting pressure on the present population is unknown but probably of considerably lesser importance than the condition of the habitat. Estimated kills in Rhode Island were 5369 in 1939 and 3361 in 1940, according to HNG, Fish and Game administrator.

Greater Prairie-Chicken (*Tympanuchus cupido*)

Status
Extinct.

Breeding
"Heath Hens," *T. c. cupido*, were common inhabitants of sandy clearings and scrubby pine–oak barrens on Long Island and in many parts of southern New England north to Boston during the early colonial era. They survived after 1835 only on Martha's Vineyard, where they became extinct about 1931 (Griscom and Snyder 1955).

In Rhode Island, no certain record survives, but a tradition of Greater Prairie-Chickens having been part of the avifauna here is indirectly supported on ecological and historical grounds. Pitch pine and scrub-oak barrens, similar to their former haunts on Long Island, Martha's Vineyard, and in the New Jersey Pine Barrens, are still found on the outwash plains in the southwest quarter of the state. These dry areas, which are thought to have been burned regularly by Native Americans as well as by natural fires, and which were probably much more extensive in precolonial times than at present, would have certainly been suitable.

In 1897, during an era when it was fashionable to liberate nonnative game birds in New England, a prairie-chicken of unknown race was shot in Rhode Island, but this individual was thought to be an introduced midwestern bird (Howe and Sturtevant 1899).

Ring-necked Pheasant (*Phasianus colchicus*)

Status
Introduced; the only currently self-sustaining population is on Block Island.

Historical
The first Rhode Island pheasants were introduced at Narragansett in 1894 by Horace Bloodgood; the birds were reportedly of English stock, and they survived and continued reproducing. Another successful liberation was made on Warwick Neck in 1902. These experiments stimulated other private releases in the next decade, and a small number were released by the state beginning in 1911. Hunting of pheasants became legal statewide in 1916. About 1500 pheasants were released annually during the period 1929–1941 by the state (Wright 1941) and about 1000 per year in the years 1954–1975 (Labiskey et al. 1964; Myers 1976). Close to the same numbers were released privately; e.g., on Aquidneck Island in the 1930s.

Breeding
Pheasants were first introduced at Block Island 6 April 1923 and became common thereafter. Forty were seen at Dickens Point in 1925 (EDi). Because of mild conditions and a wealth of habitat rich in fruiting shrubs but lacking in ground predators,

and historically aided by a very restricted hunting season, the species became particularly common at Block Island and has attained densities there as great as or greater than anywhere in New England. Their numbers may have begun to decline at Block Island in recent years, however, with the widespread succession of goldenrod and bayberry habitats into dense brush, red maple, and shadbush.

On the mainland, pheasants have been more common and persistent in the east-Bay portions of the Narragansett lowlands (particularly Aquidneck Island) than in the west-Bay area or along the South County coastal plain. They have readily adapted to suburban conditions in parts of the Narragansett Bay area and for decades have been fairly common residents in numerous localities within and around the fringe of the Providence metropolitan area from Warwick north to Pawtucket and south to Barrington. In the western and northwestern portions of the state, however, they have long been very local and often absent.

Pheasants are now common only on Block Island. Counts there have greatly exceeded others elsewhere.

Maxima. 50, Block Island, 20 June 1995 (REn); 45, Block Island, 27 June 1996 (REn); but numbers there have been much lower since then. 25, Middletown, 28 May 1990 (REn); 16, Warwick, 4 December 1982 (DWa).

GREBES (PODICIPEDIDAE)

Pied-billed Grebe (*Podilymbus podiceps*)

Status

Fairly common transient and winter resident on freshwater and brackish coastal ponds. Rare and erratic breeder; formerly more numerous.

Historical

Prior to 1899, Pied-billed Grebes were considered "common" by Newton Dexter at Long Pond, Sakonnet, where he saw adults with young and collected eggs. They were found breeding at Point Judith Pond in 1897 by Dr. George W. Field. Hathaway knew of three breeding locations. They were present for several years at Hammond's Pond, Pawtucket, where HSH saw one on 15 May 1917 and where they reportedly bred in 1916 (fide HSH). Hathaway took four eggs 22 May 1927 from a nest in "Fort Neck Pond," the nearly freshwater northern cove of Ninigret Pond. They also bred in what is now called Louttit Pond, West Greenwich (now part of Alton Jones Campus, URI), where there were two nests in 1935 and singles in 1941, 1942, and 1944. Douglas Kraus found a pair with four small young 4 September 1948 at Wesquage Pond, Bonnet Shores, during a transitory period when an impoundment caused the western portion of that salt estuary to remain relatively fresh. Richard Bowen saw an adult with a very small chick in Boyd's Marsh, Portsmouth, 24 July 1940, prior to the destruction of that salt marsh by the town dump. Elsewhere, nesting season reports are indefinite, for example, birds seen in midsummer at Tiverton, Matunuck, Easton Pond, and Scituate Reservoir. The presence of individuals or pairs during at least five summers 1937–1953 at Briggs Marsh aroused continuing suspicion of nesting, and a pair with two mottled young was seen there 4 September 1952 (DLE).

Breeding

Pied-billed Grebes were rare but regular breeders up to about 1960. Since the late 1950s, there are only a handful of reports indicative or suggestive of breeding: 1 at Long Pond, 17 July 2000 (REm); adult and 3 chicks at Trustom Pond in June 1993 (m.obs.), an individual "on territory" at Trustom Pond on 8 June 2002 (RTL), then a pair there a week later (Glen Williams), then 3 juveniles there on 31 July 2002 (CNu); also 3 adults and 2 juveniles at Trustom, 24 June 2003 (AMc), and 1 adult there on 11 June 2004 (DFi). Another pair found 7 July 1998 in West Kingston likely bred there (CRa). The decline in Rhode Island reflects a general decline in the Northeast (Veit and Petersen 1993; Levine 1998). It is possible that Rhode Island ponds have been degraded by acid rain and thus are no longer suitable for breeding by Pied-billed Grebes. Two recent summer records (26 July 2006, Jerusalem, 1 [DFi], and 28 July 2012, Moonstone, 1 [B. Sherman et al.]), suggest breeding or at least prospecting.

Nonbreeding

Pied-billed Grebes have been declining as fall transients and winter residents since the 1960s or earlier.

Fall. Migrants commonly arrive by early September, with peak numbers in October. Counts at this season have seldom exceeded 25 per day, particularly in recent years.

Fall maxima. 65, Tiverton to Little Compton, 21 October 1978 (RBo, HMP); 53 along the South Shore, 12 October 1948 (DLE); 50, Little Compton, 12 October 1970 (RBo); 50, Little Compton, 1 November 1953 (RBo); 36, Trustom Pond NWR, 13 October 2007 (STs). The highest total since 2005 is the 2007 Trustom Pond report.

Winter. Pied-billed Grebes occur through December and into early January in small numbers, and are occasionally more numerous in mild seasons, as in 1953–1954 when 121 were seen on the Newport–Westport and South Kingstown CBCs. Their presence in winter depends on the maintenance of open water on freshwater and brackish coastal ponds, but they move to tidal estuaries in severely cold weather. They survive the winter in Rhode Island, sometimes in small flocks.

Spring. The appearance of new birds in or after early March includes migrants and local winterers. Some remain into early May, and some vocalize and act in a territorial fashion.

Spring maxima. Daily totals in spring have not exceeded 5 per day. They are rare in late May; e.g., 23 May 1979 at Moonstone (JGe, RGe); 24 May 1958 at Galilee (SSD).

Horned Grebe (*Podiceps auritus*)

Status

Common migrant and winter resident along the coast; occasional inland. Rare in summer.

Nonbreeding

Horned Grebes are one of the most characteristic birds of the winter coast, where they search alone or in small groups for small fish just beyond the surf. They are relatively scarce more than a mile or so from land. Our most common grebe, they have declined substantially since their peak abundance in the 1950s.

Fall. First fall migrants arrive by late September; e.g., 3 at the East Providence Reservoir 19–21 September 1988 (RFa). They are more common by early October; e.g., 39 at Trustom Pond, 12 October 1951 (DLK).

Fall maxima. 200, Jerusalem, 31 October 1953 (DLK). Birds appear inland from mid-October to mid-November, with several counts of 15 per day.

Winter maxima. 933, Newport–Westport CBC, 21 December 1958 (DLE et al.); 400, Sakonnet Passage, 18 December 1955 (RBo). Recent totals have been much lower: 182, Newport–Westport CBC, 17 December 2011; 94, Napatree, 24 February 2011 (CRa).

Spring. Larger premigratory concentrations occur from late March to late April, with reduced numbers remaining into early May.

Spring maxima. 350 along the South Shore, 21 March 1960 (DLK); 325, Moonstone Beach, 17 April 1941 (DLK). They are less common inland in spring than during fall; e.g., 22 March 2008, Diamond Hill Reservoir, Cumberland (MLy).

Summer. There are about 25 records of summering individuals, occasionally 2 together (2, 11 June–5 July 2011, Napatree; CRa).

Red-necked Grebe (*Podiceps grisegena*)

Status

Fairly common transient and winter resident on salt water; largest numbers seen in February–April. Very rare in summer. Rare inland.

Nonbreeding

Fall. Migrants arrive in mid-October and peak numbers are found in November and December.

Fall maxima. 30, Sachuest Point, 30 November 1987 (REn); 24, Sachuest Point, 19 November 1939 (WHD).

Winter. The wintering population is usually small and fluctuates from year to year. Christmas Bird Counts seldom total more than 15, and in some years, none are observed.

Winter maxima. 64, Middletown, 14 January 1989 (SDz); 48, Sachuest, 2 January 1988 (RBo et al.).

Spring. In February–April, flocks of Red-necked Grebes can be found sitting on the water along the coast. Substantial numbers also fly past, presumably during the course of their northward migration. Unlike most other diurnal saltwater migrants that move eastward along the Rhode Island coast in spring, almost all migrating Red-necked Grebes move west or southwest, and such movement occurs almost entirely in the late afternoon and evening. These westward flights, which involve individual flocks as large as 30 birds, frequently accompany westward-moving flocks of White-winged Scoters and Long-tailed Ducks.

The westward movements along the southern New England coast are in the opposite direction of most spring migrants, which generally move north or east, and suggest that these birds may be on their way to the Hudson River via Long Island Sound.

Spring maxima. Sitting birds: 130, Sachuest Point, 16 February 1990 (GS); 110, Sachuest Point, 10 February 1989 (GS). Migrating birds: 536, Point Judith, 11 April 1998 (RLF); 445, Point Judith, 20 April 2007 (RLF); 285 moving past Watch Hill toward Long Island Sound, 13 April 1996 (RLF); 212, Point Judith, 27 April 1997. Rare in upper Narragansett Bay, a Red-necked Grebe was shot 11 October 1883 on the Seekonk River (Glezen), and there are sight records for "upper Bay" and Edgewood (see below).

Inland. Red-necked Grebes are also rare inland and are much less frequent on ponds and reservoirs than on inland waters of Massachusetts or New York. Five were at Bold Point Park, Providence, 21 March 2003 (RFa), 1 was seen 5 November 1940 at Larkins Pond (DLK) and another 30 January–3 February 1958 at the East Providence Reservoir (RLF et al.). One was picked up on golf course turf in Richmond,

7 February 1992, and released unharmed at the Weekapaug Breachway (CRa). Numbers of birds found inland in some years are likely associated with the freezing of the Great Lakes. For example, after the massive cold wave during late February 1934, which completely froze the Great Lakes (Bull 1964), 3 Red-necked Grebes were in upper Narragansett Bay off Edgewood, 23 February 1934 (DLK); 6 were at Narragansett Pier; another 10 were found dead along the South Shore, 4 March 1934 (DLK); and 9 were found dead and frozen into the ice of Watchaug Pond on 5 March 1934 (MSo). On the other hand, 1 on Watchaug Pond 6 March 1982 (SDz) and 2 on Yawgoo Pond 11–13 March 1982 (CWo, DLK et al.) may possibly have been northward-moving transients brought down by heavy rainfall accompanying the northeast storm of 6–9 March 1982.

Extreme dates. 1 September 1985, Napatree Point (CRa); 23 September 2006, Diamond Hill Reservoir, Cumberland (MLy). There are several reports of summering birds: 19 June 1977, Sandy Point Island (MPe); 30 July 2005, Ninigret Pond (v.o.); 28 August 1979, Narragansett Bay off Newport (RTL); 12 August–30 September 1985, Seapowet (DFW, RBo et al.; photo).

Eared Grebe (*Podiceps nigricollis*)

Status

Vagrant. Rare but increasing vagrant, migrant, or winter resident. The occurrence of the Eared Grebe in Rhode Island is well established by color photographs taken in 1960 and 1973 by RBo. The records, all of single birds, range from 8 December to 14 April (Buckley 1968).

Nonbreeding

LOCATION	DATE	OBSERVER(S)
Scarborough	28 January–10 March 1951	DLK, FPe, CWo
Sakonnet	13 February 1951	Williston
Narragansett	21 December 1951	DLK
Sakonnet Harbor	22 December 1951	DLK, REm
Island Park	8–26 December 1955	RBo, HEC
Warren (photos)	12–27 March 1960	RBo
Narragansett	18 January–23 February 1969	DLK, CWo, KKi
Narragansett	18 March 1972	RRV
Moonstone, Quonochontaug	27 January–5 February 1973	DLK, KKi
Seapowet (photos)	14 April 1973	RBo

LOCATION	DATE	OBSERVER(S)
Seapowet	20 January–28 February 1979	RBo, RAC et al.
Napatree Point (photos)	20 April 1985	JHW, RDe et al.
Sachuest Point	14 January 1991	Karlson
Ninigret Breachway	22–27 November 1993	PAB, DFi, DLK et al.
Sachuest Point	25 November 1993	MJM et al.
Charlestown	3 December 1993	BCG
Trustom Pond	15 October–12 November 1996	SSM et al.
Upper Narragansett Bay	5 February–22 March 1997	SRo, MJM
Quicksand Pond	13–26 September 1998	REm et al.
Trustom Pond	3–19 October 1998	CRa et al.
Trustom Pond	18 September 1999	JSJ
Napatree Point	22–24 February 2000	JSJ, SSM
Napatree Point	20 March 2001	CRa
Trustom Pond	8 October–4 November 2001	S. Wheelock, JSJ
Middletown	24 January 2002	JSJ
Napatree	7–28 March 2002	CRa
Sachem Pond	23 September–1 October 2000	STs, ASRI
Maschaug Beach	13 December 2003	GWi
Tiverton	26 December 2003	MT
Trustom Pond	17 October–19 November 2006	m.obs.
Bullock's Cove, East Providence	27 September 2013	H. Moore

Western Grebe (*Aechmophorus occidentalis*)

Status

Vagrant. About eight records.

Nonbreeding

One was seen at Jamestown 11 January 1947 (WBD), and perhaps the same individual was seen again at Narragansett 23 February 1947 and 8 March 1947 (DLK et al.). What may well have been the same bird returned to Narragansett Pier 28 December 1947 and 4 January 1948 (DLK). A remarkable group of *12* spent part of the same winter in Gloucester, Massachusetts (Veit and Petersen 1993), suggesting that an irruption to the east coast occurred at that time. One was found at Briggs Marsh 2–4 December 1956 (H. W. Harrington, SSD, RLF). Another remained at Jamestown from 4 February–10 May 1967 or later, allowing an exceptional opportunity for study by many observers (RGe, JGe, DLK, MDa, KKi, FPe, CWo et al.). The more recent observations are of 1 seen at Quicksand Pond 22 October 1973 (RBo, WBD); another at Sachuest and Sakonnet Points 5–19 December 1998 (D. Lange, RTL et al.); others at Jamestown 19 February–20 April 1999 (HWi, MT et al.); 1, Narragansett, 8–15 April 2012 (m.obs.); and 1, Sachuest Point, 14–29 January 2013 (GDe).

PIGEONS AND DOVES (COLUMBIDAE)

Rock Pigeon (*Columba livia*)

Status

Common to abundant resident, especially in urban areas.

Breeding

Numerous resident in cities and towns. They breed abundantly on office buildings, factories, and under highway bridges, as well as commonly in farming districts and even in abandoned structures on the smaller Narragansett Bay islands. A small but apparently growing population is now present at Old Harbor, Block Island. Like those that breed on the cliffs of Nahant on the north shore of Massachusetts and in similar places in Connecticut, Rock Pigeons nest in Rhode Island at a few "natural" locations. The few sites where the species breeds away from human structures include 2 or 3 sites on the Dumpling Islands, small rocky islands off Jamestown (nests examined in 1981, 1986, 1993, 1995, 1996; RLF); on The Clumps (nest, 1988) and perhaps other islands off Sakonnet Point; on the high, easterly rock seaward from Briggs Beach (up to 4 active nests through the 1980s, 1990s; RLF, JEM); in Purgatory Chasm, Middletown; and possibly on a rocky road cut along Route 295 in Smithfield (REn). Others may nest in similar sites at West Island and Little Gould Island.

Nonbreeding

Flocks in central Providence have exceeded 300; more than 500 have been found feeding in corn stubble on Aquidneck Island, and 650 have been counted on a barn roof in Matunuck. They are frequent in many parts of the western and northern interior, and the smallest numbers in the coastal zone appear to be at Block Island, where pigeons have been lacking on CBCs for whole decades; e.g., in the period 1962–1976.

Maxima. 737, South Kingstown CBC, 27 December 2008.

Eurasian Collared-Dove (*Streptopelia decaocto*)

Status

Vagrant. Two records.

Nonbreeding

One appeared at a feeder in Rumford 6 June 1999 (JSu, E. Slattery). Another was in Charlestown 14 April 2012 (N. Harvey). Eurasian Collared-Doves have expanded their range through vagrancy after having been released by humans on the Bahamas in the mid-1970s (Romagosa 2020). Some domesticated African Collared-Doves ("Ringed" Collared-Doves *Streptopelia "risorii,"* Dickinson and Remsen 2013) have occasionally been seen in the wild in North America, and hybrids with *S. decaocto* are known. However, the overwhelming majority of collared-doves in North America are clearly *S. decaocto* and it is highly unlikely that hybrids or intergrades could be identified with any certainty. Thus, we conclude (until proven otherwise) that all *Streptopelia* in Rhode Island are *S. decaocto.*

Passenger Pigeon (*Ectopistes migratorius*)

Status

Extinct; formerly common.

Historical

There are only a few Rhode Island references from the first two centuries of colonization, but these suggest that Passenger Pigeons were abundant here. On the farms of Prudence Island in 1748, there is mention of "[g]reat quantities of land fowl such as paterage, quails, pidgeons, which fly in great flocks" (Calder 1935). The Rev. Andrew Bernaby, while traveling from Rhode Island to Boston, wrote in his *Travels through the Middle Settlements in North America in the Years 1759 and 1760* (quoted in Wright 1910), "[D]uring the course of my ride from Newport, I observed prodigious flights of wild pigeons: they directed their course southward, and the hemisphere was never entirely free of them. . . . They are birds of passage, and are of beautiful plumage, and are excellent eating. . . . I met with scarcely any other food at the ordinaries where I put up, and the common people subsist almost wholly upon them during their flight." Sir Charles Blagden, writing from Newport in 1776–1780, said of the "wild pidgeon" that about 19 June "some flights . . . came over the Island, but the great flocks are not found nearer the sea than Providence" (Howe 1905). The species is mentioned occasionally in Tom Hazard's diary for Kingston (Nichols 1934), the earliest date being 5 July 1784, when he stated, "I shott four wild pigeons." Other dates range through August to 21 September. On 12 September 1785 "the pigeons flew very much. I shot at some but kild none."

Hathaway (1913:552–553) stated: "There is no definite breeding record that I can find of this bird though it formerly bred here in abundance. . . . Mr. F. T. Jencks of West Barrington . . . our veteran ornithologist, has written me as follows: 'Once in West Greenwich, a wild pigeon flew from a pine tree to another tree near by and I shot it, a beautiful male. I went and looked where he flew from and found a nest with one egg which I took. I don't remember what disposition was made of these specimens. It was close to 1880, probably in May, but more likely before 1880 than afterward. . . .' I [HSH] have in my collection a mounted adult male taken on the 'Whittaker' grounds in Cranston in October 1854, by Percia Aldrich, and mounted by him . . . [also] a young bird taken by Walter A. Angell November 2, 1886, in Cranston, now in my collection. Mr. William A. Sprague of Providence saw one sitting on a telegraph wire in Glocester on September 25, 1888." A specimen overlooked by HSH and others is one from the collection of Dr. George M. Field, collected in November 1888 at Johnston, perhaps the last recorded in Rhode Island (Schorger 1955).

Forbush (1927) stated, "T. M. Flanagan took about a dozen in Warwick in 1885–1886 according to John M. Flanagan."

According to Roger Williams, all historical Rhode Island dates can be inferred to have been between mid-June and early November.

Passenger Pigeons migrated during the day, regularly crossing water of some width, and the 1748 and 1759–1760 Prudence and Newport accounts suggest that they regularly crossed Narragansett Bay en masse, although Blagden's remarks strongly suggest that most of the fall flight passed north of Narragansett Bay. The ideas of a relatively fearless response to coastal waters by at least a minority of the population and of an early fall passage are supported by EDi of Block Island, who reported to HSH that old island residents had told her that an annual fall flight had occurred there about 10 September (see the 12 September date of a Kingston flight recorded above for 1785 in Hazard's diary). Since EDi began her long period of Block Island record keeping in 1912 at the age of 35, these flights would seem to have occurred in the general period 1850–1860 or earlier.

White-winged Dove (*Zenaida asiatica*)

Status

Vagrant. Four records; increasing continent-wide and now annual in the Northeast.

Nonbreeding

White-winged Doves have recently expanded from their original range in Texas and Central America into parts of Florida. There are at least 14 records for Massachusetts, where it has most often appeared in late June and July (Veit and Petersen 1993).

The first Rhode Island record is of 1 at a Block Island feeder 6–12 June 1999, where it was photographed by Kim Gaffett. Two were seen in 2004: 1 at Block Island, 18 June 2004 (M. Wagner; photo), and another at Watch Hill 4 November 2004

(CRa and STs independently). One remained at NBS in Middletown 13–21 December, 2017 (v.o.; photos).

Mourning Dove (*Zenaida macroura*)

Status

Common to very common resident and transient.

Historical

Mourning Doves have vastly increased since 1900, seemingly because of a decline in year-round shooting both here and on southern wintering grounds. Hathaway saw his first Mourning Dove 29 October 1887 after several years of constant field work. Howe and Sturtevant (1899) called them "not uncommon summer resident[s] in the northern and western portions of the state" but said that they were rare or absent on Bristol Promontory (despite a nest found in Warren by HSH in May 1889) as well as on most of the islands in Narragansett Bay. Arnold, however, who made daily observations at Woonsocket, 1896–1938, saw Mourning Doves in spring or summer only once (27 June 1915) and only four times in fall. Charles Doe was unable to find more than two nests during the period 1881–1930. They were absent or nearly so in Middletown in 1900, both as summer residents and as transients. Sturtevant noted two he shot 8 December 1904 as the first he had ever seen there in half a lifetime. Hathaway (1913), however, stated that they had become established in Middletown by 1910. Mourning Doves were also apparently absent or rare along the South Shore. Hathaway saw two migrants at Quonochontaug 29 October 1916 and commented on their rarity. Mary Southwick, resident at the Kimball Sanctuary, Charlestown, after 1929 seldom recorded them until the mid-1930s, when they became more regular. By the mid-1930s, they bred at least sparingly in most areas of the state except Block Island.

Breeding

In recent decades, Mourning Doves have continued a steady increase as summer residents, but this trend has been slow and gradual. Suburbanization and the proliferation of small conifer plantings also seem to have been partly responsible for their population expansion since the 1940s, especially in coastal areas. Mourning Doves are now common summer residents virtually everywhere, including locations close to downtown Providence, on all of the major islands and some small islands (e.g., occasionally territorial on Little Gould Island), and at least since the early 1970s on Block Island.

Summer maxima. 700, South Kingstown, 16 August 1971 (DWa); 700, West Kingston, 28 August 1974 (DWa); 300, Richmond, 27 August 1978 (DWa); 200, Seapowet, 4 June 1973 (RLF).

Nonbreeding

Spring. Spring arrivals traditionally appeared in interior sections from 7 March on through mid-April, but territorial males now are calling in many statewide locations in early February, almost certainly indicating residence through the winter.

Fall. Unlike most New England states, hunting has been legal in Rhode Island since 1957, and the estimated kill was 1244–2782 birds per year during the period 1958–1964 (Myers 1980).

Winter. Mourning Doves were first recorded during winter in 1920. They increased to peaks in the mid-1970s and again in the mid-1990s and have declined since (CBC data). Since the 1940s, they have become very common in agricultural areas within a few miles of Narragansett Bay and the outer coast with daily maxima to 400. Until recently, they were relatively rare in winter in the interior, but even there they are now common; e.g., 201, Wallum Lake throughout February 1992 (SMi).

Winter maxima. 1386, Newport–Westport CBC, 18 December 1976; 825, South Kingstown CBC, 27 December 2008.

Pelagic. One was seen at Block Canyon, 28 July 1995 (fide RLF).

CUCKOOS (CUCULIDAE)

Common Cuckoo (*Cuculus canorus*)

Status

Vagrant. One record.

A cuckoo found by Allen Schenck at Snake Den Farm, Johnston, on 1 November 2020 was subsequently seen and photographed by many observers and carefully identified as Common, rather than Oriental (*C. optatus*), Cuckoo during its stay until 8 November (see Photo 4). There are two records of Common Cuckoos in eastern North America: 3–4 May 1981 at Martha's Vineyard, Massachusetts, and 29 September 2013 at Blanc Sablon, Québec; there is one record from the U.S. west coast: 29 September to 2 October 2012 at Watsonville, California.

Yellow-billed Cuckoo (*Coccyzus americanus*)

Status

Uncommon to fairly common summer resident and transient. Numbers fluctuate considerably, at least partly related to the abundance of caterpillars on which they feed (Hughes 2020).

Breeding

Yellow-billed Cuckoos are most common in parts of South County away from the immediate coast, declining northward through most of Kent and Providence Counties, and common in the Lincoln and Cumberland area. Compared with Black-billed Cuckoos, Yellow-billeds seem partial to mesic, richer vegetation and repelled by dry, acidic, oak-dominated forests. Regardless of vegetation, they are uncommon in all areas east of Narragansett Bay and uncommon or rare on the major islands (though present on Conanicut). The only evidence of breeding for Block Island is 2 seen there 17 June 1987 (CRa, REn) and 2 (1 dead) 25 June 2004 (CRa). They breed at irregular intervals at many sites, even in the South County area where they are most common and best known. Thus, they may be absent in some years and numerous in others. They were unusually abundant in 1936, when they arrived at Lonsdale 3 May and remained common for the rest of the season (AEE). They were again unusually numerous in 1960.

Nonbreeding

Spring. Yellow-billed Cuckoos, more often than Black-billeds, occasionally appear very early in spring, often after storms. One was found dead in Bristol 11 April 1962 (fide REW) and another at Providence 11 April 1973 (FSt). There are several other records for April. The general arrival in spring is during mid-May.

Spring maxima. Six, Ninigret, 8 June 2002 (GW).

Fall. They are also irregularly numerous as fall transients: frequently rare or absent but at other times subject to incursions. The greatest flight in history occurred after the 1954 hurricanes, when many hundreds of Yellow-billed Cuckoos flooded the northeastern coast and were widely reported along the Rhode Island shoreline (Griscom 1955). The birds during this "flight" behaved in unusual ways, such as feeding on the ground or attempting to enter outbuildings, and were reported to "flush from the ground like quail" (Griscom and Snyder 1955). The occurrence of cuckoos was not closely linked to any of the hurricanes that fall and may have been related to the major Spongy Moth (*Lymantria dispar*) caterpillar outbreak of 1953. Some cuckoos that year remained late into the fall.

Fall maxima. 15, Block Island, 25 September 1954 (RIOC); 10 there 9 October 1954 (RCC). There was a previous late fall "flight" in 1951, with as many as 6 at Block Island on 14 October 1951 (fide RLF). Another "flight" occurred in 1964, with 10 at Block Island 26 September–2 October (SSD) and 10 there 12 October (MSl). A local breeder was still calling in Saunderstown as late as 1 November 2016 (PAB), an almost annual occurrence into October. There are 7 records for early November; 1 was collected at Middletown on 12 November 1955 (JBa, RBo), and another was seen at Matunuck 15 November 1970 (SSD, LBe, RSm).

Extreme dates. 11 April (above) to 17 November 1975, 1 found dying at Block Island (HLa).

Black-billed Cuckoo (*Coccyzus erythropthalmus*)

Status

Fairly common but erratic summer resident and fairly common migrant.

Breeding

Black-billed Cuckoos are usually fairly common in late spring and summer but are occasionally scarce (as in 1953 and 1992) or remarkably common (as in 1959, 1987, and 1993). They are fairly common in northern and western sections of Rhode Island but became rare during the agricultural era in relatively treeless southeastern sections (collected once in summer at Middletown by ESt 20 June 1889). Although Black-billed Cuckoos are usually less common than Yellow-billed Cuckoos in the interior of South County, Black-billeds are much more frequent in coastal habitats. They are now common summer residents on Aquidneck, Conanicut, Aquidneck, and Prudence Islands (10 on Prudence Island, 23 May 1946; WBD) and in the Sakonnet area. Reports of calling cuckoos in June at Block Island are ambiguous as to species, but recent field work indicates that Black-billed may breed there in some years.

Nonbreeding

Black-billed Cuckoos are rather uncommon in spring but are often common and conspicuous in fall. This is particularly true in coastal areas, especially at Block Island.

Spring. Like Yellow-billed Cuckoos, Black-billeds arrive late in spring, generally in mid-May. Early spring records include 1 found dead at Charlestown 22 April 1975 after a southerly storm (DWa) and 1 seen in Exeter 27 April 1991 (JZy). One was found in EDi's barn at Block Island 11 June 1959.

Fall. Transients or wandering birds may appear early; e.g., 13 July 1941 at Middletown (WHD). One was flushed from the dunes at Sakonnet 26 July 1955 (JBa). Transients are frequent along the coast through August and September and into early October.

Fall maxima. 12 on Block Island, 25 September 1954 (RIOC).

Extreme dates. 22 April 1975 (above) to 25 October 1959, Sakonnet (1; SSD).

GOATSUCKERS (CAPRIMULGIDAE)

Common Nighthawk (*Chordeiles minor*)

Status

Uncommon and local summer resident; uncommon spring and common fall transient. Pronounced decrease in abundance since 1990s, probably reflecting decline in roofs suitable for nesting.

Historical

Formerly known to nest on the ground around Narragansett Bay, Common Nighthawks were referred to by Howe and Sturtevant (1899) as rare summer residents. Howe and Sturtevant indicated that nighthawks nested on the ground on "Aquidneck Island," at Portsmouth, and at Providence, although we can now wonder about their nesting on the ground at the latter location. Charles H. Remington found a nest 8 June 1890 in a field at an unstated location in "Kent County." They nested regularly for at least several years near Potter's Cove, Prudence Island, where two sets of eggs were collected 30 May 1890 by F. W. Stainton and HSH. Charles Doe visited this site 15 June 1895 and found two eggs (FlMNH #56778; recently re-examined) near a rock on a small, barren piece of ground in an orchard of young trees. There are several specimens from late May to early July from Johnston during the 1890s, suggesting possible nesting in that area.

In 1903, along with Purple Martins and other insectivorous birds, nighthawks were decimated by a prolonged period of cold spring rains throughout New England. Since 1903, nighthawks have nested primarily on flat gravel rooftops in cities and towns. However, while this apparent change in habits has long been attributed to the rains of 1903, it seems as likely to have been dictated by agricultural decline, since the poorest sandy or rocky fields were being abandoned as grazing lands and were succeeding to bushes during the early part of the twentieth century. There are two records for the modern period that suggest occasional ground nesting in Rhode Island when conditions are appropriate: in June 1936, several unfledged young were found in a rocky field in Lincoln Woods (AEE). On 28 June 1951, RCC observed two males booming over a sand plain that had been swept by a forest fire that year, near Route 165 and Wood River in Exeter.

Hathaway saw a pair of nighthawks flying over the tops of buildings around Snow and Westminster Streets in Providence 17 June 1889. This observation indicates possible roof nesting at this early date, but the fact that the species escaped the notice of the many Providence taxidermists and egg collectors in that era suggests that it was at least rare if not entirely absent from Providence rooftops at the turn of the twentieth century.

Breeding

Searches for breeding nighthawks in the 1970s and 1980s located territorial birds in downtown Providence, near Hope High School and at the Capitol building. They were also in Pawtucket, Central Falls, and Johnston (12 in these areas 26 July 1982 [REn]; 7 pairs in Providence, June 1977 [STs]; 10 at the Capitol 14 July 1983 [REn; CRa]). They probably bred near Six Corners in East Providence (1950s) and in West Warwick (1973, 1982). More recently, records of possible breeders or prospectors include 3, Providence, 13 July and 24 July 2003 (DFi); 1 and 3, Providence, 5 and 11 July 2005 (DFi). None, however, have been reported on territory in the textile towns of northwestern Rhode Island, and the mills there have now largely been torn down. They have been absent from Woonsocket but probably bred during the height of the textile era, since Arnold noted them regularly there in late May, June, and July between 1901 and 1921. Nearer the coast, nighthawks have not been found on rooftops at Westerly, Wakefield, the URI campus in Kingston, Warren, Bristol, or Newport. In nearby Fall River, 1 displaying over the city 1 June 1934 (RCC) is the only suggestion of nesting. None have been proven to nest in Rhode Island since about 2005.

Nonbreeding

Spring. Nighthawks are generally very uncommon or even rare as transients in spring at nonbreeding locations, especially in southeastern sections, and there are about a dozen May records for South County. While they are rare transients along the coast during their normal arrival period in late May, there is a notable history of abnormally early arrival in coastal areas during April. Many of these instances can be correlated with strong southwesterly airflow or coastal storms. Such early records include the following: 31 March 1960 at Jamestown (fide JBa); 7 April, year and location unstated (Howe and Sturtevant 1899); 12 records for mid-April to early May. They usually appear at breeding sites during mid-May.

Spring maxima. 20+, West Warwick, 9 May 1976 (DWa); 13, Narragansett, 24 May 1984 (RLF).

Summer. Wanderers are occasionally seen in midsummer: 1 at Block Island, 19 June 1926 (EDi); 1 at Tiverton 20 June 1982 (HCa, JC); several over Kimball Sanctuary, 25 June 1935 (MSo); 1 at Green Hill, 29 June 1948 (DLK).

Fall. In the interior of New England, Common Nighthawks migrate during a highly concentrated period in the last few days of August and the first few days of September.

Fall maxima. 800–1000, Esmond, 30 August 1993 (Carpenter); 380, Burrillville–North Smithfield, 31 August 2003 (MLy); 300, North Smithfield–Woonsocket Reservoir, 2 September 2005 (MLy); 150, Chepachet, 23 August 1999 (JSJ); 81,

Providence, 4 September 2002 (DFi); 27, Block Island, 11 September 1988 (CRa et al.). One of the most unusual counts both in terms of size and location was made 2 September 1941, when Coman counted 284 and estimated 300+ at Scarborough Beach, Narragansett, where the birds were arriving from the Newport area over the water on a clear afternoon with light southeasterly winds. Late records include 1 migrating west over Napatree Point 14 October 1979 (RLF), 1 at Matunuck 16 October 1981 (v.o.), and 1 at Block Island 22 October 1977 (R.L. Norton).

Chuck-will's-widow (*Antrostomus carolinensis*)

Status

Vagrant. Increasingly frequent since the 1980s.

Nonbreeding

The first Chuck-will's-widow in Rhode Island was calling repeatedly at dusk at Block Island on 25 May 1980 (DLE, REm). Two other individuals were heard on consecutive nights at the Ruecker Sanctuary in Tiverton and at NBS in Middletown, 2–3 June 1982 (H. and J. Calhoun; GLB). One was reported in May 1982 north of Ninigret Pond (fide GLB). A series of calling birds on the west side of Wakefield began with 1 on 19 June 1984 (RAC, RBu). Another was present at the same location 30 May–20 June 1985, with a second bird calling simultaneously nearby 2–14 June 1985 (RAC, DLK et al.). One was present there 5–7 June 1987 (RAC), and 1 was at Wordens Pond, 18 May 1996 (PP et al.). The only inland report is for 1 seen at rest and in flight at Swan Point Cemetery in Providence 26 May 1992 (WBo, JHo). These records from Rhode Island occurred during a period when Chuck-will's-widows were appearing more and more frequently on Long Island and on Nantucket and Martha's Vineyard. The northward push of these birds seems to have slowed since the 1990s. An injured bird was picked up at the base of a chain-link fence in Little Compton 11 May 2010. It was taken to a rehabber in Bristol who photographed it. Two birds were present in the South Kingstown–Perryville area in summer 2011–2015 (Hope Leeson; SSM, PJL). A territorial bird was found in Coventry in 2015 and then West Greenwich (perhaps same vicinity) in 2016 (S. Brenner). One was calling at Tillinghast Pond Management Area, West Greenwich, 2 May–16 June 2020 (N. Russo; J. Mitchell).

Eastern Whip-poor-will (*Antrostomus vociferus*)

Status

Uncommon summer resident; probably declining. Uncommon and rarely detected transient.

Breeding

Whip-poor-wills are common summer residents in some of the pitch pine and scrub-oak barrens of Washington and Kent Counties, and they are widely distributed and fairly common in western and northern Providence County. Numbers fall off sharply eastward toward Narragansett Bay and the Providence–Woonsocket metropolitan axis, and they are absent from suburban areas. They were present in June at Diamond Hill in Cumberland in the late 1960s (REm), and a few may still be breeding there; e.g., 2 calling, 13–20 June 2017 (REm). There are few summer records east of Narragansett Bay between Pawtucket and Bristol. They have bred regularly in parts of northern and eastern Tiverton and have been found in eastern Little Compton (e.g., 2 males near Simmons Pond 20 June 1991 [RBo]; another there 25 June 1992 [RBo]; 1 near Quicksand Pond 6 July 1993 [RBo]); there is no evidence of breeding in that area since 1993. They have never been reported calling in the more open sections of Tiverton or Little Compton that border on the Sakonnet Passage or on Aquidneck Island. Nevertheless, small numbers of Whip-poor-wills have been present on both Prudence and especially Conanicut Islands for many years; e.g., 4 on Prudence, 18 June 2003 (REn). They have never bred at Block Island. They are now scarce along the Ministerial Highway in South Kingstown, where they were common and vociferous in past decades. The species is highly vocal on territory in late spring and early summer, particularly in the pine–oak habitats of the southwestern quadrant of the state. Recent declines may reflect nest predation by feral cats, raccoons, and other mammals and perhaps long-term effects of habitat fragmentation and pesticides.

Summer maxima. 52, Arcadia, 30–31 May 1983 (RLF).

Nonbreeding

Spring. Along with Purple Martins, Broad-winged Hawks, and waterthrushes, Whip-poor-wills are one of the earliest tropical winterers to arrive on territory in spring. The earliest records are 10 April 1941 at Bradford (MSo) and 10 April 1981 in Charlestown (Bennett). Three were reported in Charlestown 17 April 1980 (Bennett).

Spring maxima. Six, Shumunkanuc Hill, 27 May 2000 (CRa); 6, Great Swamp, 2 May 2001 and 20 May 2003 (SSM)

Fall. At the end of the breeding season, males gradually fall silent, although occasional birds have been noted calling through September (7 at Tuckertown 7 September 1973; KKi), and occasionally into the first week of October. The latest date for a calling bird is from Charlestown 12 October 1937 (MSo). Transients in fall are almost unknown, and the timing of migration is largely a mystery, but a majority of the local breeding population probably leave the state by late August; e.g., 1 banded in Middletown 29 August 1959 (JBa).

Extreme dates. 10 April (see above) to 16 October 1925, when 1 was picked up at Block Island in a weakened condition (now in the Block Island Collection; EDi).

SWIFTS (APODIDAE)

Chimney Swift (*Chaetura pelagica*)

Status
Common summer resident and common transient.

Breeding
Sir Charles Blagden's account (1776–1780) stated that the "Chimney Swallow" was "most common in the inland parts than near the seacoast." Chimney Swifts are still most common as breeders in interior towns and suburbs, where they nest in the chimneys of mills, schools, and the larger and older residences, as well as in large chimneys of isolated farms and rural houses. They are far less common close to Narragansett Bay and are especially uncommon close to the outer coast. While wandering birds are regular in June at Block Island, they do not breed. All Rhode Island nests have been in chimneys rather than in natural cavities such as hollow trees.

Nonbreeding
Spring. General arrival is in early May, with occasional single birds during April, but they are rare before mid-April. Transients commonly continue moving in early June, with occasional birds at Block Island; e.g., 30 June 1938 (EDi).

Spring maxima. 1000+, Westerly, 5 May 1953 (ESa); 300 over Providence's East Side, 13 May 1984 (REn); 200, Westerly, 24 April 1964 (ESa); 44, Napatree Point, 11 May 2014 (CRa); 40, Warren, 30 April 1983 (RBo).

Fall. The breeding population regularly disappears in August, and fall transients move through primarily in August and early September.

Fall maxima. 315, North Smithfield–Woonsocket, 2 September 2005 (MLy); 200, East Providence Reservoir, 18 August 1953 (RLF); 200, Middletown, 9 September 1939 (WHD). Rare after early September. Along with Barn Swallows, however, Chimney Swifts are often found late in fall after storms and extended southerly or southwesterly winds. A great many were noted at Block Island after the hurricane of 14 September 1944 (EDi), and greater than average numbers appeared after Hurricane Donna of 12 September 1960; e.g., 45 at Block Island 13 September 1960 (JBa). A coastal storm produced 45 at Napatree Point 25 September 1987 (CRa).

Extreme dates. 6 April 2000, Kingston (PAB), to 11 November 1982, Middletown (T. Trevor), and 21 November 1941, Middletown (Williston).

HUMMINGBIRDS (TROCHILIDAE)

Ruby-throated Hummingbird (*Archilochus colubris*)

Status

Uncommon summer resident and spring transient; more numerous in fall.

Breeding

Howe and Sturtevant (1899) called Ruby-throated Hummingbirds "common summer residents," but Hathaway (1943) considered them to be uncommon and mostly presented data on migrants. Enser (1992) recorded their presence in about a quarter of the 165 blocks on the Rhode Island Breeding Bird Atlas and confirmed them as nesting in only 3 blocks. The author found only 23 among 55,000 landbirds observed in mostly rural areas of Rhode Island, 1982–1989, or about as numerous as Eastern Bluebirds or Least Flycatchers. They are sparsely distributed and uncommon in many areas and are lacking in built-up suburbs and in many shore areas. They breed on the larger islands in Narragansett Bay but are apparently absent from Block Island.

Summer maxima. Seven, Tillinghast Pond Management Area 17 June 2012 (SSM, PJL).

Nonbreeding

Spring. Migrants or wanderers have been recorded to late May and early June; e.g., 1 moving east at Point Judith 3 June 1982 (RLF) and another migrating northeast across Narragansett Bay 4 June 1984 (RLF). A probable nonbreeder appeared at Block Island 15 June 1981 (RBo).

Fall. In fall, hummingbirds are most numerous during late August and early September. They often move westward over the dunes in coastal areas during daylight hours; e.g., at Napatree Point, from which they depart for Fisher's Island or Long Island.

Fall maxima. 40, Little Compton, 8 August 2008, at feeders (GDe); 12, Block Island, 14 September 1958 (MSl); 30 netted at Middletown, September 1956 (JBa); 15, Block Island, 4–6 October 1974 (ASRI); 8, Napatree Point, 23 September 1999 (JZy).

Extreme dates. 11 April 1952, Cranston (RCC), to 25 November 2015 at feeder, Little Compton (GDe); 9 November–25 December 2004, Tiverton, (1; GDe; photos).

Calliope Hummingbird (*Selasphorus calliope*)

Status

Vagrant. One record.

Nonbreeding

Two Calliope Hummingbirds appeared at a Little Compton feeder 21 October 2013, and 1 of these stayed until 5 November (GDe, m.obs.; see Photo 5). This record is consistent with the increasing frequency of occurrence of Calliope Hummingbirds on the East Coast.

Rufous Hummingbird (*Selasphorus rufus*)

Status

Vagrant. Substantial increase recently.

Nonbreeding

A female *Selasphorus* hummingbird was noted visiting the garden of Geoff Dennis in Little Compton 22–29 October 1995, where it was eventually identified as a Rufous Hummingbird. This was a first for Rhode Island but consistent with an accelerated increase of records in the Northeast during the early 1990s. Present daily until an unspecified date in November, the Little Compton bird was seen by P. A. Buckley, Paul Lehman, and numerous local observers and photographed by GDe. Later captured, it was transported to New York. Another Rufous Hummingbird was found by GDe at a nearby feeder the following year and was also captured and transported to New York. The Rhode Island Avian Records Committee has compiled a list of Rufous and unidentified *Selasphorus* hummingbirds that contains 21 records of 23 birds between 1995 and 2016, 11 of these from Little Compton. A partial enumeration of these includes 1 *Selasphorus* sp., Jamestown, 5 December 1998 (fide RFa); 1 Rufous, Newport, 10 December 2001–25 January 2002 (CRa et al.); 1 Rufous, Little Compton, 23–28 October 2002 (GDe); 1 Rufous, 1 November–1 December 2003 (GDe); 1 adult male Rufous, Charlestown, 29–31 October 2004 (G. Matteo et al.); 1 immature male Rufous, Little Compton, 15 October–20 December 2004 (GDe; photos); 1 immature female Rufous, Little Compton, 10 November 2004–18 January 2005 (GDe; photos); 1 immature female *Selasphorus* sp., Warwick Neck, 1 November–31 December 2005 (fide RFa); 1 adult female Rufous, Wakefield, 4–30 November 2005 (B. Chaves; photos); 1 *Selasphorus* sp., Wakefield, 4 November–19 December 2005 (B. Chavez, fide RFa); 1 adult female Rufous Hummingbird, West Warwick, 20 November 2005–19 April 2006 (C. Simas, fide RFa). The 2005–2006 West Warwick bird is the first to have overwintered in New England. A Rufous was at Snug Harbor, Wakefield, 27 October 2006 to 27 January 2007. This bird succumbed, and the specimen was confirmed as Rufous. A *Selasphorus* sp. remained at Warwick

Neck 1 November 2006–17 January 2007 (fide GDe). A Rufous was in Wakefield 17 October–20 December 2009 (v.o.). A Rufous Hummingbird was in Hopkinton late October to 30 December 2011 (Denise Gautreu et al.), and another Rufous was at Jamestown, banded 14 November 2012 (T. Mackie, A. Hill). Two *Selasphorus* sp. hummingbirds were in Little Compton 16 October 2012, 1 of which stayed to 3 December 2012 (GDe); another at Kingston 25 October to 9 November 2012 (L. Gadrel, CPe); and yet another visited a feeder in Jamestown starting about 27 November 2012 and remained for several weeks (PLE et al.; photos). Another *Selasphorus* was at Jamestown, 27 November 2012–15 January 2013 (GDe). Two *Selasphorus* sp. birds were at Little Compton 28–21 October 2015; Portsmouth from an unknown fall date to 16 December 2016; Exeter, 18 November 2017 (DPe); and North Kingstown to 1 December 2017, Wakefield 9 November–1 December 2017, and Wakefield 9–13 September 2018 (all fide GDe).

RAILS AND GALLINULES (RALLIDAE)

King Rail (*Rallus elegans*)

Status

Rare and possibly extirpated breeder and migrant. Occasionally winters.

Historical

Three records were known to Howe and Sturtevant (1899), and Hathaway (1913) gave data on eight other specimens, primarily from the Point Judith marshes. The first breeding record, previously unpublished, was of a bird incubating 10 eggs discovered by local boys in the cattail marsh near Hanging Rock, Middletown, and shown to Sturtevant before 1899. Additional birds were recorded in 1917 and 1918. King Rails were more regularly recorded after 1938, probably owing to a great increase in fieldwork. The status of King Rail in Rhode Island seems relatively little changed since, unlike that of many other southern species (see Veit and Petersen 1993, who report a decline in Massachusetts, in part due to habitat destruction). They were most often recorded during fall during the early twentieth century, and this pattern may still hold true. There seemed to be an increase in records during summer and of breeding after about 1960, but this slight increase did not persist.

Breeding

King Rails are rare and irregular summer residents in coastal freshwater and brackish marshes, as well as estuarine meadows. An immature bird, not able to fly, was found dead on the road near a small freshwater marsh north of Seapowet Inlet in August 1953 (WBD; specimen preserved but now lost), and adults were frequently seen and heard calling in summer at the same location in the mid-1950s (RBo). Just south of Seapowet Inlet on the salt marsh, an adult was seen and a nest with eggs found 1 August 1961 (ALH, REW). At the same location, adults were heard calling in May and June 1970 and 1971, and at least once were called into the road with a tape recording (SSD). In South County, 1 and sometimes 2 adults were present fairly regularly in the 1960s during May and June in a small freshwater swale northeast of Cards Pond. These were also called in to within a few feet several times (SSD), but no proof of breeding was obtained. At the lower end of nearby Moonstone Brook, a pair was present in May and June 1974 and likely bred (ESa, SSD, RSt). Two additional sites are the Winnapaug marsh near Misquamicut and Quonochontaug, where King Rails have been present along with Clapper Rails (see Clapper Rail account). The only locations north of the outer coast where territorial behavior or potential breeding have been observed are the northeast marsh at Hundred Acre Cove, Barrington, where a pair was present in May 1983 (SRi, JCB), and the Great Swamp, where 1 spent a week calling 20–27 May 2000 (DFi, CRa, SSM et al.). Birds thought to be King Rails have been heard calling in the Lonsdale marshes.

Nonbreeding

While there are records for every month of the year, most birds have been encountered during 2 periods: from 15 March through May into June and again from 15 October into late December. Ironically, the species has been recorded only once on a CBC during the long period 1950–1990 (2 birds in 1952 in Newport–Westport), but there are several records of specimens for January and February. King Rails have been rare in recent years, with the following recent records: 1, Jerusalem, 23 May 1998 (SSM); 1, Island Park, on the unusual date of 2 March 1999 (RBo); 1, Great Swamp, 20–27 May 2000 (DFi, CRa, SSM et al.); 1, Middlebridge, 29–31 May 2001 (SSM, JSJ); 1–2, Jerusalem, 5–22 May 2004 (JMa, DFi, REm, STs et al.). One at Jerusalem 25 April–31 May 2005 (JMa et al.) was seen copulating with a hybrid or intergrade on 14 May 2005 (DFi).

Extreme date. One shot at Wakefield 12 February 1889 (fide HSH).

Clapper Rail (*Rallus crepitans*)

Status

Uncommon summer resident and migrant. Very rare in winter.

Historical

Clapper Rails were never encountered by the old-time rail hunters on the Point Judith marshes, and HSH never found any in 45 years of shooting and tramping the Quonochontaug marshes. In the late 1940s, a great increase began throughout the Northeast (Griscom and Snyder 1955). Multiple sight records in Rhode Island accumulated during the short period 1950–1954, and breeding was confirmed in 1953.

An adult and several young were seen 17 July 1953 at Jamestown (MDa), a flightless juvenile was at Seapowet 22 August 1953 (CGW), and an adult with six downy young was at Galilee 23 July 1954 (DLK). Other broods were seen at Jamestown in 1958, 1959, and 1961 and at Seapowet in 1959 and 1961.

Breeding

During the the 1960s and 1970s, Clapper Rails spread in Rhode Island and appeared at Winnapaug and Quonochontaug marshes, Ninigret Pond, Potter Pond, Jerusalem, and Jamestown, in Hundred Acre Cove in Barrington (both along Runnins River and on the "Tongue"), at Adam's Point in Barrington, along the Kickamuit River in Warren, on the Colt Estate in Bristol, and in the Seapowet marshes.

At Hundred Acre Cove, 4–6 pairs were present on the "Tongue" and in the northeastern part of the cove during the period 1971–1980, but they had disappeared by 1981 (SRi). The only nest found in Rhode Island contained 14 eggs and was found 25 June 1982 at an undisclosed place in Hundred Acre Cove (W. DeRagon). At Quonochontaug, 2 males were calling 19 May 1984 (RAC); 2 were there on 1 June 2002 (STs); 3 there, 21–24 May 2003 (JSJ, STs, MJM); 2, 21 May 2005 (STs, CPe et al.). At Charlestown Breachway, they were rarely recorded until August 1992, when an adult and 3 young were found (RWa, JSJ, MJM et al.), and an adult with 4 young was there 14 August 2001 (JSJ). At Seapowet, one of the best locations in the state in the 1950s (9 calling there at the height of its statewide abundance, 31 May 1953 [RBo]; 5 on 17 December 1950 [RBo]), intensive surveys in the summers of 1981–1982 indicated that none were present (SRi). Numbers have been similarly variable in the Jerusalem and Potter Pond area, where 6–7 were calling at once in the 1970s (DLK), but usually only 0–2 pairs have been present in recent years. At Jamestown, where many early breeding records were established (9 birds 15 September 1958, including 5 young; MDa), the species has been all but unreported for decades.

At Galilee, part of the marsh that had been impounded since construction of the Escape Road in 1956, and had turned first to cattails and then eventually *Phragmites*, was restored in 1997. Three territorial pairs of Clapper Rails were present at Galilee 25 May 2014 (PP).

Fall. Fall reports are rare now but include 1, Block Island, 1–3 October 1999 (RFa), the first for the island.

Winter maxima. In total, 12 were seen on the Newport–Westport CBC in the years 1950–1956, then only 2 on the counts 1957–1972, and only 4, 1973–1990. All CBC records after 1956 have been on the South County count, paralleling what has come to be a scarcity of the species during the warmer months since the 1950s on the east side of Narragansett Bay. The following 4 records of individuals suggest some degree of winter survival: 25 February 1953, Jamestown (MDa); 27 February 1984, Middletown (v.o.); 1 March 1983, Jerusalem (GLB); 2 March 1974, Jerusalem (DLK, KKi,

FPe). One was at Watch Hill 19 December 2004 (SSM et al.). Another was recorded on the South Kingstown CBC 27 December 1998 (DLK et al.)

Spring. A bird at Jerusalem on 4 April 1982 probably constitutes an early arrival date (RAC et al.).

Mixed pairs. On 16 June 1980, a King Rail was found in the company of a Clapper on the Winnapaug marshes, but no evidence of breeding was obtained (CRa). On 7 July 1984, a nest with 7 eggs was being incubated by both a King Rail and a Clapper Rail at Quonochontaug, and 2 eggs and a chick were present in the nest 19 July (REn, CRa).

Virginia Rail (*Rallus limicola*)

Status
Locally fairly common summer resident and migrant, and fairly common to uncommon winter resident.

Breeding
Virginia Rails are fairly common during summer in freshwater marshes, especially those with cattails, and occasionally about the edges of salt marshes; e.g., at Jamestown, Seapowet, Peckham's Marsh, Galilee (eggs collected on Point Judith marsh in 1902; HSH), Napatree Point (brood seen 1980; fide ESa, CRa), and perhaps elsewhere.

Breeding maxima. Six to 8 at Lonsdale on many dates in late April and May (REn, CRa); a total of 27 located using taped vocalizations at 10 separate marshes in the spring of 1984 (CRa); 7, Cumberland, 26 May 2002 (MLy); 5, Spragueville Pond, 8 May 2005 (MLy). There is evidence of a decline. Chris Raithel has recently resurveyed several of the marshes he first surveyed during the 1980s and cannot find Virginia Rails in several areas where they were once fairly common.

Nonbreeding
Spring. Migrants appear in mid-April.

Fall. Uncommon but regular from September through early winter.

Fall maxima. 15+ flushed by dogs in the Lonsdale marshes, 2 October 1948 (WBD, RCC).

Winter. Howe and Sturtevant (1899) recorded a late date of 2 November; since then, Virginia Rails have increased and are now present regularly in early winter within a few miles of salt water as long as a vestige of open water remains. There are, however, enough scattered midwinter records to suggest that they can survive mild winters in Rhode Island: 1 in Pawtucket 23 February 1953 (RLF); 1 collected 15 February 1956 at the north end of the East Providence Reservoir (JSu, RLF); 1, Warwick, 13 March 1967 (DWa); 1, Portsmouth, 21 February 1982 (RLF); 1, Lonsdale, 24 March 1995 (REn); 1, Lonsdale, 10 February 1996 (JMu).

Winter maxima. Nine, Newport–Westport CBC, 17 December 2011; 8, Block Island CBC, 19 December 1989.

Corn Crake (*Crex crex*)

Status
Vagrant. One record.

Nonbreeding
The famous specimen, obtained by Newton Dexter "about the year 1857" in Cranston and formerly in the collection of the Franklin Society of Providence (Southwick and Jencks 1884d), is now at RWPM. The bird was apparently shot by someone else and given to Dexter. There are more than a dozen records for this upland rail on this side of the Atlantic, but most of them are more than a half-century old. Numerous British authors have discussed the decline of the Corn Crake, which was coincident with the adoption of mechanized harvesting of grain in the British Isles and Europe. The connection between the name of this species and its chosen habitat becomes clearer to North Americans when it is realized that "corn" in southern England means grains such as wheat, not the native American grain generally considered here as the sole subject of the name "corn."

Sora (*Porzana carolina*)

Status
Uncommon but likely overlooked summer resident; rare spring and common fall migrant; rare in winter.

Historical
A nest with 14 eggs was found on "the Champlin Farm," Point Judith 17 June 1906 (HSH); 9 eggs were in a nest at the same spot 29 May 1910 (HSH); another nest with 8 eggs was found at Point Judith 7 May 1911 (HSH, eggs extant); and a nest and eggs were located at Quonochontaug 4 June 1916 (HSH).

Breeding
Soras are at the southern periphery of their breeding range in Rhode Island. Prior to dedicated searches during the 1980s, there were few records for the twentieth century. An adult and 2 downy young were seen along the edge of a small pond in a pasture a half mile southeast of Seapowet Inlet 14 August 1958 (SSD), where adults

had previously been seen during summer (RBo). Soras were suspected of breeding at Long Pond, Sakonnet, in 1900 and have been seen or heard at the north end of the East Providence Reservoir (e.g., 18 June 1938; CWo); Moonstone Brook, Matunuck (many dates in May and late July); and the Middletown marshes (seen near Nelson's Pond, 30 June 1951; RBo). During May and June 1984, CRa and REn found Soras at Point Judith, Belleville Pond (4), Middletown, and northwestern Lincoln. Another was calling in northern Cumberland 16 June 1984 (RLF). The most persistent reports come from the Lonsdale marshes, where the species has been present in May and June at least since 1949; e.g., 5 there 28 May 1978 (REn). Most sites where Soras have been found in the breeding season are vegetated with cattails.

Nonbreeding

Spring. Migrant Soras are rare away from breeding sites during spring. One such record is 1 found dead at Block Island 7 April 1925 (EDi), the earliest spring record. Migrants generally arrive during late April: 4 were at Moonstone as early as 30 April 1973 (DWa). The latest record of a migrant was 24 May 1952 at Kingston (DLK).

Fall. Other than shorebirds, Soras are among the earliest non-passeriform fall migrants. There are 2 records of presumed transients at coastal points for 28 and 30 July and numerous others from early August on, with an increase in frequency after 20 August. The peak movement appears to be in mid-September, with some migration continuing into late October.

Fall maxima. 30, Boyd's Marsh, 5 September 1953 (RBo); 24+, Moonstone, 13 September 1973 (DWa); 25 heard, Middletown, 28 September 1941 (WHD); 25+ flushed using dogs, Lonsdale marshes, 2 October 1948 (RCC, WBD).

Winter maxima. One, Newport–Westport CBC, 17 December 2011; 1, Napatree CBC, 18 December 2011; 1, Napatree CBC (Quonochontaug), 19 December 2004 (SSM et al.).

Extreme dates. Two were observed walking on the shores of the north end of Briggs Marsh, Sakonnet, 18 December 1982 (DLK, RAC); another was seen 24 December 1940 at Middletown (WHD). One found dead in the road at Cards Pond 20 March 1983 (REn) could have been either a winter survivor or a very early transient.

Common Gallinule (*Gallinula galeata*)

Status

Formerly a rare spring and fairly common fall migrant and a rare breeder. Now apparently absent as a breeder and much reduced as a migrant. Very rarely lingers into midwinter.

Breeding

Common Gallinules were common summer residents at Long Pond, Sakonnet, before 1899, where N. Dexter collected eggs and young birds. Three nests but no eggs were found there in early June 1900 by Mearns. In 1935, RCC found them absent there despite the presence of cattail marsh. The cattails have since been replaced by *Phragmites,* and an adult was present there 3–4 July 2010 (REm), but breeding was not confirmed and there have been no reports at this location since *Phragmites* control was implemented by the RIDEM in 2014. There is a questionable report (in Howe and Sturtevant 1899) of breeding at Newport in the summer of 1887. After the turn of the twentieth century, breeding was unreported until 1951, when 2 broods were raised in Easton Pond, Newport (fide RCC); a pair was seen there in July 1953 (RCC). They may have bred there for many years in the formerly dense cattail beds; they were present in summer for some years before 1951 (J. Lynch, fide RCC), and broods of young were seen regularly through the late 1950s (v.o.); a pair was also present there during July 1970 (RTL).

Elsewhere, adults with a brood of 3 young were seen through the summer of 1961 at Boyd's Marsh, Portsmouth (SSD, DLE), and reports suggest they may have bred there in other years. Two birds were present in a small pond west of Tunipus Pond, Little Compton, during several summers in the 1950s and presumed to be breeding (RBo). Others have been seen on appropriate dates in early June during recent years at Westerly; in a cattail marsh near New Harbor, Block Island; in a small pond southwest of Block Island's Sachem Pond; along Moonstone Beach Road, 8 June 1996 (JMu); and near Point Judith; e.g., 1, 12 June 2004 (JSJ), and also in 1995, 1996, and 1999.

Nonbreeding

Spring. Common Gallinules are rare and irregular in spring in Rhode Island, despite their breeding in parts of southern New York and western New England. Most arrive during late April and early May, but 1 was at Green Hill 5 April 1978 (DWa). A 1st-year bird was seen at Moonstone 13 June 1972 (DWa); 1 appeared at Jamestown 7 July 1984 (CaP); another was seen in the Lonsdale marshes 29 July 1978 (REn); and there were reports from Trustom Pond in July 2002–2003.

Fall. Common Gallinules are fairly common in fall and much more frequently seen then than in spring: usually 1 or 2 per location per day.

Fall maxima. Five, Trustom Pond, 2 October 2001 (JSJ).

Winter. One was at Brickyard Pond 23–27 January 1972 (LBe). Others were observed at Moonstone 1 January 1974 (DLK, KKi, FPe, ESa, CWo) and Cross Mills 23 March 1974 (ASM). One was found recently dead at Portsmouth, 16 December 1984 (CRa, REn); 1 was on the Newport-area Christmas count on 16 December 1977 (v.o,). One was photographed in Little Compton, 16 and 26 December 2017

(GDe). One was at Point Judith, 25 December 1998 (Fred Norton). One was photographed in Barrington 29 March 1988 (DLE). A bird found on 15 February 1975 at the mill pond, Perryville (JDi), remained there until at least 6 April (SSD, DWa, KKi, FPe, ESa, CWo et al.). One at this location 7 April 1976 (DWa) and another at Green Hill 5 April 1978 (DWa) were probably successful winterers. One was at Point Judith 25 December 1998 (Fred Norton). The first for the Block Island CBC was found 20 December 2012 (CRa et al.).

American Coot (*Fulica americana*)

Status

Common to occasionally abundant fall migrant; fairly common winter resident. Rare in summer.

Historical

American Coots have fluctuated greatly in abundance over the past 100 years or so. They were generally rare in spring during the shooting era, but, at least sometimes, abundant in late fall. Flocks of up to 1000 were recorded on Point Judith Pond in 1900, for example. They were again found in lavish abundance during the 1940s (to 2000 per day on Aquidneck Island ponds). This period of abundance coincided with high numbers of many of the prairie and pond ducks in Rhode Island, suggesting that this temporary influx was primarily caused by local food conditions, perhaps coupled with some degree of success on the breeding grounds to the west.

Breeding

No nest or recently fledged young have been found in Rhode Island, but a pair was present at Easton Pond all summer in 1957 and suspected of breeding (JBa), and individuals were present in 4 other summers during the period 1940–1971.

Nonbreeding

Spring. With the return of ice-free conditions in early March, surviving winterers spread out and join northbound migrants. Most of the wintering population departs by mid-April or earlier. The largest spring counts include 90, 24 April 1948 at Middletown (RBo), and 20 there on 6 May 1950 (RBo), but late-season counts rarely exceed 1–5 birds. There are a number of records of lingerers in the last two-thirds of May, and a small number for June and July. A cluster of such summer reports came from Easton Pond in previous decades, when this artificial reservoir was choked with cattails

Fall. Coots are regular during September in small numbers, but much larger numbers appear later in fall. Peak counts are usually made from 20 October to 30 November.

Fall maxima. 2100, Newport County, 11 November 1948 (DLE); 1750, Portsmouth and Middletown, 21 November 1948 (DLE); 1550, Middletown, 11 November 1941 (RBo). Large numbers appeared throughout New England in the fall of 1994, with 1050 in Newport County 12 November 1994 (DLE, RLF), 275 at Trustom Pond 26 December 1994 (CBC), and 2320 on the Newport–Westport CBC 17 December 1994 (DLE et al.).

Winter. Totals on the mainland CBCs have ranged from 2379 to 0. The most nearly coot-less counts were during the period 1958–1964 and again in the 1980s. Even when coots are abundant in the fall, most birds move away, perhaps toward nearby wintering areas in Westport, Massachusetts, or on Long Island. The small numbers that remain into January seem regularly to survive the season. Until recently, maxima for the colder months were 150 in January, 115 in February, 170 in March, and 160 in April; most such counts are from the 1970s or earlier. The spread of *Phragmites* or purple loosestrife may be displacing coots from freshwater coastal ponds

Winter maxima. 660, South Kingstown CBC, 23 December 2006; 461, South Kingstown CBC, 26 December 2011.

During the season of 1994–1995 (see above), 800 coots at Quicksand Pond in early February dropped to 383 on 20 February 1995 (JO, RLF), including large flocks of apparently stressed birds feeding in fields and on suburban lawns more than 200 yd (183 m) from the water because of freeze-up of the pond (C. Mercer; JO, RLF; 2 emaciated birds picked up dead). A flock of 192 was behaving similarly at Wesquage Pond 19 February 1995 (RLF).

Purple Gallinule (*Porphyrio martinica*)

Status

Vagrant. Mainly in spring, but there are also three fall records and one for midwinter.

Historical

Howe and Sturtevant (1899) listed the following seven records:

1. One taken about 1857 at Westerly by N. Dexter
2. One taken in 1875 at Westerly by N. Dexter, specimen formerly in the Southwick and Critchley collection
3. One no longer extant caught alive in the middle of August 1886 in Warwick that "lived at least a year in Hardy's dime museum in Providence"
4. One found on the perhaps not-so-remarkable date of 13 January 1889 at Warren's Point, Little Compton, crudely mounted by the finder and later remounted by N. Dexter
5. A different bird killed at Sakonnet Point by N. Dexter in the fall of 1889
6. One that flew aboard a schooner at a Fox Point pier in Providence 13 May 1890 and was brought to J. M. Southwick for identification (extant at RWPM)

7. One shot at Newport about 24 May 1893 and mounted for the Newport Natural History Society (not extant)

Other records include one captured at Sakonnet Point 8 June 1900 and presented to the Boston Society of Natural History (not extant), and an adult male "collected near Newport by O. R. A. for C. E. Doe" in July 1900 (FlMNH #2986).

Nonbreeding

Spring. Aside from an adult bird killed by a cat at Block Island 12 May 1928 and now mounted in the Block Island Collection (EDi), a long hiatus in Purple Gallinule records occurred after the turn of the twentieth century. Observations in the modern period include an immature seen at Shelter Harbor, Westerly, 15–16 May 1955 (KKi, DLK et al.), and 1 near the southeast lighthouse, Block Island, 20 May 1973 (WBD), with probably the same bird a few days later near the center of the island (MSl). An adult was seen at Long Pond, Sakonnet, 28–30 May 1987 (DFi, DLE et al.). An adult male with a broken leg that was captured by a student on the grounds of the Wilbur School in Little Compton on 11 June 1998 was brought to a local wildlife rehabilitation clinic where a pin was inserted into the broken bone. Five weeks later the bird, apparently fully recovered, was released into a nearby marsh where it "flew hard before settling down in good habitat" (GDe). An adult-plumaged bird was salvaged from a snowbank in Wakefield 13–14 March 1993 (examined and photographed by CRa; specimen in South Kingstown High School collection).

Fall. The only modern fall records are 1 seen at Block Island 11 October 1991 (GTa, RDe) and an immature photographed at Wood Pond, Barrington, 24 October–9 November 1991 (RWa, JO, DLE, JCB, RBo et al.). One, apparently in its hatching year because of buff feathering around the head and neck, was picked up dead at Wordens Pond on Block Island 17 February 2017 (fide K. Gaffett). Another was found alive on Block Island 7 November 2017 (K Ga).

Yellow Rail (*Coturnicops noveboracensis*)

Status

Rare migrant, mainly in fall; 1 winter record.

Historical

Howe and Sturtevant (1899) gave four records: one collected 14 September 1894 at an unknown locality (probably Westerly) by H. O. Havemeyer Jr.; one shot 28 September 1886 at Charlestown Beach by F. L. Glezen, now mounted in the Kimball Bird Sanctuary collection; one shot by a Newport gunner 23 September 1887; and one killed on an unstated date by striking a wire in Cranston. Hathaway (1900g) recorded one that he found in a private collection in Westerly, taken near there during the last week of September 1897. Hathaway (1913) also reported a number of other early records, all taken near Point Judith, mostly in what are now called the Galilee marshes or in the surrounding hayfields, which are no longer present. These include one that was shot 10 October 1908 (Clarke) and another 15 October 1908 (Bowler); neither was preserved. One was shot 15 October 1909 (Clarke) and sent to HSH. Another four were killed in October 1910 by Bowler. On 30 September 1911, one was shot in the same area in a "fresh water run"; another was caught by a dog and two others seen 1 October 1911. Two other specimens were sent to HSH by I. R. Sheldon, who wrote (in litt.): "I do not think the Yellow Rail is anywhere near as rare as recorded. Out of 30 rails that I have seen this fall at Point Judith, four were of this species . . . they are very hard to flush and for this reason I think less rare than supposed to be. I have caught two birds with my hands." Other records include one shot at Newport 15 October 1911 by Bowler and one shot on the Quonochontaug marshes 28 September 1912 by HSH. There is one in the collection of LKi (now at NBS) taken at St. Mary's Pond, Portsmouth, 1 October 1919. The only winter record, and for years the "last" Rhode Island record of Yellow Rail, was one picked up dead on the shores of the Barrington River 20 February 1922 and brought to F. T. Jencks, the specimen now presumed lost.

Nonbreeding

Fewer Yellow Rails have been reported since the shooting era; none, in fact, were reported between 1922 and 1970. This must be at least partly due to the decline of rail hunting. Two were collected using dogs near Middle Bridge, Narrow River, 22 October 1970, and another along the west side of the Narrow River 12 November 1970, all by Robert Saunders of Narragansett; a specimen examined by the author is now at the South Kingstown High School. A bird was found dead in the road at Moonstone 13 October 1980 by T. Jarman, the specimen sent to RLF by CWo. Two recent sight records are of individuals flushed from the Palmer River salt marshes in Barrington, 13 October 1982 (SRi), and at Potter Pond, Jerusalem, 31 October 1982 (RAC). Single birds were seen at Succotash Marsh, Narragansett, 7 October 2006 (P. Rusch et al.) and 19 October 2007 (Z. Sevarino, R. Raymond, C. Nunes). There is a record for 17 December 2006, Napatree CBC, but the bird was in Connecticut (SSM, GW et al.).

Black Rail (*Laterallus jamaicensis*)

Status

Rare visitor in summer and fall.

Nonbreeding

The first Black Rail record in Rhode Island was on 28 September 1970, when the bird was collected on the East Matunuck marshes (Potter Pond) by Robert Saunders, using dogs specifically trained to hunt rails. The mounted specimen was subsequently examined and photographed by the author and is now at the South Kingstown High School (fide W. McEneaney). Another was discovered calling at Green Hill on the afternoon of 8 June 1975 by ASM, heard by several other observers through 13 June, and recorded on tape by DLK. A third bird was reportedly flushed three times on 16 September 1988 from a hayfield north of Quicksand Pond and well described from observations made as it crouched in the grass beside the shoes of the observer, C. Mercer. From 29 May to 10 June 2015, a bird was calling from shallow cattail marsh at Misquamicut Pond (GW et al.); it was heard on several nights by many observers

Black Rails are at the northern limit of their range in our area, and there are several recent summer and winter records for Long Island, Connecticut, Massachusetts, and New Hampshire.

CRANES (GRUIDAE)

Sandhill Crane (*Antigone canadensis*)

Status

Vagrant. Increasing in frequency with continent-wide population growth.

Historical

The only record for most of the historical period was one shot and another seen at Natick Hill in West Warwick, 8 or 9 October 1889 by B. Burlingame (Southwick 1889; Brewster 1890). The mounted specimen was later placed in the Smith collection and continues to reside at RWPM. This individual was listed in Forbush (1925) as the only New England record for the "Little Brown Crane."

For 72 years there were no additional records. On 23 December 1961, a Sandhill Crane was seen flying east over lower Little Compton (DLK, CWo). This bird was soon rediscovered standing on the ice at Briggs Marsh, where it remained for several days and was seen by a number of other observers.

Nonbreeding

Twenty-three years elapsed until 2 birds were discovered 22 December 1984 in fields north of Dunn's Corners, Westerly (ESa), where they had apparently been present for about 3 weeks. The birds remained at this location into January but soon began to spend most of their time in fields north of Weekapaug and remained there from 13 January on through February and March. Soon after exhibiting some courtship behavior, they were last seen 15 April (fide ESa). Their stay in Rhode Island was then among the few documented cases of overwintering in New England.

During the period 1985 to 2000, there were approximately 8 Rhode Island records, including a bird discovered and photographed near Sakonnet Point 12 October 1985 by RBo and J. W. Bowen, the day that the same or a different bird was seen in nearby Westport, Massachusetts. Others occurred at Portsmouth 1–20 April 1986 (LT, RBo, DLE et al.; photos), Galilee 15 September 1987 (AES, DFi; photo), Little Compton (an immature) 10 December 1991 (C. Mercer; photos by RBo), and Middletown 1 February 1992 (Morgan). During the period 2000 to 2005, there were 7 additional Rhode Island records. Since 2005, they have occurred almost annually in Rhode Island, sometimes overwintering; e.g., 1, Seapowet, 10 October 2015–17 March 2016 (m.obs.), and 1, Trustom Pond, 6 December 2007–29 February 2008 (D. and B. Kindseth et al.). One was found in North Kingstown, 18 March 2020 (Jeff Hall; photo) and perhaps the same bird at both Jamestown (11 April 2020; ChP) and the Great Swamp (11–12 April 2020; v.o.). A bird in Tiverton 13 May 2020 (Mark Levine; photo) is probably the latest spring date for the state.

A gradual increase in the number of records has occurred in the last 30 years along the northeastern seaboard, and Sandhill Cranes are now nearly annual, and have bred, in western New York, Massachusetts, and Maine (Kamm et al. 2013; Vickery et al. 2020).

AVOCETS AND STILTS (RECURVIROSTRIDAE)

Black-necked Stilt (*Himantopus mexicanus*)

Status

Vagrant. Increasing in frequency.

Nonbreeding

Black-necked Stilts were unknown in Rhode Island until 12 June 1966, when 3 were found in the salt marsh at Quonochontaug by ESa. These birds were examined and photographed by numerous observers until 20 June, when they departed southwestward high over the ocean (ESa; photos by RDe). A flock of 7 birds was found in the salt marsh on Conanicut Island 15 May 1972 (FPe, MDa, RLC). They were seen leaving the island in a northeasterly direction the same day. Two were at Quonochontaug 6 July 1972 (R. and C. Travisano, fide RAC), and 5 were at the same location on 25 May 1979 (Gorham, fide ESa). One was photographed at Richmond Pond, Massachusetts, on the evening of 2 July 1990 (RBo; unmentioned by Veit and Petersen 1993), after which it flew eastward into Massachusetts; it may have been the same bird seen at Briggs the next day (DLE et al.). After Hurricane Bob, 2 were present 20–23 August 1991 at Mud Pond, Moonstone (RAC; SHa et al.). A single bird was also at Block Island 20 September to 2 October 1991 (S. Madison, fide JO); 1, Great Swamp, 12 May 1995 (D. Wexler); 1, Briggs Marsh–Quicksand Pond, 23–29 July 1995 (DLE, REm, MJM, BCG et al.); 3, Charlestown Breachway, 22 May–8 June 2003 (JSJ et al.); and 1, Jamestown, 9–10 June 2008 (ChP, W. Munns).

American Avocet (*Recurvirostra americana*)

Status

Rare fall and very rare spring migrant.

Nonbreeding

The first Rhode Island record of American Avocet was 1 at Galilee on 6 October 1968 (Mr. and Mrs. K. Bates, fide RDe). Another was seen at Trustom Pond 19–20 September 1970 by DWF and Mr. and Mrs. H. Gates. Another was at Mackerel Cove, Conanicut Island, on 13 August 1972 (DRu). Others were at Briggs Marsh 16 September 1973 (DLE, REm et al.; numerous photographs), and Trustom Pond 13 October 1973 (DLK, ESa). A flock of 1 to 4 birds was present at Ninigret Pond, 22 September–1 October 1974 (DLK, DWa et al.). Two were recorded at Block Island 11–26 September 1976 (SRD), and another was there on 1 October 1977 (R. Forster). Three were present at Quicksand Pond 27 September–1 October 1981 (HWi, RBu et al.; photographed by DFi). One was seen and photographed at Quonochontaug 12–17 September 1982 (SSD, JGe, RGe et al.); 1 appeared at Quonochontaug 16 June 1989 (D. and G. Pratt, fide DLK); another was found at Quicksand Pond on 22 August 1991 (AES et al.); 1 that appeared at Cards Pond on 11 August 1992 (RFa) was probably the same as 1 that remained at nearby Quonochontaug 15–31 August 1992 (DLK et al.). One was at Ninigret, 17 August 1996 (STs, MJM, WBo); 1, Briggs Marsh, 28–30 September and 12 October 1996 (DLE, RBo et al.); 1, Cards Pond, 27 October 1996 and again at Mud Pond, 3 November 1996 (DLK, JMu). Another remained at Quicksand Pond 1–7 August 1993 (BCG, WG, RBo, DLE). One was reported at Jerusalem, 18 June 1994 (EWe). One in the "Kitchen" at Napatree Point, which attempted to dive underwater when attacked by a large falcon (probably a falconer's bird), was seized and carried off on the exceptional date of 22 August 1994 (CRa). A flock of 3 appeared at Ninigret Pond on 22 September 1994 (MJM et al.; photos) and remained through late October, with 1 bird still present 5 November (DLK); also, 1, Quicksand Pond, 14 May 1998 (RFa). More recent records include 1, Middletown, 26 July–3 August 2004 (SRi et al.); 1, Charlestown Breachway, 21–30 August 2004 (STs et al.); 2, Briggs Marsh, 1–6 October 2012 (JSJ, BCG, REm); 5, Newport, 24–30 September 2011 (RTL; m.obs.); 1, 29 November to 18 December 2015, Trustom Pond (m.obs.).

OYSTERCATCHERS (HAEMATOPODIDAE)

American Oystercatcher (*Haematopus palliatus*)

Status

Fairly common breeder and migrant on sandy beaches at the coast.

Historical

Oystercatchers have increased substantially in the Northeast in recent decades and are now breeding locally but fairly commonly along the Rhode Island coast. Oystercatchers were reported as far north as Labrador during the early 1800s by Audubon, but they then apparently disappeared from the Northeast by the early to mid-nineteenth century. With protection beginning in the early 1900s, they gradually increased and spread, returning to southern New Jersey in 1947 and breeding at Gardiner's Island, New York, in 1957 (Bull 1964). They gradually colonized much of the south shore of Long Island thereafter and jumped over Rhode Island to begin breeding in 1969 on Chappaquiddick and Nantucket Islands, Massachusetts. The push into these nearby areas, particularly into Massachusetts, was immediately reflected in a sharp increase in Rhode Island records of transients during the 1970s. They first nested on Block Island in 1976.

Unknown even as vagrants in the early part of Rhode Island's ornithological history (singles were recorded in Massachusetts in 1924 and 1936), oystercatchers were first seen at Galilee on 14 May 1938 after a violent southeasterly storm (WBD, DLK, CWo). They were subsequently unreported again until September 1954, when one appeared at Napatree Point 5 September (DLK), and others were seen at Jamestown 2–23 September (MDa) and Watch Hill and Napatree Point 12 September (DLK, CWo; ASRI). These occurrences followed Hurricanes Carol and Edna, which passed through Rhode Island 31 August and 11 September.

The next vagrant was at Napatree Point on 24 May 1958 (DLK), and another spate of records followed Hurricane Donna of 12 September 1960: on the day after the storm one was at Sakonnet (CBl), two at Napatree Point (J. Lynch), and a third at Block Island (E. Lee). An apparently different individual was seen at Block Island on 24 September 1960 (JBa).

After the 1960 flight, oystercatchers were unreported for nearly a decade, but during this time they increased considerably as breeders on Long Island. During the late 1960s, they appeared in Rhode Island as they extended their breeding range in Massachusetts. Records include one at Napatree Point on 5 October 1968 (DLK), another at Quonochontaug on 17 May 1969 (DLK), and two at Napatree Point on 27 May 1970 (DLK). The total number of individuals recorded in 1971–1975 and 1976–1980 was 25 and 68, respectively. Meanwhile, oystercatchers bred for the first time in 1974 on nearby Fisher's Island, New York (J. Bull, AMNH, pers. comm.).

Breeding

Virtually all birds nest on islands because of predation by mammals at mainland localities.

Block Island. A pair with 1 egg was discovered 6 June 1976 on a sandbar in the northern part of Great Salt Pond east of "Gunners Hill" (RLF). Present apparently without nesting in 1977, the species bred there in 1978 (1 young; ELa), 1980 (3 adults and 3 young on May 26; DLE, REm), 1981 (3 pairs and 3 young; v.o.), and 1982 (1 pair, 3 eggs). Since 2000, there have been 2–3 pairs breeding each year. They have nested mostly on West Beach from the north tip of the island to Beane Point and adjacent salt marsh.

Dyer Island. A pair was present on the south bar of this low-lying island on 27 May 1980 (RLF, JEM), and a nest with 1 egg (later unsuccessful) was found on 3 June 1982. Eggs or young have been seen almost annually since. There have been 5–6 pairs in recent years.

Big Gould Island. A pair and 2 eggs were found on the rocky southeast side on 4 June 1984 (RLF, JEM). On 3 June 1985, a pair with a nest and 1 egg was found on a small patch of sand at the southern tip of the island adjacent to the tern colony (RLF, REn). In later years, 1 (1988, 1989) or 2 pairs (1986, 1987) were present; 3 pairs in recent years.

Hog Island. A pair was found with a nest not far from the tern colony on the west side 31 July 1984 (JEM, CRa), and a pair with 3 eggs was present closer to the tern colony 3 June 1985 (RLF, REn et al.). There has been a single pair annually in recent years.

Prudence Island. A pair and 1 egg were found near Potter's Cove on 8 June 1994 (RLF, JEM). Birds were present 2015 and 2016 but are not seen every year at this site, which has much human visitation and is exposed to mammalian predators.

Hope Island. A pair was present at the north end on 3 June 1985 (RLF, JEM, CRa, REn) and was still present on 9 July 1985 (REn). There have been 3–6 pairs in recent years.

Seal Island. First seen here in 2004; a pair has been present nearly each year until at least 2010.

Spar Island. A pair was present but apparently not breeding on the northwestern-most of the 2 sandbars on 3 June 1985 (RLF, REn, JEM, CRa). There has been 1 pair annually up to the present.

Sakonnet Point islands. A pair with a nest and 3 eggs was present on West Island, 29 May 2015 (GDe).

Rose Island. A pair was present here in June 1982 but did not remain. Apparently absent in 1983–1984, pairs have been present annually 1985–1988, and 1 young was seen in 1986 (JEM, CCA, ChP). There have been 2–4 pairs present in recent years.

Napatree Point and Sandy Point Island. In 1982, a pair laid 3 eggs on the hook of Napatree Point (GLB et al.), and another pair bred near the southeast end of Sandy Point Island. Oystercatchers have unsuccessfully attempted to nest most years since then, but most chicks were lost to predation. The number of pairs has continued to increase among the breeding gulls at Sandy Point Island, where perhaps the state's largest population of oystercatchers now breeds (9 pairs raised 9 young in 2016 and 13 young in 2017; USFWS).

As can be seen, virtually all the locations where the species has bred or shown some territorial behavior are on islands. The only mainland site is Napatree Point, which would seem to constitute one of the best overall sites in the state except that the birds are exposed to foxes and skunks and to heavy human disturbance in summer. Many of the locations that have proved attractive to the species are rather rocky, and most (including Dyer Island and Napatree Point) have associated mussel beds, important for feeding during low tide. Nearly all of the locations have been adjacent to active colonies of gulls, terns, or both. Oystercatchers have not bred on the outer barrier beaches of the mainland (such as Quicksand Pond, Briggs, or along the South Shore), which provide the main breeding areas for Piping Plovers. There was a total of about 39 pairs in Rhode Island in 2016 (fide CRa).

Nonbreeding

Spring. American Oystercatchers are among the earliest shorebirds to arrive in spring; e.g., 5, Rose Island, 1 March 1997 (C. E. Johnson). There are a few records for late February. General arrival is in mid to late April.

Spring maxima. Nine, Napatree Point, 13 April 1996 (CRa); 8 passing west through the Sakonnet islands, 6 June 1984 (RLF, JEM); 7 aggressively escorted westward by 1 of the resident pairs at Sandy Point Island, 13 June 1990 (RLF, JEM).

Fall. In late summer and early fall, concentrations are larger than in spring and confined to the outer coast. The largest numbers on record appeared after Hurricane Gloria on 27 September 1985. On the day after the storm, approximately 70 were present in mainland areas including 15 at Briggs, 10 in the Point Judith area (including several sleeping with gulls in parking lots), 13 at Quonochontaug, 26 at Napatree Point (including many migrants moving west along the shore), and numerous others elsewhere (DLE, DLK, DFi, RLF et al.). Still at Napatree Point were 21 on 5 October 1985 (CRa), and 10 were at Block Island the same day (ASRI). Three remained at Napatree Point in 1985 after Hurricane Gloria as late as 23 November (CRa).

Fall maxima. 111, Napatree Point, 2 September 2016 (CRa).

Extreme dates. 27 February 1990, Charlestown Breachway (MJM), and 27 February 2012, Napatree (CRa); to 10 December 2016, Seal Island, Bristol County (2; L. Mendes), and 27 December 2015, Napatree Point (SSM, GWi et al.).

PLOVERS (CHARADRIIDAE)

Northern Lapwing (*Vanellus vanellus*)

Status
Vagrant. Two records.

Nonbreeding
The first Rhode Island record was a bird shot on Block Island on 20 November 1932 and brought to EDi. It is now mounted in the Block Island Collection (Dickens 1934). Its presence in our region may have been related to the "Great Atlantic storm" of 17–19 November 1932 (Murphy and Vogt 1933). Two lapwings appeared at Little Compton, 1–7 February 2013 (J. Park, m.obs.), following the occurrence of 3 birds in Massachusetts during November–February 2012–2013.

Black-bellied Plover (*Pluvialis squatarola*)

Status
Common transient in spring and fall. Uncommon in summer and rare winterer.

Nonbreeding
These plovers never became as rare during the shooting era as some other shorebirds, although they were much reduced in numbers. They are now regularly found in concentrations of 50 or more at such favorable places as Galilee, Quicksand Pond, the Quonochontaug flats, Napatree Point, and elsewhere.

Spring. Probable spring transients appear during early April, although some March records may represent late-winter wandering from nearby wintering areas. Peak movement is typically from mid to late May. Some linger into mid-June; e.g., 1, Block Island, 22 June 2004 (CRa); 1, Block Island, 22 June 2005 (CRa).

Spring maxima. 300 along the South Shore, 27 May 1962 (DLK); 163 and 88, Matunuck and Galilee, 26 May 1978 (DWa); 150 along the South Shore, 19–20 May 1956 (DLK). Spring records inland include 1 or more at the Great Swamp, 27 April–25 May 1994 (SSM); 3 migrating northeast over Kingston, 21 May 1962 (DLK), and 1 at Fiskeville, 5 May 1951 (JGe, RGe).

Summer. Small flocks of up to 20, mainly one-year-old birds, occur through the summer.

Fall. Migrants appear in mid-July, with a protracted migration through November.

Fall maxima. 400, Little Compton, 12 October 1970 (RBo); 200, Newport, 16 October 1954 (SSD); 175, Briggs and Quicksand Ponds, 16 September 1973 (DLE, REm); 150 at Galilee, 26 July 1975 (R. Arbib, SRD). Since 2005, there are only 11 counts of 50+ birds, all from fall at Block Island; e.g., 79, Block Island, 16 October 2015 (REm et al.); 99, Block Island, 10 November 2011 (SSM et al.). Fall records inland include individuals that put down on low-water flats at the East Providence Reservoir 25 September 1989 (ILe), 3 there 17 October 1985 (DLE), and 1 there 27 October 1984 (DLE). One was at Arnold Mills Reservoir on 7 November 1993 (DMK). Other records include 15, Wood River Junction, 27 September 2000 (REm); 4, Woonsocket Reservoir, 18 August 2002 (MLy); 1, Diamond Hill Reservoir, 7 September 2002 (MLy); 2, Wood River Junction, 4 September 2003 (JSJ); 2, Arnold Mills Reservoir, 14 August 2005 (MLy); 2, Woonsocket Reservoir, 11 September 2005 (MLy); 7, East Providence Reservoir, 23 September 2010 (JSu); 42, Heaton Orchard Road (turf farm), Richmond, 2 October 2010 (RTL); 15, Heaton Orchard Road, Richmond, 29 September 2011 (RTL); 7, Wood River Junction (turf farm), Richmond, 2 September 2012 (RWa); 3, Diamond Hill Reservoir, Cumberland, 1 November 2008 (MLy); 8, Wood River Junction (turf farm), Richmond, 18 August 2013 (SSM, PJL).

Winter. Formerly unknown in winter; there are now innumerable winter records of this species along the outer coast and north regularly in Narragansett Bay to Island Park and occasionally Riverside and Watchemoket Cove.

Winter maxima. 34, Block Island CBC, 20 December 2012 (SCo et al.); 31, Newport–Westport CBC, 20 December 1980 (DLE et al.); 18, South County CBC, 29 December 1973; 11, Napatree, 17 December 1997 (CRa). Probably regularly survives winter at Block Island; e.g., 30, Block Island PDC, 20 February 2012 (SSM et al.).

American Golden-Plover (*Pluvialis dominica*)

Status

Uncommon to fairly common fall migrant, very rare in spring.

Historical

The largest numbers of golden-plovers usually occur during or following storms with easterly winds, as in other conditions the birds pass well offshore, or perhaps high overhead. They were formerly abundant under such conditions (1850–1870) and were shot over decoys or by "stalking" at such formerly famous shooting grounds as Sachuest Point, Coddington Point, the south ends of Prudence and Conanicut Islands, and elsewhere in what were then coastal sheep pastures. H. W. H. Powel (unpublished), recalling shooting shorebirds on barren Aquidneck promontories in the 1870s, mentioned "flights of Green Head plover when [1 to 3] heavy flocks would be in sight at the same time; coming steadily for half a day at a time." With constant shooting they became rare by 1899 and remained so until at least 1925. They subsequently increased, and the first sizable flock was of 39 at Middletown on 3 September 1940 (WHD). Despite continued increase, they have not regained their former abundance. Part of the recent scarcity is probably due to lack of suitable habitat as most pastures have become overgrown or developed.

Nonbreeding

Spring. Unreported in spring for many decades, golden-plovers began to appear again in the 1940s away from their more usual northward route through the western Mississippi River valley. One appeared very early at Napatree Point 15–29 March 1986 (CRa), and another or the same bird was there 16 April 1986 (REn). Other spring dates range from 3 April 1966 at Napatree Point (DLK, CWo) and 4 April 2009, Sachuest (RTL), to 6 June 1970 at Quonochontaug (ESa).

Summer. Occasional individuals are seen in midsummer; e.g., 6 July 1991, Napatree Point (CRa).

Fall. Peak numbers appear from late August through early October. Most birds now occur on coastal mudflats, as their preferred wet pastures have mostly disappeared. Modern sod farms, such as at Richmond, provide a seemingly suitable alternative.

Fall maxima. 202, Easton Pond, 17 October 1954, when the reservoir was artificially drained following saltwater flooding during hurricanes (v.o.); 116, Napatree Point, 5 September 1991, after a strong southerly rainstorm (CRa); 89 in a Middletown field, 2 September 1963 (SSD). Substantial numbers have recently foraged on the Richmond turf farms, about 8 mi (13 km) inland: 70, 3 September 2013 (m.obs.); 52, 30 August 1986 (DFi, DLK); 46, 9 September 1986 (RBo).

Inland. They appear inland where conditions are suitable: 1 shot at Pawtucket, 28 August 1891, formerly in the URI collection; 1 on the East Providence Reservoir flats, 14 September 1989 (DLE); 1 at Cumberland Reservoir, 28 September 1980 (REn); and 1 at a reservoir in Smithfield, 3 November 1951 (RCC). More recent records include 1, Woonsocket Reservoir, 6 October 2001 (MLy); 1, Diamond Hill Reservoir, 27 October 2002 (MLy); 1, Diamond Hill Reservoir, 16 October 2004 (MLy); 2, Arnold Mills Reservoir, Cumberland, 20 October 2007 (MLy).

Extreme dates. Regular into late November; 1 remained at Moonstone, 6 November–12 December 1979 (v.o.), and another was photographed on the Block Island CBC 18 December 1995 (DFi).

Killdeer (*Charadrius vociferus*)

Status

Fairly common summer resident and common migrant; uncommon in winter.

Historical

Killdeers were reduced to rare summer residents during the shooting era, despite a wider availability of breeding habitat then than at any time since. Around the turn of the twentieth century, it was exceptional to see even a single Killdeer at any season, and two shot in 1887 merited a note in *Forest and Stream*.

Most unusual was a major incursion of Killdeers that reached southern New England in fall 1888, after the late-season hurricane of 23–27 November. During this storm, the barometer fell below 29 in (74 cm) and winds of 84 mph (135 kph) were recorded. After this storm, large numbers of Killdeers appeared in southern New England coastal areas, particularly Rhode Island and Massachusetts (Chadbourne 1889). Large flocks were still present that year at least into late December, and numbers wintered on Cape Cod, which they had not done before.

Howe and Sturtevant (1899) stated that Killdeer had been seen in nearly all parts of the state by 1899 and listed at least five breeding records 1894–1899. Another record is given in Remington and Commissioners (1908) for 1895. Killldeers bred in Warwick in 1903 (HSH); near Wickford, circa 1902–1904 (Howe 1904); and in Cranston in 1910 (HSH) but continued to be scarce until the 1920s. With protection, Killdeers rapidly increased. They bred at Narragansett in 1917 (HSH), and migrant flocks were seen at Block Island in 1917 and 1918, followed by breeding in 1919. They nested commonly statewide in the 1920s.

Breeding

Killdeers now commonly breed in agricultural and athletic fields, pastures, and gravel pits and other barren areas, even on the median strips of expressways and the gravel roofs of buildings. Killdeers are most common in the populous eastern and southern sections of the state. Even at Block Island, where the percentage of open ground once exceeded any other part of the state and the species nested for many decades, suitable habitat in recent years has declined to the point at which breeding is only sporadic.

Nonbreeding

Spring. Spring transients regularly appear in late February or the first week of March. In a record mild winter, a migrant appeared in Cumberland on 22 January 1989 and was still present 5 February 1989 (DMK). Other early transients have been seen on 8 February 1937 at Block Island (EDi) and 13 February 1960 at Middletown (JBa); a flock of 8 arrived at the East Providence Reservoir on 21 February 1949 (RBo), and 6 appeared at Block Island on the same day. Similarly, 8 appeared 1 February 2006, Matunuck (RTL), and 6, 18 January 2013, Sachuest NWR (RTL). Killdeers are common transients through March and early April, and late migrants occur until early May or later.

Spring maxima. 40, North Kingstown, 21 March 1989 (LAH); 34 along the South Shore, 21 March 1986 (Swanson); 28, West Kingston, 10–12 March 1995 (DLK); 24, Narragansett, 1 April 1978 (DLK).

Fall. Flocks of postbreeding birds and migrants commonly assemble along muddy shores and in agricultural fields during late summer and throughout the fall.

Fall maxima. 165 on the Wood River Junction turf fields, 1 September 1997 (STs); 138+, Diamond Hill Reservoir, 23 October 1994 (DMK); 120, Portsmouth, 17 September 1949 (DLE); 115, Switch Road, 15 August 2004 (STs); 110 migrating in several flocks past Napatree Point, 8 October 2009 (CRa).

Winter. Small flocks of migrants are commonly seen into late November and some remain into winter in milder years.

Winter maxima. 54, Newport–Westport CBC, 26 December 1955; 23, Jamestown, 1 December 1957 (SCr); 20, Tiverton, 17 December 1984 (DLE).

Common Ringed Plover (*Charadrius hiaticula*)

Status

Vagrant. One record.

Nonbreeding

A juvenile discovered by P.A. Buckley appeared on the Ninigret flats on 15 September 1991, but it eluded other observers until rediscovered on 22 September by Bob Conway. Noted were the wide, slightly separated breast band and characteristic vocalizations. The bird appeared to lack use of 1 foot, which was held out at an angle, allowing very close observation of its unwebbed toes. The only previous New England record was also of a juvenile at Monomoy 5 September 1990 (Veit and Petersen 1993), and there has been a slight increase in frequency of occurrence in the Northeast since about 2010.

Semipalmated Plover (*Charadrius semipalmatus*)

Status

Common migrant in spring and very common in fall, rare in early winter. Fairly common inland in suitable habitat.

Nonbreeding

Unlike many species of shorebirds, Semipalmated Plovers have increased in abundance over the past 50 years and occur later in fall, with numerous records in December.

Spring. Migrants regularly appear during early May and peak numbers occur during the last half of May. Early arrivals include an immature at Napatree Point 2 April 1983 (CRa), 2 at Weekapaug 5 April 1991 (GTa, ATa), and an adult at Napatree Point 11–17 April 1983 (CRa).

Spring maxima. 450 at Galilee/Jerusalem on 18 May 1940 (DLK); most spring totals are 200 or fewer.

Summer. Groups of 5 or fewer occur through June.

Fall. After the Short-billed Dowitcher, Least Sandpiper, and Lesser Yellowlegs, this species is among the earliest of the returning Arctic shorebirds. The first migrants appear in late June and early July; peak counts of adults are in late July and juveniles in late August. They are less numerous inland than at the coast but still fairly common: 15, Diamond Hill Reservoir, 14 September 2003 (MLy); 24, Diamond Hill Reservoir, Cumberland, 18 August 2007 (MLy); 40, East Providence Reservoir, 11 August 2012 (J. Sweeney). Rather late were 3, Diamond Hill Reservoir, 27 October 2002 (MLy).

Fall maxima. 920, Quicksand Pond, 14 August 2004 (REm); 625, Quicksand Pond, 4 August 2001 (REm); 500, South County, 16 August 1983 (RAC).

Extreme dates. 2 April 1983 (see above); numerous records in November; crippled bird shot at Middletown on 13 December 1913 (ESt); 1 seen at Narragansett on 24 December 1976 (STs, DLK); 10, Block Island, 22 December 2011 (DFi, STa).

Piping Plover (*Charadrius melodus*)

Status

Uncommon and local summer resident along the sandy beaches of the outer coast, as well as a fairly common spring and common early fall transient on coastal mudflats. Very rare in winter.

Historical

These characteristic birds of beach and dune were formerly common summer residents before the height of the shooting era but became rare by the turn of the twentieth century. Howe and Sturtevant (1899) called them "uncommon summer residents and uncommon migrants," but could cite only a few breeding records along the Westport and Little Compton beaches, plus an adult and two young that were collected at Charlestown Beach in 1895. Remington and Commissioners (1908) provided data on a nest "in Washington County" in 1908. A flock of nine was seen at Third Beach, Middletown, on 17 April 1912 (ESt). Hathaway (1913) called them "extremely rare summer residents and migrants" and referred to a single pair that bred from 1908–1912 at Briggs Beach. Doe collected eggs on various South County beaches, mainly Narragansett, Matunuck, Moonstone, occasionally Charlestown, Green Hill, Galilee, and near Point Judith. These numbered five sets of eggs in 1913, six sets in 1920, nine sets in 1921, and 25 additional sets 1922–1927. There are eggs in HSH's collection from Weekapaug for 1918 and 1920, and the species returned to Block Island as a breeder in 1917. Hathaway (1943) stated that they became relatively common in the state after 1920. He knew of 10 breeding pairs in Rhode Island in 1936 and estimated that many others were present as well.

Breeding

The 1938 hurricane created considerable habitat for Piping Plovers through the destruction of houses and removal of beachgrass. Following these drastic changes, there was a short period circa 1940–1945 when gasoline rationing and other wartime restrictions kept the public from rebuilding on beaches. Local conditions therefore encouraged a substantial increase in the Piping Plover population, and a peak in numbers lasted at least to the end of World War II. During this 5–7-year period, fragmentary data suggest that breeding occurred on nearly every barrier beach on the outer coast; e.g., 37 adults at Napatree Point 16 June 1943 (Stoddard). In 1954, Hurricane Carol rejuvenated the habitat by removing hundreds of beach cottages along the South Shore, flattening the dunes, and creating many wash-over strips suitable for nesting. A vast new area of sand at Galilee, created by dumping dredging spoil on the marsh in the mid-1950s, added considerable new habitat. This area, now covered by *Phragmites*, remained unvegetated long enough to cause statewide numbers to reach a second peak; e.g., 40 at Galilee in July 1960 (RLF).

Numbers gradually dropped to 10–14 pairs by the 1970s. Piping Plovers disappeared by 1972 from the Galilee/Jerusalem area, where they had been common, as well as from many other barrier beaches along the South Shore.

During the period 1981–2004, the number of breeding pairs in Rhode Island increased from 10 to 70 (USFWS, TNC; Cairns and McLaren 1980). Productivity ranged from 0.5 to 1.7 fledged young per territory during the period 1981–2003, with 113 fledglings in 2002, 72 fledglings in 2003, and 105 fledglings in 2004. This increase in nesting pairs and their productivity resulted mainly from enforcement of regulations on beaches and provision of nesting exclosures. More recently, 3 pairs nested at the mouth of Narrow River in 2008, but none were there in 2016 (TNC); 1–2 pairs returned to Second and Third Beaches in Middletown in recent years, an area where there were several historical records of breeding (USFWS); 2 pairs nested on Block Island in 2009 but none since (CRa); and 24 pairs nested at Quicksand and Briggs Beach in 2016 (TNC).

Summer maxima. 42, Napatree Point, 18 July 2009 (RTL).

Nonbreeding

Spring. Piping Plovers generally arrive in mid to late March. Local breeders are commonly augmented by transients in spring on mudflats along the outer shore.

Spring maxima. 56 along the South Shore, 4 April 1954 (DLK); 40, Jerusalem, 7 April 1946 (DLK); 35, Quicksand Pond, 16 April 1973 (ASi).

Fall. Groups of adults and juveniles congregate at the end of the breeding season from mid-July to late August.

Fall maxima. 44, Little Compton, 21 July 1944 (fide RLF). Rare after mid-September, but there are at least 10 records for November and 4 for December.

Extreme dates. 9 March 2008 and 2017, Quicksand Pond (GDe), and 11 March 1956, Jerusalem (DLK), to 13 January–4 February 1962, Napatree (DLK).

Lesser Sand-Plover (*Charadrius mongolus*)

Status

Vagrant. One record.

Nonbreeding

An adult male was found by L. Ferraresso and MJM on Ninigret Flats in the morning of 24 July 1999. It remained until 26 July, was seen by many, and was both photographed and videotaped (*NAB* 53:445). There are other eastern North American records from New Jersey, Ontario, Louisiana, and Florida (Curtis 2007).

Wilson's Plover (*Charadrius wilsonia*)

Status

Vagrant. Occurs in spring and fall.

Nonbreeding

The first undoubted record was of 1 seen at Jerusalem, 16–17 May 1954 (RCC, SSD, CWo, DLK, GRa, RLF et al.), although 1 was reported collected near Newport in 1883–1884 by Col. Powel. Another was seen shortly thereafter at Briggs Marsh, 1 September 1954, the day after Hurricane Carol (which brought many other Wilson's Plovers to the Northeast; DLE). Others appeared at Napatree Point on 27 May 1962 (DLK, KKi, FPe, CWo); 2 were there 17 May 1964 (GBi, RDe, Knox, Bates); and 1 was there on 11 September 1965 (GBi, Bates). One appeared after a storm at Galilee on 28 April 1979 (L. McGraw, RLF), and another was seen there 11 August 1982 (JGe, RGe). One was observed at Ninigret 6 August 1983 (RAC). A decade passed before another was seen at Quicksand Pond 10–11 June 1992 (R. Deegan; photo). One was at Easton's Beach 22 April 1999 (RTL); 1, Block Island, 13 and 20 April 2003 (Corrie Heinz); 1, Napatree, 15 June 2005 (CRa). On 4–5 May 2018, a male was at Goosewing Beach, which was joined 6–7 May by a second, less-well-marked bird (GDe; photos). On the last day, both were noted taking off over the ocean to the southwest. One was at Charlestown Breachway, 4 May 2019 (S. Miller, CPe, D. Heitzmann). One was at Scarborough Beach, 3–16 March 2020 (J. Kiamco et al.; photos).

Snowy Plover (*Charadrius nivosus*)

Status
Vagrant. One record.

Nonbreeding
A single Snowy Plover was found and photographed at Goosewing Beach, Little Compton, 17 April 2017 (GDe; see Photo 6). It consorted with Piping Plovers, departed late in the day, and was not seen again. Others have occurred on Cape Cod in June 1994 and in New Haven, Connecticut, in October 2004.

SANDPIPERS (SCOLOPACIDAE)

Upland Sandpiper (*Bartramia longicauda*)

Status
Rare breeder and uncommon migrant.

Historical
Upland Sandpipers were probably common summer residents near the coast during the agricultural era, but their status inland during this period is unknown. Howe and Sturtevant knew of no certain breeding records as of 1899, and only suspected the species of nesting, facts that suggest that the large numbers shot in the early shooting era may have been primarily migrants from farther north rather than local breeders.

Sir Charles Blagden (1903) said of the "grass plover" at Newport in 1776–1780: "They come in great numbers and are very tame at first but grow extremely shy and hard to shoot in a few days." He collected one on April 29, year unstated (Howe 1905). Giraud (1844) called them "common" in Rhode Island, and Baird et al. (1884) mention that they had been found "breeding near Narragansett Bay." Hathaway regularly heard them migrating over his shooting blind at Quonochontaug in late summer but almost never saw them alight there. His highest count of such overhead migrants was 19 on 23 July 1911. Prior to the 1930s, this species had not been seen in sizable flocks for many decades, but HSH's notes state that in 1843 "they were so plentiful at Sakonnet and on Conanicut Island that the gunners tried to get the protective law removed." A kind of golden age occurred in the period 1930–1965, when late-summer flocks of 20–30 birds were not uncommon at traditional locations (maximum 43 at Jamestown on 14 August 1941; HNG). Later counts in this golden

FIGURE 5. Shorebirds at Charlestown Breachway, © Margaret La Farge.

era included 31 at Matunuck on 7 August 1965 (CWo) and 30 at Middletown on 19 August 1963 (RLF).

Still called "grass plovers" by nineteenth-century gunners, they were very common in late summer at such favorite shooting grounds as Prudence Island, Beaver Neck on Conanicut Island, portions of southern Aquidneck Island, and elsewhere, but with constant shooting they had become scarce by 1885. With protection, there was some increase after about 1915, but they never recovered their former abundance because by that time (early twentieth century) their habitat was disappearing because of the decline of sheep and cattle farming.

Breeding

Upland Sandpipers were first confirmed nesting beginning in 1915 at a number of coastal locations, where breeding birds were sporadically and locally present into the 1950s. More recently, they have declined again because of the disappearance of hayfields and the transformation of most open habitats into housing or industrial developments. A history of known breeding locations is as follows:

Block Island. Eggshells of Upland Sandpipers were found at Dickens Point in 1915, and in 1916, young that could not fly over a stone wall were seen (EDi). No other nests or young were found for many years, but birds were seen regularly at Dickens Point, and these were suspected of breeding on southern portions of the island. One was flushed from a nest with 4 eggs near Dickens Point on 14 June 1981 (RBo; photos), 1 was seen there 17 June 1986 (REn), and 2 were heard there on 19 June 1990 (CRa). The Lewis-Dickens property, historically home to Upland Sandpipers and Grasshopper Sparrows, was acquired by ASRI and TNC in 1983 and has been managed as a natural agricultural landscape with priority given to the continued breeding of its grassland fauna and flora.

Little Compton. Birds were seen periodically during early summer in the area northeast of Briggs Marsh, especially 1932–1939, but no proof of breeding was obtained. The habitat has long since succeeded to brush and housing developments.

Aquidneck Island. Migrants or "staging" local breeders were common in late summer in the Middletown area, and birds were reported in June (and presumed nesting) near Green End and Paradise Avenues, all during the period 1946–1962. Additional observations of individuals in June during the 1960s and early 1970s at 3 other Middletown and Portsmouth locations suggested a more widespread breeding distribution. After about 1975, with accelerating destruction of much of the island's open land by housing developments together with a cessation of mowing in many remaining hayfields, and the planting of others to nursery stock, Upland Sandpipers probably ceased breeding on the island. The only sign of their continued presence since then is a bird seen south of Sandy Point, Portsmouth, in June 1984 (CRa).

Conanicut Island. First noted 15 July 1929, when a young bird was caught and banded by M. Wood. In 1936–1937, A. M. Clark reported them as common and breeding near Jamestown and found young birds in 1939. In 1940, 4 breeding pairs were noted, and 1 nest located; young were seen in 1941 and 1943 (HNG). On 17 June 1947, WBD found 15 adults and 3 young. Thereafter, the species gradually declined in summer, although 1 or more pairs may have persisted into the mid-1960s. A calling male in the U.S. Navy's Beavertail antenna fields 9 June 1972 (RLF) is apparently the last record.

Quonset Point. Four birds in fields at the northern end of the airfield 14 May 1985 (CRa) and 2 there 18–19 May 1987 (CRa, REn) possibly bred or attempted breeding.

West Greenwich. Breeding was suspected in a large field in the northeasternmost part of the town adjacent to I-95, where individuals and occasional pairs were reported consistently in May during the 1980s, and an agitated bird was present 20 June 1985 (DFi, DLK). This field was destroyed for industrial development in the mid-1990s (fide CRa). This is the only location away from the coast in Rhode Island.

Matunuck area. An "anxious" pair of "field, or whistling plovers" was found at Moonstone on 14 May 1921 (CED); 4 were seen there and thought to be breeding by HSH in late July 1938. On 1 July 1946, DLK found a pair with 3 young and later a second pair with a single chick on 14 July of the same year. Birds were seen there in spring until about 1965 (DLK). The last potential breeding there was a pair observed 4 June 1972 (ESa).

Nonbreeding

Spring. Upland Sandpipers normally arrive during late April and early May. A remarkable record was an Upland Sandpiper that alighted in a tree on a foggy morning at Swan Point Cemetery 26 April 1990 (DFi, DLE).

Fall. This species is among the earliest fall migrants, even for a shorebird.

Fall maxima. 20, NBS, Middletown, 24 August 1955 (JBa); 11–13, Richmond turf farms, 12–19 August 1986 (v.o.). Counts have not exceeded 3 birds per day per locality since 2005.

Extreme dates. 5 April 2000 (JSJ, MT et al.) to 4 November 1995 (MHC, SMC), both at the Richmond turf fields. Rare before late April and after mid-September.

Whimbrel (*Numenius phaeopus*)

Status

Uncommon spring and fairly common fall transient to coastal areas, rare on the shores of Narragansett Bay and extremely rare inland.

Nonbreeding

Spring. Numbers in spring are always small (maximum 6 per day), and in some years, Whimbrels are unreported during spring. Most occur between mid-April and mid-May. One was seen at Quicksand Pond 6 June 1987 (RBo).

Fall. The earliest occurrences include individuals at Matunuck on 29 June 1943 (WBD) and 2 on 29 June 1988 at Quonochontaug (DLK). Most records fall into 2 well-marked periods: 1 from about 20 July–10 August (mostly adults) and another peak 25 August–20 September (primarily immatures).

Fall maxima. 48, Charlestown, 2 September 1976 (DLK); 40, Little Compton, and 48, Trustom Pond, both 18 September 1949 (WBD, DLK); 40, Middletown, 18 September 1941 (WHD); 35, Little Compton, 6 September 1940 (DLE); 12, Ninigret, 21 August 2004 (STs et al.). There have been no groups of more than 20 birds in the past 20 years.

Whimbrels are only rarely observed on or near the shores of Narragansett Bay: 1 was seen on 5 May 1945 on Poppasquash Point, Bristol; another was shot at the mouth of the Pawtuxet River in 1876 by F. T. Jencks. One was observed flying south over Nag's Pond, Prudence Island, on 2 August 1944 (HNG), and 2 were seen at "Merriman's Pond," Barrington in August 1943 (HNG). There are only 2 records well inland: 1 at Fiskeville on 30 April 1979 (JGe, RGe) and 1 at Arnold Mills Reservoir, 30 August 1993 (DMK).

Extreme dates. 6 April 1946 (DLK) and 8 April 1984 (RAC), both at Galilee, to 3 November 1970, Trustom's Pond (RRi), and 8 November 1959, Block Island (MSl).

Eskimo Curlew (*Numenius borealis*)

Status

Probably extinct; formerly an erratic but occasionally numerous fall migrant.

Nonbreeding

There are 3 specimens in RWPM, 1 of which is the specimen listed in the museum's catalog as collected in October 1842 on the "Payne Farm, Johnston." Giraud (1844) stated that it was "seen every [fall] season" in Rhode Island. Eskimo Curlews are mentioned in Rives (1883–1884) among birds shot near Newport. By 1899, Howe and Sturtevant called them "rare migrants" and stated that they were "formerly quite abundant, migrating with [golden-plovers]." Six were shot in 1886 at Little Compton by H. T. Marshall (fide HSH), and 1 was taken by LKi at Fort Adams on 4 September 1888, now on display at NBS in Middletown. The only twentieth-century record, and one of the last specimens collected in New England, is a male shot on 5 September 1913 at Little Compton by C. B. Clarke and purchased by HSH, now in the possession of ASRI.

Long-billed Curlew (*Numenius americanus*)

Status

Vagrant. More regular as fall migrants before about 1850.

Historical

While data are sparse, Long-billed Curlews apparently occurred regularly as fall migrants before the market gunning period. They were extirpated from eastern portions of their breeding range (as far east as Illinois) during the last half of the nineteenth century. There is a previously unrecorded specimen in the RWPM collection, shot by Newton Dexter at Point Judith in August 1859 and examined by RLF. Newton Dexter wrote to Howe that, "the last one he killed or saw in Rhode Island was about 1862." According to the shooting diary of Robert L. Dring, a Newport market gunner (Howe 1901a–c), Long-billed Curlews were shot in southern Rhode Island (very likely on the Sachuest marshes) 15 July 1870 and 29 July 1874, and eight were shot 2–14 August 1873. There is a specimen shot on 8 September 1897 at Jamestown by Thomas R. Stetson, acquired by RHH and placed in the collection of William Brewster (Howe 1899a; formerly in BMS, now presumed lost).

Two Long-billed Curlews were seen in the fields at Dickens Point, Block Island, 3 August 1912 by Elizabeth Dickens, who wrote the following to HSH: "[S]tanding on a knoll a hundred yards [91 m] distant in a pasture, were 2 birds, cinnamon brown, fully as large as bitterns but in a shape like hudsonians [Whimbrels, or Hudsonian Curlews, were recorded regularly in the same fields], and with bills not less than ten or twelve inches [25–30 cm]. Of course, I frightened them, and they flew away, neighing like colts." Again, on 5 September 1946, six were reported by EDi in her west meadow.

Nonbreeding

While other individuals have been observed on Long Island and the Massachusetts coast during the modern period, there is only 1 recent Rhode Island record: an individual impeccably described from the Ninigret flats 9 August 1991 (RAC).

Hudsonian Godwit (*Limosa haemastica*)

Status

Rare but regular fall migrant; very rare in spring.

Historical

Hudsonian Godwits were apparently not rare before the turn of the twentieth century; for example, 50 were shot at Charlestown Beach on one occasion in the 1880s "after a September blow" (Marshall, fide HSH), and 18 were seen and six shot at Point Judith during a heavy northeaster on 30 August 1903 (W. A. Dring, LKi). Thereafter they seem to have declined and then remained generally rare for the next half century, long after protection in the early 1900s. Counts during the period 1903–1971 never exceeded three to four per day, but numbers have increased since then.

Nonbreeding

Spring. The "Ring-tailed Marlin" of the old-time Rhode Island gunners is very rare in Rhode Island during spring because the bulk of the population migrates north through central North America. There is a record of 3 "in the grey plumage" shot in early May 1873 at West Island, off Sakonnet, by George H. Mackay (Mackay 1911; see also Phillips 1929). Others in full alternate plumage were seen at Jerusalem 27 May 1981 and 29 May 1986 (JGe, RGe).

Fall. In Rhode Island, the earliest fall records are for individuals at Quonochontaug on 9 July 1977 (MPe) and 14 July 1986 (DFi, DLK); thereafter, peak numbers have been largely of juveniles in September, often during storms or heavy rain. Large counts of adults, such as those that appear at Monomoy and Newburyport, Massachusetts, have not often appeared in Rhode Island, probably owing to the lack of extensive tidal flats.

Fall maxima. 112, Quicksand Pond, 2 September 1974 (DLE, REm); 50, Trustom Pond, 3 September 1972 (SSD, DLK et al.); 38, Ninigret Pond, 5 September 1988 (RAC et al.). All these flocks were seen during rainstorms, some associated with tropical storms. Hudsonian Godwits occasionally appear inland: 1 was caught at Woonsocket Reservoir by a Peregrine Falcon 3 November 2001 (MLy et al.); a late bird remained at Arnold Mills Reservoir 7–12 November 1993 (DMK, GV). Six were in a field along Matunuck School House Road in South Kingstown 21 October 1989 (JO). Hudsonian Godwits are now rare in Rhode Island; there have been a total of 14 birds seen since 2005.

Extreme dates. "[E]arly May" (see above) to 26–28 November 1988, a bird in a pool of water in a pasture at West Kingston (GTa, ATa, DLK, RAC).

Marbled Godwit (*Limosa fedoa*)

Status

Rare fall transient. Very rare in spring and early summer.

Historical

This spectacular shorebird was apparently more common in the early shooting era than at any time since. A single gunner, R. L. Dring, shot 26 in Rhode Island between 1867 and 1874 (Howe 1901e), but by the turn of the twentieth century they were considered to be rare to very rare and were then unrecorded until 1934, even by Hathaway, who was constantly in the field during late summer.

Nonbreeding

In the modern period, Marbled Godwits have been seen nearly annually at coastal locations and have increased, totals per year of 1–7 being reported in recent decades.

Spring. Individuals have been recorded 22 May 1977 at Jamestown (REn), 23 May 1991 at Jerusalem (SHa), 23 May 2010 at Napatree (RDe), and in a field at Little Compton 1 June 1969 (SSD). One was at Napatree Point 27 June 1968 (DLK).

Fall. The earliest fall record is for 2 at Trustom Pond 25–26 July 1972 (DLK, KKi, FPe).

Fall maxima. Five, 28 July 1973, Ninigret Pond (DLK, KKi, FPe); 5, Charlestown Breachway, 3 September 2005 (JSJ et al.); 4, Briggs, 8 September 1948 (DLE); 4, Jerusalem, 25 August 1999 (DLK); 1 was at Green Island off Warwick in upper Narragansett Bay, 29 August 2004 (v.o.).

Extreme dates. 22 May 1977 (above) to 14 October–26 November 2002, Napatree (CRa); 23 October 1948, Quicksand Pond (DLE); 2 November 2001, Trustom Pond (PP); and 14 November 2007, Napatree (RTL).

Ruddy Turnstone (*Arenaria interpres*)

Status

Common migrant in spring and fall, locally fairly common in midsummer and winter.

Ruddy Turnstones remained fairly common during the shooting era and are still common migrants on pebbly beaches, rocky coasts, and breakwaters. They are perhaps the most common spring shorebird on the shores of upper Narragansett Bay at Bristol, Warwick, and Prudence Island, but they have not been recorded on freshwater shores inland.

Nonbreeding

Spring. The peak of spring migration is in late May. Turnstones are often more numerous in upper Narragansett Bay in late May–early June than they are along the South Shore.

Spring maxima. 500, Napatree Point, 31 May 2010 (CRa); 400, Napatree Point, 26 May 2012 (CRa); 350, Napatree Point, 27 May 2013 (CRa).

Summer. Small numbers, mainly 1-year-olds, often spend the summer; e.g., 11, Napatree, 7 July 2001 (CRa).

Fall. The peak of southbound birds occurs in mid to late August.

Fall maxima. 234, Napatree Point, 7 August 1989 (CRa); 180, Napatree Point, 15 August 1989 (CRa); several counts of about 150 at Napatree. Counts later in

September seldom exceed 50, and smaller numbers are reported through October to the end of November.

Winter. Turnstones were first recorded in winter 7 January 1891, when 4 were seen at Middletown (ESt). Thereafter, there was only 1 other recorded occurrence (1949) prior to 1960. Since then, they have clearly increased, mainly at Napatree but elsewhere as well. The most regular winter location during the 1990s was the south end of Third Beach and around Sachuest Point, where a flock of 20–25 have wintered annually since the 1970s.

Winter maxima. 71, Newport–Westport CBC, 16 December 2000 (DLE et al.); 30, Third Beach, Middletown, 2 January 1994 (SSM); 19, Sachuest Point, 22 February 2016 (RTL).

Red Knot (*Calidris canutus*)

Status

Fairly common spring and fall migrant in small numbers along the outer coast; rare in Narragansett Bay and very rare inland. Rare in winter on the coast.

Nonbreeding

The "Robin Snipe" of nineteenth-century gunners favors sandy tidal flats along the South Shore.

Spring maxima. 60+, Napatree Point, 30 May 1946 (DLK); 60, Napatree Point, 2 June 1982 (CRa, RLF); 50, Quicksand Pond, 26 May 1988 (DLE); 43, Galilee, 27 May 1962 (DLK); 28, Napatree, 4 June 2002 (CRa).

Summer. Several records of single birds spending the summer; e.g., 22 June 1992 at Napatree Point (CRa), 23 June 1985 at Quicksand Pond (RBo), 24 June 1955 at Galilee (fide RLF), 25 June 1992 at Ninigret Pond (PAB), 29 June 1937 at Quonochontaug (HSH), and 1 July 1993 at Quicksand Pond (RBo).

Fall. The largest fall counts of Red Knots in Rhode Island are of immatures in September and October; peaks of adults in July such as occur in Massachusetts do not appear here, so fall counts are not much larger than in spring, unlike most shorebirds.

Fall maxima. 118, Ninigret Pond, 10 August 1976 (RAC); 72, Napatree Point, 21–22 September 1989 (CRa); 47, Briggs Beach, 10 August 1991 (JO); 42, Napatree Point, 3 September 2011 (H. Golet). There are only 9 counts of 20+ birds per day per location since 2005, all from Napatree Point.

Winter. The tendency for Red Knots to overwinter in Rhode Island has substantially increased. A flock of 6–7 was seen at Napatree Point 23 December 1981, and 2 were still present on 9 April 1982 (CRa, DLK). Again, 5 birds stayed at Napatree Point for most of the winter of 1983–1984, with 2 remaining into April (CRa). One or 2 birds wintered at Napatree 2007–2008, last seen 16 April 2008 (CRa et al.). At locations other than Napatree Point, the much scarcer winter records include 1 at Jerusalem on 9 January 1966 (DLK), 2 at Weekapaug to 10 January 1976 (ESa, DLK, CWo), and 1 at Galilee on 30 March 1958 (DLK).

Ruff (*Calidris pugnax*)

Status

Rare migrant.

Nonbreeding

There were 3 specimens taken during the late shooting era: Sakonnet Point 3 July 1900 by N. Dexter; Point Judith marshes on 31 August 1903 by W. A. Dring (obtained by LKi; see Hathaway 1913; specimen in MCZ); and Point Judith on 7 September 1909 by W. T. Bowler (obtained by HSH and now in possession of ASRI). There have been about 30 records of single birds since the 1950s, with most during July–October (e.g., Weaver 1961). Spring records include 1 female at Trustom Pond 28 May 1966 (DLK, KKi, FPe); a male with black ruffs, Quonochontaug 19 June 1968 (FPe, KKi, DLK); 1 female, Napatree 4–15 May 2003 (B. Nikula et al.); 1 female, Great Swamp Management Area 12 April 2006 (DFi et al.); 1 female, Napatree Point 31 May 2007 (CRa, RLF); and Charlestown Breachway 22 May 2010 (fide R Fa). Fall records include 1 male, Sisson Pond, Middletown, 11–15 August 1999 (PAB et al.); 1 male, Charlestown Breachway 30 June 2004 (JMu); 1 female, Charlestown Breachway 8 August 2006 (G. Gerdts et al.); 1 at Quicksand Pond, Little Compton, 16 August 2011 (REm), 1 at Barrington, 23–27 October 2012 (B. Shamgochian); and a male at Charlestown Breachway 8 July 2020 (Patrick Felker et al.). There have been 7 Ruffs in the state since 2005, 3 of these in spring and 4 in fall.

Sharp-tailed Sandpiper (*Calidris acuminata*)

Status

Vagrant. One record.

A juvenile Sharp-tailed Sandpiper was found at Galilee 13 November 2021 by Sam Miller and Liam Corcoran (see Photo 7). It was clearly photographed and seen by numerous other observers until 20 November. This is 1 of about 8 records for New England.

Stilt Sandpiper (*Calidris himantopus*)

Status

Rare spring and uncommon fall migrant.

Historical

Known as the "Mongrel" or "Bastard Yellowleg" to nineteenth-century Rhode Island gunners, Stilt Sandpipers were probably more common as fall migrants during the post–Civil War era (likely due to better habitat) than at present. *Two hundred and nineteen* were shot near Newport by market gunner R. L. Dring between 1867–1874, including up to 69 birds in a year. Howe and Sturtevant (1899) considered them to be "not uncommon transients" and included information that suggests that they may have persisted better than some other species through the shooting era. At present, however, they are generally rather uncommon and irregular in fall, with numbers seemingly smaller than those known in the nineteenth century, possibly due to a lack of adequate habitat. Stilt Sandpipers, perhaps even more than dowitchers, are specialists of salt pannes. Many Rhode Island salt marshes had not yet been ditched in the nineteenth century and contained numerous pools that filled at the highest tides and only slowly drained, thus providing optimum habitat. In addition, some of the most favored locations for Stilt Sandpipers during the nineteenth century were at Briggs Marsh (until early in the twentieth century primarily a salt marsh and not a pond as it is today) and in the Middletown marshes (mostly obliterated by a reservoir and dump in the twentieth century, although the former dump section has been restored to salt marsh with tidal flow).

Nonbreeding

Spring. There are only a handful of spring records: 1 collected at Sakonnet Point on 9 May 1895 by Newton Dexter (formerly at RWPM; no longer extant); 1, Galilee, 4 June 1955 (DLK); 1, Galilee, 11–20 May 1958 (DLK et al.); 1, Jerusalem, 23 May 1959 (RMc); 1, Newport, 12 May 1966 (RSt); 1, Quonochontaug, 21–24 May 1975 (JC et al.); 1, Galilee 29 April 1979 (RLF, SSD); 2, Trustom Pond, 5 May 1982 (DVT); 3, Weekapaug, 23 May 2006 (RLF).

Fall. The earliest record in fall is of 2 at Briggs Marsh 2 July 1941 (DLE). Most adults pass through in late July; immatures pass through in August and early September. Stilt Sandpipers have been recorded inland at a small pond in Smithfield, where single individuals (perhaps the same) were present on 2 and 17 September 1949 (DLK), and another was at the Richmond turf fields 8 September 1999 (DLK).

Fall maxima. 80, Quonochontaug, 26 July 1973 (DLK); 60, Napatree Point, 14 September 1980, where they are rare because of a lack of grassy pools (RBu et al.); 52, Briggs, 11 August 1990 (JO et al.); 50, Quonochontaug, 18 August 1969 (DLK et al.); 45, Charlestown Breachway, 8 August 2007 (PCa).

Extreme dates. 29 April 1979 (above) to 16 November 1957, Gardiners Pond (JBa).

Curlew Sandpiper (*Calidris ferruginea*)

Status

Vagrant. About 12 records, possibly increasing.

Nonbreeding

Curlew Sandpipers were included in a list of birds shot near Newport by Col. J. H. Powel, a well-known local sportsman (Powel 1884), but details underlying this assertion are lost. Otherwise, the first for Rhode Island was 1 in "defective" spring plumage discovered at Quonochontaug on 26 May 1973 (DLK, CWo, FPe, KKi; photographed by S. Ross). An identifiably different individual appeared 2 days later at Cards Pond, Matunuck, 28–29 May 1973, during a period of low water (DLK, SSD, FPe, DWa; photo). Others were soon recorded, including 1 well-studied at Ninigret Pond 27 August 1977 (RAC), 1 in partial alternate plumage at Quicksand Pond 23–28 August 1986 (DLE, REm; photographed by RBo), 1 in alternate plumage photographed at Galilee, 10 August 1987 (M. Fitzpatrick), 1 1st-year bird at Quicksand Pond on 11 May 1990 (DLE, REm), 1 at Napatree Point on 31 August 1991 (CRa), 1 at Ninigret on 12 August 1992 (RAC), 1 in alternate plumage at Charlestown Breachway 15–16 July 2004 (JSJ et al.), 1 at Charlestown Breachway 21 May 2011 (PLE et al.), and 1, Napatree, 1 June 2018 (DPe; see Photo 8).

Red-necked Stint (*Calidris ruficollis*)

Status

Vagrant. One record.

Nonbreeding

An alternate-plumaged Red-necked Stint was found at Napatree Point 5–7 July 2020 (Bill Thompson and many others; eBird) and remained for 3 days (see Photo 9). There are about 20 records for the east coast of the United States (Mitra et al. 2021).

Sanderling (*Calidris alba*)

Status

Very common spring and fall migrant at the coast and fairly common lingerer in midsummer; very rare inland. Fairly common winter resident.

Historical

Sanderlings apparently did not winter in Rhode Island until the early part of the twentieth century.

Nonbreeding

Spring. Migrants occur mainly from 5 May–10 June, and the peak is ordinarily in mid to late May. Counts of such transients in May and the first week of June (rarely more than 50 per day) are much smaller than counts in fall.

Spring maxima. 200, Sachuest Point, 19 May 1991 (LAM); 150, Sachuest Point, 14 April 2013 (RWa); 80, Napatree, 5 May 2015 (CRa); 50, Napatree Point, 13 June 1991 (CRa).

Summer. Small numbers of probable nonbreeders are common from mid-June into early July.

Fall. Early fall birds appear by 15 July or earlier and increase rapidly in late July. Contrasting with counts on Cape Cod (Veit and Petersen 1993), Rhode Island counts of adults in late July and early August have rarely exceeded numbers of immatures that pass through later in summer and early fall. Instead, migration is greatly protracted, with greatest numbers from late August to late September.

Fall maxima. 900, Quicksand Pond, 5 October 1991 (DLE); 620, Easton's Beach, 28 September 2011 (RTL); 450, Napatree, 29 July 2008 (CRa); 500, Scarborough, 17 September 1966 (DLK). Sanderlings slowly decline in late fall, but moderate numbers are now regular on the outer coast and along the shores of the Sakonnet Passage into winter. The records suggest that they are generally scarcest in April, prior to the appearance of new migrants.

Sanderlings are fairly common in upper Narragansett Bay but uncommon inland: 12, Diamond Hill Reservoir, 1 November 2008 (MLy); 9, Wood River Junction (turf farm), 27 August 2016 (eBird, m.obs.).

Dunlin (*Calidris alpina*)

Status

Common spring and fall migrant and common to uncommon winter resident on coastal mudflats. Rare inland, and rare anywhere in summer.

Historical

During the shooting era, this highly gregarious species was depleted and was an uncommon fall and rare spring transient during Howe and Sturtevant's era (to 1899); for decades afterward, they did not occur in Rhode Island during winter at all. The latest fall record known to ESt (to about 1905) was 29 October. The first record during midwinter was of 40 at Sakonnet on 29 January 1933 (WBD). Others were seen again in 1937 and 1939, and they became more regular in occurrence thereafter. They have gradually increased and are now regular winter residents at certain coastal mudflats, with midwinter counts since the 1950s to 100 or more, and still higher numbers more recently. Since the 1950s, they have steadily increased and are now one of our most numerous shorebirds.

Nonbreeding

Spring. Counts include a mixture of birds overwintering in or near Rhode Island plus migrants.

Spring maxima. 500, Quonochontaug, 22 May 1971 (KKi, FPe, ESa); 500, Quicksand Pond, 22 May 1980 (JCB); 460, Quicksand Pond, 26 April 1980 (RBo); 450 there, 17–18 May 1975 (DLE, REm); 305, Napatree, 21 April 2003 (CRa).

Summer. Numbers of migrants taper off during the last week of May.

Fall. Dunlins are usually the last of the mudflat sandpipers to appear in fall, normally after mid-September or later. An early migrant or nonbreeder was 1, Ninigret Pond, 5–6 July 2003 (STs).

Fall maxima. All at Quicksand Pond: 1000, 17 December 1983 (DLK et al.); 600, 2 November 1958 (RBo); 500, 15 October 1973 (DLE, REm); 500, 24 October 1973 (KKi, FPe); 500, 5 November 1989 (DLE). Counts at other Rhode Island locations have rarely exceeded 300, perhaps owing to a lack of extensive mudflats.

Winter maxima. 500, Quicksand Pond, 1 January 1983 (RBo); 425, Quicksand Pond, 30 January 1979 (RBo); 1405, Newport–Westport CBC, 17 December 1983 (DLE et al.); 401, South Kingstown CBC, 22 December 2007 (STs et al.). Flocks often become smaller and more scattered in late winter.

Inland. Dunlins are uncommon inland: 12, Scituate Reservoir, 23 October 1949 (WBD); 5, Cumberland Reservoir, 24 October 1992 (DMK), and 2 there 12 November 1993 (GV); 1–2, Woonsocket Reservoir, 6–20 October 2001 (MLy); 1, Woonsocket Reservoir, 16 October 2004 (MLy); 19, Diamond Hill Reservoir, Cumberland, 1 November 2008 (MLy).

Purple Sandpiper (*Calidris maritima*)

Status

Common migrant and winter resident on rocky islands, headlands, and breakwaters along the outer coast; very rare in similar habitat along the shores of Narragansett Bay, and rare anywhere on mudflats and sandy beaches.

Nonbreeding

These "rock" shorebirds are found most often on the rocks and islands of Sakonnet, Easton's, and Sachuest Points; along the rocky shores of Newport, Beavertail, and Narragansett; on the Point Judith and Block Island breakwaters; on rocky sections of the South Shore at Matunuck and Weekapaug; and on the rocky flats at the end of Napatree Point. This is the hardiest of all the Arctic-breeding shorebirds of eastern North America.

The earliest transient date is for 2, Napatree Point, 2 September 2009 (RTL). There are a few other September (see "Extreme dates" below) and October records, but general arrival is rarely before mid-November. Maximum winter counts rarely exceed 200 and usual counts are 10–50 per day.

Maxima. 500, Newport, 24 February 1960 (JBa); 330, Conanicut Island CBC, 28 December 2002 (ChP, CaP et al.); 294, Sachuest Point, 14 December 2008 (L. Parmelee). Maximum counts at Block Island have not exceeded 25 birds.

Purple Sandpipers are rare away from the rocks and jetties of the outermost coast. Birds collected 7 November 1899 at Bullocks Point, Riverside (in RWPM), and at Hope Island 15 November 1902 (LKi; now in NBS collection), and also 8 seen at Bristol (Colt State Park) 20 November 2002 (DFi) and 8 at Colt State Park 22 November 2003 (DFi), suggest overland arrival in fall. Several recent records in Narragansett Bay suggest overland departure in spring: 10 at Despair Island 28 May 1997 (RLF, JEM) and 9 moving between Hope Island and Despair 26 May 1998 (CRa, RLF, JEM). There are several records for Rose Island including 2 there 6 May 1999 (JEM). The only winter record in Narragansett Bay is for 1 at the south end of Prudence Island 1 January 1946 (WBD). They have been noted along the shores of the Sakonnet Passage along the Middletown shore north of Third Beach. Birds have been noted at least once feeding on the flats with Dunlins in such atypical locations as the Great Salt Pond at Jerusalem (DLK, RLF), and they frequently feed on the flats in the "Kitchen" at Napatree Point.

Extreme dates. 2 September 2009, Napatree Point, and 13 September 1898, 1 shot at Cormorant Rock (ESt, in RWPM), to 14 June 2000, Newport (RLF, CRa), and 16 June 1986, Napatree Point (CRa).

Baird's Sandpiper (*Calidris bairdii*)

Status

Uncommon fall migrant.

Nonbreeding

Baird's Sandpipers prefer the grassy margins of freshwater puddles, golf courses, turf farms, wrack lines, and sometimes open sand flats. Virtually all individuals reported are juveniles.

Fall. From 1 to 8 birds have been reported annually in most years since the 1940s; a major "flight" in 1974 produced reports of 36 birds. The earliest records include 1 at Briggs Marsh 18 July 1982 (RBo) and 1 at Ninigret 29 July 1993 (DLK et al.). Both these birds must have been adults, as juveniles would not have yet fledged. Other individuals have been seen at Tiverton 1 August 1957 (RBo) and Trustom Pond 1 August 1976 (DLK).

Fall maxima. Eight, Richmond, 29 August 1983 (GLB); 7, South County, 11 September 1967 (DLK); 6, Quonochontaug, 6 August 1974 (DWa); 5, Easton Pond, 21 August 1990 (CRa, REn). They are regularly reported into mid-October but are rare after that.

Away from the outer shore, Baird's Sandpipers are relatively rare. They have been found on saltwater shores away from the outer coastline at Barrington, Island Park, Nonquit Pond, and Seapowet. The relatively recent development of turf farming in South County has created a wealth of suitable habitat, especially in the immense tracts along Heaton Orchard Road in Richmond. Recent maxima there include 8, 29 August 1983 (see above), and 7, 30 August 1981 (KKi, FPe, TAd, STs). Seven were found on a West Kingston turf farm 30 August 1990 (CRa). Thus, given suitable ground on which to alight, Baird's Sandpipers seem equally common at inland locations. Other records at nonturf farms include 1, Diamond Hill Reservoir, 14 September 2003 (MLy); 1, Woonsocket Reservoir, 11 September 2005 (MLy).

Extreme dates. 16–25 July 2003, Charlestown Breachway (adult; RFa, DFi et al.), and 18 July 1982, Briggs Marsh (adult; RBo), to 11 November 1997, Seapowet (DSp).

Little Stint (*Calidris minuta*)

Status

Vagrant. One record.

Nonbreeding

A Little Stint was found and photographed 4 July 2012 at the Charlestown Breachway (CPe; see Photo 10). A second was found at the same location 4–7 July 2020 (Tim Metcalf, m.obs.; eBird). There are about 15 records for the east coast of the United States (Mitra et al. 2021).

Least Sandpiper (*Calidris minutilla*)

Status

Very common to abundant migrant coastally, generally preferring the grassy portions of salt marshes to open mudflats.

Nonbreeding

Spring. Peak migration is generally in early May, earlier than for Semipalmated Sandpipers, and flocks of 10 to 20 are often seen in mid-April.

Spring maxima. 700, Quicksand Pond, 14 May 1988 (DLE); 600, Cards Pond, 13 May 1956 (DLK).

Summer. Least Sandpipers are rare after about 20 May, and they are less common in June than Semipalmated Sandpipers. However, southbound adults often appear in numbers during the last week of June; e.g., 5, Ninigret, 27 June 1993 (PAB, DLK et al.); 55, Quicksand Pond, 30 June 1989 (CRa, RBo); 95, Ninigret Pond, 30 June 2001 (STs).

Fall. The fall migration is more protracted, and smaller flocks continue into early September.

Fall maxima. 850, Ninigret Pond, 10–18 July 1998 (STs); 715, Charlestown Breachway, 9 July 2005 (STs, CPe et al.); 700, Ninigret Pond, 12 July 2003 (STs); 600, Quonochontaug, 9 July 1937 (HSH); 430, Seekonk River, 14 August 2002 (MLy).

Inland. Least Sandpipers occur more frequently in the interior than do Semipalmateds, particularly in spring.

Maxima inland. 118, Diamond Hill Reservoir, 26 August 2006 (MLy); 84, Woonsocket Reservoir, 7 September 2002 (MLy); 80, Richmond turf fields, 16–17 August 1999 (CRa; SSM); 75, Richmond, 6 September 2000 (STs); 65, Arnold Mills Reservoir, 15 August 1993 (DMK).

Extreme dates. 23 March 1990, Trustom Pond and Briggs Marsh (MJM, DLK), and 30 March 1957 (SSD), to 14 December 1946, Jerusalem (DLK), and 26–30 January 2013, Cards Pond.

White-rumped Sandpiper (*Calidris fuscicollis*)

Status

Uncommon spring and fairly common fall migrant in small numbers at coastal locations; very rare inland.

Historical

White-rumped Sandpipers were unknown in spring and uncommon in fall during the shooting era, but since 1920 have greatly increased. They are now regularly found among other "peeps" on saline mudflats and in salt marshes in both spring and fall.

Nonbreeding

Spring. The peak of migration is in late May, and White-rumpeds are rare before early May.

Spring maxima. 75, Quonochontaug, 23 May 1971 during a great flight of other shorebirds (CWo, SSD, LBe, REm); 40, Quonochontaug, 9–13 May 1978 (RAC et al.); 36, Napatree Point, 4 June 1990 (CRa); 28, Napatree, 3 June 2004 (CRa); 20, Galilee, 11 June 1960 (SSD). Most spring counts are for fewer than 10–15 birds per day per locality.

Summer. Usually only single birds are seen during June, but an unusually large number was present in June 1988, when a maximum of 44 birds was present at Quicksand Pond 12 June; 20 were still there on the 16th (JO, REm).

Fall. This is one of the later migrating shorebirds; numbers often do not appear until late August. They occur regularly through October and early November. There are numerous records through mid-November; e.g., 10 at Gardiners Pond, 11 November 1957 (JBa), and a flock of 12 appeared at Portsmouth 28 November 1986 (DLE et al.). A few single birds have appeared in late June and early July; these may be early migrants or failed or nonbreeders.

Fall maxima. 100, Briggs Marsh, 4 September 1995 (JO); 75, Briggs Marsh, 12 September 1950 (RBo); 58, Napatree Point, 20 October 1989 (CRa); 40, Briggs Marsh, 26 August 1995 (JO); 40, Little Compton, 8 September 1946 (DLE).

Inland. 43, Heaton Orchard Road, 30 August 2005 (PAB).

Extreme dates. 27 April 1958, Galilee (DLK, CWo et al.), and 28 April 2007, Quonochontaug (DFi), to 23 December 1950, Newport–Westport CBC.

Buff-breasted Sandpiper (*Calidris subruficollis*)

Status

Uncommon fall migrant; one spring record.

Nonbreeding

Buff-breasted Sandpipers usually alight and feed in grassy areas and harvested potato fields, as well as on turf farms, during a very restricted period in early fall. They have been observed most frequently on potato fields near Moonstone Beach and South Kingstown and also on the golf course at Misquamicut. They are less often seen in locations lacking extensive fields, such as Seapowet, Quicksand Pond, Briggs Marsh, and Napatree Point. There are only 3 records from Block Island (1–2 October 1982 [ASRI]; 28–30 September 1984 [ASRI]; 1, 7 August 1991 [CRa]). This species became exceedingly rare during the shooting era and remained rare for many years, but there has been a considerable increase in numbers of birds seen in Rhode Island since the 1960s.

Fall. Buff-breasted Sandpipers arrive late in fall for a sandpiper, probably because most individuals that put down on the Atlantic coast are juveniles, which migrate later than adults. There are 2 records of birds that must be adults based on date: 1 shot at Little Compton on 22 July 1906 by C. B. Clarke (fide HSH), and another seen at Briggs Marsh 25 July 1987 (DLE).

Fall maxima. 27, Richmond turf farms, 29 August 1983 (GLB, RAC); 14, Matunuck, 12 September 1971 (DLK, ESa); 14, Matunuck, 16 September 1977 (ESa, CWo, DWa); 8, Briggs Marsh, 10 September 2002 (JSJ).

Extreme dates. 19 May 1986, Quonochontaug, 1 of few spring records from the Northeast, impeccably described by RGe and JGe; to 5 November 2008, Colt State Park (DFi).

Pectoral Sandpiper (*Calidris melanotos*)

Status

Uncommon spring and fairly common fall transient to grassy portions of marshes, flooded fields, and other open coastal habitats; occasional inland. Numbers fluctuate widely from year to year.

Historical

The "Creaker" of the old-time gunners, Howe and Sturtevant (1899) referred to Pectoral Sandpipers as "abundant migrants," a status contrasting with the then-reduced numbers of most other shorebirds. They were, however, rare in spring at that time. A sense of their abundance in fall can be obtained from Howe (1901e) of 2337 shot in the Middletown area by a single market gunner (W. A. Dring) during the period 1867–1874, with totals of up to 668 per year.

Nonbreeding

Spring. Pectoral Sandpipers are among the earliest migrant shorebirds in spring, often arriving in mid-March. A count of 12 on 23 March 1978 at Jerusalem (STs, DRo, JMu) was early for so many. It is more regular as a transient in early to mid-April.

Spring maxima. 50, Galilee, 13 April 1970 (A. Schroeder); 45, Cards Pond, 2 May 1981 (RAC, DRo); 30, Weekapaug, 15 April 1981 (GLB); 30, Galilee, 29 April 1981 (DLK).

Summer. Rare in June, individuals have been seen at Seapowet on 3 June 1973 (DLE) and Warwick on 7 June 1967 (DWa). One remained at Quonochontaug 5–22 June 1978 (v.o.); 1 was at Potter Pond 20 June 1995 (JCB), and 1 at Galilee 28 June 1959 (DLK).

Fall. Pectoral Sandpipers have a very protracted fall migration and may appear at any time from mid-July to late November. Larger than average numbers are sometimes found during storms and severe rainfall that creates sizable puddles on grasslands.

Fall maxima. 150 at Block Island after a 2-day northeast storm, 16 September 1933 (EDi); 110 with a large number of American Golden-Plovers at Napatree Point after a southerly rainstorm, 5 September 1991 (CRa, RWa); 60, Quonochontaug, 14 September 1913 (HSH et al.); 58, Newport, 14 August 1955 (WHD); 50, Galilee, 30 July 1960 after Tropical Storm Brenda (DLK).

Inland. Prior to the late 1970s, there were only 3 records at interior locations, all in fall: 1 shot at Hammond's Pond, Pawtucket, 18 August 1894 (H. Cash and HSH; extant); 6 in Lonsdale 15 October 1934 (AEE); and 15 at Smithfield 17 September 1949 (DLK). Recently, there has been a considerable increase in inland records, in part reflecting increased coverage: 1, Great Swamp, 26 July 2004 (SSM); 15, Cumberland, 12 November 1993 (GV); 11, Slocum, 16 September 2005 (PAB); 8, Arnold Mills Reservoir, Cumberland, 20 October 2007 (MLy). Less common in spring: 2, Great Swamp, 10 April 1991 (DFi et al.); 12, Cumberland, 8 May 1982 (REn).

Extreme dates. 15 March 1981, Quicksand Pond (DLE, REm); to 3 December 1972, Matunuck (RLF), 17 December 1950, Seapowet (RBo), and 5 January 2010, Goosewing Beach (GDe; photos).

Semipalmated Sandpiper (*Calidris pusilla*)

Status

Abundant spring and fall migrant through coastal areas, very rarely alighting in the interior. Counts in Rhode Island have only rarely approached numbers regularly recorded elsewhere in southern New England or Long Island, probably because of the relative scarcity of extensive mudflats.

Nonbreeding

Spring. The usual period of spring migration is from late April through the first few days of June, later than Least Sandpipers. Peak migration is about 12–31 May or slightly later.

Spring maxima. 5000, Cards Pond, 29 May 1973, when the pond was artificially drained (DWa, FPe, KKi); 1500 at Galilee, 30 May 1954 (DLK, RBo); 1000+, Quonochontaug, 24 May 1971 (KKi); 800+, Ninigret, 27 May 1984 (RAC); 550, Napatree Point, 2 June 1982 (CRa, RLF).

Summer. 40, Galilee, 25 June 1954 (DLK); 24, Napatree Point, 26 June 1986 (CRa). There are numerous records of smaller flocks in late June, many undoubtedly nonbreeding immatures.

Fall. Adult numbers peak in late July, juveniles in late August to early September.

Fall maxima. 1950 at Quicksand Pond, 28 July 1990 (RBo); 1500, Quicksand Pond, 6 August 1990 (RBo, DLE); 1500, Briggs Marsh, 25 August 2002 (DLE, REm); 1500, Charlestown Breachway, 22 July 2006 (STs). Migration tapers off in late August but continues through September and occasionally October, with scattered reports even in early November (see Phillips 1975).

Inland. 430, Seekonk River, 14 August 2002 (MLy); 9 shot at Hammond's Pond, Pawtucket, 6–9 September 1892; 5 shot from a large flock at the same location, 18 August 1894 (H. Cash); 19, Arnold Mills Reservoir, 12 August 1991 (DMK); 13, Diamond Hill Reservoir, Cumberland, 27 August 2006 (MLy); 2, Wood River Junction turf farm, 20 October 2007 (STs).

Extreme dates. Early May (earlier records prone to misidentification of Least Sandpipers) to 16 November 1986 (DFi, DLK); 17 November 1956, Tiverton (2; SSD); 25 November 1950, Warren's Point (1; RBo); and 22 December 1951, Middletown (1; WHD).

Western Sandpiper (*Calidris mauri*)

Status

Uncommon fall migrant on coastal mudflats; very rare in spring.

Nonbreeding

This species of the northwestern Arctic usually consorts with Semipalmated Sandpipers, which it resembles relatively closely, but it often feeds in deeper water than the latter species and occurs later in fall.

Spring. Most have appeared in late May and early June, but earlier records are anticipated given that Western Sandpipers regularly winter in New Jersey and appear to be extending their winter range northward. Records are all of single birds: 28–29 May 1984, Napatree Point (RAC; CRa); 5 June 1983, Napatree Point (RBu, RAC, HWi); 6–11 June 1984, Napatree Point (CRa et al.). Exceptional was 1 closely observed at Trustom Pond on 5 April 1952 (DLK), a remarkable date (but see above about northward range extension).

Fall. The earliest fall records are for individuals at Ninigret Pond, 1 July 1991 (PAB); Napatree Point, 1 July 2004 (CRa); Sachuest Point, 1 July 2006 (RTL); 4, Ninigret, 9 July 1994 (PAB); and 8 appeared at Block Island as early as 23 July 1976 (SRD). Peak migration, largely of immatures, is from late August to late September.

Fall maxima. 25, Quicksand Pond, 31 August 1974 (REm, DLE); 21, Quonochontaug, 15 September 1975 (DWa); 20, Little Compton, 2 September 1960 (RLF); 20, Ninigret Pond, 14 August 2004 (STs); 17, Easton Pond, 6 October 1974 (RLF); 16, Briggs Marsh, 4 September 1995 (REm); 15, Block Island, 13 September 1960 (JBa);

8, Charlestown Breachway, 26 August 2015 (DFi); 7, Block Island, 12 October 1964 (RTL).

Inland. One, Heaton Orchard Road, 30 August 2005 (PAB); and 1, Switch Road, 8 September 2005 (JSJ).

Extreme dates. 5 April (see above) to 5 November 1950, Little Compton (DLE); 5 November 1983, Napatree Point (CRa); 6 November 1948, Little Compton (DLE).

Short-billed Dowitcher (*Limnodromus griseus*)

Status

Common to abundant migrant.

Nonbreeding

Spring. The average arrival of Short-billed Dowitchers is during the first week of May, and peak numbers occur about 20 May. There are now many records from the last half of April, but this was not the case before about 1970. Twenty-four were at Quicksand Pond 26 April 1980 (RBo).

Spring maxima. 1500+ resting on the Quonochontaug marshes, 24 May 1971 (KKi); 500, Quonochontaug, 22 May 1976 (DLK et al.); 300, Napatree Point, 20 May 1980 (MPe).

Fall. Short-billed Dowitchers continue as common transients in smaller numbers through late September, and there are occasional reports of dowitchers (some specifically identified as Short-billeds) to mid-November. Remarkably, there were no reports of this common coastal species at any inland location until 15 July 1996, when 35 were noted flying over the Great Swamp (CRa).

Fall maxima. 1000, Ninigret, 9 July 1987 (DLK, DFi); 1000, Ninigret, 9 July 1990 (DFi, DLK); 500, Napatree Point, 23 July 2008 (CRa); 420, Briggs Marsh, 13 July 1972 (DLE); 420, Ninigret, 13 July 1992 (PAB).

Extreme dates. 8 April 1991, Quonochontaug (5; CHa), to 2 November 1986, Quonochontaug (7; CHa), and 11 November 1987, Quonochontaug (30; CHa). One dowitcher at Napatree Point 22–29 November 1993 was not identified to species (CRa).

Long-billed Dowitcher (*Limnodromus scolopaceus*)

Status

Uncommon fall migrant; rare in spring and early winter, occasionally survives the winter.

Nonbreeding

Fall. Most occur during late August–November. There are 2 specimens: 1 collected by C. B. Clarke on 28 September 1908 at Point Judith and obtained by HSH (extant and recently examined), and another shot by LKi on 16 September 1913 at Point Judith, in the NBS collection. Occasional adults have appeared in July, including singles on 14 July 2000 at Ninigret Pond (SSM, PAB), 18 July 1947 at Trustom Pond (DLK), 21 July 1990 at Briggs (RBo), and 24 July 2012 at Charlestown Breachway (DFi). There is 1 record of this dowitcher inland: Wood River Junction, 25 October 1947 (DLK).

Fall maxima. Eight, Quonochontaug, 22 September 1988 (DLK).

Winter. Recently, Long-billed Dowitchers appeared during winter on Long Island and eventually New England. The first such birds included 1 carefully observed at Galilee on 12 December 1971 (DLK) and possibly the same individual rediscovered at Quonochontaug, 26–27 December 1971 (ESa, DLK, FPe, RSt). On 25 November 1979, a flock of 4 appeared at Weekapaug Inlet (ESa), and 3 of those continued through the winter to 11 May 1980 (ESa, DLK, KKi, FPe, RLF et al.); also, 1, Weekapaug the following winter, 6 November 1980–12 April 1981 (ESa et al.). There are also 3 CBC records: 1, Galilee, 27 December 1998 (DLK et al.) continuing through 18 January 1999 (STs); possibly the same or another at Jerusalem, 21 and 28 March 1999 (STs) and again 8 April 1999 (PAB); 2, Newport–Westport CBC (Briggs Marsh), 17 December 2005 (REm, GLB et al.); 1, Block Island CBC, 22 December 2005 (CRa et al.).

Spring. An individual at Seapowet on 15 March 1953 that "had quite a long bill" very likely belongs here, as do observations of dowitchers reported as this species from Jerusalem on 7 April 1984 (RAC), Seapowet on 14 April 1973 (RBo), and Jerusalem on 29 April 1973 (ASM, CWo, DLK et al.). There was also an unidentified dowitcher (a flyover) from Moonstone 17 March 2007 (CNu) and a Long-billed at Napatree 8–11 May 2004 (CRa).

Eurasian Woodcock (*Scolopax rusticola*)

Status

Vagrant. One record.

Nonbreeding

There is 1 oft-cited record in Baird et al. (1884), accepted by C. J. Maynard (1896) but doubted by G. M. Allen (1901). Quoting Baird et al.: "Mr. George N. Lawrence cites an instance in 1866 where a friend of his shot, near Newport, Rhode Island, a large woodcock which weighed fourteen ounces." The fact that American Woodcocks rarely reach, and never exceed, 9 ounces (255 grams), while the mean weight of the

European bird is 14 ounces (397 grams), naturally suggests that the specimen was *rusticola*. Allen (1901) wrote: "[A]s the identification is based alone on weight, this species must be included as a Rhode Island bird with some misgivings." Mendall and Aldous (1943) listed 260 grams (9.17 ounces) as their greatest recorded weight for the American Woodcock in fall. There are, however, a number of well-established records of Eurasian Woodcock for the eastern United States, the most recent of which was in Cape May, New Jersey, in 1956 (Boyle 2011).

American Woodcock (*Scolopax minor*)

Status

Fairly common resident and migrant. Rare in winter.

Historical

Sir Charles Blagden (1903), in the era 1776–1780, declared that "there are Woodcock's nests in woods of most of the islands round Rhode Island," for example, islands in Narragansett Bay. By 1899, however, Howe and Sturtevant described it as "a formerly common summer resident, now becoming almost rare as a breeding bird." This decline, coincident with an all-time high in the amount of local as well as regional breeding habitat, was almost certainly due to year-round hunting pressure on the Rhode Island breeding grounds. Hathaway took considerable interest in the known breeding locations of this species. He knew of very few in the late-nineteenth century (among them Warwick and Cranston), and CED found it nesting in Bristol in 1891. By 1920, however, and especially by the early 1930s, coincident with protection during the breeding season, woodcock had recovered and vastly increased as Rhode Island breeders. By the period 1934–1942, HSH knew of more than 35 locations where they were breeding and thought that they had become common and widespread summer residents. A high count of territorial birds just after this era was for 35 males on Prudence Island (mainly near the south end) on 29 April 1946 (WBD).

Breeding

Block Island has had one of the highest densities during the breeding season in recent years, a status probably enhanced by the lack of ground predators (an estimated 200 pairs in 1978 [JEM, CCA]; 42 were recorded on 11 May 1999 [SCo]). Even at Block Island, an evidently downward change in numbers has been in progress in recent years; the decline is inversely proportional to the increasing size of the island's vegetation (CRa, REn). As in Blagden's time, woodcock breed on all of the larger islands in Narragansett Bay and have even been present during the breeding season on such tiny outposts as 110-acre (45-ha) Dutch Island (2 nests in 1978; CCA) and 95-acre (38-ha) Hope Island (June probings, 1978; RLF).

Nonbreeding

The largest kills known here were made by the market hunter R. A. Dring, who shot 126 and 276 in the Newport area in the 2 years 1872–1873, despite his main attention to mudflat shorebirds.

Spring. Recent counts include 30, northern Cumberland, 7 March 2002 (MLy); 14, Ninigret NWR, 27 April 1988 (LAH); 13, Trustom Pond, 29 April 1988 (LAH).

Fall. Wandering birds appear where they do not breed in August. Peak movement occurs in late October and first half of November (to 50 per day in November 1932 in "Cherry Valley," West Glocester, according to hunters in litt. to HSH).

Winter. There are several January and February records of single birds, and overwintering survival is likely in milder winters.

Winter maxima. Eight, Newport–Westport CBC, 16 December 2006; 6, South Kingstown CBC, both 2011 and 2012; 5, Great Swamp, 21 December 1991 (REn, CRa); 3 displaying at dusk on Sachuest Point, 26 January 1982 (GLB et al.); singles at Foster, 22 January 1980 (Spencer), and Limerock, 2 February 1991 (REn).

Wilson's Snipe (*Gallinago delicata*)

Status

Common migrant and fairly common winter resident.

Historical

In the early shooting era, snipes were killed in remarkable numbers, both spring and fall, and the species would seem to have been several times more numerous than in the modern period. The shooting records of the market gunner Robert A. Dring list a total of 1586 killed by him in the Newport region between 1867 and 1874, 509 of them in fall. In his best spring season in 1873, he shot 349 between 27 March and 28 April. H. W. Hare Powel (1899) recalled that his uncle John Hare Powel's average fall kills on Aquidneck Island in the 1870s and 1880s ranged between 400 and 450 per year.

Nonbreeding

Fall. General fall arrival is in late August or September, but there are many earlier records. Some of these may be of nonbreeding summering birds. The earliest of these was at Briggs Marsh, 16 July 1941 (DLE).

Fall maxima. 120, Easton Pond, 13 November 1955, when the water was artificially lowered (JCB, DLK, CWo, SSD, RLF et al.); counts at other locations have rarely exceeded 30.

Winter. Snipe are common in the more open parts of southern Rhode Island in December and commonly linger into winter in small groups in wet pastures and along streams, even flushing from ponds and watercourses in wooded areas. They appear to survive the season regularly, although data from the late-winter period are fewer than for the pre-Christmas period, and the extent to which they remain in such upland situations through the late-winter period is poorly known. Nearly all of the winter records come from milder locations in southern sections. One of the best locations to see numbers in midwinter has been at Green End Pond, Middletown.

Winter maxima. 33, Middletown, 20 December 1979 (v.o.); 30, Green End Pond, 7 December 1979 (DFi).

Spring. Peak migration occurs in mid-April.

Spring maxima. 36, Trustom Pond, 2 April 1990 (MJM); 34, Sisson Pond, 17 April 1999 (REm); 33, Moonstone, 6 April 2002 (SSM); 27, Great Swamp, 8 April 2012 (PLE); 23, Jamestown, 24 March 2007 (CNu). Two later records are Moonstone Brook, 24 May 1973 (DWa), and another flushed from the dunes at Charlestown Beach 9 June 1974 (RLF).

Prospecting and Breeding

Snipe have increased as breeders in central and western Massachusetts in modern times and might be expected to occupy cattail or boggy habitats in northern sections of Rhode Island. During early morning in the Great Swamp 27–30 April 2002 (SSM, PJL), a snipe was observed giving its "booming" display flight and also a series of rail-like "kek-kek-kek" notes on the descent from its display. One was also "winnowing" predawn at the Great Swamp on 13 May 2006 during an ASRI "birdathon" (REm, PLE et al.).

These behaviors certainly constitute prospecting and suggests that breeding may occur in the future. There is a questionable, though plausible, record of a nest, eggs, and 2 birds alleged to have been collected at Little Compton on 10 May 1870 by Dr. Henry F. Marshall, which seems yet more plausible in light of the above observations.

Terek Sandpiper (*Xenus cinereus*)

Status

Vagrant. One record.

Nonbreeding

A Terek Sandpiper was found at Napatree Point 28 June 2020 (JSJ) where it remained through 1 July (see Photo 11). This is the third record for eastern North America, with others from Massachusetts and Virginia (eBird; Mitra et al. 2021).

Spotted Sandpiper (*Actitis macularius*)

Status

Uncommon breeder and fairly common migrant.

Historical

Spotted Sandpipers were substantially more numerous before 1900 than they are now, and the decline seems in part related to changes in agricultural practices in Rhode Island, especially the decline of cattle and sheep farming. Bare ground and shortgrass habitats once maintained by grazing animals are now covered by dense brush and *Phragmites*. For instance, Talbot (1884b), working Warwick Neck in 1882–1883, described Spotted Sandpipers as "everywhere abundant . . . nesting in great numbers along these shores." Charles Doe's egg collection contains 10 clutches from Warwick as well as two from Green Island off Edgewood. William C. Rives called them very common in 1884 in the Ocean Drive area. Hathaway (1900b), in studying lower Little Compton 2–6 June 1900, called them "everywhere abundant along the shore, especially on Sakonnet Point, where several sets of eggs were found in the grass back of the beach."

Breeding

Spotted Sandpipers are now uncommon to rare summer residents about the larger ponds, reservoirs, rivers, and marshes of the interior and on barrier beaches, coastal ponds, Narragansett Bay islands, and elsewhere along saltwater shores. There are probably fewer than 30 pairs statewide. Spotted Sandpipers have been absent along Ocean Drive and at Sakonnet Point, although they still breed at Quicksand Pond and Briggs Marsh (DLE, REm). At Napatree Point, CRa (in litt. to DLE, 1994) stated that the species had declined drastically in the previous decade from perhaps 7 pairs to 1. At Charlestown Beach, DLK found 5 pairs in 1947; there have been none reported for years.

Spotted Sandpipers have bred on some of the smaller, well-vegetated islands in Narragansett Bay that lack most ground predators; e.g., Hope, Rose, Dutch (2 pairs in 1981), Big Gould, and Fox. Even on these islands, observations declined through the period 1970–1999, and even there the species may now be absent (RLF, JEM). Along the mainland shores of Narragansett Bay and the margins of the larger Narragansett Bay islands, all indications suggest that they are vastly less numerous as breeders than in former times and may now be either very rare or absent. Inland, CED collected 2 clutches in West Gloucester in 1915–1917, but information on nesting status in the interior is limited, and fieldwork since the 1980s suggests they are absent from acidic upland ponds of western Rhode Island.

Nonbreeding

Spotted Sandpipers are still fairly common as migrants in Rhode Island.

Spring. General arrival is in early May, with transients to the end of May or later. The largest numbers occur at Napatree Point.

Spring maxima. 13, Napatree Point, 26 May 1982 (GLB).

Fall. They become relatively uncommon by mid-September and still scarcer by late September. There are occasional records through October, in 1 instance with individuals at 3 separate locations, all on 14 October 1990 (v.o.). There are more than a dozen recent Napatree Point counts for 15–30 birds during July (CRa, DLK).

Fall maxima. 38, Napatree Point, 7 August 1989 (CRa); 33, Napatree Point, 22 July 2010 (CRa); 25, southern Aquidneck Island, 25 July 1970 (RTL); 15, Block Island, 7–9 August 1991 (REn, CRa); 16, East Providence Reservoir, 23–24 July 1953; and 15 there, 9 September 1953 (RLF). There are only 2 counts of 20+ birds since 2005.

Extreme dates. 13 April 1978, Charlestown (DWa), to 2 January 2003, Seekonk River (1; RFa, DFi), and 5 January 2013, Beavertail (JMa, CPe). Rare after October. One at Jamestown 26 February 2012 (ChP, CaP) suggests overwintering.

Solitary Sandpiper (*Tringa solitaria*)

Status

Fairly common spring and common fall migrant.

Nonbreeding

Spring. Solitary Sandpipers favor freshwater marshes and farm ponds. Peak numbers appear late April to mid-May.

Spring maxima. Five in an Exeter farm pond, 14 May 1994 (RLF); 4, York Pond, Providence, 17 May 1977 (v.o.).

Fall. More numerous in fall, Solitary Sandpipers have appeared at Little Compton 7 July 2005 (RFa), at the Great Swamp 9 July 1973 (DWa), and Galilee 12 July 1957 (v.o.); most appear in late August and early September.

Fall maxima. 14, Block Island, 9 September 1960 (JBa; ASRI); 12, Metcalf Mills, Providence, 22 August 1946 (WBD); 8, South County, 3 August 1965 (DLK); 8, Scituate, 23 October 1949 (WBD), a remarkable number for so late.

Extreme dates. 25 March 2000, Burdickville (STs, DFi), and 12 April 1981, South Kingstown (RAC), to 30 November 1993, Lonsdale (REn).

Lesser Yellowlegs (*Tringa flavipes*)

Status

Uncommon spring and common fall migrant; very rare in winter. Uncommon inland.

Nonbreeding

Spring. Numbers are quite variable from year to year, in part reflecting suitability of local habitats. Lesser Yellowlegs are more inclined to use freshwater habitats and grassy marshes than are Greaters. Peak counts of up to about 15 birds per day occur from mid-April to early May.

Spring maxima. 60, Cards Pond, 2 May 1981 (RAC, DRu); 6, Matunuck, 30 March 1968 (DLK, CWo) is early for so many. There are scattered records for small numbers through mid to late May. They are rare in early June, and the latest records of migrants or lingerers are for individuals at Briggs on 10 June 1934 (RCC) and 11 June 1939 (DLE).

Fall. Lesser Yellowlegs are one of the earliest and most conspicuous migrant shorebirds in summer, sometimes (presumably nonbreeders and failed breeders) arriving back from their boreal breeding grounds in late June (8, Quonochontaug, 25 June 1977; ESa, CWo). A more protracted migration of juveniles passes through from early August through September. Migrants regularly appear into late October, occasionally November, and there are several December records.

Fall maxima. 350 between Quonochontaug and Jerusalem, 9 August 1982 (DLK); 300 flying southwest over Galilee, 28 July 1942 (DLK); 250, Quonochontaug, 10 August 1933 (HSH); 250 flying past Quonochontaug, 30 August 1963 (DLK); 210, Briggs Marsh, 4 September 2012 (GDe).

Inland. Lessers are uncommon at inland locations but have been occasionally numerous when conditions are favorable; e.g., 26, Wood River Junction, 4 September 2003 (JSJ); 23, Heaton Orchard Road, 30 August 2005 (PAB); 18, Diamond Hill Reservoir, Cumberland, 18 August 2007 (MLy); 20, Sisson Pond, 8 September 2007 (WM); 20, Scituate Reservoir, 23 October 1949 (WBD).

Extreme dates. 27 February 2008, Quicksand Pond (GDe), and 12 March 1986, Quonochontaug (DFi, DLK), to 4 January 1981, Weekapaug (DLK et al.).

Willet (*Tringa semipalmata*)

Status

Common and increasing breeder; fairly common spring and fall transient. Rare in winter.

Subspecies

Two subspecies occur in Rhode Island: "Eastern Willet" *T. s. semipalmata* breeds at the immediate coast, mainly in salt marshes, whereas "Western Willet" *T. s. inornata* occurs as a migrant in fall on beaches and tidal flats. As the 2 forms can be distinguished in the field, and indeed are likely to be separate species (Oswald et al. 2016), the status of each is treated separately below.

Eastern Willet (*Tringa (semipalmata) semipalmata*)

Historical

Willets bred in salt marshes along the Atlantic coast in colonial times but were extirpated from New England by about 1875 (Forbush 1929). At that time, market gunning had formed a gap in the breeding distribution between Nova Scotia and New Jersey. Gradually, this gap was filled on Long Island (1966; Bull 1976), Connecticut (1978; Zeranski and Baptist 1990), and eastern Massachusetts (1976; Veit and Petersen 1993). By the mid-1970s, Willets were common summer residents in the salt marshes of southern Long Island with 75–100 pairs nesting by 1975 (PAB, pers. comm.). Rhode Island, with its less-extensive salt marshes, was among the last parts of the Atlantic coast to be colonized by breeding Willets, which were confirmed to be breeding in 1987.

Howe and Sturtevant (1899) called it rare but gave no specific records. Hathaway (1913) discussed the relationship of the two subspecies here and stated that he had neither seen an Eastern Willet nor knew of any in Rhode Island collections. Yet, on 8–9 May 1913, 2 were shot at Block Island from a flock of 10, and another flock of 26 was seen on 19 May 1913 at Block Island (EDi). There were no additional records until the early 1930s, after which flocks appeared irregularly on the outer coast and, much more rarely, in Narragansett Bay, for example, one at Prudence Island 6 June 1946 (WBD) and another there 28 June 1983 (CRa, JEM). Reports of prospecting birds in summer began increasing during the early 1960s, for example, one at Briggs Marsh 24 June 1961 (REW), another there 28–29 June 1969 (ESa), one at Quonochontaug 25 June 1974 (DLK), and two there 22 June 1978 (DLK, KKi, FPe). During the 1970s, Willets were observed several times during June performing territorial displays at various South Shore locations (e.g., one calling over Potter Pond 2 June 1979; RLF), and three nonbreeding pairs were present at Napatree Point through June 1983 (CRa).

Unusually large numbers of Willets occurred during the falls of 1912 and 1913. In 1912, HSH saw 18 during August at Quonochontaug, and I. R. Sheldon saw 25 on 23 August at Point Judith. Hathaway saw 15 on 10 August 1913 at Quonochontaug, and 17 were shot along the South Shore in early August of the same year, but no specimens are extant. Given the decline of *semipalmata* during the nineteenth century, it seems more likely these were *inornata*; however, the rather early fall dates suggest *semipalmata*.

Breeding

An enumeration of the colonization of breeding sites in Rhode Island appears below:

Hog Island. A displaying bird was seen near the tern colony at the western marsh on Hog Island 4 June 1985 (RLF, CRa), and the first breeding record for the state was established here 11 June 1987, when a bird was accidentally flushed from 4 eggs (JEM). A pair was present, and a bird flushed from eggs at the same location, on 8 June 1988 (RLF, JEM et al.). None were present 1989–1991, but 1 pair was seen in 1992 (JEM), none in 1993–1994, and 1 was flushed from 4 eggs on 1 June 1998 (RLF, CRa; photo).

Jerusalem/Potter Pond. Willets were first seen displaying over the Potter Pond marsh 2 June 1979 (RLF), but they did not breed. A pair was present all summer in 1986, and 4 territorial birds were present in the vicinity of the tern colony on Potter Pond 19 June 1987 (JEM). Three were present on the Jerusalem marsh 7–28 June 1988 (RLF). On 15 July 1989, a pair of adults and 2 half-grown young were present near the Jerusalem flats (RAC). Two pairs occupied the Jerusalem (Succotash) marshes in 1990 (DLK). On 11 June 1991, 7 birds, mostly in pairs, were in violent contest over the marshes southeast of Potter Pond (RLF, JEM). Two pairs were at Potter Pond in 1992 (JEM). A nest and 5 eggs were observed on the marsh south of Potter Pond 18 May 1993 (JEM et al.).

Ninigret Pond. Displaying birds were noted in the 1980s, and breeding in the mid-1990s seems likely; e.g., 6 birds, including a territorial pair northeast of the breachway 29 June 1996 (JEM).

Quonochontaug. An excited adult was unexpectedly and closely flushed from under the marsh grass near the old channel 25 May 1991 (DLK et al.). A nest with 2 eggs was found 24 May 1994 (CWo).

Sandy Point Island. On 13 June 1990, 1 repeatedly dive-bombed the author on the southeast end of the island, and 3 additional birds were present. On 11 June 1991, a territorial pair was present at the same site (RLF, JEM). One was photographed incubating 4 eggs there 9 June 1994 (RLF, JEM). This island lacks any salt marsh.

Winnapaug. Willets apparently have occupied this extensive marsh later than some of the other South Shore marshes; territorial birds were present during the late 1990s, and 1 was flushed from 4 eggs along the driveway to a house in the central marsh 24 May 2000; 3 pairs were estimated to be nesting (RLF).

Seapowet marshes. An estimated 3–4 pairs annually since about 2005, a territorial pair in Peckham's Marsh (REm), and 2 territorial pairs in Peckham's Marsh (Fogland), 5 July 2013 (REm).

Since about 2000, the Willet population has continued to grow and expand, and they now breed at Block Island, Napatree Point, Prudence Island, and virtually all locations with suitable habitat.

Nonbreeding

Spring. Nova Scotia breeders may ordinarily pass by at sea in spring, given the irregularity of their occurrence and frequency of records on Block Island.

Spring maxima. An exceptional count of *150* at Jerusalem 11 May 1981 (GLB); 47, Block Island, 26 April 2014 (DFi); 37, Brenton Point, 1 May 2008 (RTL); 36, Galilee, 17 May 1958 (SSD); 35 flying past Point Judith, 1 May 1993 (RLF); 30, Block Island 6 May 1975 (ELa).

Fall. Most *semipalmata* depart by late August, so migrants from September on may be mostly *inornata*, but this deserves closer study. Before September, *inornata* are substantially outnumbered by *semipalmata*.

Fall maxima. 64, Napatree, 13 July 2014 (CRa); close observation of 2 significantly smaller and darker birds seen in direct comparison with a typically pallid Western Willet at Napatree Point, 6 September 1975 (RLF et al.).

Extreme dates. 15 April 1950, Trustom Pond (DLK). Late-fall dates for Eastern Willets are not determined; see below for late dates of Western Willets.

Western Willet (*Tringa (semipalmata) inornata*)

Nonbreeding

There is a record of presumed *inornata* for Warwick 29 July 1896. This specimen, from the Brewster collection (BMS #15487), is now in the MCZ. There was 1 in the old URI mounted collection, no longer extant, labeled Providence, 30 September 1880, without further data. Recent fall maxima have rarely exceeded 5 per day, and most occur from late August to early October. Late records include 1 at Napatree Point 18 October 1964 (ESa); 1, Ninigret, 20 October 1979 (TAd, KKi, FPe); 1, Weekapaug, 21 October 1956 (DLK). The only records later than these were of individuals that overwintered at the Jerusalem/East Matunuck/Potter Pond beach and marsh complex 21 November 1985 through 21 April 1986 (DFi, DLK et al.), and 1 at Block Island, 20 December 2012–17 March 2013 (SSM et al.).

Inland. There is an inland record of an individual of unknown race well seen in a plowed field in Exeter 12 May 1953 (JGe, RGe), and another seen and heard at the Great Swamp 2 May 2001 (SSM).

Spotted Redshank (*Tringa erythropus*)

Status

Vagrant. One record.

Nonbreeding

On 30 May 1955, D. Burger of Bristol found a blackish, yellowlegs-like shorebird in the Seapowet marshes in Tiverton. He called RCC, who correctly identified it from its description and sent RBo to investigate. That afternoon RBo easily found the bird, which was in full alternate plumage, and he was afforded excellent observational opportunities to eliminate the possibility of oiled or melanistic Greater Yellowlegs. It apparently disappeared in the night, judging by the inability of L. Griscom and a busload of Massachusetts birders to find it the next day.

Greater Yellowlegs (*Tringa melanoleuca*)

Status

Common spring and fall migrant. Nonbreeding birds occur uncommonly during summer; rare in winter.

Nonbreeding

Spring. Greater Yellowlegs are the more frequently encountered yellowlegs in spring and fall, though they are outnumbered in a few prime locations by the more aggregative Lesser Yellowlegs. Greaters are also hardier birds than Lessers, having peak counts later in fall and a larger number of winter records.

The general arrival is in mid-April, with peak numbers in late April and early May.

Spring maxima. 54, Jamestown, 26 April 1976 (MDa); 52, Fort Getty, 4 May 2006 (DFi); 47, Westerly, 27 April 1983 (ESa); 40, Cards Pond, 2 May 1981 (RAC). Numbers decline after mid-May, but small numbers occur regularly into the first week of June and occasionally later (11, Quonochontaug, 13 June 1983; ASM).

Fall. Fall migrants normally arrive in mid-July. Fall migration is greatly protracted, often lasting into mid-November. Most appear at the coast, but they are regularly inland after heavy rain or when lowered water levels in ponds and reservoirs provide foraging habitat.

Fall maxima. 150, Quonochontaug, 31 October 1971 (RSm); 110, Seapowet, 22 October 2002 (RBo, DLE); 100 between Point Judith and Green Hill, 3 November 1957 (DLK, CWo); 40, Scituate Reservoir, 23 October 1949 (WBD); 18, Diamond Hill Reservoir, 22 August 2002 (MLy); 28 at Block Island on 28 September 1997 (STs) was a maximum for the island.

Winter. Small numbers regularly survive through December and into January. Winter survival has never been proven, however.

Winter maxima. Six, Weekapaug, 19 January 1980 (TAd, KKi, FPe, ESa); 6, Jerusalem, 7 January 2000 (STs); 6, Matunuck, 29 December 2001.

Extreme dates. 24 January 2008, Sachuest Point (1; RTL); 18 February 1995, Green Hill (RLF); 21 February 2000, Jerusalem (SSM); 23 February 1974, Ninigret Pond (2; DWa); 24 February 2007, Block Island (2; SSM et al.). About 10 other January records since 2005; also 1 at Jerusalem 26 January 2002 and 2 there on 16 February 2002 (STs).

Wood Sandpiper (*Tringa glareola*)

Status

Vagrant. One record.

Nonbreeding

A Wood Sandpiper, the first for New England, was found at the Marsh Meadows Wildlife Refuge, Jamestown, 13–30 October 2012 (CPe and hundreds of others; see Photo 12). This Palearctic sandpiper has been recorded 3 other times on the U.S. east coast, in New York (twice: 1907 and 1990) and Delaware (2008).

Wilson's Phalarope (*Phalaropus tricolor*)

Status

Uncommon migrant on coastal mudflats in fall. Rare in spring.

Historical

There are four records from the late nineteenth century, between 1880 and 1899, all during August and September (Rives 1887b; Howe and Sturtevant 1899).

Nonbreeding

Wilson's Phalaropes surged into the Northeast beginning in the 1960s, and there were records of courting and breeding birds in Massachusetts and New York in the 1970s and 1980s. Concurrent with these sporadic breeding attempts was a substantial increase in records from Rhode Island. Wilson's Phalaropes favor pools or salt pannes at the upper reaches of salt marshes as well as freshwater pools, especially flooded fields. They are not pelagic, like the other 2 species of phalarope.

Spring. More birds were seen during fall than in spring, but they were more or less annual from the late 1950s (13 June 1959, Galilee; SSD) to 1993 (5 May 1993, Quicksand Pond; PAB). There were about 27 spring records during that time period; they have decreased since then.

Fall. Through most of the modern period, Wilson's Phalaropes have been regular migrants in small numbers on coastal mudflats with other shorebirds in fall. Totals in recent decades have never exceeded 10 birds per year, and usually average 2–6. There have been 19 records since 2005.

Fall maxima. Four, Quonochontaug, 1 September 1974 (SSD; photos); 4, Quicksand Pond, 1 September 1980 (RBo); 3, Briggs Marsh, 7 September 1995 (J Osborn); 3, Charlestown Breachway, 21 August 2012 (JSJ, BCG).

Inland. One, Woonsocket Reservoir, 11 August 2002 (MLy).

Summer. There are several records of early transients or summer lingerers: singles 4 July 1976 at Sakonnet (P Carter), 7 July 1976 at Ninigret (DLK, DRo), 2 July 1980 at Quonochontaug (DFi), and 6 July 1996 at Charlestown Breachway (BCG).

Extreme dates. Late April to 9 October 1974, Jerusalem (AMc).

Red-necked Phalarope (*Phalaropus lobatus*)

Status

Fairly common migrant in spring and fall over offshore waters, occasionally appearing in massive "wrecks." Rare but regular on tidal flats among other shorebirds.

Historical

"Wrecks" or "flights" of phalaropes have a long history of occurrence in the Northeast. They are probably caused by the coincidence of a storm with peak migration and perhaps a reduced food supply (Underwood and Stowe 1984). The first record of a flight of Red-necked Phalaropes in Rhode Island was on 15 May 1895, when Newton Dexter observed several thousand of these birds passing *over* Sakonnet Point during a southeast gale. After a southwest gale on 27 April 1931, there was a big flight at Block Island (EDi), and they abounded in "every pond hole" at Sakonnet (HNG). After a severe southerly storm, they were numerous along the Rhode Island coast on 14 May 1938, with 100 seen at Little Compton (DLE). During 25–27 May 1940, an estimated 10,000 phalaropes, including many of this species, were seen off "West Beach," Block Island after a "long northeaster" (EDi). During the phalarope "extravaganza" of 9–11 May 1969, many hundreds and perhaps thousands of Red-necked Phalaropes accompanied the more numerous Reds all along the shore from Napatree Point eastward at least to Newport (v.o.).

Nonbreeding

Spring. Red-necked Phalaropes have a less pelagic distribution than Reds and are generally more frequently seen from shore than the latter during nonstormy weather.

Spring maxima. 70, south of Block Island, 26 May 1973 (RLF et al.); 47, Cox Ledge, 24 May 2007 (STs et al.); 27, Cox Ledge, 30 May 1992 (DLE, REm). There are now several early June records, the latest being 10, Cox Ledge, 6 June 1976 (STs); 2 there 7 June 1975 (RLF et al.); and 1 there, 11 June 2016 (m.obs.).

Fall. There are records for early migrants or summering birds, including 1, Charlestown Breachway, 16 July 2005 (STs, CPe et al.); 6, 10 mi (16 km) southwest of Block Island, 22 July 1993 (DFi); 1 seen from the Newport–Jamestown ferry, 24 July 1954 (RBo); another at Trustom Pond, 3 August 1976 (DLK, FPe); and 1, Charlestown Breachway, 5 August 1999 (JSJ).

Fall maxima. 250, Harbor of Refuge, 28 August 2011 (RLF); "hundreds" of unidentified phalaropes seen at night 82 mi (132 km) southeast of Block Island 18 September 1993 (CAv) may have been largely this species; 105, Cox Ledge, 9 September 2006 (STs et al.).

Inland. There are few reports for inland locations for any season, but in fall they have been observed in Narragansett Bay or the Sakonnet Passage as far north as Spar Island in Mount Hope Bay, at Seapowet (twice), Island Park (twice), near the Jamestown Bridge, and among the Dumpling Islands in Jamestown harbor, usually after storms. There is a specimen collected in "Providence County," 28 August 1908.

Extreme dates. 14 April 1934, 15 mi (24 km) south of Block Island (14; fide EDi), to 10 November 1937, Middletown (WHD).

Red Phalarope (*Phalaropus fulicarius*)

Status

Uncommon and highly irregular spring and fall migrant, usually migrating well offshore and only occasionally seen along the coast. At lengthy intervals, they have appeared in great flights after easterly gales or under conditions of extensive fog in spring, when thousands have appeared in outer coastal locations or lower Narragansett Bay.

Historical

As with Red-necked Phalaropes, large numbers of Red Phalaropes occasionally appear at the coast, either after storms or during fog. These "flights" may be triggered by food shortages, as many storms in spring do not result in wrecks. Wrecks of Red Phalaropes in Rhode Island include 23 May 1892 (specimens collected at two Newport locations); 17 May 1930, flight of unknown extent off Block Island (EDi); 25–28 May 1940, 10,000 phalaropes off Block Island, mostly Reds with some Red-necked (fide EDi); 10 May 1956, 500 noted at sea off Rhode Island (*AFN*); one dead at Block Island 30 May 1958 (MSl), and no others seen, but 4000 observed off nearby Long Island. The best-documented flight occurred 9–11 May 1969 after heavy rain and southeast winds all day 9 May. Large numbers were seen the first day from Napatree Point to Newport, with huge numbers off Beavertail and some in lower Narragansett Bay (6000 in the West Passage near Jamestown Bridge; MDa). In clear weather the next day, 1000 were in Ninigret Pond (ESa), 1000 were off Narragansett (DLK), and there were 15 still off Jamestown 13 May (MDa). A smaller flight of 200 occurred 17 May 1970 along the South Shore after a day of fog and rain with 30-knot (56 kph) southeast winds (DLK, ESa, CWo). In sum, large "flights" have occurred from 30 April to 30 May, later than those of Red-necked Phalaropes.

Nonbreeding

Spring. Red Phalaropes are seen occasionally from the Block Island ferry or from such coastal promontories as Point Judith. Even at Cox Ledge they are not seen with great regularity or in any numbers in spring. There is a record for the remarkably early dates of 12–13 March 1955, when 1 was found at Cards Pond on a day "when the temperature reached only 10° f [–12 °C]" (DLK). A flock of 7 appeared at Point Judith 16 March 1993, after the formidable northeast snowstorm of 14 March 1993 (BCG, MJM). Most occur between late April and late May. The latest spring records are for 1 off Block Island 2 June 1938 (fide EDi) and 3, 8 mi (13 km) east of Block Island 8 June 1983 (RBo, HMP, RKl).

Spring maxima. 50, Cox Ledge, 19 May 2003 (CPe).

Fall. In fall, Red Phalaropes are only occasionally seen from land and never in numbers remotely comparable to those seen in spring; e.g., 1, South Kingstown, 30 July 2001; 3 after a storm at Briggs Marsh, 15 August 1953 (JGe, RGe); 1, Third Beach, 31 August 2005 (RTL); 1, Moonstone Beach, 5 November 2005 (STs et al.). Offshore, they are found more regularly and are seen in about the same abundance as during spring. They are most often reported from late August through October and are more likely to be seen later in the season than are Red-necked Phalaropes.

Fall maxima. 68, Galilee, 6 September 1979, the day of Hurricane Frederick (DLK); 32, Narragansett, 8 September 1941 (v.o.). Up to 20 per day have been seen in fall at Cox Ledge; e.g., 20, Cox Ledge, 7 October 1996 (MJM).

Inland. The only occurrences away from the outer shore are for the West Passage near the Jamestown Bridge in 1969 (see above) and for 1 shot in the fall of 1898 off Gaspee Point in upper Narragansett Bay (E. A. Armstrong, fide HSH).

Extreme dates. 12 March 1955, Cards Pond (DLK), to 20 December (no year given; Allen 1909).

SKUAS AND JAEGERS (STERCORARIIDAE)

Great Skua (*Stercorarius skua*)

Status

Rare winter resident offshore.

Nonbreeding

A skua was seen repeatedly in fog about 6 mi (10 km) south of Block Island 23 January 1972 (G. Rosenband and W. Krawiec), and 3 to 5 birds were seen 3 February 1978 75 mi (121 km) south of Block Island (B. Morris, *AB* 32:328). Based on patterns of occurrence that have subsequently emerged, these early records almost certainly pertained to Great Skuas and not South Polar Skuas (see below). Another individual attending a dragger at the Fish Tail, 6 mi (10 km) north of Block Canyon, 4 March 1995, was unquestionably this species (40+ observers; PAB, DLE, RLF et al.).

South Polar Skua (*Stercorarius maccormicki*)

Status

Uncommon migrant and summer resident offshore.

Nonbreeding

South Polar Skuas have been found at Cox Ledge and elsewhere offshore between late May and October, with a peak in observations during June. Among 35 observations during the warmer months through 1990, 15 were observed in May and June and 20 from July through November. Devillers (1977) outlined the transequatorial migration of South Polar Skuas to the North Pacific and a similar pattern occurs in the Atlantic (Veit 1978; Harrison et al 2021).

Maxima. One South Polar and two unidentified skuas, 10 mi (16 km) east of Block Island, 24 September 1978 (DRn, STs, CWo); 2, 36 mi (58 km) south of Block Island, 25 September 1996 (BCG); 2, 7 mi (11 km) east of Block Canyon, 31 May 1997 (RLF, DLE, PAB, RBo et al.); 2, Cox Ledge, 15 June 1974 (G. S. Raynor); 2, 30 mi (48 km) south of Block Island, 14 September 1992 (RAC); 1, Cox Ledge, 11 July 1995 (DFi et al.); 1, 10 mi (16 km) south of Block Island, 11 August 1995 (MJM); and 2, Mud Hole, 24 May 2007 (STs, CPe, PLE). It is unusual to see any skua from land, even during storms: 2 were seen near the north end of Block Island 29 August 1995 (DFi, JO et al.).

Pomarine Jaeger (*Stercorarius pomarinus*)

Status

Common transient offshore; uncommon within sight of land.

Historical

There were only five spring records of Pomarine Jaegers before the 1960s: 3 May 1954 off Warren's Point (NPH), one shot 9 May 1913 off Block Island (Hathaway collection), sight records for 17 May and 25 June (HSH), and an adult freshly dead on the Quonochontaug Marsh 26 June 1916 (HSH). There was a great inshore flight in late summer 1910, when HSH saw three to four and as many as eight every day from the beach at Quonochontaug. Griscom, who happened to be traveling by sloop through Block Island Sound 21 August 1910, and again 2 days later, confirms HSH's impressions of unusual numbers by describing jaegers there as "common." Two Pomarines were collected off Point Judith in that period and sent to HSH. Such rare appearances of numbers close inshore probably result from unusual abundance of baitfish (herring, sand lance), as described for the 1886 flight of jaegers and Cory's Shearwaters in Buzzards Bay (see Veit and Petersen 1993 and Cory's Shearwater account).

The status of both jaeger species was poorly known before the early 1970s, when regular trips to Cox Ledge began. Pomarine Jaegers are less frequently seen from shore than are Parasitic Jaegers, but numbers of this species exceed those of Parasitic Jaegers at Cox Ledge and farther offshore. Observations at Cox Ledge and vicinity during the period May through October 1966–1974 yielded 174 Pomarine, 119 Parasitic, and three Long-tailed Jaegers.

Nonbreeding

Spring. Observations before 20 May are few, and the majority appear to pass offshore during the last week of May and the first week of June. More offshore observations are needed in late April and the first half of May, when occasional early transients are likely.

Spring maxima. All for Cox Ledge and vicinity: 9, 15 May 1986 (DFi et al.); 8, 12 June 1975 (DLK, RAC, KKi, FPe); 6, 31 May 1980 (DLE et al.); 6, 8 June 1975 (RLF); 4, 31 May 1975 (DLK, CWo).

Summer. Small numbers of both Pomarine and Parasitic Jaegers have been present throughout the summer, at least in recent years. In addition to the late June records mentioned above, observations for this species during the breeding season (at Cox Ledge unless otherwise indicated) include a flock of 9 off Sakonnet Point, 24 June 2001 (REm); 1, 30 mi (48 km) south of Block Island, 27 June 1984 (RBo, HMP); 6, 27 June 1975 (DLK, DWa). There are 2 old inshore observations of perhaps the same (possibly summering) individual 16 July 1941 along the Middletown shore of the Sakonnet Passage (WHD, Williston) and at Sakonnet Point the next day (DLE). Three were seen in the Sakonnet Passage 17 July 1948 (RBo). See Veit and Petersen's (1993) discussion of variations in levels of nonbreeding based on fluctuations in lemming populations. Lean lemming years may provide an ample source of summer lingerers, failed breeders, and early fall migrants.

Fall. There is much variation in abundance from year to year, but in general Pomarines tend to occur somewhat later in fall than Parasitics, with peak numbers appearing in late September and October. Pomarine counts in waters south of Block Island are usually greater than along the immediate coast; jaeger numbers are more heavily weighted toward Pomarine over Parasitic there than closer inshore.

Fall maxima. A "steady migration" of 168 Pomarines (and only 4 Parasitics) was observed moving southwest over Block Canyon 17–18 September 1992 (DLE, NPH); 60, 50 mi (80 km) south of Block Island, 9 September 1995 (DLE, WRP et al.); 56 feeding at fishing trawlers 40 mi (64 km) south-southeast of Block Island, 27 September 1990 (RAC); 36, Cox Ledge, 16 September 1973 (THD, RLF); 34, 40 mi (64 km) southeast of Block Island, 12 September 1993 (WRP, DLE et al.); 20, Cox Ledge, 20 September 1979 (DLK); 15, off Sakonnet, 6 September 1944 (RBo); 9, Cox Ledge, 26 October 1974 (THD); 10 flying over Harbor of Refuge, Galilee, after the passage of Hurricane Irene 28 August 2011 (CRa).

Extreme dates. The latest inshore fall records, all for Point Judith, are 3, 15 November 1982 (DLK); 1, 20 November 1982 (RAC); and a dark-morph bird, 29 November 1984 (DLK, CRa). A large jaeger, thought to be this species, was seen during a southerly storm at Point Judith 3 December 1986 (CRa); 12, Cox Ledge, 20 November 2004 (STs, GW et al.).

Inland. There is an unusual inland record of a wounded and emaciated subadult found resting on the turf fields north of Wood River Junction 31 July 1993 and captured alive (S. Carr; specimen prepared by CRa; AMNH #831186). No unusual weather preceded its appearance.

Parasitic Jaeger (*Stercorarius parasiticus*)

Status

Fairly common spring and common fall transient offshore, regularly seen in small numbers from coastal promontories, and generally commoner inshore than Pomarine Jaeger.

Historical

Not mentioned by Howe and Sturtevant (1899), Parasitic Jaegers were first recorded when HSH Hathaway 1901d)) recorded three shot at Charlestown Beach 2 September 1897, one of which is extant. Hathaway, writing soon after the great flight of Pomarine Jaegers that occurred along the New England coast in 1910 (Hathaway 1913), considered Parasitics to be much less frequent than Pomarine Jaegers. He recorded one shot on the late date of 27 November 1909 at Easton Pond by C. B. Clarke and sent to HSH (still extant). Data from recent years show Parasitic Jaegers to be at least twice as numerous as Pomarines along the mainland coast but considerably less numerous offshore (see Pomarine Jaeger account).

Nonbreeding

Spring. Individuals were seen flying east over the Galilee Escape Road in the early evening of 11 May 1980 (RLF), and 1 was chasing a kittiwake off Point Judith 12 May 1984 (RLF); most shore-based counts are of 1–3 birds.

Spring maxima. Ten, Point Judith, 15 June 1975 (JZe); 8, Cox Ledge, 15 June 1974 (RAC).

Summer. Regular during midsummer both offshore and within sight of land: 1 off Scarborough 23 June 1972 after a storm (KKi); 1 at Cox Ledge 25 June 1981 (DLK); 1, 10 mi (16 km) southeast of Point Judith 28 June 1977 (DLK); an adult off Napatree Point 29 June 1987 (CRa); 1, 30 June 1984 at Briggs Beach (DLE); 2 at Cox Ledge 1 July 1995 (DFi, CAv et al.); and 4 there 8 July 1977 (DLK).

Fall maxima. 20, Harbor of Refuge, 28 August 2011 (RLF); 11, Sakonnet Passage, 18 August 1940 (RBo); 11 past Point Judith into a 20-mph (32-kph) southwest wind, 27 October 1985 (RLF); 8 at the same location, 15 November 1982 (DLK). On 27 September 1985, during Hurricane Gloria, 36 were counted at Point Judith, mostly moving eastward past and over the point late in the day on the 50–90 mph (80–145 kph) south-southwest winds of the southerly quadrant of the storm (RLF, CRa). Also, 14, Cox Ledge 16 September 1974 (RLF et al.).

Inland, Parasitic Jaegers have occasionally been seen in upper Narragansett Bay; e.g., 1 off Poppasquash Point, Bristol, 4 September 1937 (HNG); 1 near Conimicut Light, 25 October 1937 (HNG); 1 in Mount Hope Bay 17 September 1938 (RBo); 1 off Saunderstown "well within Narragansett Bay" 13 July 1938 (A. M. Clark); 1 off Bristol 12 September 1948 (RBo); 1 chasing terns off the north end of Prudence Island 4 July 1987 (DFW); 3, 13 September 2004 (GDe); 1, 6 July and 15–17 August 2005 (GDe). There is only 1 spring record north of the outer coast: a dark-morph adult and a light-morph subadult together chasing Common Terns at the Spar Island colony in central Mount Hope Bay 1 June 1998 (RLF, CRa).

Extreme dates. 16 April 2004, Point Judith (RLF), to 27 November 1909 at Easton Pond; 9 December 1986, Point Judith (DFi, DLK); and 19 December 1993, Charlestown (MJM).

Long-tailed Jaeger (*Stercorarius longicaudus*)

Status

Uncommon migrant far offshore; very rare within sight of land.

Long-tailed Jaegers are the most pelagic of the jaegers in the North Atlantic, and the least likely to be seen from land in Rhode Island. They were unrecorded in Rhode Island until the 1950s, presumably owing, in part, to the lack of pelagic coverage before then. The first state record was of a moribund adult that remained at Napatree Point 7–14 September 1952 (DLK, FPe, CWo). Another sick bird, this time an immature, was discovered near Beavertail lighthouse, Conanicut Island, and approached to within 20 ft (6 m) 24 September 1956 (RLC, MDa). One was seen in Block Island Sound on 20 June 1965 (WHD), and another or the same was seen on the same date flying along Moonstone Beach (CWo).

Nonbreeding

Spring. All records are from Cox Ledge or other offshore waters. Two adults (1 photographed) 8 June 1974 (RLF); 1, 14 June 1975 (DLK), and 4 the next day (JZe); an adult 22 June 1975 (RAC); an adult 6 June 1976 (NHBC); 1, 4 June 1978 (DLE, CWo et al.).

Fall. (All from Cox Ledge, unless otherwise indicated) A remarkable 5 were seen 24 September 1978 (DRn, STs, CWo); 1, 1 September 1968 (M. Kleinbaum); adults 21 September 1974 and 30 August 1975 (both DLK). A dark-morph bird was seen 7 September 1983 (RBo); adults were seen 40 mi (64 km) south-southeast of Block Island on 10 September 1990 (RAC), and at Block Canyon 17–18 September 1992 (DLE, NPH). Other records include 1, Cox Ledge, 30 August 1993 (DFi, DLE et al.); 2, 40 mi (64 km) southeast of Block Island, 12 September 1993 (WRP, DLE et al.); 2, 40 mi (64 km) southwest of Block Island, 22 September 1993 (DLE); 1, 1 July 1995 (CAv, DFi et al.). A light-morph juvenile Long-tailed Jaeger at Point Judith on 9 September 2009 was so close to shore as it moved northeast that its rounded central tail feathers were easily visible (PAB). Also, subadult, 31 May 1997, Block Canyon (PAB, m.obs.); 1, 2nd-summer plumage, 17 June 2000, Cox Ledge (STs, PAB, SSM); 2 juveniles, Mud Hole, 8 September 2004 (STs et al.); 1–2 juveniles, Mud Hole area, 24 August 2005 (JMa, GW); 4, Block Canyon, 3 September 2013 (CPe et al.).

Extreme date. An immature jaeger seen from Point Judith 15 January 2001 (DFi, CNu) was identified as Long-tailed. There is a winter specimen of Long-tailed Jaeger from the Gulf Coast of Florida in 1889 (Lee 1989) and observations of Long-taileds off northeastern Florida in December and January (Wallace and Wigh 2007), but Long-tailed Jaegers are otherwise unrecorded in the western North Atlantic during winter, as North American breeders winter mainly in the Falkland Current area off Argentina (Veit 1985). The latest record in Massachusetts waters is for 10 November

(Veit and Petersen 1993). This Rhode Island record may signify changing distribution related to climate.

AUKS, MURRES, AND PUFFINS (ALCIDAE)

Dovekie (*Alle alle*)

Status

Uncommon and erratic visitor in fall and winter; occasionally appears in very large numbers, usually in November.

Historical

Howe and Sturtevant (1899) referred to Dovekies as rare and irregular winter visitors, both along the shore and in Narragansett Bay. The most notable wreck in history occurred in 1932 (Murphy and Vogt 1933) when thousands were seen during and after a storm 17–19 November all along the East Coast south to Florida and Cuba. Elizabeth Dickens stated it "fairly rained Dovekies" on Block Island. About 20 were captured alive by boys from St. George's School, Middletown, and birds were found inland at Glocester, North Smithfield, Lincoln, Johnston (five), Providence (four), East Greenwich, Wickford (two), Usquepaugh, Peace Dale, and elsewhere. Numerous others were stranded on land in coastal districts, seen in coastal ponds, or washed up dead along the shore. On 22 November, a number were still swimming in Watchaug Pond (MSo).

Nonbreeding

There have been no subsequent irruptions of equal magnitude, but a number of lesser wrecks have been recorded, as well as a number of records of large numbers along the shore. There was a large flight in 1944: 115 were seen off Sakonnet on 11 November 1944 (RBo), and 1 was caught alive near Wordens Pond on 27 December (fide HNG). A still larger flight occurred in 1948: several were blown ashore in Melville and a number noted off Sakonnet about 28 November. The possibility of numbers having been blown into Long Island Sound was suggested by the steady easterly passage of 500 in less than 1 hour on 4 December at Quonochontaug (DLK), as well as 100 each at Sakonnet and Point Judith the next day, also flying east (WBD, DLK). Another flight occurred along the shore 11–23 December 1949, but no Dovekies were reported inland. One was picked up at Kingston 24 January 1953 (DLK), and many others were seen along the shore about the same time. November flights occurred in 1959 and again in 1965, when birds were observed all along the shore, and 1 was picked up alive in Coventry on 11 November 1965. Possibly the most sizable wreck since 1932 occurred during a southeast gale 2–3 November 1966, when birds were scattered all over the state. Thirty reports were received, with birds inland at Pawtucket, Barrington (5), Bristol, East Providence, Providence, Wordens Pond, Middletown, Newport (6), Matunuck, and North Kingstown. Another flight occurred in early November 1969, when 1 was seen at Wordens Pond on 2 November (DLK et al.). In 1970, a very impressive flight occurred after a gale on 12 November, with birds inland at Barrington (preserved; RLF), Coventry, and Cranston. Many small flocks occurred along the shore 19–26 November 1971, and birds appeared in Winnapaug Pond and ashore in Avondale. Veit and Petersen (1993) indicate that this species' fall flights in Massachusetts are independent of those of other alcids and often not associated with storms. Dovekies move southward in NAO-negative years; this is probably related to food scarcity farther north (Veit and Manne 2015).

Apart from these irregular incursions, Dovekies are most often found offshore from November to February, but small numbers are sometimes seen from shore, especially at Point Judith and Block Island, and there have been occasional unseasonal records. One was found a half-mile inland at Ninigret NWR 12 March 1989 (SHa). Late-spring occurrences include 2 in alternate plumage resting on the water off Weekapaug 1 May 1955 (DLK) and later reported by fishermen to have *summered* there; 1 at Middletown 5 May 1940 (fide WHD), and 1 at Block Island 16 May 1930 (fide EDi).

Maxima. 230, pelagic cod boat, 21 January 2013 (PLE); 220, Block Canyon, 10 March 1996 (RLF, RBo et al.).

Extreme dates. 27 October 1930, Block Island (EDi), to 17 July 1955, Quonochontaug (RCC).

Common Murre (*Uria aalge*)

Status

Formerly a very rare winter visitor; has increased substantially since about 1990 and is now regular in winter in near offshore waters, such as Cox Ledge. Rarely seen from land, even during storms.

Historical

Howe and Sturtevant (1899) recorded one taken at Point Judith on the authority of F. T. Jencks, and another allegedly taken at Sakonnet in 1894 by Newton Dexter. The earliest extant specimen, previously unpublished, is in the Doe collection (FlMNH #2881) labeled "Thick-billed Murre, Narragansett Bay, January 1909" (its identity as *U. aalge* was discovered by T. Webber and confirmed at AMNH by J. Bull). The first modern record was for an oiled bird found at Point Judith 10 February 1930 and obtained by Hathaway; the specimen's present whereabouts are unknown. Another oiled bird was found at Davis Point, Newport, 20 April 1934 by J. J. Lynch. One was found at Block Island, again oil-soaked, 7 April 1940; now in the BISC (H. Madison). On 23 March 1941, another oiled bird was found at Warren's

Point (RCC, RBo et al.; MCZ #291103). One was found dead on Block Island on a CBC 26 December 1944 (EDi)

Nonbreeding
Since 1950, sight records include Sachuest Point 31 January 1953 (NPH); 2 off Easton's Point, Middletown, 22 March 1958 (JBa, RCC, RLF); individuals at Jamestown 25 November 1960 (MDa) and 30 November 1968 (RLC); off Newport 10 December 1970 (RTL); an oiled bird at Narragansett Pier 9 January 1977, perhaps oiled at the *Argo Merchant* sinking on the Nantucket Shoals (DLK, RAC, RBo et al.); at Cox Ledge 12 March 1977 (RAC, DLE, DLK, CWo); and at Warren's Point 9 November 1982 (RBo, HMP). There is a recent specimen salvaged from Napatree Point 4 March 1989 (UConn #09930, courtesy of Prof. G. A. Clark, Jr.) and 2 were seen at Point Judith, 22 January 1995 (DFi). An unprecedented *13* were seen roughly midway between Montauk Point and Block Canyon 4 March 1995 (PAB, DLE, RLF et al.). Over the next decade, Common Murres were carefully documented by DFi, STs, SSM, and others during many trips to Cox Ledge and other waters east and south of Block Island: 1 (plus 3 unidentified murres) 28 December 1998 (SRC, C. Hempstead, SSM); 8, 4 February 2001 (DFi, SSM); 28, 9 February 2002 (STs, SSM, S. Keeble). Recent counts offshore have routinely broken double digits (see "Maxima" below), and birds are regularly encountered from the Block Island ferry. These data, in contrast with the almost complete absence of offshore records from prior decades, make it clear that Common Murres have substantially increased in abundance in Rhode Island waters since about 1990 (Veit and Guris 2009; Veit and Manne 2015).

Maxima. 45, Cox Ledge to 20 mi (32 km) south-southeast of Block Island 28 December 2003 (STs); 44, Cox Ledge, 6 January 2016 (PLE, Keith Mueller et al.); 21, Cox Ledge, 19 March 2011 (CPe et al.); 7, pelagic cod boat, 29 January 2013 (PLE); 7, Block Island PDC, 25 February 2013 (SSM et al.). There are 2 records from summer: 1, Cox Ledge, 5 June 2003 (JSJ et al.), and 1, Galilee, 12 July 2017 (CRa); possibly the same bird at Quonochontaug 27 July 2017 (Stephanie Vasquez).

Thick-billed Murre (*Uria lomvia*)

Status
Rare and erratic winter visitor. Many winters go by with none reported.

Historical
Bull (1964) and Zeranski and Baptist (1990) list flights noted in the New York–Connecticut area in 1883–1884, 1890–1891, December 1894, 1901–1902, December 1926, December 1932, and 1976. Jencks (1884) mentioned one shot between Warwick Neck and Patience Island 26 December 1883 and 10 others taken later the same winter. Howe and Sturtevant (1899) considered them to be irregular winter residents

along the outer shore, and listed a bird shot in Bristol in the winter of 1883–1884. There was a well-documented but no longer extant inland specimen taken from a small flock on the Blackstone River above Pawtucket 16 December 1894, formerly in the collection of Harry Cash. There is no corroboration in Rhode Island, however, for what Zeranski and Baptist (1990) term the largest flight in Connecticut, when thousands were seen, and hundreds shot on the western half of that coast between mid-December 1890 and mid-February 1891.

In Rhode Island, the many records near the turn of the twentieth century during flight and non-flight years suggest that Thick-billed Murres were once more frequent in Rhode Island waters than were Razorbills. This numerical superiority prevailed until the early 1940s. Only nine records accumulated during the period 1950–1976, despite a vast increase in coastal field work, while in the same period there were five records of Common Murres and more than 50 coastal observations of Razorbills.

Nonbreeding

Winter. This quarter-century of rarity was abruptly if temporarily ended during the winter of 1976–1977, with a flight along the New England coast that may have exceeded all others in recorded history. This "flight" produced a minimum of 8 Rhode Island records from inshore locations during late December, mostly of small numbers. A maximum of 40 was noted at Point Judith 23 December 1976 (STs), another was picked up injured in Providence 11 January 1977 after a severe northeast storm the previous day, and thousands were still present off Cape Cod into mid-January. Remnant birds were noted from Cox Ledge in February and March. This incursion may have been related to unusual oceanographic conditions that affected the murres' food supply farther north; the winter of 1976–1977 was a strongly negative NAO year (Veit and Guris 2009, Veit and Manne 2015).

Inland. Some Thick-billed Murres have turned up well inland. One was noted floating northward past Seapowet at Jack's Island, Tiverton, 17 March 2001 (RFa). One was swimming in the northernmost part of Ninigret Pond 23 January 1972 (DLK, FPe, KKi, ESa). Another inland record involves a specimen at RWPM picked up dead in Woonsocket 30 November 1931, a date that correlates with a flight off Block Island the same year. Also, 1, Cox Ledge 18 January 2002 (JSJ), and 8 on Napatree CBC 22 December 2002. There were 2 reports of 2 birds in January 2005, and 5 reports of 6 birds in February 2005, all South County (plus 1 at Beavertail). One was captured at the Westerly train station 19 January 2020 (v.o.).

Summer. An oiled bird was found along the South Shore 23 May 1993 (skin prepared by CRa; AMNH #831191). An individual with a broken wing was at Bonnet Shores 14 June 1954 (DLK), and another thought to be this species seen fleetingly off Second Beach, Middletown, 13 August 1966 (DLE, REm). A photograph of the latter individual could not be specifically identified (courtesy L. Tuck), but the probability of its being this species is heightened by the capture of 7 birds alive at various Long

Island and Cape Cod locations at the same time (25 July–30 August 1966) (Bull 1974; Veit and Petersen 1993).

Maxima. 40 off Block Island, 6 December 1931 (fide EDi); 40, Point Judith, 23 December 1976 (STs); 10, Cormorant Rock, 27 November 1937 (WHD).

Extreme dates. 9 November 1982, Warren's Point (RBo, HMP), to summer.

Razorbill (*Alca torda*)

Status

Uncommon to fairly common winter resident; substantial increase since about 1990.

Historical

Howe and Sturtevant (1899) referred to Razorbills as rare winter visitors and mentioned that a number had been taken along the outer shore. From 1900 up to the present, they have increased, dramatically so since about 2005 (Veit and Manne 2015). In the 1920s, Elizabeth Dickens received reports by fishermen of counts of 20–50 birds and even "great numbers" in waters south of Block Island. During January 1930, many were killed by an oil spill from a tanker and washed ashore on Block Island (see also Atlantic Puffin and Common Murre accounts).

Nonbreeding

Peak numbers appear from December through March. Razorbills are regularly reported into April and 2 were at Block Island 20 May 2012 (DFi).

Inshore maxima. 546 off Point Judith, 5 January 1992 (RLF, DLK et al.); 318, South Kingstown CBC, 21 December 2002 (STs); 290 passing west off Point Judith, 26 December 1999 (RLF); 200 seen from Beavertail, 9 January 2015 (WM); 155, Napatree CBC, 23 December 2007 (SSM, GW et al.).

Offshore maxima. 249, Block Island CBC, 8 January 2004 (CRa et al.); 248, Block Island PDC, 19 February 2001 (SSM et al.); 210, Cox Ledge, 11 March 2006 (JSJ); 137, Sharks Ledge, 9 March 1985 (DFi, DLK).

There are few reports from locations north of the mainland coast. An oiled bird that subsequently died was captured in Narragansett Bay off Barrington 22 February 1961 (REW; in MCZ). Another was blown ashore during a gale and found on Indian Avenue, Middletown, 10 April 1961 (REW). One alive but ill near Island Park, Sakonnet River, 13 January 2013, was brought to rehab but died (AMNH). One, Sakonnet River south of old stone bridge abutments, 26 February 2004 (CRa et al.).

Extreme dates. 24 October 1960, Block Island (JBa), to 20 May 2012 (see above), to summer. One was reported through June 1960 at Block Island (fide ALH), and 1 spent the entire summer of 1931 there (EDi).

Great Auk (*Pinguinus impennis*)

Status

Extinct. Occurred in precolonial and perhaps early colonial times.

Historical

Eaton (1898) listed numerous bones of this species found in Niantic shell heaps at three locations near New Harbor, Block Island. He stated that these heaps were definitely established prior to the colonization of the island by European settlers about 1661. Eaton listed bones of this flightless bird as often or more often than those of many other aquatic species still common in coastal Rhode Island, suggesting that the Great Auk may have been a fairly common bird near Block Island. If so, they probably appeared along the mainland coast as well. There are numerous records of bones in shell heaps from Massachusetts northward and others for Long Island and Florida.

Black Guillemot (*Cepphus grylle*)

Status

Rare winter visitor. Very rare in summer.

Historical

A bird shot in Mount Hope Bay in 1859 by Newton Dexter and presumed not extant by Howe and Sturtevant (1899) was exhumed and re-examined at RWPM by Ferren. Another was shot in upper Narragansett Bay off Bullocks Point, Riverside, by S. W. Thayer and sent to F. B. Webster, the date (about 1881) unknown. A bird that was formerly in the collection at URI labeled "18 January 1887 from Newport" may have been the same as one said to have been shot at Newport by Newton Dexter (fide HSH). One was shot by Dexter off Sakonnet 18 November 1892 and acquired by HSH, present whereabouts unknown. One was shot by a fisherman off Second Beach, Middletown, in February 1900 (LKi). Hathaway (1913) mentioned one that was shot off Sakonnet 1 January 1906, as well as two shot at Cormorant Rock 30 December 1906 and 28 November 1909, all collected by C. B. Clarke. The latter two specimens are still extant. Four others were shot in 1909 at Cormorant Rock by Clarke, the dates now lost.

Nonbreeding

Black Guillemots have increased and now appear regularly in winter off the north end of Block Island. There are 15 or more records since 1980 between Point Judith and the Massachusetts border. Two records at Cox Ledge, 26 February 1977 and 12 March 1977 (CWo, DLE, RAC, DLK), suggest that the species may not be as confined to littoral habitats as is often thought (Veit and Petersen 1993) and may occur occasionally well offshore. The only records during the warmer months are for an immature south of Fort Mansfield, Napatree Point, 1 May 1990 (CRa), 1 in alternate plumage closely observed near the Point Judith breakwater 5–7 June 1981 (RLF), and 1 at Sachuest Point, 12 May 2017 (RTL). Since 1998, Black Guillemots have been recorded regularly on the Block Island February counts, most frequently off the Clay Head bluffs, but also off Mohegan Bluffs.

Maxima. 16, Block Island, 9 March 2015 (Mike Resch); 9, Block Island, 17 February 2014 (SSM et al.).

Extreme dates. 6 November 2004, Block Island (DFi, REm et al.), to 23 April 2001 (Evelyn Rhodes) and 7 June.

Long-billed Murrelet (*Brachyramphus perdix*)

Status
Vagrant. One record.

Nonbreeding
One was found a few hundred feet off the sea wall at Narragansett Pier 27 November 1998 by SSM; it remained until 9 December and was clearly photographed and observed by many. There have now been >80 records, including about 10 on the U.S. Atlantic coast, of this Asian species in North America, most since 1979 (Sealy and Carter 2012; Howell et al. 2014). Despite identification difficulties and fairly recent (1997) separation of this species from Marbled Murrelet, it seems clear that there has been a substantial increase in vagrancy of Long-billed Murrelets to North America since the 1970s.

Atlantic Puffin (*Fratercula arctica*)

Status
Uncommon fall and winter visitor, usually far offshore.

Historical
There are several old reports of puffins from Block Island and vicinity, mostly by fishermen who reported to EDi. Her diary listed two birds for 2 February 1921, two for 16 January 1922, and one for 26 January 1922. On 15 February 1922, two puffins were said to have bobbed up under the stern of a boat near Block Island (Hathaway 1943, citing a letter from EDi, but this observation is not mentioned in her diaries).

On 10 January 1930, after perhaps the largest oil tanker disaster in Rhode Island waters, one came ashore along the south coast but soon swam off again (EDi). Another record is for the unlikely date of 1 June 1931, when HNG, fishing off Sakonnet, saw one "by the bell buoy southeast of West Island. . . . I was very close to it and particularly remember the red-yellow and I think blue bill."

Nonbreeding

Thereafter almost a half century passed until the memorable alcid winter of 1976–1977, which produced the first modern sight records. Singles were seen at Cox Ledge 6 and 12 March 1977 (DLK; RAC, DLK, CWo). There is a report of 2–3 birds seen in mid-March 1984 at a point 50 mi (80 km) south of Block Island (W. Davis, fide RAC). A remarkable 58+ puffins were seen southwest of Block Island on a track between Montauk Point and Block Canyon 4 March 1995 (PAB, WBo, DLE, RLF et al.). Nearly all of these birds were seen from 30–60 mi (48–97 km) southeast of Montauk, some nearly within sight of Block Island. One was seen offshore 22 April 2010, pelagic (B. Harris), and there are 2 summer records, the one listed above from 1931 and 1 adult off Fox Island in Narragansett Bay 3 June 2003 (CRa, RLF). Two were seen at Cox Ledge 4 February 2001 (SSM, DFi). There are 8 reports since 2005 (all since 2009), all pelagic. There are 2 reports of birds seen from shore: 1 at Block Island from 22 March 2003 (Alex von Arx, K. Gaffett) and another Middletown, 31 January 2013 (CRa).

Recent maxima. Three, 19 November 2009, Cox Ledge (B. Harris).

GULLS AND TERNS (LARIDAE)

Black-legged Kittiwake (*Rissa tridactyla*)

Status

Common to abundant fall migrant and winter resident offshore.

Nonbreeding

Spring. Only the weakest sign of any return movement has been detected in March and early April in inshore waters, and there are remarkably few reports of kittiwakes seen from the mainland coast in spring.

Spring maxima. 90, Sharks Ledge, 9 March 1985 (DFi, DLK); 110, Cox Ledge 29 March 1983 (DFi). Rare after early May. Late inshore records include 6 migrating eastward past Point Judith and 2 at Beavertail, 3–6 June 1984 (RLF, RAC; REn); 1 at Galilee, 8 June 1983 (RBo); and 1 on the flats at Napatree Point, 12 June 1983 (CRa, REn). There are no records for midsummer.

Fall. While common in offshore waters during the colder months, Black-legged Kittiwakes are not often seen from shore except from Point Judith and at Block Island. Most have been seen during strong southerly or easterly winds, especially in late fall. More than a few hundred kittiwakes in Rhode Island waters is distinctly unusual, and the rare appearance of large numbers, particularly near the mainland, appears to be correlated with food conditions. Veit and Petersen (1993) have connected the remarkable numbers of seabirds, particularly kittiwakes, along the eastern Massachusetts coast in the late 1970s and early 1980s with a vast increase in the population of the American Sand Lance (*Ammodytes americanus*). Thousands of kittiwakes appeared in 1983 in nearby Buzzards Bay, as well as in Rhode Island (see below), again in response to an unusual abundance of sand lance. Most are seen from late October through early January.

Fall and winter maxima. 7500, passing east off Sandy Point, Block Island, 18 December 1977 (RLF, SRD); 4826, passing west past Point Judith, 26 December 1982 (RAC); 4400 there 25 November 1983 (DLE, REm); 1525, Block Island, 28 December 1980 (CBC); 1000, Cox Ledge, 20 November 2004 (STs, GW et al.).

Inland. There is only 1 record from inland fresh water, an adult carefully examined during a period of fair weather at the East Providence Reservoir 25 October 1956 (RLF, JSu). After alighting on an overhanging tree branch along the reservoir shore, this bird was later noted spiraling upward and westward toward the Seekonk River.

Ivory Gull (*Pagophila eburnea*)

Status
Vagrant. Three records.

Nonbreeding
An immature Ivory Gull was discovered on the runway of the old air base northwest of Ninigret Pond on 29 December 1985 by New York birder Donald Mawhinney, who submitted 2 excellent photographs. Another immature was present at Briggs Marsh 7–8 December 2007 (REm, m.obs.), and an adult bird was photographed by GDe at Quicksand Pond in Little Compton 23 January 2010 (see Photo 13).

Sabine's Gull (*Xema sabini*)

Status
Rare but regular pelagic transient, mainly in late summer and early fall.

Nonbreeding
The records are as follows: adult, Cox Ledge, 31 August 1975 (Jenks, Hagen, Crandall); basic adult, Cox Ledge, 28 September 1975 (R. Radis, fide THD); "subadult," Cox Ledge, 8 July 1976 (DLK et al.); immature, south of Block Island, 11 September 1976 (THD; photo); alternate adult, Cox Ledge, 18 August 1979 (B. Spencer); immature, Cox Ledge, 3 September 1987 (DFi); immature, Cox Ledge, 5 September 1980 (RAC, DLK, GLB et al.); immature, 54 mi (87 km) south of Point Judith, 10 September 1990 (RAC); 1, 40 mi (64 km) south-southeast Block Island, 5 September 1995 (MJM); 1, 36 mi (58 km) south of Block Island, 25 September 1996

(BCG); adult, Mud Hole, 8 September 2004 (STs et al.); adult, 25 mi (40 km) south of Block Island, 11 September 2005 (STs, JSJ). The only nonpelagic records are a well-described bird found resting at Sandy Point Island 15 August 1978 (MPe); 1 found on the beach at Jerusalem 29 August 1992 (NF); adult, upper Narragansett Bay, 29 August 2004 (GDe); 1 at Harbor of Refuge 28 August 2011 (RLF); 1 juvenile was photographed at Napatree Point 6 September 2012 (DFi, STa).

Bonaparte's Gull (*Chroicocephalus philadelphia*)

Status

Common to abundant migrant and winter resident in inshore waters, in Narragansett Bay, and in the largest estuaries. Has declined substantially since 1980. Uncommon in summer and rare at any time inland.

Historical

Numbers increased since 1902, when Mearns (1902:8) considered Bonaparte's Gulls to be rare winter visitors. They first wintered in large flocks in 1949 (RCC) and have since been found in flocks exceeding 1000 birds in Narragansett Bay. They have been especially numerous near the sewer outlet at Field's Point in the Providence River, and they often drink and bathe on freshwater ponds close to Narragansett Bay. A period of abundance from the 1940s to the 1980s was followed by a substantial decline.

Nonbreeding

Fall. Flocks of transients occur in September and October, but large numbers do not generally appear until November. Numbers have declined substantially since the 1980s.

Fall maxima. 3000, upper Narragansett Bay through December 1956 (CBl); 1500, Narragansett, 21–28 December 1947 (DLK) and again 21 January 1983 (KKi); 1000, Easton Pond, 27 November 1983 (v.o.); 650, Block Island, 28 December 1980 (m.obs.); 405, Napatree CBC, 21 December 2003 (SSM et al.); 24, Cox Ledge, 20 November 2004 (GW). These gulls are not especially hardy and sometimes depart Narragansett Bay with the onset of very cold weather. In warmer winters they regularly remain in Narragansett Bay all winter, and counts of 500 have been obtained in January on 3 occasions. The February maxima are 550, Easton Beach, 20 February 1998 (JEM, Linda Jacobson), and 400 at Jamestown 14 February 1953 (MDa).

Inland. Eight, Woonsocket Reservoir, 3 November 2001 (MLy).

Spring maxima. 350, Watchemoket Cove, 28 March 1976 (CWo). As with scoters and grebes breeding in northwest Canada and Alaska, there is some evidence of a westward spring migration along the Rhode Island coast; e.g., 50, Point Judith, 15 May 1982 (RLF).

Summer. First recorded in summer in 1927; there are now numerous records of lingering individuals and flocks of 2–5 birds (usually immatures) between late May and August. Unusual numbers remained through the late spring of 1983, resulting in unprecedented counts of 120 and 80 at Trustom Pond, 4 and 10 June 1983 (DVT).

Summer maxima. Twelve, Galilee, 13 July 1997 (STs); 11, Briggs Marsh, 3 July 1990 (DLE); 9, Newport, 29 June 1927 (J. T. Nichols); 2 juveniles, Charlestown Breachway 9 August 1999 (SSM).

Black-headed Gull (*Chroicocephalus ridibundus*)

Status

Regular winter resident in small numbers at traditional sites; occasional at other seasons. Formerly found mainly in association with Bonaparte's Gulls; now more often with Ring-billed Gulls.

Historical

Black-headed Gulls expanded their breeding range from continental Europe to Iceland by 1911 and reached Massachusetts in 1930. They bred in Greenland in 1969, Newfoundland in 1977, the Magdalen Islands of Québec in 1982 (Aubry 1984), and in Massachusetts and Maine in 1984 and 1985. The first Rhode Island record was of one at Jerusalem 22–28 November 1941 (Coman, DLK). They appeared annually in winter after the 1950s. A banded bird was seen in 1955 at Field's Point (a 1967 Netherlands-banded bird appeared in Connecticut in 1985), but birds from the growing Newfoundland and Labrador colonies are probably supplying the bulk of birds on our coast (Montevecchi et al. 1987). Black-headed Gulls are partial to polluted urban situations, and in Rhode Island have occurred at the Field's Point wastewater treatment facility in Providence. A pair nested unsuccessfully at Monomoy, Massachusetts in 1986 (Holt et al.1986).

Nonbreeding

Fall–winter. Black-headed Gulls have aggregated since the early 1950s near the sewer outlet at Field's Point, Providence, and on the nearby flats at Watchemoket Cove, East Providence, where the small but fairly dependable winter population has been roosting.

Fall and winter maxima. Counts never exceeded 3 per day for many years until *14* appeared at Watchemoket Cove 4 March 1967 (RBo) (Bowen 1967); 14 on 4–5 February 1973 (R. Spicer, CWo); 12, 12 March 1974 (CWo); 15, 17 March 1982; 15, 18 April 1977 (all Watchemoket, FPe et al.). Since 1990, numbers at Watchemoket Cove have rarely exceeded 3 birds, rising to 5 on 5 February 2000 (SSM, PAB) and 6 on 20 January 2006 (m.obs.). In eastern Massachusetts, a similar decline has accompanied municipal sewage cleanup (Veit and Petersen 1993); improved municipal wastewater treatment efforts in upper Narragansett Bay may be having a similar effect.

The only location other than the Providence River that has attracted even the transitory loyalty of a wintering population is Weekapaug Inlet, where as many as 5 birds wintered during the 1980s; a maximum of 6 was present there 30 March 1985 (ESa). They no longer winter at Weekapaug but are regular in winter at Scarborough Beach, near the sewer plant. The earliest fall records include 3 at Seapowet 4 September 1978 (RBo), and individuals at Watchemoket Cove 7 September 1979 (CWo) and 11 September 1976 (STs). Lingering birds are occasionally reported into May; e.g., 3, Seekonk River, 13 May 1996 (RBo).

Summer. A subadult was at Briggs/Quicksand Pond, 15 June–10 August 1990 (DLE, RBo). An adult was seen at Field's Point 4 July 1965 (SCr). One was seen at Galilee on 13 July 1997 (STs); 1 was at India Point 22 July 2001 (MLy); a subadult was at Ninigret Pond 9–31 August 1982 (DLK); 1, Briggs Marsh, 2 August 2003 (REm, DFi, STa); 1, Quicksand Pond, 20 July 2004 (GDe, RFa); 1 adult, Galilee Harbor, 29 August 2005 (STs); 1, Little Compton, 7 July 2016 (GDe; photo).

Little Gull (*Hydrocoloeus minutus*)

Status

Rare spring and fall migrant; occasional at other times.

Nonbreeding

Little Gulls first appeared in Rhode Island during the 1940s and 1950s. An immature bird was at Briggs Marsh 25 August–8 September 1946 (RCC, DLK, DLE, RBo). An adult was at Field's Point 6–7 February 1954 (FPe, EWe, DLK, CWo), and an immature was there 12 February–30 March 1955 (FPe, EWe). Since that time, about 75 records have accumulated, with a peak during the 1970s and 1980s. More than half of the records are for the winter (November–March), unlike Massachusetts, where peaks of migrants appear in April and August (Veit and Petersen 1993). Many have been seen at or near the Field's Point sewer outlet in the Providence River, which has been inaccessible to birders since about 1970. There are also a few records from the coast and even over offshore waters. They are less tied to polluted waters than are Black-headed Gulls.

Maxima. Two adults, Briggs Marsh, 26 May 1984 (DLE); 2 adults, Watchemoket, 31 January 1987 (DLE); 2 adults during storm, Camp Cronin, 16 April 2007 (PAB). There is a total of 12 reports since 2005, 3 of these from May.

Laughing Gull (*Leucophaeus atricilla*)

Status

Common spring and abundant fall migrant; common and increasing during summer; rare in early winter.

Historical

Laughing Gulls have never bred in Rhode Island, despite a long history of doing so in southeastern Massachusetts (Nisbet 1971). They were first recorded in Rhode Island at Sakonnet in 1884 by Newton Dexter and were next seen at Sakonnet in August 1903 by LKi. The third record was of two adults shot at Point Judith in 1908 by C. B. Clarke. Numbers increased dramatically during the next decade, for example, 17 per day in 1911, 100 in 1912, 200 in 1915 and 1916. By 1921, Laughing Gulls were common, and there is one estimate of 1000 birds at one locality in 1926. Numbers seem to have stabilized between the early 1920s and the late 1940s or early 1950s, corresponding to their pattern of abundance at the primary New England nesting colony at Muskeget Island in Massachusetts. From the mid-1950s through the late 1970s, numbers of spring migrants declined sharply in Rhode Island, coincident with the abandonment of nearly all the New England breeding colonies. Numbers increased again during the 1980s and 1990s.

Nonbreeding

Spring. In spring a few migrants appear in mid-April, and general arrival is during the last half of April. Spring birds have traditionally been noted in upper Narragansett Bay locations earlier than in more logical but less hospitable outer shore locations. They are always found in smaller numbers on the outer coast in early spring and are most in evidence during the last third of April and the first half of May, after which numbers dwindle until early June, when only a few birds remain.

Spring maxima. 1000, Second Beach, Middletown, 28 May 1926 (ESt); more recently 200 there 23 May 1951 (DLE).

Summer. 500 at Second Beach 1–2 July 1936 during a herring run (ESt).

Fall. By mid-July, adults and juveniles begin to arrive. Since the 1920s, Laughing Gulls have been common in August along the outer shore, in Narragansett Bay north to the Providence River, and on adjacent fresh water. Their abundance varies from year to year, partly in response to the abundance of such schooling fish as herring and predatory bluefish (*Pomatomus saltatrix*).

Fall maxima. Big increase since 1970 or so; 3550 moving west past Point Judith in less than 2 hours, 10 November 1979 (RAC); 2500, Watch Hill, 7 November 2007 (CRa); 1850, Seekonk River, 18 September 2001 (RFa, DFi); 1500, Point Judith, 2 October 1983 (DLE); 1450 moving west there in 2 hours, 16 October 1976 (RLF); 1500 in 1 hour at Galilee, 18 December 1948 (DLK); 75, Galilee, 18 December 1948

(DLK). After Hurricane Gloria, 1000 were seen 5 October 1985 in Block Island Sound (CRa, REn). The highest offshore count is for 300 on 22 September 1993, 40 mi (64 km) southwest of Block Island (DLE).

The U.S. east coast breeding population between Maine and New Jersey increased from 31,000 pairs in 1977 to 92,000 pairs in 1994–1995 (Nisbet et al. 2013), and increasing counts of transients in Rhode Island reflect this growth.

Extreme dates. 3 March 1995, Warwick (*FNRIB*), and 9 March 1997, Easton Pond (BCG), to 2 January 2011, South Kingstown CBC (1; STs et al.), and 5 January 1950, Newport (5; WHD).

Franklin's Gull (*Leucophaeus pipixcan*)

Status

Vagrant. About 10 records.

Nonbreeding

One was at Tiverton, 20 September 1959 (P. William Smith, L. Vaughn). On 27 May 1977, 2 adults were seen by Noble Proctor, DWF, and others at Cox Ledge, a seemingly unlikely location to find this western vagrant (excellent photograph by Proctor on record). Another adult appeared at the same location on 9 June 1979 (RRV). There is a winter record of an immature at Weekapaug Inlet 10 December 1990 (CRa), and 1, Sachuest, 27 July 1995 (Lyle Brinker). More recent records include 1, Jerusalem, 3 November 2005 (JSJ); and 1, Block Island, 8 July 2009 (T. Magarian). In the eastward flight that occurred during the fall of 2015, there were 3 reports from Rhode Island: 4 at Block Island 11 November (SSM, PJL), 1 at Quonochontaug 14 November (DFi), and 1 at Napatree (flying west) 15 November (CRa).

Black-tailed Gull (*Larus crassirostris*)

Status

Vagrant. One record.

Nonbreeding

An adult Black-tailed Gull was found in the Third and Second Beach areas of Middletown 24 June 1995, where it remained until at least 15 October (POn et al.; O'Neill 1995) and was seen and photographed by hundreds. What was assumed to be the same bird was again present on the same beaches 25 July–5 September 1996 and 30 August–18 October 1997 (m.obs.).

Mew Gull (*Larus canus*)

Status

Vagrant. Three records.

Nonbreeding

An adult was seen at Point Judith 9 April 1974 (RAC). Another was at Sachuest 8 October 1995 by Nelson Roy et al., who submitted a detailed description. Neither of these 2 birds were identified to subspecies (European *L. c. canus*, western North American *L. c. brachyrhynchus*, or Siberian *L. c. kamtschatschensis*). An adult *L. c. kamtschatschensis* was at East Providence 6 January to 11 February 2006 (CPe; SSM et al.; see Photo 14), and a fourth bird of unspecified race was at Little Compton 26 March 2010 (fide RFa). The AOS does not recognize the "Kamchatka Gull" *L. c. kamtschatschensis* as having occurred in North America, and we therefore follow Dickinson and Remsen (2013) in referring to these taxa as above. Numerous observations in northeastern North America seem likely to be *L. c. kamtschatschensis* on criteria summarized by Adriaens and Gibbins (2016).

Ring-billed Gull (*Larus delawarensis*)

Status

Abundant migrant and winter resident. Locally common nonbreeding summer resident.

Historical

Ring-billed Gulls were scarce in Rhode Island at the turn of the twentieth century. They have greatly increased in accord with their increase as breeders in eastern Canada; for example, from 1000 to 80,000 pairs at their largest colony on Little Galloo Island, eastern Lake Ontario, during the period 1945–1967 (Bull 1974), and from 20,000 to 700,000 pairs in Michigan between 1960 and the 1980s (Brewer et al. 1991). There was an average annual increase of 11% in the Great Lakes and Upper St Lawrence River between 1976 and 1984 (Blokpoel and Tessier 1986). By the mid-1990s, there were 15,000 pairs breeding nearby on Lake Champlain, and scattered inland colonies were in New Hampshire, Maine, Massachusetts, and the Canadian Maritimes.

According to Merriam (1877), this gull was "not rare" in Connecticut in winter, but a vast decrease took place in the next several decades due to egging on the Great Lakes (Bent 1921). Howe and Sturtevant (1899) did not know of a single record for Rhode Island; the first record was an 1891 specimen discovered later by HSH in RWPM. There were several later reports just after the turn of the twentieth century (Mearns 1901, 1902) that suggest that by 1899–1902, Ring-billed Gulls were rare winter visitors.

Nonbreeding

Ring-billed Gulls visit ponds and reservoirs and frequently rest on golf courses, shopping mall rooftops, and empty parking lots. Favored habitats at all seasons are the mudflats of coastal ponds, fields near the coast, and such harbors as Galilee and Providence. Unlike the larger breeding gulls of Rhode Island, they rarely frequent dumps or directly scavenge with other gulls at such facilities as the fish-packing plant at Galilee but are often present in numbers at upstate shopping centers and in the vicinity of fast-food restaurants. Unlike the larger gulls, Ring-billed Gulls are generally scarce over offshore waters and almost never attend fishing vessels. They are relatively uncommon at Block Island.

Fall and winter maxima. 2476, Newport–Westport CBC, 17 December 2011 (REm et al.); 2070, Seekonk River, 16 October 2005 (MLy); 1500, East Providence Reservoir, 21–23 February 1983 (RAC); 1500, Diamond Hill Reservoir, 7 March 2002 (MLy); 1000, Matunuck, 25 November 1982 (DLK, RAC). The earliest arrival of a juvenile was at Galilee 17 July 1962 (DLK).

Summer. Flocks of up to 250, mostly immatures, are now present throughout the summer along the South Shore and in Narragansett Bay, especially near the Providence sewer outlet at Field's Point and in Watchemoket Cove.

Herring Gull (*Larus argentatus*)

Status

Abundant resident and transient near salt water and common on fresh water throughout the state from Wallum Lake southward. Nests on coastal islands. In 2016, 1371 pairs of Herring Gulls nested at 13 localities in Rhode Island (CRa).

Historical

Herring Gulls colonized New England in the late 1800s, and the first Rhode Island nesting occurred at Narragansett Bay in 1927. Sturtevant told Drury (*Bulletin Massachusetts Audubon Society* 44:4) that during his youth in the Newport–Middletown area (ca. 1870s), he spent a period of 4 years without seeing a single Herring Gull, though HSH's journal describes them as being very common during winter and spring in the Providence River and Narragansett Bay in the 1890s. Howe and Sturtevant (1899) described them as abundant winter residents but indicated that they were uncommon in summer. They increased during the early part of the nineteenth century to such an extent that ESt saw 1000 at Third Beach, Middletown, 1–2 July 1936 during a herring run.

Breeding

Herring Gulls bred in Massachusetts in 1912 and on Long Island after 1931. The first breeding in Rhode Island dates to 1927, when HNG found unfledged young on

"Hope Island" (more likely on nearby islets) in Narragansett Bay. They then bred regularly on the islands off Sakonnet Point from 1937 on. The colonization and spread of Herring Gulls into New England is extensively documented by Kadlec and Drury (1968) and Drury (1973, 1974) and updated by Nisbet et al. (2013).

Sakonnet islands. The first Rhode Island Herring Gull colony appeared on The Clumps, a 1-acre (0.4 ha) island between East Island and West Island, where RCC found 2 downy young 18 July 1937. David Emerson found 10 nests there in 1938 and 41 nests in 1940, when an additional 9 nests were found on nearby West Island. There were 67 nests on these 2 islands in 1941, and 144 nests on the Sakonnet islands, including East Island, in 1944 (DLE, RCC). The 3 islands had 315 nests in 1946 (RBo, RCC), but in 1948 there was a temporary decline to 150 pairs. About 400 and 500 pairs were present in 1951 and 1952 according to Clement and Woodruff (1962), but data on the Sakonnet colony from an egg-spraying trip conducted on 24 May 1952 by USFWS personnel indicates that 3759 eggs were sprayed with oil, suggesting a population of about 1300–1500 pairs. A total of 4042 eggs were sprayed in 1953, and the population was estimated at 1400–1600 pairs. There was no further record until 1961, when REW estimated 2300 pairs on these 3 islands. Similar numbers, remarkably large by later standards, were present on the Sakonnet islands through the mid-1960s or later, and during this period 2 or more pairs regularly nested (usually unsuccessfully) on the (mainland) tip of Sakonnet Point itself. These birds fed at the Middletown dump, before it closed in the late 1960s. With closure of this dump, numbers on the Sakonnet islands began to dwindle. Estimated nesting pairs on the 3 main islands and 2 rocks north of West Island varied between 335 and 604 nests, with an average of 470 (RLF, JEM) in 1976–1990 but had declined to 196 pairs in 2016 (CRa).

Sandy Point Island, Watch Hill. This long sandbar was formerly connected to Napatree Point until it was broken through during the 1938 hurricane. Gulls started nesting there "slightly after 1938" according to the owner of the island, even though none were noted breeding by DLK in 1940. By 1946, 150 pairs were present (DLK), and these increased to 250–300 pairs in 1947 and 350 pairs in 1950 (DLK). Later estimates of breeding pairs include 1250 in 1962, 750 in 1963, 2500 in 1965, and 1900 in 1972 (aerial estimates; WHD). There were 1000 pairs there in 1995, but these declined to 327 pairs nesting in 2016 (CRa).

Block Island. The colony in the dunes west of Sachem Pond has been one of the largest in Rhode Island and witnessed one of the fastest expansions of any state colony. Twenty adults, and a nest and 1 egg, were found 8 June 1958 (B. Littlefield, fide EDi). By 1961, this number had increased to 250–300 pairs (MSl). Estimates of breeding pairs include 750 in 1962, 450 in 1963, 1300 in 1965, and 2000 in 1971 (WHD, MSl). By 1970 and perhaps earlier, this colony had expanded southward to the barrier dunes of Great Salt Pond, where there were about 100 nests in 1973 (MSl,

RLF). There were about 1000 pairs nesting in 1988, but these declined to 119 pairs nesting in 2016 (CRa), almost surely a consequence of the cessation of open dumping at the landfill.

Rose Island. Breeding was first noted in 1972 when more than 150 pairs were present (RLF). This colony peaked at 520 pairs in 1992; thereafter, the colony fell to 154 pairs in 2016 (RLF, CRa et al.).

Gull Island. This tiny island adjacent to the east end of the Newport–Jamestown Bridge, once used by the U.S. Navy, contained an active colony of Common Terns in the 1970s and early 1980s and later attracted a pair of Great Black-backed Gulls and up to 5 pairs of Herring Gulls. The colony was abandoned in 1981.

Dutch Island. One nest was found on the west side of the island in 1981, but none was present with Great Black-backed Gulls on the island's south end in that year or other years (JEM, RLF). Continued checks in later years have proved negative.

Ninigret Pond. There are 2 records (1976 and 1985) of single nests behind the barrier beach on a 2-acre (0.8 ha) marshy island often used by nesting Common Terns. This is the only known site for breeding in the mainland portions of western Rhode Island between Napatree Point and Narragansett Bay.

Hope Island and associated islets. This was one of the largest Herring Gull colonies in Rhode Island. It peaked at 1500 nests in 1980 but subsequently declined to 174 nests by 2016. About 100–150 pairs of Herring Gulls nested on Little Gooseberry Island during the 1960s but have since disappeared; the island was recolonized in 1995 and has been continuously occupied since that time, with a peak of 15 nests in 2007.

Dyer Island. There were 300–500 pairs of Herring Gulls nesting during the period 1961–1972 (REW, WHD) and 140 pairs in 2016 (CRa).

Dumpling Islands. There are 6 rocky islets and 6 smaller rocks located along the east side of Jamestown that have been used as breeding sites for gulls and terns. Some of these islands may have been colonized by Herring Gulls as early as the mid-1940s, but the first definite count is 25 adults nesting in 1950 (DLK, MDa, RCC); 385 pairs were estimated in 1972 (WHD) and 17 pairs in 2016 (CRa).

Rocks off Land's End, Newport. The granite rocks immediately south of Land's End, Newport, supported a maximum of 6 pairs during the period 1978–1981, after which none have been present.

Ocean Drive, Newport. Gulls have recently colonized a group of islets west of Price's Neck that had formerly been inhabited by terns. On an island east of Price's Neck, where terns have been present intermittently, single nests of Herring Gulls were present in 1978 and 1980 (RLF). There were 94 nests in 2005, down to 40 nests in 2016.

Napatree Point. A nest with 2 eggs was discovered on the "hook" 8 June 1984 (RLF, JEM). Other nests were at the same location in 1988 and 1990 (CRa).

Hog Island. Single pairs nested on the southwest beach in 1985 and 1988 (RLF, JEM), but none have nested there since.

Nonbreeding

New England Herring Gulls probably reached peak abundance in the 1970s (Veit and Petersen 1993; Nisbet et al. 2013) and have been declining since about 1980. The decline is mainly due to the closing of open garbage dumps and the decline of fisheries, thus the consequent reduction in food supply for gulls. Christmas Bird Count data for the years 1995–2004 suggest that 260,000+ Herring Gulls winter along the U.S. east coast between Massachusetts and Virginia (Nisbet et al. 2013).

Winter maxima. 9621, Newport–Westport CBC, 18 December 2010; 9000, Greenwich Bay, 30 January 2006 (DFW); 7500, Moonstone Beach, 10 November 2007 (STs).

Thayer's Gull (*Larus thayeri*)

Status

Rare winter visitor.

Nonbreeding

A single 1st-winter bird, probably a Thayer's Gull, was at Belleville Pond 28 January 1965 (RLF). Close study of birds in attendance at the Galilee fish plant have yielded most of the other reports, all made since 1983 (when Thayer's Gull was given species status by AOU). These include 2 regularly seen there from 8 January to 27 March 1983 (RAC, DLK; photos); at least 2 and probably 3 in the Galilee area from 2 February through 28 April 1984 (DLK, RAC); a 1st-year immature 26 December 1987 at Galilee (RAC, DFi; photo), and another 1st-year immature 30 January 1988 at Galilee (SG). A 1st-winter bird was at Block Island 11 November 2010 (PJL, SSM). One was at Cox Ledge, 20 December 2014 (PLE).

Note: The AOU (Chesser et al. 2017, 2020) determined Thayer's Gull to be conspecific with Iceland Gull, largely based on inferred hybridization in central Canada (Weir et al. 2000). Nevertheless, this taxonomic situation is not demonstrably different from several other species pairs of large gulls still treated as separate species; e.g., Western and Glaucous-winged in the Pacific Northwest and Herring and Glaucous in Iceland and Alaska. Therefore, we follow Dickinson and Remsen (2013) and Browning (2022) in retaining Thayer's Gull as a separate species, include records of this distinctive form, and encourage future documentation of the occurrence of Thayer's Gulls in Rhode Island.

Iceland Gull (*Larus glaucoides*)

Status

Uncommon winter resident, mainly at the coast. Rare in summer.

Historical

Iceland Gulls were unrecorded in Rhode Island until 1914, when an immature was shot at Block Island and obtained by HSH. As with most large gulls, they increased very substantially during the twentieth century, probably mostly because of increased food supply at garbage dumps and commercial fishery operations. As far as can be determined, all Iceland Gulls in Rhode Island are of the North American race *L. g. kumlieni*.

Nonbreeding

Iceland Gulls are more numerous than Glaucous Gulls in Rhode Island and occur both earlier in fall and later in spring. Iceland Gulls regularly appear during November and remain into April. At any season, this species is very much less common in our area than north of Boston or on Nantucket (Veit and Petersen 1993).

Winter. Most occur around dumps and fishing operations, especially at Galilee.

Winter maxima. 13, Galilee, throughout January 1983 (RAC); 7, Galilee, 16 January 1988 (SG); 5, Galilee, 16 February 1984 (DLK); 5, East Providence Reservoir, 8 February 2005 (RFa); 5, Greenwich Bay, 15–23 January 2009 (DWi).

Summer. There is a summer specimen of an immature collected at Field's Point 28 June 1965 (RLF). An adult was seen at Point Judith 18 May 1941 (RCC, RBo, WHD). There are several records from 2003 on from early May to mid-June.

Extreme dates. 23 October to 28 June.

Lesser Black-backed Gull (*Larus fuscus*)

Status

Rare but increasing fall and winter resident in coastal areas. Increasingly seen during summer as well.

Nonbreeding

The first observation was of 1 that was photographed at Warren 18 November 1979 by RBo. Three other records in the Warren River–Touisset area through 1982 may pertain to the same bird. As the rate of occurrence increased through the 1990s and 2000s, many reports pertained to wintering adults faithful to particular sites, most notably a bird that returned to Grassy Point, Ninigret NWR, every year from December 2002–2016. They have now occurred in every month of the year, with peak numbers in October–November. As of 2012, up to 10 individuals per year have been found. Lesser Black-backed Gulls have expanded their breeding range to Iceland (late 1960s) and Greenland (late 1980s), and the Greenland population exceeds 3000 pairs. Despite widespread speculation to the contrary, North American Lesser Black-backed Gulls seem to derive from the Greenland nesting population (Zawadzki et al. 2021).

Maxima. 13, Cox Ledge and surrounding waters, 4 November 2015 (K. Mueller); 8, Block Island, 2 October 2011 (DFi, STa); 6, Block Island, 7 October 2005 (DFi); 3 immatures, Galilee, 26 February 1983 (GLB).

Glaucous Gull (*Larus hyperboreus*)

Status

Uncommon and irregular winter visitor and rare transient. Very rare in summer.

Historical

This species was not listed by Howe and Sturtevant (1899), who overlooked a record of one found dead on the top of Cormorant Rock, Middletown, by G. H. Mackay 18 April 1892. Mearns (1902) found Glaucous Gulls at Newport in 1900 and 1901 and claimed these as first state records. They were then nearly unreported again until about 1938.

Nonbreeding

Glaucous Gulls are most often found in harbors and other places, such as Galilee, where fish are processed or at dumps near the ocean. They are less common than Iceland Gulls inland. Glaucous Gulls are seldom seen in the fall before late December and occur most frequently in January and February, decreasing by early April. They are rare after April, but there are about 15 records from June through early September.

Maxima. Five, Galilee, through January 1983 (RAC); 4, Galilee, 30 January 1981 (RAC); 4, South Kingstown CBC, 22 December 2007.

Hybrids. Glaucous × Herring Gull hybrids have been reported at Oakland Beach 20 March 2007 (CNu) and on the Newport–Westport CBC 3 January 2009 (REm et al.).

Great Black-backed Gull (*Larus marinus*)

Status

Abundant and conspicuous resident in coastal areas and regularly common inland. A total of 1343 nesting pairs was counted at 15 locations in 2016 (CRa, RLF).

Historical

Howe and Sturtevant (1899) called this bird a "not uncommon winter resident" along the outer coast (26 August–23 April) but stated that they were seen in Narragansett Bay only occasionally. They were apparently not present during summer. Mearns (1902), making observations at Newport in 1899–1902, called them "rather rare winter residents" and mentioned single birds that occurred regularly near Newport harbor. Hathaway saw 12 over Roger Williams Park 25 November 1889, which may be the highest count from the turn-of-the-twentieth-century era. Allen (1909) called them "common winter residents," a surprising statement based on their status in Massachusetts at the time. One was shot at Block Island in 1911 and shipped to RWPM, where $10 was paid for it and the fact noted that it was only the "second" (actually the third) specimen for the state. A gradual increase occurred in the next two decades, with maxima at Block Island of 10, 9 November 1915, and 45, 8 November 1922 (EDi). Lynch noted that by 1934 they occurred commonly in winter with Herring Gulls on large freshwater ponds. A few began breeding in Essex County, Massachusetts, in 1931 (Eaton 1931) and in the early 1940s at Boston Harbor, Martha's Vineyard, and elsewhere. In Rhode Island, Great Black-backed Gulls first bred in 1944, but the breeding population remained small prior to about 1960. Since then, the species has spread more rapidly to as many or more islands as the Herring Gull.

Breeding

Sakonnet islands. On 30 June 1944, DLE and RCC found 2 pairs and 1 nest with eggs among nesting Herring Gulls on The Clumps off Sakonnet Point. Two pairs were again found there 23 June 1946 by RCC and RBo, and 1–2 pairs bred there through 1951 (DLE). Twelve pairs were found on West Island in 1961 with "several" more pairs on nearby East Island (REW). The number of nests in 1976–2016 averaged 213 (RLF, JEM).

Briggs Beach rocks. Single nests were located on the most westerly of these 3 rocks in 1989 and 1990 (RLF, JEM), but were subsequently abandoned.

Sandy Point Island, Watch Hill. Several young were found 12 July 1961 (REW), long after the island had been colonized by Herring Gulls, suggesting that a few may have been breeding as early as the late 1950s. There was a peak of 539 nests in 1997, reduced to 159 nests in 2016 (CRa).

Pawcatuck River islands. A single nest was found with Common Tern nests on Horace Island in the Pawcatuck estuary in 1990, and in the same year a pair was noted with 2 young in a nest on a 15-foot Osprey platform on an island to the south of Horace (RLF, JEM). The platform has since disappeared, and there has been no nesting since then.

Dyer Island. Several pairs were first noted on this low island in 1961; they had probably been breeding here for some time (REW, RLF). There was a peak of 247 nests in 1993, down to 88 nests in 2016 (CRa).

Dumplings. One or 2 pairs nested in 1961 (REW). There were 138 nests on all the Dumpling Islands combined in 1986 and 39 nests in 2016 (CRa).

Block Island. Breeding began in the north dunes in 1961, and by 1971, 100 pairs were estimated to be breeding there (MSl). There was a peak of 756 nests in 1994, down to 560 nests in 2016.

Little Gould Island. Herring Gulls began nesting on Little Gould Island, known for its heron colony, in 1976 and a nest of Great Black-backeds was found in 1978 (RLF, JEM). Only 1 or 2 pairs were present for nearly a decade, after which numbers climbed to 25 nests by 1990, 35 nests in 2001, and 23 nests in 2016.

Big Gould Island. First found breeding with a single nest in 1977 (JEM). Numbers gradually increased to 200 pairs by 1990 (JEM et al.), 250 nests in 1990, and 125 nests in 2016.

Little Gooseberry Island. Only a few pairs were present among numerous Herring Gulls during the period 1961–1962 (REW, RLF), but with later defoliation of the island by guano, this species totally replaced Herring Gulls (which have since disappeared). There was a peak of 26 nests in 2004, then 13 nests in 2016 (CRa).

Rose Island. Probably present by 1972 or earlier; 50 pairs were estimated 10 June 1976 (RLF), 85 nests in 1995, and 67 nests in 2016.

Seal Rock, Mount Hope Bay. A single nest was noted in 1988 (RLF), marking for a time the northernmost Narragansett Bay location, but there has been no nesting in recent years.

Seekonk River. A pair had a nest among those of Double-crested Cormorants on a platform associated with a series of isolated pilings opposite the mouth of the Ten Mile River 13 June 1995 (RLF, JEM). This is currently the northernmost Narragansett Bay location and is well to the north and inland of any known nests of Herring Gulls. They were still nesting here in 2016.

Nonbreeding

With a vast increase as a local breeding bird in the last 50 years, these gulls are now abundant or at least common year-round in most coastal areas, and moderate or sometimes sizable aggregations of young birds accumulate in summer at points all along the coast and in Narragansett Bay. They are regular at dumps and ponds in eastern inland sections at all seasons but are proportionately less common than Herring Gulls at inland locations (once an exceptional 45 at Scituate Reservoir, 29 January 1983; REn). Great Black-backeds generally outnumber Herrings offshore in winter; e.g., 200 Great Black-backeds and only 10 Herring Gulls at Block Canyon 4 March 1995 (RLF, DLE, PAB et al.). Along the mainland coast, the largest winter aggregations occur in the Sakonnet and Point Judith areas, where large numbers roost on offshore islands or the Point Judith breakwater.

Winter maxima. 2421, Block Island CBC, 20 December 1989; 1388, Block Island CBC, 21 December 2006 (CRa et al.).

Brown Noddy (*Anous stolidus*)

Status

Vagrant. Three records.

Nonbreeding

A Brown Noddy was picked up dead on Block Island a week after the hurricane of 19 September 1928, but unfortunately the specimen was not preserved (EDi). On 12 September 1960, at the height of Hurricane Donna, 3 were seen flying over the center of Block Island by JBa. An adult was seen at Point Judith during Hurricane David 6 September 1979 (RAC). Brown Noddies appear much less frequently in the Northeast following hurricanes than do either Sooty or Bridled Terns.

Sooty Tern (*Onychoprion fuscatus*)

Status

Vagrant. Often, but not exclusively, after tropical storms. One winter record.

Nonbreeding

Summer/fall. There were 3 old records of this pelagic tern known to Howe and Sturtevant (1899). These records are 1 shot at Point Judith in September 1876 after a great storm (F. T. Jencks), 1 found dead in Newport, 1877, and an adult male taken inland at Woonsocket on 16 July 1897 and placed in the Smith collection at Roger Williams Park. None of these are extant, although a mounted adult at RWPM without data may be the Woonsocket bird. An immature was shot by C. B. Clarke at Coddington Point, Newport, on the astonishing date of 8 January 1908 and acquired by HSH; the specimen is now in the possession of ASRI.

The afternoon after the hurricane of 14 September 1944, 12 birds were seen north of Gould Island in the East Passage (RBo, in a boat). There were no further reports until Hurricane Donna 12 September 1960, when an unprecedented flight occurred. At the height of this storm, 25 birds "in tight formation" were seen struggling against the wind at Touisset, Massachusetts, less than a mile from the Rhode Island line (RBo). Two were also seen passing over central Block Island during the storm, and an immature with 7 adults was resting there the next day (JBa). Two others were seen at Newport, also on 13 September (REW), and an adult was picked up dead at Howlands Pond, Acoaxet, about a half-mile from Quicksand Pond (RLF; specimen in Peabody Museum of Salem). On 14 September, 1 was picked up at Block Island in a weakened condition (MSl, JBa); it had been banded *during the storm* on Long Island. This or another bird (an immature) was still present at Block Island on the 15th (JBa).

There were no further reports until the day after Hurricane Belle 10 August 1976, when 1 was seen at Cox Ledge (F. Purnell, RAC, DLK et al.). All records, however, paled in comparison with events during Hurricane David, whose inland path created southwesterly winds over Rhode Island on the afternoon of 6 September 1979. In all, 62 were seen that day along the South Shore, mostly passing Point Judith to the eastward, and an additional 15 were reported from the South Shore and at Cox Ledge the next day. During Hurricane Bob on 19 August 1991, 1 was seen at Point Judith (AFr), and 1 was at Sakonnet Point on 17 September 1999 after Hurricane Floyd (REm).

One was at Harbor of Refuge after the passage of Hurricane Irene 28 August 2011 (CRa). A juvenile passed Point Judith heading southwest on 6 September 2016 during Hurricane Hermine (PAB). Singles were seen in the Seekonk River, 4 August 2020 (Nemes, Kneidel; photo), and the next day at Charlestown Inlet (DFi, STa).

Spring. Four days after the passage of "a southerly storm of hurricane proportions," 2 were reported at Block Island 23 May 1928 (fide EDi). One of the very few occurrences not related to a storm, and one of very few spring records, is a bird seen off the Sakonnet breakwater 18 May 1986 (DLE); 3 adults were also seen from Fort Cronin during a coastal storm 16 April 2007 (PAB, MT), and an adult male was found dead at Cockeast Pond 17 April 2007 and taken to AMNH (GDe, CRa). Other occurrences that lacked any connection to storms are 1 at Cox Ledge on 14 July 1987 (DFi), and 1 reported from Napatree Point 11 August 1992 (RFl).

Bridled Tern (*Onychoprion anaethetus*)

Status

Vagrant. From tropical waters; mainly, but not entirely, during and after tropical storms.

Nonbreeding

On 12 September 1960, during Hurricane Donna, an adult struck a parked automobile and was picked up dead at Block Island by an island resident and brought to EDi. It is now mounted in the Block Island Collection. On the remarkable date of 7 January 1962, after a storm that deposited more than 4 in (10 cm) of rain, an immature was found dead at Block Island and brought to EDi. This bird is now in the RWPM mounted collection, where it was recently examined. There is a sight record of an adult seen from a boat among the Sakonnet Point islands on 23 July 1963 by J. Mahoney and others. This bird circled the boat on a sunny day and was clearly differentiated from the similar Sooty Tern. During Hurricane David in the afternoon and evening of 6 September 1979, 5–7 adult Bridled Terns were seen with many Sooty Terns passing Point Judith (RAC, RBu, GLB et al.), and 2 adults were seen at the same location the next day (RAC et al.). Hurricane Bob of 19 August 1991 produced 1 at Newport and 1 at Point Judith on the day of the storm (SRs; AFr). Eleven birds were reported statewide after the passage of Hurricane Irene 28 August 2011 (m.obs.; see Photo 15), and 5 were seen and photographed at Block Canyon 3 September 2015 (PLE et al.), supporting the notion that Bridled Terns ordinarily disperse northward over Gulf Stream and slope waters, even in the absence of storms.

Least Tern (*Sternula antillarum*)

Status

Locally common breeder and fairly common spring and fall migrant.

Historical

Sir Charles Blagden (1903), writing during the Revolution (1776–1780), described the "mosquito gull" as "very frequent about ponds near the beaches" on Aquidneck Island, where "on July the 2nd I shot [1] with a minnow in its mouth, which it was taking to the young then hid in the sand near the beach." By 1870, Least Terns were extirpated in Rhode Island. Newton Dexter considered them to have been common in earlier times (1850s and 1860s), but by 1899 they had become so rare that Howe and Sturtevant (1899) not only knew of no suggestion of breeding, they listed only two records of transients. The species then remained totally unrecorded at any season through 1913. Hathaway's monitoring of the Westerly and Charlestown beaches through the 1930s from his cottage at Quonochontaug, as well as CED's successful efforts to collect Piping Plover eggs (without mentioning Least Terns) between Point Judith and Matunuck during the period 1913–1927, all suggest a lack of Least Terns along the South Shore for at least the first third of the twentieth century. With protection, a regional increase occurred, and the first Rhode Island breeding in perhaps a half century or more was discovered by HNG on Briggs Beach, where there were two nests on 4 June 1929. In 1930, two pairs were again on territory at the same location. There is no further record until 1936, when 8 to 10 pairs were present at Briggs (HSH) and a pair was located on Sakonnet Point (then a substantial sandbar; RCC).

After the great hurricane of 21 September 1938, hardly a house was left standing on the low-lying barrier beaches from Napatree Point to Jerusalem, and these newly cleared beaches provided ideal nesting habitat for Least Terns. In 1954, Hurricane Carol then renewed the open habitat on the beaches almost as effectively as in 1938 by re-breaching most of the coastal ponds, flattening dunes, eradicating beachgrass, and removing many structures that survived, or were rebuilt after, 1938.

Breeding

One of the most important forms of disturbance has come from vehicles driven on beaches and over open sandy areas, but this practice has mercifully been curbed or eliminated in many areas by various regulations (all of Little Compton very early; Napatree Point and the Trustom–Matunuck area after about 1960). The entire stretch of beaches between Watch Hill and Weekapaug breachway was still open to off-road vehicles until the late 1990s, and some areas were even beginning to have traffic problems. Beaches between Misquamicut and Watch Hill and the beach west of Charlestown Breachway have been managed such that vehicles are restricted during tern and plover nesting seasons. Similar management began to get underway in Rhode Island in the late 1980s with permanent exclusion of the public from Moonstone Beach and temporary fencing of breeding areas at Quicksand, Briggs, East Matunuck, Misquamicut, Napatree, and elsewhere. Seasonal exclusion of vehicles from additional South County beaches in the late 1990s has already begun to make an important contribution to improved nesting success of both Least Terns and Piping Plovers. From 1986 to 1994, the state population ranged from 250 to 430 birds.

Napatree Point. Before the 1938 hurricane, 39 summer houses and military buildings extended over most of the length of the peninsula. The hurricane washed all of these away. Yearly counts of the numbers of pairs include 10 and 20 in 1941 and 1943 (Stoddard); 30 in 1944 (HSH); 12 in 1945, a sharp decrease attributed by HSH to disturbances from the 1944 hurricane; 30 and 20 in 1946–1947 (DLK); 6 in 1954 (DLK); 25 in 1959 (DLK); 10 in 1960 (DLK); and none in 1961 (REW). There was a peak of 140 adults nesting in 1985 (CRa) and 100 adults in 1998. The colony has not exceeded 48 adults since then, and there was no nesting in 6 years since 1998. Chronic mammalian predation limits recruitment at this site.

Sandy Point Island. This site was connected to Napatree Point until 1938 and only sporadically visited by observers before the 1970s. There were 20 pairs in 1946 and 1947 (DLK), none in 1961 (DLK), 15 pairs at the west end in 1965 (RLF), and 4 pairs in 1969 (M. Erwin). In general, success has been very poor because of masses of bathers using the island. Thirty adults nested in 1985 (CRa). In 1998, 400 adults appeared and began to nest on a dredge spoil deposition. There was no follow-up to determine success, but birds did not reappear the following year, which suggests that they failed. There was no nesting again until 2010, when 800+ adults abruptly appeared. Ninety-one young fledged from this colony (USFWS). They have nested in 3 out of the last 6 years in more modest numbers, but no young were fledged. Gull predation of chicks limits reproduction. The site is now managed and monitored by USFWS in agreement with the Connecticut-based group Avalonia Conservancy.

Misquamicut. There were 7 pairs in 1961 (REW, DLK), 130 adults 1–28 June 1988 (CRa), and 149 adults and 75 young there through July 1988 (CRa). This has been one of the more consistently occupied colonies in recent years, with nesting every year and a peak of 214 adults in 2006. The colony has produced young every year since 2004, when monitoring by USFWS began.

Quonochontaug. Least Terns bred irregularly in the late 1960s, perhaps earlier; there was still 1 pair in 1971 east of the breachway (DLK, CWo). There was breeding on the beach near the west end of Quonochontaug Pond in 1949 and 1954 (ESa) and perhaps in other years. In 1971, several pairs bred in dunes southwest of "Nokes Island" (ESa). Nesting was sporadic and probably limited by off-road vehicle use before USFWS took over management in 2004. Since then, terns have nested consistently, with a peak of 162 adults in 2012. Habitat was improved by Hurricanes Irene (2011) and Sandy (2012). In 2012, 111 young fledged.

Ninigret Pond. Ten pairs were on territory near the west end in 1974 (RLF). As at Quonochontaug, this site allows off-road vehicle use during summer. Least Terns consistently nest here, and they also benefit from USFWS management and monitoring. There was a peak total of 150 nesting terns in 2008. Forty-six young fledged in 2014.

Green Hill Beach. Ten pairs were present in 1961 (CWo), but none have nested since then.

Trustom Pond. A varying colony of up to 18 pairs, usually 3–6 pairs, has been present at least from the early 1960s (v.o.). Moonstone Beach, the barrier beach for Trustom Pond, has been a federal refuge since the early 1980s and has had the most intensive and consistent Least Tern management of any Rhode Island site. A maximum of 222 adults nested in 2015 and fledged 142 young.

East Matunuck. A colony of 5–20 pairs was present annually from 1950–1977. There were 20 pairs in 1976 and 37 birds at Potter Pond on 9 June 1971 (RLF). This site was marginal for many years but has improved slightly with USFWS oversight. The best recent year was 2014, when 54 adults produced 23 young.

Galilee. A colony formed in the early 1950s, first on fill dumped onto the marsh for the Escape Road, later on dredging spoil pumped onto the marsh's west side. Douglas L. Kraus estimated 100 birds here throughout July 1952, and 40 were present on 11 June 1954 (DLK). They were finally forced to abandon the dredging spoil because of the growth of beachgrass and *Phragmites*.

Narragansett. This site has been occupied during several recent years, with a peak of 92 adults in 2007 and 38 adults in 2014. Twenty-two young were fledged in 2014.

Middletown. Other than a nest found in the Second Beach area in mid-July 1938 (WHD), no other signs of territoriality have ever been reported on Aquidneck Island.

Seapowet. Two pairs were present in 1951 on the outer beach north of the main inlet into the Sakonnet Passage (DLE). Two pairs were present on the beach north of a small tidal creek on 6 June 1990 (JEM), and 8 pairs had eggs and young there 20–25 June 1991 (JEM, RLF). Forty adults were present in 2015.

Sakonnet Point. A pair was on territory here in 1936 (RCC), but there has not been any further evidence of breeding, and most of the point has washed away.

Briggs Beach. Least Terns were first found on territory here in 1929 (HNG); a variable but usually dependable colony is still present. Counts by David Emerson since 1937 have usually varied from 4 to 15 pairs, with 40 pairs in 1960. Little Compton had 65 pairs in 1946 (RCC, RBo). The site is private property and is managed for terns and Piping Plovers by TNC through a grant from RIDEM. The site is not occupied every year, but 112 adults nested in 2003 and 66 adults in 2016. Best productivity was in 2016, when 24 young fledged.

Quicksand Pond. There have been as many as 20–30 pairs (generally 5–15 pairs) in recent years, and 60 birds were here 10 July 1988 (DLE). This is a TNC preserve and is managed with help from a RIDEM grant. The site is more consistently occupied than Briggs, with a peak of 132 adults in 2007. Twenty young fledged in 2013.

Block Island. About 3 pairs were present at Sandy Point 1963–1966 (MSl); reported again in 1970 or 1971, but none since.

Barrington. A few pairs may have bred from time to time on Rumstick Point, the northernmost Rhode Island breeding location. Two pairs were on territory 9 June 1977 (RLF).

In 1982, there was a relatively sudden increase in the statewide population, which temporarily rose to an estimated 200 pairs (CRa, JEM, RLF et al.). This sudden increase would have been impossible without movement of birds from adjacent states.

Nonbreeding

Least Terns arrive somewhat later in spring and depart somewhat earlier in fall than other breeding terns. The species is normally rare before early May. The earliest counts of larger groups include 10 at Galilee 5 May 1965 (KKi), 25 along the South Shore 6 May 1962 (DLK), and 59 at Galilee 7 May 1961 (CWo). The great majority of the breeding population and the few transients are gone by 1 September.

Least Terns do not generally stray inland within New England. There are, however, several Rhode Island records of birds at 2 locations within a few miles of the coast and Narragansett Bay. One was seen over Belleville Pond on 7 August 1982 (CRa, REn). Since 1991, the species has been recorded regularly from mid-May to mid-June at the Great Swamp impoundment: 4 on 25 May 1991 (RDe, MJM et al.); 1 on 5 and 17 June 1992 (SSM); 1–2 from 26 May–1 June 1993 (SSM, CRa); 1–2 from 19–31 May 1994 (SSM); 2 on 22 May 1998; 1 on 2 June 2005 (SSM et al.); and 4 on 26 May 2014 (M. Pagliarini, M. Schenk).

Extreme dates. 25 April 1982, Point Judith (RLF), to 22 September 1941, Little Compton (DLE).

Gull-billed Tern (*Gelochelidon nilotica*)

Status

Vagrant. Very rare in spring and rare in fall. Most records are in late summer and early fall.

Nonbreeding

Spring. Gull-billed Terns are much rarer in spring than in fall, with only 4 records. One was at Quonochontaug 27 May 1970 (DLK), and an adult was moving east past Point Judith 16 May 1982 (RLF). An adult seen on the Galilee flats during a driving rainstorm on the exceptionally early date of 15 April 1983 (DLK), and later relocated there 19–20 April (CRa, REn, RAC), was feeding on earthworms washed onto the roadway. One at Watchemoket Cove 7 May 1984 (RAC) constitutes the only occurrence away from the outer coast.

Fall. Many but not all records of Gull-billed Terns follow hurricanes and tropical storms. After Hurricane Bob 19 August 1991, the largest influx on record occurred, with 5 birds statewide, including 3 at Quicksand Pond and 1 at Cards Pond the day after the storm (DLE; RAC) and 1 at Napatree Point the next day (CRa). The first Rhode Island record was after the hurricane of 14 September 1944, when 1 was seen on 15 September in lower Narragansett Bay off Newport (RBo) and another was found on 17 September at Briggs Marsh (DLE). On the day after Hurricane Carol, 1 September 1954, 1 was seen at Quicksand Pond (DLE). Another at Misquamicut 6–12 September 1954 (RCC, DLK et al.) may or may not have been the result of Carol. Two appeared at Galilee 15–19 August 1958 (DLK) without the influence of any storms. After Hurricane Donna on 12 September 1960, 2 were found at Block Island 16 September by the Brookline Bird Club, and 1 was collected the next day (JBa; MCZ #261401). Other individuals probably driven by Hurricane Donna appeared at Galilee 18 September 1960 (RTL) and at Block Island 3 October 1960 (JBa). Other records include a bird at Block Island 30 September 1967 (RTL), an adult at Quonochontaug 14 August 1971 after an earlier southerly disturbance (GRa), and 1 at Trustom 4 August 1973 (CWo). After Hurricane Belle on 10 August 1976, singles were seen at Quonochontaug and Ninigret Ponds, 11 and 12 August (STs, DWa), and perhaps another bird at Briggs Marsh, 14–15 August (DLE, REm, CWo). Further records, none during or after storms, are for an adult at Quonochontaug 7 July 1978 (DLK); an immature at Jerusalem 26 July 1977 (DWa); 1 at Quonochontaug 21 July 1991 (RWa); and 1 in basic plumage (and possibly a hurricane relic) at Galilee 6 September 1979 (RAC). One was seen on 2 consecutive days passing westward off Watch Hill 10–11 October 1981 (RLF, CRa, Ren). There were 5 observations at 5 South Shore locations, 19 August–9 September 1983, perhaps involving as few as 2 birds (RAC, DLK et al.). Single birds were seen at Napatree Point, Block Island, and Ninigret Pond, 3–14 August 1991 (CRa; Brinker; JGe, RGe). One was at Sachuest marsh, 11 September 2019 (RTL); and another at Charlestown Breachway, 20 June 2020 (Alan Kneidel; photo).

Caspian Tern (*Hydroprogne caspia*)

Status
Uncommon spring and fall coastal migrant, sometimes more numerous after tropical storms.

Historical
Howe and Sturtevant (1899) knew of only three records but quoted Newton Dexter as saying that he had seen it many times [*sic*] off Sakonnet, where it "usually [flies] high and [is] rarely killed." These nineteenth-century records and those from modern times seem most likely to pertain to transients from inland colonies; those that have appeared in modern times after tropical storms likely originate from the southeastern United States.

Nonbreeding
Spring. Caspian Terns move through quickly in late April and early May.

Spring maxima. Seven, Trustom Pond, 10 May 1984 (CRa, DLK et al.); 7, Quicksand Pond, 26 April 1980 (RBo et al.); 4, Trustom Pond, 10–17 June 1982 (DVT).

Summer. Rare in midsummer; usually single birds seen.

Summer maxima. Three, Trustom 16 July 1983 (DVT); 2, Trustom, 24–30 June 1986.

Fall maxima. 14 moving west off Warren's Point, 21 September 1952 (RBo); 7, Briggs Marsh, 13 September 1992 (REm); 5, Napatree, 20 September 2002 (CRa).

After the hurricane on 14 September 1944, "several" were at Block Island (EDi), 15 in lower Narragansett Bay on 16 September (RSt), and 1 at Briggs Marsh on 17 September (DLE). After Hurricane Carol on 31 August 1954, about 10 birds were seen at various outer shore locations to 12 September, with a maximum of 4 in Little Compton the day after the storm (DLE). The day after Hurricane Donna of 12 September 1960, about 10 birds were noted along the coast (v.o.). Approximately a half dozen appeared after Hurricane Gloria in 1985 and a similar number after Hurricane Bob in 1991.

Inland. Caspian Terns have been reported once in lower Narragansett Bay (15 September 1963; RMa), once at Tiverton (12 May 1963; M. Gardler), once off Conimicut

Point (24 April 1995; REn), and several times on the Seekonk River (fide RLF). Two appeared at the East Providence Reservoir 14 September 1960, the day after Hurricane Donna (DLE, REW, RLF). Probably different individuals were at the Great Swamp impoundment 28 April 1994 (SSM) and Wordens Pond 7 May 1994 (CRa). Three were at the former site on 2 May 1995 (SSM); 1, Swan Point Cemetery, Providence, 7 May 2015 (RFa); 5, St. Mary's Pond, Middletown, 26 April 2006 (JSJ).

Extreme dates. 11 April 2004, Napatree (CRa), to the exceptional date of 1 December 1973, 1 found resting in a field near Trustom Pond (DLK, FPe).

Black Tern (*Chlidonias niger*)

Status

Uncommon spring and common fall migrant.

Historical

Newton Dexter stated (in Howe and Sturtevant 1899) that they were "very abundant off Sakonnet at times, usually after a heavy southeaster in early September." Sturtevant noted a great many in the Newport area 20–31 August 1900. Hathaway (1913) recorded a notable flight at Point Judith 3 September 1906 that lasted all day during a 30-mph (48-kph) storm accompanied by rain. In this instance, hundreds were seen coming from the northeast and flying into the wind, and later they appeared in "great numbers" over the waters between Point Judith and Newport.

Probably the greatest flight in history occurred in September 1955, unrelated to any storm. On 4 September there were 50 in upper Narragansett Bay. On 10 September there were 250 at the latter location, 300 at Napatree Point, 250 at Point Judith, and 150 in Block Island Sound (v.o.). On 18 September 1955, *2000* were estimated in the vicinity of Block Island (Austin, Isleib).

Because of poor coverage after the hurricanes of 1924, 1928, and 1944, few or no Black Terns were observed, but 20 at Block Island on 20 September 1928 were undoubtedly connected with the hurricane of the day before. After Hurricane Carol of 31 August 1954, about 70 were seen along the outer coast 1–12 September 1954. After Hurricane Donna of 12 September 1960, a great many were seen along the South Shore by J. J. Lynch, and 20 remained for a week at the East Providence Reservoir (RLF). During Tropical Storm Doria on 28 August 1971, more than 100 were seen at Point Judith (SSD), and more than 220 were along the South Shore, mainly at Moonstone, 29 August–4 September 1971 (CWo, SSD et al.). Other recent storms, particularly those of 1976, 1979, and 1985, were not accompanied by large numbers of Black Terns. Thirty or more that appeared after Hurricane Bob 19 August 1991 were also relatively few by earlier (and more recent) standards. After Hurricane Irene on 28 August 2011, 200+ were found along the coast after the storm, including 90 at Trustom Pond NWR and 58 at Napatree Point 29 August. These high counts included many juveniles and may have been the result of high reproductive success as well as displacement or concentration by the storm.

Nonbreeding

Spring. Most spring records are of 1–4 birds, usually in May. Veit and Petersen (1993) discuss spring occurrences in Massachusetts, both coastally and in the interior, and note that these often occur during rain.

Spring maxima. 15, Jerusalem, 27 April 1977 (DWa); 5, upper Narragansett Bay, 10 May 2012 (GDe).

Summer. One was seen south of Block Island 16 June 1987 (DLK et al.). There are several other summer records: 18 June 1961 at Seapowet (DLE) and adults seen inland at Beach Pond 18 June 1932 and 24 June 1928 (A. A. Green, in litt. to HSH).

Fall. In fall, Black Terns are normally irregular transients in small numbers and are sometimes unreported. Early transients, probably subadults or failed breeders, have occurred at Trustom 2–4 July 1974 (CWo) and at Little Compton 8 July 1951 (WBD) and 9 July 1939 (DLE). There are many other reports for later in July, but records rise to a definite peak in late August and early September. Observations decline again through September for non-storm years, and there are late occurrences for 29 September and 16 October (1, Sachuest Point, 16 October 1996; DLE). The latest record is for 2 struggling into a 50-mph (80-kph) southeast storm over Wesquage Pond 24 October 1959 (DGa).

Fall maxima. 2000, Block Island, 18 September 1955 (Isleib, Austin); 43, upper Narragansett Bay, 1 September 2012 (GDe). Note: 1 inland fall record, a bird conservatively reported as *Chlidonias* sp., at Woonsocket Reservoir, 3 August 2003 (MLy).

Extreme dates. Late April–24 October 1959 (above).

Roseate Tern (*Sterna dougallii*)

Status

Common spring and fall transient and fairly common summer visitor. Formerly a regular local summer resident; has not bred since 1982.

Historical

Roseate Terns were uncommon nonbreeding summer residents at the turn of the twentieth century, when all terns were greatly reduced in numbers (Howe and Sturtevant 1899). Charles E. Doe, however was the first to suspect Roseates of nesting in Rhode Island. On 26 June 1897 and again on 12 June 1898, he rowed out to "Gull Island" "in the Atlantic Ocean" off Land's End, Newport, and found a colony of 300 terns nesting in the vegetation of a sandy hollow and about the rocks on the north side of this tiny island. His notes indicate that he thought both species were breeding but despaired of separating Roseate nests from those of the more numerous Common Tern.

Roseate Terns gradually increased after the shooting era, and the first confirmation of nesting was on the Sakonnet Point islands, where HNG found an unstated number breeding in 1929 and 1930 in grassy portions of The Clumps. David Emerson continued these observations after 1938, counting 31 pairs at this location in 1941. However, the increasing numbers of breeding Herring Gulls (which began nesting there in 1937) were soon to eliminate the terns. In 1944, there were only two pairs (and 144 gull nests), and only three pairs of terns in 1946. Numbers fluctuated from two pairs in 1949 to as many as 10 pairs in 1951, but Roseate Terns disappeared after 1952 when 500 pairs of gulls were breeding on the three main Sakonnet Point islands. Roseate Terns also bred at Peckham's Marsh along the Sakonnet Passage in northwesternmost Little Compton, where on 27 June 1936 RCC found a single nest with five eggs in the salt marsh amid a small colony of Common Terns. This is an atypical site for Roseate Terns, and the five eggs indicate two or more females laying in one nest, as the usual clutch size is two.

Except for breeding locations discovered in the last few years, most of the older historical locations lack proof. In 1937, WHD found about 30 pairs on the largest of the Island Rocks, a small group of rocky islets to the east of Sachuest Point in Middletown. In 1938, Karsch reported 25 pairs of Common and Roseate Terns nesting here but did not indicate the relative numbers of the two species. This location may possibly have supported Roseate Terns for many years previously, but its rocky, exposed condition does not provide satisfactory habitat for this species. In 1940, WHD noted that both species were more numerous than usual about the "tern colonies" near Brenton Point, Newport, where he observed 40 Roseates on 25 June 1940. While breeding at this location seems likely, no proof was obtained. An even more likely location for suspected breeding (because of grassy conditions) is on the Dumpling Islands off Jamestown, where a few pairs were in attendance with Common Terns (prior to the massive breeding of gulls) in July 1951 (MDa).

Breeding

After the early 1950s, the Roseate Terns seem to have been extirpated as breeding birds from the Rhode Island coast. In early June 1977, however, 1–4 birds attended 4 outer coastal tern colonies, with 1 pair continuously present and apparently incubating at Horace Island in the Pawcatuck estuary (RLF). Thereafter, visits to this island confirmed the presence of 1–2 nests per year during the period 1978–1980 (RLF, JEM). They have been absent from this site since 1981. The only other recent location for confirmed breeding is on a low island east of Price's Neck, Ocean Drive, Newport, where 2 pairs were present in 1979 and 2 nests were located under beach-peas in 1981 (RLF, JEM).

Nonbreeding

Spring. Roseate Terns generally arrive during mid-May, later than Common Terns. Roseates are much less numerous during spring than in fall. Concentrations (of nonbreeders?) occur at Napatree Point in early June.

Spring maxima. 589 passing west, Point Judith, 27–28 May 2003 (RLF); 150, Middletown, 24 May 1960 (JBa); 150, Green Island, 29 August–3 September 2004 (DWa); 80, Napatree Point, 9 June 1993 (RLF, JEM); 70, Napatree Point, 9 June 2000 (RLF, CRa); 60, Napatree Point, 1 June 1993 (CRa).

Summer. Numerous postbreeding adults and juveniles appear from nearby colonies, particularly from late July through early September, and it is common to see adults (often colorbanded) from the Great Gull Island and Buzzards Bay colonies feeding young at such locations as Napatree Point. Numbers frequently enter Narragansett Bay in late summer, particularly when baitfish are particularly numerous: 150, Green Island, 29 August–3 September 2004 (DWa).

Fall. General departure is early relative to the disappearance of Common Terns and is usually in early September. By mid-September, very few are reported, and there are only a few records for late September. During the last 10 years there have been only 2 counts in excess of 200 birds, and neither of them is more recent than 2007. At Napatree, the best local staging area, Roseate Terns are now scarce in both spring and fall.

Fall maxima. 1000+ at Jamestown through August 1959 (MDa); 500, Napatree, 16 August 2001 (CRa); 450, Napatree, 11 August 1993 (CRa); 400, Napatree, 21 August 2000 (CRa); 250, Galilee, 1 August 1954 (DLK). Unlike Common and Least Terns, Roseate Terns have never been reported away from salt water. They are also proportionately much less frequent than Common Terns in Narragansett Bay and over waters south of Block Island.

Extreme dates. 21 April 2000, Charlestown Breachway (2; D. Abbott et al.), to 12 October 1937, Little Compton (HNG).

Common Tern (*Sterna hirundo*)

Status

Common breeder and very common transient both inshore and offshore.

Historical

Howe and Sturtevant (1899) referred to Common Terns as "common summer residents" and knew of their breeding at two locations, Cormorant Rock and Dyer Island, but undoubtedly these two islands were not the only locations where they bred. As was the case with many other local breeding species, Howe and Sturtevant had a very limited understanding of specific breeding sites, as evidenced by the "discovery" of 125 pairs in 1902 on a rock east of Rough Point, Newport, reportedly a long-standing colony and a location that should have been well known to ESt. This period at the turn of the twentieth century followed a half century of depredation and egging in New England, such that the New England population of Common Terns was at its lowest historical level in the 1890s and early 1900s (Nisbet 1973; Arnold et al. 2020). Hathaway (1913) stated that the Common Tern had made a notable increase since 1907, judging mainly by the size of transient flocks at Quonochontaug in late summer, and he noted 1000 there 17 August 1910 with an additional 6000 elsewhere along the South Shore the same day. During 1911–1916, HSH routinely obtained counts of 3000 at Quonochontaug in late August. During the same period, EDi often observed large numbers at Block Island. She regularly obtained late-summer counts in excess of 1000 and noted 15,000 there on 29 August 1916.

By then, Common Terns had recovered from hunting and egging and were perhaps at an all-time peak in abundance in northern New England during the approximate period 1910–1920. In Massachusetts, the peak was about 1920–1930, at which time there were 30,000–40,000 breeding pairs compared with 10,000 today (Veit and Petersen 1993; Arnold et al. 2020). Common Terns, as well as Roseates, were soon to begin a gradual decline due to increasing pressure from breeding gulls. With a reduction in egging at the much-reduced gull colonies well before 1900, and despite a short but destructive period of millinery shooting of gulls in the closing years of the nineteenth century, the long-depressed breeding gull population began a southward expansion along the Maine coast. The increase of gulls on the Maine coast was slow to accelerate during the period 1900–1921, but the population there was rising sufficiently quickly that controls began to be introduced as early as 1934 (Palmer 1949). In Massachusetts, significant breeding of gulls began in the 1920s, and massive increases were noted from the 1940s onward. Many of the early gull colonizations in Rhode Island occurred during 1937–1945. The takeover of certain islands by Herring Gulls immediately began to push terns toward smaller and less optimal nesting locations, and this process of displacement, continued by Great Black-backed Gulls even with the recent decline in Herring Gull numbers, continued in Rhode Island into the 1990s.

With the reappearance of breeding gulls from Maine to Massachusetts, the size of transient tern flocks along the Rhode Island coast began to decrease after the 1920s, and since 1935 very few counts exceeding 1500 have been recorded (see "Recent maxima" below). Common Terns continued to decline as Rhode Island transients through the general period 1940–1960 and later. This decline in transient numbers has unquestionably resulted from the vast increase in the larger gulls in other New England states, where many large colonies of terns have been overrun. An important theme in the history of breeding Common, Roseate, and Arctic Terns in Massachusetts (see Veit and Petersen 1993) is their displacement by gulls from safe, offshore sites such as Muskeget and Nantucket to insecure, inshore locations where predation by Great Horned Owls, Black-crowned Night-Herons, and even terrestrial predators is more prevalent.

While competition for breeding space on islands seems to have been an important factor in reducing tern numbers throughout our region in the last 50 years, other reasons have also been offered for their decline, including pesticides, predation, water pollution, lack of food, and others. Currently, heavy metals or pesticides appear not to be important factors in Common Tern breeding efforts, even at the Providence River colony (Custer et al. 1985, 1986).

Breeding

Common Terns breed north to within a half mile of the Washington Bridge in the Providence River, and they are regular visitors to some locations a very short distance inland, such as the East Providence Reservoir. There is still only 1 record for any point in the more distant interior (Lonsdale, 21 June 1988; REn), and they have not been recorded at the Scituate Reservoir or at any other point in western Rhode Island.

Below is a list of past and present locations where Common Terns have been known to nest in Rhode Island, 33 of which were listed by Clement and Woodruff (1962). Rhode Island tern colonies were not thoroughly surveyed again until 1977, when Ferren and Myers began annual or near-annual counts.

Little Compton.

1. Briggs Marsh. Six to 10 pairs bred on the salt marsh to the east of the breachway in the early 1940s and continued to about 1955, absent thereafter (DLE). Others have been noted nesting on a rock in the north end of Briggs Marsh at least since 1977, with circa 10 pairs in most years (JEM, DLE). There were 40 adults and 11 nests on this island in 2012 but not since. No nesting in salt marsh in recent years.
2. Briggs Beach Rocks. Common Terns have nested on several sizable rocks off Briggs Beach since the 1970s; e.g., 7 pairs on "Great Rock" June 1977 (REm); 40 adults and 16 nests on Great Rock June 1984 (REm); and 15 nests on "Third Rock" 8 August 2005 (REm). Since about 2008, however, only 1–2 pairs have attempted nesting on these rocks, usually unsuccessfully, likely due in part to increasing human disturbance.
3. Sakonnet Point islands, including three main islands to the south of Sakonnet Point and the smaller "Clumps," including Sakonnet Light, to the east. Apparently absent before 1900 (N. Dexter), the species appeared about 1920 according to HNG, especially on rocks to the northwest of the 3 main islands (60+ pairs on 1 rock in the late 1920s; HNG). A few pairs bred sporadically on these rocks until 1950 (DLE), but in the early 1960s these sites supported only breeding gulls (REW, RLF). They were also present on The Clumps during the

early 1930s and thereafter. David Emerson noted them breeding at the Clumps in 1941, but by then there were already 67 Herring Gull nests on the 3 main islands, and Common Tern abundance appears to have dwindled rapidly thereafter. It is interesting to note that Roseate Terns withstood pressure from gulls on the main islands until 1951 by breeding in the island's vegetation, a habitat almost completely eradicated by gull excrement after 1970.

4. Sakonnet Harbor. Said to have nested on a rock in the harbor in various years prior to 1960 (REW), but none were present in the 1970s or since.
5. Peckham's Marsh, on the Tiverton/Little Compton line. On the salt marsh, with some perhaps in Tiverton, there were 16 nests in 1936 (RCC) and 17 nests in 1961 (REW), but none since.

Tiverton/Portsmouth. Fewer than 5 nests in this general area up to the 1950s. A few pairs used to nest on abandoned barges along the Tiverton shoreline, but these derelict vessels were removed.

Mount Hope Bay. Probably present for decades on Seal Rock south of Mount Hope, where there were 20 and 30 pairs in 1961–1962 (REW). Consistently occupied in recent years, peak of 100 adults in 2004, 50 in 2016. Spar Island, a small sandy island in Mount Hope Bay, has been Rhode Island's largest colony for a few years (e.g., 251 nests, 25 May 2001), with 240 adults present in 2016.

Providence River. Showing the tolerance of this species for polluted waters, a few pairs bred in the mid-1940s on rocks near "the Socony docks" along the East Providence shore (RCC). They were common in the Providence River and on nearby fresh water through the 1950s; there were 50 pairs nesting in 1988 on a sunken barge just south of the Washington Street Bridge (JEM, RLF). This site was consistently occupied, but the decking of the barge collapsed, and the structure was then set on fire by vandals. Forty adults nested in 2008, and 7 adults in 2016. Elsewhere in the Providence River, Common Terns have nested on navigational structures and on an abandoned oil off-loading platform. Nesting Great Black-backed Gulls are now limiting tern nesting in this area. Common Terns nested on the roof of the Conimicut Point lighthouse (Warwick) in 2011 (35 adults) and in other recent years. A colony was discovered on Rock Island, off the shore of Cranston near the mouth of the Pawtuxet River, in 2015; 40 adults were present in 2016.

Southern Aquidneck Island.

1. Gull Rock, in the East Passage northeast of Rose Island, at least from 1972 on (RLC). There were 24 nests 10 June 1976 (RLF, JEM), but no more recent nesting.
2. A rocky island west of Price's Neck, Newport, probably the same as one called "Seal Island" by Mearns at the turn of the twentieth century. Terns were breeding here in some numbers in 1902. Also likely the site of a tern colony in

1940 (WHD), it contained 68 pairs in 1961 (REW). About 25 adults and some young were present in 1972 (MDa), and 15–20 nests were present in 1976 (Erwin). Thereafter, the island was claimed by gulls, and there has been no further nesting.

3. Green Bridge Pond, Ocean Drive. In 1961, 13 pairs bred on the rock in the salt estuary, and others were then said to nest on rocks in the adjoining tidal estuary. None bred for some years afterward because of the roosting of gulls, but in 1986, approximately 10 pairs were again breeding (RLF). Through the late 1980s and 1990s, as many as 12 pairs have nested there. There were 25 adults nesting at this site in 2016.
4. Lily Pond, Ocean Drive. In 1961, 7 pairs attempted to nest on a small rock (REW). No further data until 1994–1995, when 12 and 44 pairs were nesting (RLF, JEM). There were 12 adults in 2014, and 1–2 adults in 2016.
5. "Big" Gooseberry Island, Ocean Drive. Terns were said to nest here prior to colonization by gulls (Clement 1956; owner of the island, pers. comm.). There has been no recent nesting.
6. Bailey's Beach, Ocean Drive, in 1961, there were 6 pairs on a rock to the southwest of the beach. Infrequently occupied in recent years, but 4 adults in 2016.
7. Gull Rock, northeast of Land's End, Newport, contained probably the largest shooting-era colony in the state. Doe took a number of egg sets from a large colony in 1897 and 1898 and estimated the numbers present in the latter year at 300 adults, when they were "as numerous as last year." A colony of 125 pairs was breeding in 1902 (King 1902d), and a correspondent of ESt noted that they had bred there well before 1900. Robert E. Woodruff found 73 nests in 1961, and in talking to local fishermen learned that Common Terns had bred there for many decades previously. When the site has been examined in 1976–1977 and later years, only gulls have been in residence (RLF, JEM). There has been no recent nesting.
8. Cormorant Rock, off Sachuest Point, is undoubtedly the best-known historic colony. The species was abundant here throughout the shooting era, and the colony was estimated at a remarkable 175 pairs in 1899 (Sturtevant). Because of the vulnerability of this location to high seas, numbers have varied over the years: 12 pairs in June 1900; 50 pairs in 1902; "small colony" in 1903. Then there was no further record until 1946, when only a few eggshells were found (RBo, RCC). There were only 2 pairs in 1961 (REW) and perhaps 10 pairs in 1972 (RLF).
9. Island Rocks, east of Sachuest Point. Terns were nesting here in 1905 (Sturtevant) and again noted in 1937–1938 (WHD, Karsch) but absent in 1940 (WHD). There were 11 pairs in 1961, and 10 adults were there in 2016 (CRa).
10. Dyer Island, west of Portsmouth. Another historic tern colony mentioned in Howe and Sturtevant (1899), where (usually) small numbers have bred.

Charles E. Doe collected eggs here on 5 occasions between 1887 and 1897 and found about 50 pairs breeding in the first 4 years but said the colony had "diminished fast" to only 50 birds on the last occasion. The birds were nesting on the central marsh as well as widely over the whole of the then-unvegetated island. Reginald H. Howe landed here in 1896 and found 20 birds and 1 nest. Roland C. Clement estimated 10 pairs in 1946. There were 2 pairs in 1961, together with a large colony of Herring Gulls (REW). There was 1 pair in 1977 but none in 1979–1980, then 80 nests on 7 June 1983 (RLF, JEM). There were 8 adults nesting in 2008 but none since.

Prudence Island. On 14 June 1946, RCC found 7 eggs and 20 pairs attempting to nest at "Gull Point," near Potter's Cove, a difficult location because of potential predation and frequent human disturbance. There has been no more recent nesting.

Hope Island. Common Terns were said to have bred here in 1934–1938 (fide REW), but undoubtedly on offshore rocks rather than on the main island. These satellite islets include Round Rock, on the east side of Hope, where there were 10 pairs in 1961 (REW) and several present in 1976 (Erwin) but none after 1978 (RLF, JEM). Doe visited the Despair Islands north of Hope Island on 20 June 1915 and found 60 nests on the larger, more northern, islet and 15 nests on the smaller, more southern, islet. Estimates of nests on 6 July 1919 were 50 and 10, respectively. In the mid-1960s, terns were lacking, and gulls were nesting on the northern islet (RLF). With the abandonment of Hope Island by the U.S. Navy and its occupation by gulls in the early 1970s, terns have bred or tried to breed on the northern islet in most years, at least since 1978, while the southern islet has been too wave-washed for occupancy. Despair Island is the largest Rhode Island colony in some years, with 150 adults in 2013. There have been few individuals in recent years, however.

Wickford/Quonset/Davisville. On the Wickford harbor marker, east of the Wickford Harbor jetties, is a cement platform 30 ft (9 m) in diameter owned by the U.S. Coast Guard. The terns nesting in this small colony were used in contaminant studies by Custer et al. (1985). The Coast Guard replaced this structure with one that is not suitable for tern nesting, and only a few birds have been intermittently present since.

Conanicut Island. Known to breed at 3 locations:

1. For many years off Fort Wetherill on a high, rocky island: 9 pairs in 1961 (REW) and 1972 (MDa); 12+ pairs in 1976 (RLF). Now infrequently occupied, but 1 nest in 2015.
2. By tradition through the 1950s or earlier on the Dumpling Islands in Jamestown harbor; e.g., 37 nests in 1961 (REW). After the mid-1970s, terns were mostly forced off by gulls, but 3 pairs attempted to nest on the southernmost of the Dumpling Islands in 1976, and another pair bred on one of the smallest rocks in the group in 1978 (RLF, JEM). No more recent nesting.

3. At Sheffield Cove, west side, where terns have attempted to breed occasionally in recent years on a low sandbar. There were 13 nests on 10 June 1976 (RLF), but no nesting since the 1990s.

Narrow River. Common Terns were said to have bred on a rock at the mouth of the inlet in 1936 (Russell). No recent nesting.

Trustom Pond. Depending on water levels, a low, rocky, and sometimes grassy island may be present in the middle of the pond. Terns have nested more or less regularly here at least since 1935 (CWo et al.), varying in numbers and success with the height of the water level. Sample counts include 100 and 60 pairs in 1946 and 1947 (DLK), and 100 pairs in 1961 and 1962 (REW). During most of the period since 1970, this colony has not been present because of higher water levels, but occasional colonization has been made possible by artificial breaching of the barrier beach by refuge personnel. There were 20 nests in 1976 (SRi) and 28 nests in 1983 (DVT), but none more recently.

Green Hill Pond. A small colony was seen in June 1975 inhabiting a tiny island at the east end of Green Hill Pond (DLK). Territorial behavior has been reported by residents in other years. The colony had a maximum of 40 adults in 2004 but was abandoned as of 2008 (CRa).

Quonochontaug Pond. Probably a few have nested here for many years on rocks in the southeast marsh; e.g., "several pairs on a ledge," 23 June 1938 (HSH). Thirty nests were found at probably the same location in 1970, but all failed (ESa). As many as 19 nests were there on 9 June 1974 (RLF). There was 1 nest and a few other birds 9 June 1976 (RLF) but none in 1983 (ESa). High tides frequently flood this site and human disturbance is a factor. No recent nesting.

Sandy Point Island, Watch Hill. Terns formerly bred from about 1938 (just after the island was separated from Napatree Point in the hurricane) through perhaps 1960, but thereafter they were unable to compete with gulls. Douglas L. Kraus made the following counts: 20 pairs in 1946 and 1947, 50 pairs in 1949, 30 pairs in 1950. Their date of disappearance here is unknown; none were present in the 1960s or later. Only recent nesting was 1 nest in 2010 (USFWS).

Nonbreeding

Spring. Normal arrival is 1–10 May with general arrival in mid-May. There is a considerable migration in May and early June past coastal promontories and through coastal waters.

Spring maxima. 550, Point Judith, 6 May 1995; 500 flying east, Warren's Point, 9 May 1954 (RCC); 315, Point Judith, 14 May 1995. Transients at Cox Ledge have been noted to 11 June.

Since 1st-year immatures mostly remain within the winter range during their first summer and are therefore uncommon at this latitude, 800 presumed adults off Little Compton on 20 June 1937 (RCC) and 1000–1200 along the Middletown shore 21–28 June 1937 (WHD) are unusual numbers at a time when the breeding population is usually scattered in small colonies.

Fall. Concentrations along the mainland coast increase after midsummer with the aggregation of local breeders and an influx of large numbers of juveniles from such nearby colonies as Great Gull Island, New York, and Buzzards Bay, Massachusetts, but no recent counts approach those known many decades ago. Terns are highly opportunistic in relation to food, particularly in fall, and unusual concentrations suggest good baitfish conditions. Numbers routinely remain at favorable locations along the outer shore until the first week of September, but there is often a mass exodus along the whole coast thereafter.

Fall maxima. 3000 at Beavertail through August and September 1959 (MDa); 2500, Green Island, Warwick, 7 September 2006 (DFW); 2200, Green Island, 29 August 2004 (DWa); 2000, Briggs Beach, 12 August 2006 (REm); 2000, Charlestown Breachway, 12 August 2006 (STs).

Extreme dates. 8 April 1981, Napatree Point (CRa), to 17 December 1983, Acoaxet (DLE, REm).

Arctic Tern (*Sterna paradisaea*)

Status

Vagrant. Fewer than 10 records.

Nonbreeding

The most southwesterly breeding location of this circumpolar tern is Noman's Land, southwest of Martha's Vineyard, Massachusetts, where up to 20 pairs bred into the 1990s (WHD, USFWS). Noman's Land is a mere 35 mi (56 km) from Block Island. Despite the presence of varying numbers of nesting pairs during the historical period at such Massachusetts locations as Muskeget, Nantucket, and outer Cape Cod, none have ever been suspected of breeding on the Rhode Island coast. In Rhode Island, there are at least 11 sight reports (many of them listed by Conway 1992) and a single specimen, a bird found dead on Main St., Block Island, on 16 May 1928, which is now in the Block Island Collection (EDi). Other sight records include 1, Cox Ledge 7 July 1987 (DLK, DFi); 1 at Ninigret 15 July–4 August 1993 (PAB, DLK et al.); 1, 48 mi (77 km) southwest of Block Island 12 September 1993 (WRP, DLE, CAv et al.); 1, Block Canyon, 31 May 1997 (PAB, RLF et al.); 1 1st-summer plumage, Charlestown Breachway, 4 July 2000 (CRa); 1, Napatree, 20 July 2020 (A. Kneidel; photo).

Forster's Tern (*Sterna forsteri*)

Status

Uncommon fall and rare spring transient. Very rare in winter.

Nonbreeding

Spring. One, perhaps a very early migrant, was present at Weekapaug and Quonochontaug during the period 15–24 March 1980 (GLB et al.; photograph on file). A juvenile was seen on the Seekonk River 24 June 2002 (MLy), suggestive of local nesting; 1-year-old birds were seen at Briggs Marsh, 28 May 2011, and at Quicksand Pond, 30 June 2010 (GDe).

Fall. Forster's Terns are regular visitors in small numbers in late summer and early fall among flocks of Common and Roseate Terns and more rarely in pure flocks. The earliest record of fall arrival is for 1 at Third Beach, 5 July 1995 (RLF, DLE et al.) and also 5 July 2003, Ninigret (STs et al.).

Fall maxima. 354, Seekonk River, 16–23 September 2001 (RFa, DFi); major flight in 2001 with about 100 others distributed along the south coast; 54, Briggs Marsh, 24 October 1994 (DLE); 44, Beavertail Point, 12 September 1959 (MDa).

Large numbers have on occasion been brought northward by tropical storms. On 1 September 1954, the day after Hurricane Carol, a flock of 25 was found at Briggs Marsh, and about 20 were reported from coastal areas during the following week. After Hurricane Gloria (27 September 1985), Forster's Terns were moving westward after the storm. On 30 September, 50 birds were counted on a 1-hour ferry crossing to Block Island, and they were constantly streaming past. More than 130 appeared after Hurricane Bob of 19 August 1991, with 56 between Point Judith and Charlestown, 17 in Little Compton on 20 August, and 40 in Middletown on 22 August (RAC, DLE et al.). Numbers remained for weeks thereafter including 41 at Ninigret 7 September 1991 (RWa, PAB) and a remarkable 172 in the Seekonk River 13 September 1991 (JO; DLE; DLE et al.); 100, Green Island, 29 August–3 September 2004 (DFW); 164 in Seekonk River 25 September 2006 (RFa et al.). In the months following Hurricane Irene (2011), about 150 birds were observed statewide.

Forster's Terns have been appearing and remaining much later in fall. The first November records were for 1–5 birds at Napatree Point 5–12 November 1986 (CRa) and 21 at Charlestown Inlet 11 November 1989 (RLF). Up to 7 birds were present statewide in early November 1991 (v.o.), and 4 were at Sachuest Point and Ninigret 7–24 November 1993 (PAB; DLK). One stayed at Trustom Pond to 20 November 2001 (PPa; JSJ).

Winter. Winter occurrences include 1 in Tiverton 14 December 1991 (JHo, MJM); 1, South Kingstown CBC, 21 December 2003 (STs et al.); 1 at East Matunuck on 23 December 1962 (DLK); 2, Galilee (South Kingstown CBC), 23 December 2006 (STs); 1 at Jerusalem 27 December 1981 (DLK, DLE, REm, STs); 1, Conanicut

Island CBC, 28 December 2003 (ChP, CaP et al.); 1 at Third Beach, Middletown, 6 January 1960 (Baird 1961); 1 at Point Judith, 6 January 1988 (RBo); 1, Sachuest, 7 January 1995 (D. Speicher); 1 at Ocean House Marina, Charlestown, 3 January 1999 (STs).

Inland. Five, Diamond Hill Reservoir, 29 September 2001 (MLy).

Royal Tern (*Thalasseus maximus*)

Status

Formerly rare vagrants from the south, now greatly increased to uncommon or even fairly common visitors.

Historical

A single pair of Royal Terns bred at Nantucket in the 1860s (Griscom and Snyder 1955), but there are no records of their early presence on the Rhode Island coast, even as vagrants. They were among the terns most heavily devastated by the plume trade at the turn of the twentieth century, but with rigid protection they have recovered. They now breed in southern New Jersey and wander in small numbers annually to southern New England. Unrecorded in Rhode Island until 1944, they then began to increase. By the 1960s, they visited annually in non-storm years, punctuated by larger incursions during and after tropical storms.

Royal Terns were first recorded after the hurricane of 14 September 1944, when one appeared at Bonnet Shores on 15 September (Ball) and another was found at Briggs Marsh on 17 September (DLE). Single birds at Napatree Point 14 August 1952 (DLK) and 5 July 1954 (Gates) were not associated with any storm. After Hurricane Carol on 31 August 1954, very limited coverage of the coast produced six at Little Compton and one at Tiverton the day after the storm (DLE, C. Gardner), with lingering birds at Misquamicut and Napatree Point until 12 September (DLK, RCC). In following years, other birds appeared without the aid of storms, for example, 27 July 1957 at Napatree Point (SSD), 28 July 1958 at Galilee (RLF), and 28 June 1959 at Galilee (DLK).

After Hurricane Donna of 12 September 1960, the largest influx on record occurred in Rhode Island and elsewhere in the Northeast. On 13 September, 26 were resting along the southern shores of Little Compton (RLF), while five were at Newport (REW), six at Jerusalem (DLK), nine at Trustom (J. J. Lynch), and eight at Block Island (JBa). At the last location, JBa counted 31 on 14 September, 30+ on 15 September, and 10 on 16 September; a single bird lingered to 18 September. Remnant storm birds were noted along the mainland coast until 18 and 22 September, and even to 17 October at Jamestown (RLC), while the last individual was seen at Block Island 24 October 1960 (JBa). A well-mummified bird that was found the following winter in Middletown became the first state specimen (REW; in MCZ).

Nonbreeding

Since 1960, hurricanes have usually brought smaller numbers. Tropical Storm David of 6 September 1979 brought only 5 birds. The wholly pelagic Hurricane Gloria of 27 September 1985 deposited about 25 birds in Rhode Island (compared with 170 at Martha's Vineyard and Nantucket), and approximately 12 were present through the second week of October. Hurricane Bob of 19 August 1991 produced 35 or more. After Hurricane Irene, 15 were seen on Block Island 30 August 2011 (DFi, STa), and at least 15 others were seen across the state 28–30 August.

Spring. Royal Terns are rare in May and June, but increasingly observed as their population increases and they spread northward.

Fall. After 1960, Royal Terns were seen virtually annually in Rhode Island (to 28 birds per year in 1979). Unlike many terns, this species is very much at home in coastal waters as well as waters farther offshore; there are now several records for Cox Ledge. There is a gradual increase in reports through June, and the peak comes in late July, with diminishing reports through August. For decades there was a persistent lack of September reports for non-hurricane years; September is the month when most significant storm-driven maxima have occurred.

Fall maxima. 15, Block Island, 30 August 2011 (DFi, STs); 12, Trustom, 3 July 1983 (DVT).

Extreme dates. 31 March 2014, Newport (1; P. Champlin; photo), to 18 November 1993, Napatree (CRa), and 23 November 1974, Moonstone Refuge (A. Magee). Unprecedented in New England is the midwinter record of a banded bird that was blown ashore in southeastern Middletown during a blizzard, 27 January 1987 (LT, prepared by CRa; now AMNH #831189).

Sandwich Tern (*Thalasseus sandvicensis*)

Status

Rare summer visitor. Increasing in frequency of occurrence, perhaps because of warming climate.

Nonbreeding

After the passage of Hurricane Donna during the afternoon of 13 September 1960, 1 was photographed the following morning resting with a flock of Royal Terns on the tip of Sakonnet Point (RLF), and a different bird was photographed later the same day among Royal Terns on the flats at Quicksand Pond (RBo; RLF). Another large tern with a black bill was seen resting on a buoy at Newport the same day but was too distant for positive identification (REW et al.).

During the morning of 6 September 1979, under conditions of strong southeast winds associated with a much-weakened Hurricane David, an adult was seen at

Jamestown (GLB), and 2 adults were found the next day at Galilee and Point Judith (RAC) after the remarkable southwest winds of the previous afternoon.

One of the largest influxes on record was associated with Hurricane Gloria of 27 September 1985. On the day after the storm, 5 flew west past Point Judith in a 1-hour morning watch (RLF), 1 was photographed in a flock of gulls in the parking lot at Scarborough Beach the same morning (RLF), 3 were at Jerusalem (DLK, DFi, RLF), 2 were at Briggs Marsh (RBo et al.), and 5 were in the vicinity of Napatree Point, mostly passing west offshore in the late afternoon (RLF, REn, DLK, DFi). While the number of Sandwich Terns in Rhode Island associated with Gloria was unprecedented, counts were trivial compared with numbers in southeastern Massachusetts, where more than 75 were present at Martha's Vineyard and Nantucket. Hurricane Bob of 19 August 1991 produced about 16 birds in Rhode Island. Four were at Scarborough Beach and 1 at Sand Hill Cove the day of the storm (RAC et al.). On the following day, 1 was at Briggs (DLE), 9 were between Point Judith and Charlestown (RAC et al.), and 1 or 2 were at Napatree Point (CRa). Two remained at Trustom, 22–24 August (RAC, RWa).

Hurricane Emily, which passed offshore on 2 September 1993, did not directly affect the Rhode Island coast, but may have brought a Sandwich Tern to Napatree Point on 4 September 1993 (CRa). One was at Quicksand 17 September 1999 following Hurricane Floyd (REm), and 1 was at Napatree 24 September 2003 in the wake of Hurricane Isabelle (CRa).

After Hurricane Irene, there were about 16 birds statewide, including 5 at Block Island 30 August 2011 (DFi, STs) and 1 on the Sakonnet River near Little Gould Island 30 August 2011 (GDe). Other records not related to storms are 2, Ninigret, 15 August 1997 (STs); 1, Napatree, 24 September 2003; 1, Ninigret, 31 July 2004 (STs); 1, Ninigret, 30 July 2005 (PLE); 1, Warwick, 2 August 2006 (v.o.); 1, Middletown 5–6 July 2008 (RTL); and 1, Middletown, 23 May 2013 (RTL).

Black Skimmer (*Rynchops niger*)

Status

Fairly common visitor during the warmer months, greatly increased from past decades. One recent breeding record.

Historical

Since 1951, Black Skimmers have occurred annually in late summer and fall. Skimmers may well have bred along the Rhode Island outer coast in early historical times, as they apparently did on Cape Cod, but no record remains. The first Rhode Island occurrences were after the hurricane of 26 August 1924, when two were seen at Quonochontaug 1 September (fide HSH) and two were reported at Block Island 10 September (fide EDi). They were then unreported until three were seen at Briggs Marsh 20 September 1936 (RCC).

Breeding

A pair of skimmers was present annually during June 1983, 1984, and 1985 at the Horace Island tern colony in the Pawcatuck estuary (REn, CRa, JEM, RLF). In 1985, their nesting attempts were foiled by Mute Swans. A nest was found on Sandy Point Island in 1992 (JEM, RLF). There has been no suspicion of nesting since.

Nonbreeding

During and after the hurricane of 14 September 1944, the entire Northeast was deluged with skimmers. In the first 2 days after this storm, an estimated 1000 birds were reported statewide, including 275 at Newport, "hundreds" at Block Island, 157 at Briggs, 50 at Bonnet Shores, 75 at Touisset, and numbers at Matunuck and Point Judith Pond. There were 64 still at Nayatt Point 1 October (HNG), and smaller numbers remained throughout October, including 1 accidentally collected at Riverside on 20 October by a surf-casting fisherman. The last birds were seen at Block Island 28 October (EDi) and Portsmouth 29 October (RBo, DLE). The next spring, a single bird was seen fishing at Sandy Point, Block Island during "the last week of March" (Beebee, fide EDi). This is a remarkable time of year, but this record's plausibility is enhanced by the fact that more than 200 hurricane-driven skimmers remained into early January 1945 in the area of western Long Island (Bull 1964).

After 1944, skimmers appeared more frequently. Such early records include 1, 30 May 1946 at Galilee (DLK); 3 adults at Sandy Point Island 30 June 1946 (DLK); and 1 at Jerusalem, 30 August–3 September 1951 (FPe, EWe, DLK). In 1953, they were more numerous and widespread than any time since 1944, with reports from 5 locations of 1–5 birds each.

After Hurricane Carol of 31 August 1954, the coast was again visited by large numbers but not as many as in 1944. There were initial reports from at least 10 locations of a minimum of 440 birds, and large numbers remained through the fall; e.g., 375 at Quicksand Pond 26 September–3 October (RBo); 450 at Conimicut Point 31 October (GRa); and 40 still at Conimicut 11 November (BRi). The latest individual in 1954 provided the latest current fall date for the state; it was also found at the most unlikely (essentially inland) location: Watchaug Pond 27 November 1954 (DLK).

In 1955, 1 appeared as early as 26 May at Little Compton (Hendershee). Later that year, 32 were recorded at Pawtuxet 1 November 1955 and a very late flock of 45 was at Bullocks Point 25 November (both CBl).

Since 1955, Black Skimmers have been reported annually, with from 1–15 reports of 1–10 birds and non-storm flocks of up to 30 birds. After Hurricane Donna of 12 September 1960, only about 33 skimmers appeared. After Hurricane Gloria, 61 skimmers were at Napatree Point 29 September 1985 (CRF, RLF) and at least 30 others were noted in the state in the days after the storm.

Extreme dates. 4 May 1995, Quonochontaug (REn), to 27 November 1954, Watchaug Pond (see above); 1 March (1945) record (see above).

TROPICBIRDS (PHAETHONTIDAE)

White-tailed Tropicbird (*Phaethon lepturus*)

Status

Vagrant. Two records.

Nonbreeding

An adult White-tailed Tropicbird was observed passing west by Point Judith (20 yd [18 m] away) 28 September 1985, the morning after the passage of Hurricane Gloria (RAC). The weather had cleared, and there was a modest west wind. Two others appeared in Massachusetts at the same time (Veit and Petersen 1993). In July 1972, a tropicbird with a yellow bill and long, drooping tail feathers was seen flying over the beach and marsh at Quonochontaug by GRa.

Red-billed Tropicbird (*Phaethon aethereus*)

Status

Vagrant. Three records, none associated with a tropical storm. The status of both species of tropicbirds in the Gulf Stream off Cape Hatteras, North Carolina, has been clarified by Lee et al. (1981).

Nonbreeding

An immature Red-billed Tropicbird was found in a weakened condition on a 10th floor window ledge of the Hospital Trust National Bank in downtown Providence 3 July 1973. No unusual weather conditions preceded its appearance. It died shortly after its discovery and is now MCZ #331,370. Another bird, this time an immature female, was found dead on Green Hill Beach by NCl 19 July 1975 (MCZ #330,456). During a flurry of records from the New England coast 2002–2011, a Red-billed Tropicbird was seen at Little Compton 30 May 2008 (RLF, CRa). This Rhode Island bird was identifiably different from bird(s) seen at Seal Island and Matinicus Rock, Maine, during the same year.

LOONS (GAVIIDAE)

Red-throated Loon (*Gavia stellata*)

Status

Common to abundant transient and common winter resident at the coast.

Nonbreeding

Fall. Red-throated Loons migrate south through Rhode Island mainly during late October and November. They are often far more numerous than Common Loons.

Most of the fall flight seems to pass far south of Block Island and is therefore only partially noted by land-based observers (171 past Mohegan Bluffs on a sunny day, 11 November 1997; RLF). Migration continues through December and early January in smaller numbers; e.g., 279 west past Point Judith, 29 December 2001 (RLF); 73 doing the same, 5 January 1992 (RLF). While common in the Sakonnet Passage, Red-throated Loons are usually scarce at any season in upper Narragansett Bay areas (14, Warren, 16 November 1985; RBo). Red-throated Loons are much more numerous than Common Loons over shelf waters during winter; e.g., 30 north of Block Canyon, 10 March 1996 (v.o.).

Fall maxima. 627 past Point Judith in 3 hours, 23 November 1941 (DLK); 438, Little Compton, 21 November 2007 (J. Osborn); 261, Narragansett, 28 November 2014 (SSM, PJL).

Winter maxima. 371, Newport–Westport CBC, 14 December 2014; 335, South Kingstown CBC, 22 December 2001.

Spring. Some begin moving north in late February; e.g., 25 flying north at Narragansett, 22 February 1951 (RBo). Most birds move through Block Island Sound and over offshore waters between late March and mid-May, earlier than do Common Loons.

Inland. One, 1 November 1940, Watchaug Pond (DLK); 1, 17 November 1945, Wordens Pond (DLK); 1, Burrillville, 24 November 2006 (MLy).

Spring maxima. 509, Point Judith, 27 April 1997; 474, Point Judith, 1 May 1993 (RLF). Totals of 4–8 migrants, including birds in alternate plumage, have been seen at Point Judith during early June. Inland, there are more than a dozen spring reports of individuals on fresh water ranging from 7 March to 18 May.

Summer. There are about 18 records (up to 1995) of individuals through summer, and 1 or 2 are found almost annually, though much less often than Common Loons.

Extreme dates. Early fall transients appear in late August; e.g., 5 at Napatree Point 29 August 1971 (ESa, FPe) and 5 at Quonochontaug 29 August 1912 (HSH).

Pacific Loon (*Gavia pacifica*)

Status

Vagrant. Now almost annual, mostly at Block Island and other rocky shores.

Nonbreeding

Pacific Loons have fluctuated in abundance as vagrants to the east coast, with peak numbers occurring in the 1950s, then again more recently (since about 1990). They appear disproportionately frequently in coastal Rhode Island compared with the immediately adjacent states, perhaps because of the abundance of rocky coastlines.

Most reports come from Block Island and other rocky shores along the outer coast. There were 7 records 1951–1960 that span the dates of 12 November to 17 May, 2 in 1961–1970 from 2 January to 11 March, 9 in 1971–1980 from 17 December to 12 June, 4 in 1981–1990 from 5 November to 15 February, and 11 in 1991–2000 from 13 December to 1 June. More recent records include 1, Block Island, 2–6 October 2007 (J. Williams et al.); 1, Middletown, 15 February–6 March 2013 (P. Champlin et al.); 2, Block Island CBC, 20 December 2012 (CRa) (see Photo 16).

Maxima. Pacific Loons have been reported in 9 of 11 years since November 2005. Four, Block Island, 6 May 2009 (DFi et al.); 4, Block Island, 12 November 2010 (SSM, PJL).

Extreme dates. 2–6 October 2007, Block Island (J. Williams et al.), to 12 June 1974, Point Judith.

Common Loon (*Gavia immer*)

Status

Common transient and winter resident on salt water; fairly common spring and uncommon fall transient inland. Fairly common nonbreeding summer resident along the coast.

Nonbreeding

Winiarski et al. (2013) showed maximum abundance of wintering Common Loons off the south end of Block Island and along the Rhode Island coast west of Point Judith.

Fall. Southbound Common Loons arrive in late August or early September; e.g., 3 moving west through Block Island Sound, 8 September 1963 (DLK); 16 flying southwest over Block Island, 17 September 1960 (DLK, JBa). Migration continues to mid-December (Powers and Cherry 1983).

Fall maxima. 350 past Block Island's southeast light, 11 November 1996 (RLF); 280 past Block Island's north end, 17 October 1976 (RLF); 250 resting on water off Quonochontaug, 27 November 1991 (CRa); 245 past Block Island, 30 October 1960 (JBa); 235, Ninigret Beach, 16 November 2014 (PP).

Up to 4 per day have been seen during October and November on larger bodies of fresh water such as the Scituate Reservoir.

Winter. In winter and early spring, large numbers of Common Loons often concentrate along the sandy shores of South County. The largest densities found by Winiarski et al. (2014) were at sea southwest of Block Island

Winter maxima. 1342, Watch Hill to Quicksand, 18–20 February 1995 (RLF; see below); 547, Napatree CBC, 18 December 2005. To compare wintering abundance east and west of Point Judith, a loon census was conducted between Napatree Point and Quicksand Pond 18–20 February 1995, with the following results: Napatree to Quonochontaug—527; Quonochontaug to Green Hill—656; Green Hill to Point Judith—91; Point Judith to Quicksand Pond—75 (RLF).

Spring. Numbers passing coastal points increase in late March and eastward movement begins in early April, including birds inland (1 in Barden Reservoir, 4 April 1983; JGe, RGe).

Spring maxima. The highest counts of coastal migrants have been observed in all-day sea watches at Point Judith, where birds pass northeasterly over the peninsula (especially in early morning), as well as through Block Island Sound: 1015, 16 May 1982 (RLF); 988, 28 April 1985 (RLF); 955, 14 May 1995 (RLF); 65 heading northeast over Sachuest Point into the interior of Tiverton–Little Compton, 15 May 1983 (REn, RTL); 19 flying northeast over Watchaug Pond, 14 May 1940 (WBD, DLK).

Summer. Summering nonbreeders and wanderers are regularly reported on salt water, rarely in flocks to 21 and 25 birds, but individuals are rare at that season away from salt water; e.g., 1 flying southeast over Apponaug, 19 July 1974 (KKi); 2 at Larkins Pond, 3 August 1955 (DLK); 1 at Arnold Mills Reservoir, 9 July 2005; and 1 immature at Woonsocket Reservoir, 9 July 2005, and again 31 July and 14 August 2005 (MLy).

Common Loons could possibly nest in Rhode Island, though no evidence of breeding has been reported. After nearly a century of absence from southern New England, they have occupied the Quabbin and Wachusett Reservoirs in north-central Massachusetts in recent years, while others have bred in Connecticut. Three pairs were breeding in Worcester County, Massachusetts, in 1987, and there were 9 pairs statewide in 1991 (Veit and Petersen 1993). Conditions on Rhode Island's sizable Scituate Reservoir might prove to be suitable should the new southern New England breeding population continue to expand.

ALBATROSSES (DIOMEDEIDAE)

Yellow-nosed Albatross (*Thalassarche chlororhynchos*)

Status

Vagrant. Four records.

Nonbreeding

The first Rhode Island record is of a subadult seen at Cox Ledge on 21 August 1976 by CWo. This bird, which approached to within 15 ft (4.6 m) of the boat and was kept under observation for an extended period, possessed a narrow border of black along the edges of the largely white underwings. On 21 August 1979, exactly 3 years after CWo's record, another immature bird appeared at Cox Ledge and was examined and photographed by THD and numerous New York observers. There have been 2 additional observations of unidentified albatrosses: 1 reported by fishermen in the area of Cox Ledge in May 1979, and another seen distantly about 11 mi (18 km) east of Block Island's southeast lighthouse on 7 July 1979 by Kevin D. Powers of the Manomet Bird Observatory. Most Yellow-nosed Albatrosses in eastern North America have appeared in March–June (McDaniel 1973; Finnegan 2001; Veit 2005), so the 2 Rhode Island records from August are atypical in this respect.

Black-browed Albatross (*Thalassarche melanophris*)

Status

Vagrant. One record.

Nonbreeding

An adult Black-browed Albatross was studied at close range and photographed 5 December 2012 about 50 mi (80 km) southeast of Point Judith (B. Gervelis). There were 9 records of Black-browed Albatrosses in the northwest Atlantic during the years 2009–2013 (Davis 2014).

AUSTRAL STORM-PETRELS (OCEANITIDAE)

Wilson's Storm-Petrel (*Oceanites oceanicus*)

Status

Common austral migrant and resident offshore.

Nonbreeding

Wilson's Storm-Petrels are transequatorial migrants from their breeding grounds in the Antarctic and adjacent waters. They are one of the most abundant seabirds off the North American coast during the boreal summer.

Inshore. Wilson's Storm-Petrels enter inshore waters of Narragansett Bay during fog and haze. Some have wandered into the East Passage between Newport and Jamestown harbors (8 in fog west of Goat Island 27 July 1995; RLF, DFi, PAB et al.). They are sometimes forced over land during storms; e.g., 1 observed flying west across the Jerusalem marsh toward Potter Pond 28 August 1971 during Tropical Storm Doria (DLK). During Hurricane Donna on 12 September 1960, 40 were seen flying low over the center of Block Island (JBa) and 10 were seen over Rumford being blown backward (RLF). During Hurricane Gloria on 27 August 1985, more than 100 were seen being pushed inland to the west of Point Judith (RLF, CRa), while 1 remained all morning in Galilee Harbor the next day (RLF, DLK, DFi). Seven were seen passing over Wakefield during Hurricane Bob, 19 August 1991 (RAC), and 1 was found dead on the URI campus in Kingston the same day (CRa; AMNH #831180).

Maxima. Counts from shore usually do not exceed 150, but there are a few exceptions: "thousands," Point Judith, 14 August 1899 (ESt, LKi); 600, Quonochontaug, 2 July 1983 (C. Perry); 452, Point Judith, 13 June 2000 (RLF).

Offshore maxima. Counts at Cox Ledge routinely exceed 300; counts from there (unless otherwise indicated): 4000, 16 July 1987 (DFi, DLK, RBo et al.); 2000, 8 mi (13 km) south of Block Island, 11 August 1987 (DFi); 2000 between Block Island and the wreck *Suffolk*, 35 mi (56 km) to the south, 27 June 1984 (RBo, HMP); 1500, 3 mi (5 km) south of Block Island, 23 August 1997 (DLE, DFi). Shipboard surveys conducted in 2009–2010 by URI (Paton et al. 2010) show highest densities of Wilson's Storm-Petrels somewhat south, or farther offshore, of the highest densities of Great and Cory's Shearwaters.

Extreme dates. The earliest offshore record is for 20 at Cox Ledge 7 May 1980 (RBo); inshore records are 1, Sakonnet, 13 May 1989 (DLE et al.), and another off Weekapaug, 19 May 1972 (ESa). The latest inshore date is for 8, Block Island Sound, 27 September 2002, but there are later counts at Cox Ledge of 41, 28 September 1974 (m.obs.), and 4, 77 mi (124 km) south of Block Island, 9–10 October 1993 (CAv).

White-faced Storm-Petrel (*Pelagodroma marina*)

Status

Vagrant. Four records.

Nonbreeding

An immature was collected 18 August 1953 by M. S. Gordon and R. H. Backus from a Woods Hole oceanographic vessel at lat 39°48'N, long 71°02'W (Gordon 1955), not at lat 39°40'N, long 71°02'W as stated by Watson et al. (1986). This location is approximately 54 mi (87 km) south-southeast of Block Island. Robert Cushman Murphy referred this individual to *P. m. hypoleuca*, the breeding form of Madeira and the Salvage Islands. According to Watson et al. (1986), this specimen has mysteriously disappeared from the AMNH alcohol collection.

Another individual was 54 mi (87 km) south-southeast of Point Judith on 10 September 1990 (RAC), remarkably near the location of the 1953 observation. One landed aboard a fishing boat 80 mi (129 km) south of Block Island 30 August 2001 and was seen and photographed by Kim Gaffett (fide CRa). *Four* were seen at Block Canyon 18 August 2013 (A. Wilson et al.), and 1 was there, 3 September 2018 (CPe et al.) (see Photo 17).

NORTHERN STORM-PETRELS (HYDROBATIDAE)

Leach's Storm-Petrel (*Hydrobates leucorhous*)

Status

Uncommon migrant on the outer continental shelf; rare inshore.

Nonbreeding

Leach's Storm-Petrels are rarely seen close to shore in Rhode Island despite the presence of very small nesting colonies on Penikese Island and Noman's Land of southeastern Massachusetts. Nesting birds from the Gulf of Maine feed far offshore and, even during migration, tend to stay near the continental slope except when diverted by storms. Spring and summer records selected from a considerably larger number are 1 caught alive and banded at Block Island, 7 May 1962 (MSl); 1 collected while it sat in the water off Nayatt Point, Barrington, 9 June 1889 by J. M. Southwick, CED, and Stainton; 3 in Block Island Sound, 15 June 1975 (SRD); 1 about 8 mi (13 km) southeast of Point Judith, 15 June 1976 (RAC); 1 seen 18 June 1950 about 3 mi (5 km) off Newport (RBo); individuals at Cox Ledge, 14 June 1980 (DFi; photo) and 25 June 1981 (DLK); 1 found dead in the water near Castle Hill lighthouse at the mouth of Narragansett Bay 25 June 1989 (fide RLF); 1 brought to the NBS by a Newport resident 21 July 1971 (LOG, SSD, LBe); 2 at Cox Ledge 25 August 1973 (Axtell and Tozzi); 3, 60 mi (97 km) southeast of Block Island 25 August 1984 (RBo); and 3 about 40 mi (64 km) south-southeast of Block Island 10 September 1990 (RAC).

Storm-driven birds occur much later in fall than do Wilson's Storm-Petrels. The period of peak occurrence is October, with some birds lingering into early winter. Such records are 1 found alive on Briggs Beach, Little Compton, 22 August 1991, probably a victim of Hurricane Bob of 3 days earlier (REm et al.); 1 in the Sakonnet Passage off Tiverton 1 September 1954, the day after Hurricane Carol (CGa); 1 found after Hurricane Donna 12 September 1960 in a garage in Newport (REW); and 7 birds in the East Passage, north of Big Gould Island, Narragansett Bay, 15 September 1944, the day after a severe unnamed hurricane (RBo). During the height of Hurricane Gloria on 27 September 1985, when winds exceeded 70 mph (113 kph) from the south, approximately a dozen were seen with numerous Wilson's Storm-Petrels being pushed due north into the interior, to the west of Point Judith (CRa, RLF). Other possible or probable storm-driven birds include 1 seen off Point Judith 29 September 1938, a week after the 1938 hurricane (WBD); 1 found 8 October 1901 in a weakened condition on Potters Avenue, Providence, 2 days after "a gale that swept Cape Cod" (Southwick 1902); 1 collected 14 October 1891 in Narragansett Bay, the specimen acquired by HSH but subsequently lost; 1 seen during a southeast storm off Point Judith 24 October 1971 (CWo, ESa); another seen there under similar conditions 30 October 1973 (DWa); 1 seen *northwest* of the fort at Napatree Point 31 October 1991, the day after a hurricane-force northeast storm (CRa); 1 thought to be this species off Point Judith during a southeast gale 10 November 1990 (PAB); another unidentified storm-petrel there during another southeast storm 29 November 1980 (RAC); 1 at East Providence 29 October 2012 during Hurricane Sandy (T. Seiter); 1 identified as Leach's off Point Judith during an easterly storm 11 December 1993 (MJM); 10 seen off Warren's Point, 2 January 1939 (Gairloch).

Maxima offshore. 12, Block Canyon, 18 August 2013 (A. Wilson et al.).

Extreme dates. Early May–2 January 1939, Warren's Point (Gairloch).

Band-rumped Storm-Petrel (*Hydrobates castro*)

Status

Rare summer visitor offshore. Four records.

Nonbreeding

One was seen and photographed at Atlantis Canyon 28 August 1994 (W. Petersen, RBo) and another was at the shelf break about 100 mi (161 km) south of Block Island 13 July 1997 (RRV, PAB). These observations fit a pattern of July and August occurrence between the continental shelf break and the Gulf Stream south of New England and New York during about 1990 to the present. One was photographed at Block Canyon 10 September 2010 (P. L'Etoile et al.), and 2 were seen there 3 September 2015 (C. Pedro and P. L'Etoile; see Photo 18).

SHEARWATERS AND PETRELS (PROCELLARIIDAE)

Northern Fulmar (*Fulmarus glacialis*)

Status

Rare to uncommon but regular winter visitor to offshore waters; has substantially increased since 1970s.

Nonbreeding

The first state record (and still the only instance of 1 seen on or from the mainland) was of a weakened bird caught alive 2 October 1965 as it rested on Second Beach, Middletown, after a southerly storm. It was photographed, banded, and later released by REW. The first sight records of birds at sea were obtained soon after the earliest trips by birders to Cox Ledge. One was at Cox Ledge 16 September 1973 (THD, DWF, RLF et al.) and another was there 23 September 1973 (RLF, DLK, CWo). Four were at Cox Ledge 24 May 1974 (THD) and then 25 on 8 June 1974 at the same location (RLF). Since these initial observations, the species has been seen annually and numerous records have accumulated for the period September–June, almost all at Cox Ledge or in waters south of Block Island. Peak occurrence in September–December and again in May and June suggests a passage of migrants and nonbreeders.

Northern Fulmars often arrive in late September; e.g., 1 at Block Canyon, 5 September 1992 (CAv). There is substantial variation from year to year, and in some years, none are seen. The winter of 2015 and 2016 seems to have been a big season for them; there were several counts in the 30s and 40s between November 2015 and January 2016. There are a few records from east of Block Island north to within 10 mi (16 km) of Sakonnet, and there are 3 records from Block Island Sound, all seen from the Block Island ferry: 23 June 1990 (CRa, REn); 1, 1 March 2003 (SSM et al.); 2, 10 November 2011 (SSM et al.), and 22 December 2011 (DFi, STa). Other than the 1965 bird, the species has so far never been seen on or from the mainland, and only 1 other individual has been observed from land in Rhode Island: 10 June 1981 from the Mohegan Bluffs at the southeast corner of Block Island (RBo).

Maxima. All for Cox Ledge unless otherwise noted: 150–250, 15 May 1986 (DFi et al.); 90, 20 mi (32 km) southeast of Block Island, 30 December 2015 (CPe); 59, off Block Island, 24 September 1977 (THD); 58, Block Canyon, 4 March 1995 (PAB, DLE, RLF et al.); 50, 8 June 1983 (RBo, RAC, DFi); 37, 13 June 1982 (RAC, DLK, DLE); 19, 20 mi (32 km) south of Cox Ledge, 11 March 1989 (JMa).

Black-capped Petrel (*Pterodroma hasitata*)

Status

Vagrant. Four records.

Nonbreeding

On 27 September 1985, during the height of Hurricane Gloria, 2 *Pterodroma* petrels were closely observed a short distance to the west of the Point Judith lighthouse by RLF and CRa. They were first noted just east of the Harbor of Refuge jetties struggling low over the water against the 70 to 90 mph (113–145 kph) southerly winds, but after a short period they were forced over the beach and parking lot, where a brief but unobstructed look was obtained before they disappeared westward over the dunes and along the shore of Sand Hill Cove. Among other field marks clearly noted were the "three-piece bill," white underparts (including underwing), dark gray mantle, and an extensive white, high, rump patch. These features made clear their identification as either Black-capped Petrels (*Pterodroma hasitata*) or Bermuda Petrels (*P. cahow*). The identification as Black-capped Petrel was confirmed by David B. Wingate (pers. comm.), who ruled out Bermuda Petrel on the basis of the written descriptions. Specimens of *P. hasitata* have been obtained after hurricanes in Connecticut and New Hampshire, and there are at least 5 records from New York and 3 from eastern

FIGURE 6. Shearwaters and storm-petrels at Cox Ledge, © Margaret La Farge.

Massachusetts. Another Black-capped Petrel was seen over Wakefield in the eye of Hurricane Bob 19 August 1991 (RAC). There are 2 other more recent records from Block Canyon, at the continental slope south of Rhode Island: 10 September 2010 (m.obs.) and 3 September 2015 (CPe et al.; see Photo 19).

Cory's Shearwater (*Calonectris diomedea*)

Status

Rare spring and common late-summer and fall visitor.

Historical

Cory's Shearwaters were not recorded in Rhode Island before the great "flight" of 1886 that brought hundreds or thousands to Buzzards Bay (Baird 1887, Rives 1887a). They were first collected off Point Judith and Newport during this remarkable herring extravaganza, when "enormous schools of Atlantic Herring (*Clupea harengus*)" (Veit and Petersen 1993) occurred in the bay. Other inshore incursions apparently took place under similar conditions in 1935, 1949, 1955, and 1969. They have even entered such estuaries as the Sakonnet Passage and lowermost Narragansett Bay on rare occasions, such as 100 off Seapowet 11 October 1969 (HMP).

Nonbreeding

In most years, these warm-water shearwaters are rarely seen from land, but they do occasionally enter Block Island Sound. They have sometimes appeared in large numbers in Buzzards Bay (as in 1886 and during many more recent flights, including 2017), and in major flight years they are not infrequently observed from the Block Island ferry, but westward they are always rare along the South Shore and do not enter Long Island Sound. They are rare off Watch Hill, consistent with the fact that they are unrecorded from Connecticut except for 1 inland specimen (Zeranski and Baptist 1990).

Spreading westward after the breeding season from the Azores, Canary Islands, and other east Atlantic breeding colonies, Cory's Shearwaters usually appear in late summer along the coast between Cape Cod and Cape Hatteras. They are rare within sight of the Rhode Island mainland before mid-July, and more often do not appear before August. The earliest inshore records are 2 at Sakonnet Point 24 June 2001 (REm) and 19 in Block Island Sound 29 June 1982 (REn). One was captured alive at Charlestown Beach 3 July 1970 (SSD, ESa).

Farther offshore, they often appear earlier in spring. In 1975, nearly every spring trip to the Cox Ledge area reported them, beginning with 1 east of Block Island 25 May 1975 (RAC, DLK). Other counts that year included 83 at Cox Ledge 21 June (v.o.) and 400 there 14 July. These counts are supported by shipboard surveys conducted 2009–2010 by URI (Paton et al. 2010) that showed Cory's Shearwaters and Great Shearwaters being concentrated along the 20-fathom (37 m) depth contour, which includes the Cox Ledge area. They have remained into late October and even early November in years when feeding conditions and perhaps water temperatures have been appropriate. In years when they are numerous, individuals have regularly been seen to mid-November.

Maxima. *1100,* 3 mi (5 km) east of Block Island to 35 mi (56 km) south of the island, 27 June 1984 (RBo, HMP); 980, Cox Ledge, 3 July 1979 (v.o.); 800, Block Island Sound, 3–5 October 1969 (ASRI); 340 off Sakonnet Point–Briggs Beach, 3 September 2017 (REm).

Extreme dates. 25 May 1975 to 20 November 1935, *350* off Block Island (fide EDi). The 1935 record and that of 1 seen and photographed at Barrington 1 November 2012 (B. Shamgocian) are reminiscent of the massive incursion of Cory's Shearwaters into Buzzards Bay, Massachusetts, during the fall of 1886 (see above).

Sooty Shearwater (*Ardenna grisea*)

Status
Common nonbreeding visitor.

Nonbreeding
Inshore. Sooty Shearwaters appear during late May and June and are more common and regular in waters south of Block Island than farther inshore. They usually decline in abundance later in the summer, but in some years are common again in fall. They are more frequently seen from shore than Great or Cory's Shearwaters, especially during southeast winds and fog. They are most often seen from Point Judith eastward, but occasional birds have been seen as far west as Napatree Point, 26 May 1992 (CRa), and Quonochontaug 26 May 2000 (G. Williams). They usually decline by late June and become scarce after mid-July. Sooties are usually the least numerous of the 3 larger shearwaters in our waters, although they sometimes surpass the others in numbers during late spring and early summer.

Early inshore records include individuals off Point Judith 12 May 1974 (HWi) and Matunuck 15 May 1976 (RBu). Two were seen passing Point Judith during a southeast storm 28 November 1980 (DLK). There are 2 records from uppermost Narragansett Bay off Riverside 21 May 1964 and 1 July 1962 (both CBl). Another was found in the northern portion of the Sakonnet Passage, where it was swimming within a few feet of Little Gould Island's south "beach" on the morning of 30 May 1995 (DLE, RLF).

Maxima. 472, Point Judith, 24 May 1980 (RLF); 100+ each resting in the surf off Sandy Point, Block Island, 6 June 1976, and Point Judith, 9 June 1976 (RLF).

Offshore. Generally Sooty Shearwaters are found earlier in spring and later in fall offshore than they are at inshore locations. Offshore, the earliest spring dates are

1 on 15 April 2010 off Block Island (B. Harris); 30 April 1982, 10 mi (16 km) south of Block Island (DFi); and another at Cox Ledge 2 May 1979 (fide RAC). They are relatively scarce in summer; the maximum August count is 10, Cox Ledge, 19 August 2000 (SSM et al.); 4 were 10 mi (16 km) south of Block Island 9 September 1996 (WRP, DLE et al.); and 12 were 40 mi (64 km) southeast of Block Island on 12 September 1993 (WRP, DLE et al.).

Maxima. 600, Cox Ledge, 10 June 1978 (THD, DLE, CWo); 600, Cox Ledge, 17 June 2000 (SSM et al.); 390, off Block Island, 7 June 2015 (CPe et al.); 250, south of Block Island, 27 June 1984 (RBo, HMP).

Winter records. One dead, Block Island, 26 December 1940 (EDi); 1, flying east off the north end of Block Island, 20 December 1999 (REm, REn, DFi, RLF, CRa et al.); 1, 75 mi (121 km) south of Block Island, 2 February 1978 (B. Nikula); 1, Cox Ledge, 27 March 1983 (DFi); 1 shot by C. B. Clarke off Point Judith, 28 March 1902 (formerly in the collection of LKi but no longer extant).

Great Shearwater (*Ardenna gravis*)

Status

Common nonbreeding visitor to offshore waters during June–October, occasionally to December.

Nonbreeding

Inshore. Near land, Great Shearwaters are the least numerous of the regularly occurring shearwaters and were not seen from the mainland coast until 1935. There are still less than 2 dozen inshore records, most of 5 or fewer birds, and half of these records have been from the Block Island ferry. Four were picked up dead along Moonstone Beach 28 June 1998 (RFa). The highest counts from shore are 11, Sandy Point, Block Island, 28 June 1982, and 12 off Point Judith during the offshore passage of Hurricane Lili, 13 October 1990 (PAB, RLF, REn). One seen at Gaspee Point, Warwick, 13 August through 5 September 2004 (GDe), is the most inland occurrence for Rhode Island.

Offshore. Counts at Cox Ledge during the warmer months commonly range from 30–200 birds. Shipboard transects (Powers 1983; Paton et al. 2010) show that Great Shearwaters reach maximum density during summer in an area corresponding approximately to the 20-fathom (37 m) depth contour, which indents shoreward south of Sakonnet Point and includes the Cox Ledge area. Data from a very limited number of pelagic trips after mid-October suggest that Great Shearwaters may often remain in small numbers through late October and November. Six were seen off Beavertail Point 21 November 1995 (RLC), and 5 were seen passing Point Judith during a southeast storm 29 November 1980 (RAC); also, 83, Block Island Sound, 10 November 2011 (SSM et al.); and 3, Mud Hole, 7 December 2015 (K. Mueller). Great Shearwaters arrive for breeding at their main colony at Tristan da Cunha in late October, so birds off New England in October and November must be prebreeders or nonbreeding adults.

Maxima. 1500, 40 mi (64 km) south-southeast of Block Island, 12 September 1993 (WRP, DLE); 1040, Cox Ledge, 4 August 2009 (JSJ et al.); 1000, Cox Ledge, 14 July 1978 (DLE, CWo, HCJ); 950, south of Block Island, 27 June 1984 (RBo, HMP).

Extreme dates. 9 May 2009, southeast of Block Island (CPe et al.), to 16 December 1979, Block Island (G. Lockhart).

Manx Shearwater (*Puffinus puffinus*)

Status

Uncommon but increasingly numerous visitor and migrant; rare in winter. Commonest inshore.

Nonbreeding

Rhode Island's first Manx Shearwater was seen 21 July 1940, when WHD reported 1 off Land's End, Newport. Next were 2 other records of small shearwaters that likely belong here, both of them close inshore: 1 *inside* the Point Judith breakwater 18 September 1951 (RCC) and another 14 July 1964 in Narragansett Bay north of Goat Island, Newport (REW). Singles were seen off Point Judith 29 April 1967 (DLK, CWo), Narragansett 24 October 1971 (RRV), and Point Judith 13 March 1972 (RRV), and 4 more were off Sakonnet Harbor 6–7 June 1970 (JAG, DLK, CWo, SSD et al.). Thereafter, a veritable flood of observations began, coincident with the increase of the species elsewhere in New England, including the 1973 discovery of the first North American nest under a plank on nearby Penikese Island, Massachusetts (Bierregaard et al. 1975), and increasing numbers breeding in Newfoundland after 1976 (Nisbet et al. 2013).

The continued presence of Manx Shearwaters during spring and summer near Beavertail and Point Judith in the 1970s suggested that they might have been breeding nearby. Searches of the most likely sites on Hope, Dutch, and other islands in Narragansett Bay and along the coast since 1976 by the author and JEM have so far been fruitless. After the mid-1980s, a sharp reduction in the presence of Manx Shearwaters was noted off mainland promontories such as Point Judith and Beavertail, although occasional birds continue to pass coastal promontories and the species remains regular in offshore waters. Records along the Rhode Island coast have been concentrated along the central coast, particularly off Point Judith and in the Beavertail Point area, where individuals and small flocks have been seen. They are rare after early October; 1 was seen 20 mi (32 km) south of Block Island 24 October 1993 (CAv).

Maxima, prospecting birds. 17, Beavertail, 2 June 2012 (RWa); 12, Beavertail Point, 21 May 1976, and 11 there 1 June 1979 (RLC); 23 "passages" of individuals there, 26 April 1980 (GLB); 10, Block Island Sound, 6–9 June 1976 (RLF). In addition to the 1964 record near Goat Island mentioned above, there are now 2 additional records for lower Narragansett Bay between Goat and Rose Islands for July 1996 (C. Johnson). Most unusual was a small black-and-white shearwater, thought to be a Manx, seen off Sabin Point, East Providence, 1 April 2001 (RFa). Two passing very close to Point Judith near dusk on 1 July 2015 and headed for the West Passage (PAB), and 1 in Narragansett Bay north of Big Gould Island 25 May 2016 (CRa), suggest that Manx Shearwaters continue to prospect and may breed in Rhode Island.

Maxima offshore. Eight, Cox Ledge, 17 June 2000 (SSM et al.); 6, off Block Island, 7 June 2015 (CPe et al.); 5, Cox Ledge, 4 August 2009 (JSJ et al.).

Extreme dates. 4 March 1980, Point Judith (GLB), to singles on 29 November 1980, Point Judith (RAC), 13 December 2009, off Block Island (Paul L'Etoile et al.), and 4 January 1992, Point Judith (RLF).

Audubon's Shearwater (*Puffinus lherminieri*)

Status

Rare summer visitor.

Nonbreeding

Rhode Island's first Audubon's Shearwater was found dead at Briggs Beach 27 July 1938 by DLE (MCZ #291307). There were no more records until the 1970s. The following recent sight records were made by observers thoroughly familiar with Manx Shearwaters and are believed correct: 1 at Cox Ledge, 13 August 1977, coincident with the intrusion of a warm-core eddy of Gulf Stream water close inshore (Dieter, Emmons, fide THD); 1 or 2 birds 10 mi (16 km) south of Block Island, 26 August 1978 (THD et al.); 1 at Cox Ledge and another 12 mi (19 km) east-southeast of Point Judith, 4 August 1979 (RBo, RAC, DLK, DRo, CWo). Others at Cox Ledge include 1 September 1979 (DFi, KKi, FPe, CWo), 4 September 1983 (DLK, STs, DRo), and 10 September 1985 (DLK, DFi). Other records include 1, 8 mi (13 km) southeast of Block Island, 28 July 1987 (DFi); individuals at Cox Ledge and Block Canyon, 21 and 28 August 1992 (DFi; DLE); 2 at Cox Ledge, 8 July 1993 (DFi, CAv et al.); 1, 10 mi (16 km) southwest of Block Island, 29 July 1993 (SSM et al.); 1, 11 mi (18 km) southeast of Block Island, 26 August 1993 (CAv, DFi); 5 and 2 at Atlantis Canyon, 12 and 25 September 1993 (WRP, DLE et al.; CAv); 3 near Block Canyon, 28 July 1995 (PAB, DFi, RLF et al.); 1, 10 mi (16 km) south of Block Island, 9 September 1995 (WRP, DLE, RBo); 1 in Block Island Sound, 3 August 1996 (WBo, DLK); 1 at Cox Ledge, 25 July, 31 July, and 22 August 1996 (CAv, DFi, RTL); 1, 15 mi (24 km) south of Block Island, 31 August 1996 (DFi); 1, 20 mi (32 km) south of Block Island,

24 August 2005 (STs et al.); 1, Block Island Sound, 23 August 2002 (DFi); and 1 at the Mud Hole, 24 August 2005 (STs).

This warm-water shearwater and several other tropical pelagics such as Bridled Tern and Black-capped Petrel have lately been shown to be much more regular in the Gulf Stream off Cape Hatteras than previously suspected (Lee and Booth 1979; Lee 1984, 1986), and tracking data suggest that Audubon's Shearwaters regularly enter the Gulf of Maine (P. Jodice, Clemson University, pers. comm.).

Maxima. 47, Block Canyon, 3 September 2015 (CPe et al.); 26, Block Canyon area, 19 August 2009 (JSJ et al.).

STORKS (CICONIIDAE)

Wood Stork (*Mycteria americana*)

Status
Vagrant. Two records.

Historical
For more than a century, the only Rhode Island record of a Wood Stork was of an immature shot as it was flying northward over the Barrington River in company with a Great Blue Heron, 8 August 1896 (Hathaway 1896). The specimen is in the RWPM, where it was recently re-examined. Another individual had been shot 21 days earlier in nearby Seekonk, Massachusetts (Brewster 1896), very likely having passed through Rhode Island, and a third was collected at York County, Maine, 18 July 1896 (Vickery et al. 2020), indicating that the Rhode Island record was not an isolated occurrence.

Nonbreeding
The only modern record is for 1 at Block Island 3 November 2019 (Clara Cooper-Mullin; photo). Episodic northward dispersal by multiple Wood Storks has occurred in 1955 and 1958–1963, including a flock of 11 at Plymouth Beach, Massachusetts, 18 November 1963 (Veit and Petersen 1993; Levine 1998).

FRIGATEBIRDS (FREGATIDAE)

Magnificent Frigatebird (*Fregata magnificens*)

Status
Vagrant. About 15 records, usually not associated with storms.

Nonbreeding

The first frigatebird for Rhode Island was a female found dead on the Block Island docks 16 November 1932 and brought to EDi (Dickens 1934). Harold N. Gibbs later interviewed a man who stated that he had shot this bird as it sat in a tired manner on a beach in company with gulls. The specimen is now in the Block Island Collection. A second was well seen on the surprisingly early date of 21 March 1958 at Block Island's north end by Eileen and John Lee; the latter had a previous familiarity with frigatebirds in the Pacific (fide RCC). On 21 May 1959, 1 was reported sailing along the Newport and Middletown shores by a Newport resident and hastily verified by JBa (Baird 1959b). Another was seen 30 June 1960 soaring over the NBS in Middletown (J. Mahoney). On the exceptional early winter date of 21 December 1973, a female appeared in the area of Second Beach, Middletown, and remained the next day to be listed on the Newport–Westport CBC (LOG, RMc, TSG). Another female, seen 2 mi (3 km) south of Block Island 15 October 1988, appeared 30 minutes later at Cox Ledge, where it attended the fishing boat *Supersquirrel* for a short period (m.obs.; thoroughly photographed by F. Gallo). Another bird with some white in the plumage was seen along Ocean Drive 26 or 27 June 1990 during a storm (L. Steere, fide CRa). Two seen in 1995 were thought to be different birds: 14 October 1995 at Sachuest Point (JDa) and 28 October 1995 at Galilee (photographed and specifically identified as *magnificens*; PP, PAB, MJM). Most remarkable is a midwinter record of an adult female seen at close range as it flew over Sachuest Point heading for Sakonnet 4 February 1991 (JGe, RGe). Singles were seen at Providence 13 October 1999 (RFa, DFi), Matunuck 14 November 2005 (RTL), and Block Island 9 June 2011 (DFW). Another was photographed at Little Compton 30 October–1 November 2012 (G. Dennis; see Photo 20), following Hurricane Sandy. A white-faced immature was sailing west over Point Judith, 22 December 2018 (GWi et al.; photos), and the same(?) was seen by others at 3 other locations west to Matunuck, from where it headed southwest over the sound toward Montauk. One was at Beavertail Point, 10 June 2020 (S. Roberts; photo).

Many of these records were and are presumed to be Magnificent Frigatebirds on the grounds that this is the most likely species to appear, given that it is the only one nesting in the North Atlantic. Nevertheless, Lesser Frigatebird (*Fregata ariel*) has been photographed in Maine in May 1960 (Snyder 1961), Great Frigatebird (*F. minor*) in California and Oklahoma (Howell et al. 2014), and Ascension Frigatebird (*F. aquila*) in England in July 1953 (Harrop et al. 2014). Therefore, the scrutiny of future frigatebirds for specific identification is justified.

GANNETS AND BOOBIES (SULIDAE)

Brown Booby (*Sula leucogaster*)

Status
Vagrant. Three to four records.

Nonbreeding
An adult, apparently exhausted, Brown Booby landed on the upper deck of the party fishing boat *Supersquirrel* at Cox Ledge 19 August 1989. It was unable to take off because of the boat's railings. Finizia and Kraus were called by Al Jarman, captain of the *Supersquirrel*. Kraus identified the bird, and it was photographed by Conway and Finizia, then released the next day outside the Point Judith breakwater, when it flew off toward the southeast. Another or the same adult bird was seen flying east over Harbor of Refuge 26 August 1989 (RAC). Other records include 1 at Cox Ledge 5 May 2001 (K. Mueller) and another 3 mi (5 km) south of Sakonnet Point 27 July 2012 (M. Schrimpf).

Northern Gannet (*Morus bassanus*)

Status
Common to abundant migrant and winter resident. Rare in summer.

Nonbreeding
Along the mainland outer coast, gannet counts off Point Judith are generally about twice as large as those off southern Little Compton, and substantial counts are seldom made along the South Shore to the west of Point Judith.

Spring. Counts of migrating gannets over inshore waters are smaller in spring than in fall.

Spring maxima. 2000, Block Island, 4 April 2009 (DFi); 568 in 1 hour, Block Island, 16 April 1959 (A. M. Bagg); 501, Point Judith, 29–30 April 1995 (RLF).

Summer maxima. 80, Point Judith, 6 June 2000 (RLF); 75, Block Island, 7 June 1995 (RLF, AES)

Fall maxima. 3000 off Napatree Point, 21 November 1991 (CRa); 2800, Point Judith, 25 November 1983 (DLE, REm); 2000+, Point Judith, 10 November 1991 (RLF); 1800, Point Judith, 16 November 2005 (DFi).

Winter maxima. 1996, South Kingston CBC, 26 December 1999 (DLK et al.); 1672, South Kingstown CBC, 21 December 2002 (STs); 903, South Kingstown CBC, 22 December 2001 (STs et al.).

A count of 270 during fog off Napatree Point 28 November 1985 (CRa) was a prologue to an "unprecedented flight" of similar size the same day along the Connecticut coast of Long Island Sound (Zeranski and Baptist 1990). During the period 10–25 November 1991, unprecedented numbers were present in Block Island Sound, apparently in response to an abundance of sand lance. On 20 November 1991, 2000+ were diving furiously off Matunuck (RWa, SHa et al.), and 3000 were present off Napatree Point the next day (CRa). These numbers are unusual in western Rhode Island and perhaps reflect localized concentrations of baitfish such as sand lance or menhaden.

Inshore records. Gannets are uncommon away from the outer coast: 12 were in the Sakonnet River off Seapowet 31 March 2001 (RFa), another reached Little Gould Island 13 February 2005 (GDe), and 2 were observed in Greenwich Bay 15 April 2005 (GDe). The only inland record away from salt water is 1 seen over Johnston on 26 November 1950 after a severe gale the day before, when winds over southern New England reached 100 mph (161 kph) (RCC).

ANHINGAS (ANHINGIDAE)

Anhinga (*Anhinga anhinga*)

Status
Vagrant. Two records.

Nonbreeding
An Anhinga was seen soaring over North Smithfield 11 August 1996 by BCG and W. Gearhart. They described a long-necked bird with a long, spreading tail soaring and drifting northward, hawk-like, over their home. Another was seen flying north over the Seekonk River at Swan Point 12 May 2002 (DFi, STa).

CORMORANTS (PHALACROCORACIDAE)

Great Cormorant (*Phalacrocorax carbo*)

Status
Common migrant and abundant but declining winter resident on salt water. Rare inland.

Historical
Great Cormorants were rare at any season on the Rhode Island coast before 1900 (Jencks 1886a; Mackay 1894; Howe and Sturtevant 1899), during a period when

they were nearly exterminated from North America (Bent 1922). One of the most persistent hotspots is Cormorant Rock off Sachuest Point in Middletown, where the top of the rock appears to have been worn smooth from the feet of cormorants. On a map dated 1776 and surveyed by Des Barres, these same rocks are correctly located and called "Cormorant Rocks" (Mackay 1894), and Sir Charles Blagden (1776–1780) reported that cormorants "harbour in large quantities on a rock about 2 miles [3 km] from Sachawest [*sic*] Point, thence called Cormorant Rock" (Howe 1905).

Nonbreeding

Winter. The rocky coastline between Martha's Vineyard and Point Judith holds some of the largest numbers of Great Cormorants anywhere on the northeastern coast of the United States, and substantially higher densities than adjacent parts of Massachusetts, Connecticut, or New York. Since about 1990, the Great Cormorant population in southeastern Rhode Island has been in decline. The current (2007–2016) 10-year average on the Newport–Westport CBC is a paltry 74.5, with the maximum count being 108 on 14 December 2014—a far cry from the 1980s when this CBC once represented the national maxima.

Winter maxima. 3256, Sakonnet, 16 December 1984 (RAC et al.); 3100, Sakonnet, 20 December 1986 (RAC et al.); 1000, Little Compton, 31 January 1947 (RCC); 200, Block Island, 20 February 1946 (EDi). More recently: 175, 8 November 2008 Block Island (CRa, SSM et al.).

Migrants continue to arrive into January; e.g., 255 flying west past Point Judith 5 January 1992 (DLK, RLF).

Fall. Great Cormorants arrive in late September and reach peak abundance by December.

Spring. Lingerers remain through April (40 at Jamestown, 25 April 1972; MDa), and Point Judith sea watches in recent years indicate almost daily passage of a few transients through most of May (e.g., 13 moving east 15 May 1982; RLF), with occasional birds still moving in early June. A northward overland flight in spring by some of the population is suggested, however, by the observation of 60 adults in 4 flocks flying northeast toward the Taunton River at Stone Bridge, Tiverton, 6 March 1971 (RBo). Records of transients among overland flocks of Double-crested Cormorants include 1 with 15 of the latter heading east-northeast over Chapman Pond, Westerly, 28 April 1979 (RLF), and another seen in a flock of 310 moving southwest over Arcadia 18 September 1982 (REn).

Summer maxima. 12, Block Island, 8–20 June 1981 (RBo); 11, Galilee, 19 August 1984 (DWa).

Inland records. There is an increasing number of inland records: 1, Diamond Hill Reservoir, 18 December 1983 (REn); 2, Lincoln, 9 February 1992 (DMK); 1,

Lonsdale, 11 February 1992 (REn); 1, Diamond Hill Reservoir, 7 November 1993 (DMK); 1, Tucker Pond, 26 December 1994 (SSM). Small groups of <10 occur on the East Providence Reservoir.

Maxima. 44, Seekonk River, 13 March 2005 (MLy).

Double-crested Cormorant (*Phalacrocorax auritus*)

Status

Common and increasing breeder. Abundant transient, vastly increased since about 1980.

Historical

Sturtevant recorded Double-crested Cormorant only once during summer, and Howe and Sturtevant (1899) listed no other summer records. Twenty-five summering birds at Middletown 23 June 1937 (WHD) were considered to be an exceptional number. Counts in the 1950s rarely exceeded 50, and a flock of 170 was on the Sakonnet islands 20 June 1971 (RLF, DLE). A minimum of 1000 was present in 1980. During 1977–1979 they colonized Fisher's Island, New York, and the Norwalk Islands of western Connecticut and spread elsewhere through southern New England (Buckley and Buckley 1984; Hatch 1984).

Breeding

Double-crested Cormorants first nested at the Sakonnet islands in 1981. Nest counts at Sakonnet rose rapidly thereafter and a meteoric spread to other islands soon began. An island-specific history of their colonization follows:

Islands off Sakonnet Point (West Island, East Island, The Clumps). Double-crested Cormorants first bred in 1981. Seventeen nests were located on the southern extension of West Island, and 1 nest on The Clumps, in June 1981 (RLF, JEM). West Island was the largest colony in Rhode Island, rising to a peak of 954 in 1989. Thereafter, the West Island colony declined as colonies on nearby islands grew. Most of the breeding activity has been on bare rocks of the southeastern parts of West Island, with growing numbers on a cliff face to the west and occasionally on the chimney-like ruins of the old fishing lodge. High and lofty East Island took the longest to be colonized, but that colony has grown rapidly from an initial 9 nests in 1985 to 444 in 1994. East Island was abandoned as a nesting site after 2011, and birds probably moved to West Island.

Little Gould Island. This forested island off Tiverton has nesting herons as well as gulls and cormorants. It was not colonized by cormorants until 1987. A single nest in the southern cedars in 1987 grew to 628 nests by 1994. The predicted death of most of the island's ancient forest (Ferren and Myers 1990) was well on its way to completion in 1994. Here, the species has nested entirely in the trees, moving gradually from the stunted cedars on the southern end into the tall trees of the interior of the island and seriously threatening the island's remarkably diverse flora. This site has declined since reaching a peak of 753 nests in 1996. Recent nest counts have been in the mid-400s.

Little Gooseberry Island. Once thickly vegetated in the 1960s, this now-barren island was initially devegetated by gull guano and then eroded by rainfall, thereby setting the stage for cormorant occupancy. One cormorant nest appeared in 1984, 12 in 1987, then a peak of 181 in 1991. Thereafter, the colony declined and apparently transferred to better-vegetated Hope Island. There were 4 nests in 2004, but there has been no nesting since then.

Hope Island. This 25-acre (10-ha) island in central Narragansett Bay had 16 nests in 1991; they were all on the cross-arms of old telephone poles along the island's west side. In 1994, all but a handful of the 286 nests were in the central island trees. They continued to increase to 528 nests in 2015.

Big Gooseberry Island. Two nests were present in 1985, 1 in 1992, and 3 in 1993 (JEM, RLF). This island has a seasonally occupied private house, and efforts to inhibit birds such as gulls and cormorants from nesting have been made by the owners and caretakers. They were apparently successful in discouraging nesting birds and there has been no nesting since 2003.

Dumpling Islands. A single nest, later abandoned, was found on the northeastern Dumpling in 1986. No further breeding was noted until 1994, when 19 nests were located adjacent to the largest roosting site on the northwestern Dumpling. This site's colony has increased in recent years. In 2016, there was nesting on 3 of the Dumpling Islands, for a total of 227 nests.

Seekonk River. Two nests were on pilings off the mouth of the Ten Mile River along the East Providence shore on 23 June 1994 (RLF, JEM) and had apparently been there for 2 years previously.

Big Gould Island. This former navy torpedo testing facility is located north of the Newport Bridge. Cormorants began to nest in trees on the west side of the island during 1998. This site has been stable at 25–30 nests.

Gull Rock. This is a craggy island off the eastern shoreline of Newport. Cormorants began to nest here in 1995, and the colony reached a peak of 44 nests in 2013.

Price's Neck. This is an island off the south coast of Newport. Cormorants began to nest here in 2003 and the colony reached a peak of 188 nests in 2010.

There are other pilings and little islands where small numbers of cormorants have nested. The statewide total was 2364 nests in 2016 (there were 18 in 1981, 181 in 1990–1991, 2217 in 1995, and 1735 in 2004).

Nonbreeding

The majority of the northeastern breeding population crosses southern New England spring and fall by an overland route (Dorr et al. 2020).

Spring. Migration of adults in alternate plumage peaks during late April, and immature birds continue a protracted migration through early June.

Spring maxima. On 25 April 1981, RAC encountered a flock moving along the Charlestown shore that he estimated to be 1.5 to 2 miles (2–3 km) long, 0.25 mile (0.4 km) wide, and 150–200 ft (46–61 m) deep, containing approximately *40,000* cormorants. Also, 4277, Point Judith, 28 April 1985 (RLF); 40 flying northeast, Richmond, 22 June 1996 (CRA).

Fall. The southwestward movement of small flocks along the coast becomes obvious in September, but migration does not reach its spectacular peak until mid or late October.

Fall maxima. 22,000 past Point Judith, 10 October 1982 (RLF); 13,000, Watch Hill, 10 October 1981 (RLF, REn, CRa); 7640, Napatree, 23 October 2001 (CRa); 4500 past Block Island, 12 October 1987 (REn).

Winter. Unlike Great Cormorants, Double-crested Cormorants are partial to shallow estuaries, harbors, and salt ponds. Double-cresteds have dramatically increased as winter residents since the 1970s.

Winter maxima. 63, South County CBC, 14 December 1991 (DLK et al.); 55, Galilee 17 February 1992 (DLE, REm); 50, Galilee, 2 January 1986 (m.obs.). Elsewhere, counts have also increased; e.g., 38, Newport–Westport CBC, 15 December 2012 (REm et al.).

PELICANS (PELECANIDAE)

American White Pelican (*Pelecanus erythrorhynchos*)

Status

Vagrant. Increasing in recent years.

Nonbreeding

The first Rhode Island record was of an immature discovered 28 September 1946 at Bonnet Shores by DLK and others, who were able to approach to within 100 yd (91 m). This individual was also reported from Narrow River the same day and may have been 1 of several which spent most of October of the same year on Long Island (Bull 1964). Another was first found by DVT and remained for most of the period 11 November–6 December 1978 at Easton Pond, Newport, where it roosted nightly on the flats. On most days, it commuted eastward past Sakonnet toward the Westport River area or fed occasionally in Quicksand Pond or along Newport's Ocean Drive, returning to Easton Pond in the afternoon. Early in the morning of 3 July 1989, a third bird was found by RBo resting on a sandbar at Quicksand Pond, from which it promptly departed eastward through Acoaxet toward Westport. Two discovered at Narragansett 4 January 1994 (S. Howden) remained until 5 March, thus surviving one of the bitterest winters on record. These birds often frequented the Galilee docks to receive handouts from fishermen. Single birds were present at Allens Harbor 3–6 December 1994 (D. Havens), Warwick 1 August 1996 (fide WBo), Block Island 27 September–6 October 1996 (MSl, ASRI et al.), Trustom 7 November 1996 (MJM), Napatree Point 18 October 1997 (CRa), and the same or another at Trustom the next day (Al Richards). Also, 1, Prudence Island 13–20 December 2004 (M. Rehor), then possibly the same bird on Winnapaug Pond 29 December 2004–20 January 2005 (JGa et al.); and 1, Narragansett 8 June 2005 (T. Seiter), then possibly the same bird at Charlestown Breachway 2 July 2005 (PLE). One was on the Palmer River in Barrington 18 September 2008 (B. Lombardi), and another at Westerly, 16 June 2010 (Kate Pisano). One was at Trustom Pond NWR, 24 July 2010 (RDe), 1 at Jamestown 4–5 January 2011 (Pam Long et al.), and another at Trustom Pond, 20–24 November 2012 (JMu, SSM, PJL et al.). One was at Nonquit Pond, Tiverton, 17–19 January 2020 (Pagarini, Palmer, Schenck). This increase since about 1985 is consistent with an overall increase of records in the Northeast during those years.

Maxima. Three, Trustom Pond, 19 January 2014 (Hans Bucht).

Brown Pelican (*Pelecanus occidentalis*)

Status

Vagrant. Increasing in frequency.

Nonbreeding

One was shot at Block Island 17 April 1921 and examined by EDi, but not preserved. On 21 June 1927, an immature was seen resting on the water about a mile off Narragansett Pier and 2 hours later seen flying in a northeasterly direction over Newport harbor into Narragansett Bay. Its presence may have been related to a gale of near hurricane strength that passed off Cape Cod on the 20th (Nichols 1927). One was seen off Sakonnet Point 10–12 April 1954 by Williston, CWo, and RBo. An immature was seen flying west along the shore past the Point Judith breakwater 17 August 1958 by SCr. After Hurricane Donna on 12 September 1960, 6 pelicans of uncertain species were reportedly seen off Poppasquash Point, Bristol, by local residents.

None were seen in Rhode Island between 1960–1988, the period of DDT-related decline (Shields 2020). On 7 July 1988, 3 were noted fishing off Sandy Point, Block Island (Anderson). A single bird was photographed as it circled over the ocean off

Ninigret Pond 18 November 1988 (AFr). On 20 August 1991, the day after Hurricane Bob, 1 was seen off Matunuck and presumably the same bird was later over Ninigret Pond (both RAC). In July 1992, several were present near Stonington Point, Connecticut, and in neighboring Little Narragansett Bay. At least 1 of these was resting on Sandy Point Island 21 July 1992 (RDe), 1 was sitting among cormorants on pilings off Napatree Point 23 July 1992 (CRa), another was seen there 10 August 1992 (NF), and local residents reported up to 5 diving into Little Narragansett Bay during the same period. Several appeared during 2000–2002: 1, Block Island, 24 June 2000 (CRa); 2–3, Sakonnet Point, 25 July 2000 (fide GDe); 1, Block Island, 21 December 2000 (CRa, SSM, PJL); 1, Ninigret Pond, 29 September 2001 (L. Landon; photo); 1, Jamestown, 15 January 2002 (ChP). A weakened bird was captured 18 February 2000 at Galilee and was put on an airliner to Florida for rehabilitation. In 2011, there were several reports spanning 30 August to 3 September from the South Shore of Rhode Island to Block Island. All of these may have been of the same bird. After Hurricane Sandy in 2012 there were at least a dozen reports ranging from 31 October to 19 November. At least 3 of these were brought to rehabilitation clinics; 1 had come from a fishing vessel 100 mi (161 km) south of Block Island. One was photographed at the north tip of Block Island 20 June 2014 (C. Amaral). One was at Galilee, 4–7 January 2019 (M. Macleod, B. Glemboski, CPe) and 1 in Block Island Sound, 22 September 2019 (DFW).

HERONS AND EGRETS (ARDEIDAE)

American Bittern (*Botaurus lentiginosus*)

Status

Rare and local summer resident and uncommon to fairly common transient. Uncommon in winter.

Historical

During the early twentieth century, American Bitterns were common summer residents in cattail marshes. They declined during the 1940s, and this decline steepened during the 1970s. By 1980, the only territorial birds were found at Block Island, the Westerly Cedar Swamp, and the Lonsdale marshes.

Breeding

During extensive explorations of Rhode Island marshes during the breeding bird atlas survey, 1981–1986, American Bitterns were not confirmed breeding anywhere in Rhode Island. The only territorial birds were found at the Lonsdale marshes and Belleville Pond. American Bitterns have actually been outnumbered by Least Bitterns. During the 1990s, American Bitterns seem to have fared best in the relatively predator-free environment of Block Island, where birds responded to tapes for several years at Peckham's Pond, but have since disappeared. Elsewhere, summer records have been limited to scattered observations at former breeding locations and occasional strays; e.g., 1 at Galilee 4 June 1990 (DLK) and another in Barrington 14 June 1983 (JCB). A recently fledged chick was found on the east side of the road at the Jerusalem marshes 15 July 2002 (CRa). Some disappearance must be the result of habitat destruction, including replacement of cattails with *Phragmites* and purple loosestrife, but in other areas birds have disappeared from marshes that seem not to have changed much. Possibly, bitterns have suffered from reductions in abundance of their invertebrate prey, as have American Black Ducks (Longcore et al. 2020).

Nonbreeding

Spring. Early migrants, perhaps from nearby wintering areas, trickle into coastal areas from mid-March onward, but general arrival is during the last 10 days of April and the first few days of May.

Fall. In recent years, 2–3 per day have been seen at the outer coast, most often in salt marshes.

Winter. Since 1940, individuals have occasionally lingered into early winter; e.g., 1, Seapowet, 2 February 2001 (REm); 1, Trustom Pond, 15 February 2002 (Glenn Williams); 1, Block Island, 18 February 2002 (CRa, SSM et al.); total of 2 reported during February 2002.

Maxima. Three, Moonstone (Mud Pond), 21 September 2007 (WM, RTL); 3, Napatree CBC, 27 December 2015 (SSM, GW).

Least Bittern (*Ixobrychus exilis*)

Status

Rare summer resident and rare, seldom detected, transient.

Breeding

Howe and Sturtevant (1899) stated that they bred "in all the reedy marshes of the state," and this assessment is supported by data from turn-of-the-twentieth-century oologists as well as recent use of tape-recorded vocalizations. They are nevertheless restricted to marshes with narrow-leaved cattails. According to CRa (pers. comm.), Least Bitterns require open patches of water among cattails rather than unbroken stands. Historical stations for confirmed or suspected breeding are as follows (see Jencks 1880c, 1881):

Quonochontaug. The most reliable location for many years was in cattails at Fresh Pond, between Quonochontaug and Ninigret Ponds. Two were shot here, 1–4 June 1897 (HSH); 2 nests found, 22 and 29 June 1913, and another 4 June 1916 (HSH); and a pair was with 4 young, 4 days old, 1 July 1917 (HSH). The species was noted

FIGURE 7. Heron nesting colony, © Margaret La Farge.

regularly in May and June until about 1936 (HSH, DLK) and more sporadically thereafter; a pair 11 June 1984 (CRa) and 1 seen 17 June 1987 (v.o.) suggest continued occupancy, but the habitat has been succeeding to *Phragmites* and shrubs. Chris Raithel surveyed this site by taping from a canoe in spring 2015 and found no birds but did find 1 in 2017.

Moonstone Brook, South Kingstown. Territorial birds were present regularly in summer from 1940 through 1970 and intermittently since then (CWo, DLK et al.). One was there 23 July 2012 (RWa), and they were still nesting there in 2016. This spot and the fringe marsh around Trustom Pond may now be the only nesting locations in Rhode Island.

Point Judith/Galilee area, Narragansett. Many or most records come from a cattail pond a half mile west of the Point Judith lighthouse. Four eggs were collected at Point Judith 1 June 1886, perhaps at this marsh (Holland). There is a set of eggs labeled "Point Judith marshes" dated 29 May 1910 collected by HSH, on which date he found 3 active nests in the same marsh. Several young were collected 8 June 1915, and a nest and 4 eggs 24 June 1916 (CED). No further proof of nesting was obtained there, but individuals were occasionally seen in the salt marsh south of the present Escape Road from 1952 on. They may have bred in the small cattail marsh and sea of *Phragmites* that appeared after a massive dumping of dredge spoil on the marsh, replacing the *Spartina alterniflora* at the site. They were rediscovered at the small cattail marsh a half mile west of the Point Judith lighthouse 6 June 1982 (RLF). Two were seen there 7 June 1984 (CRa); 2 on 19 May 1987 (CRa); and 1 on 14 June 1989 (REn) and 30 June 1999 (JGe, RGe). Raithel surveyed this site in June 2016 and 2017 and found no birds.

Belleville Pond, North Kingstown. An incubating bird was discovered and photographed there in 1975 (F. Golet). Four were counted 6 June 1984, including a pair and 2 other calling individuals, possibly amounting to 3 pairs (CRa). Three were present 15 May 1985 and 3 at 2 locations 1 June 1986 (CRa, REn). Raithel surveyed this site in 2015 and 2016 and found no birds.

Hanna's Marsh, East Greenwich. A bird flushed here by canoe 28 June 1972 (DWa) suggests possible breeding.

Lonsdale marshes, Lonsdale. One was seen 7 June 1974 (RLF), and 4 (3 possible pairs) 12 June 1984 (CRa, REn). Raithel surveyed this site in 2015 and 2017 and found no birds.

Almy Pond, Newport. Eggs were collected here in 1884 by John Snowdon Howland; Doe found 2 nests within 20 ft (6 m) of each other on 16 June 1895, and other nests in 1898 and 1915. A recently fledged chick was seen about 1982 (TTr, fide CRa), and 5 birds were present on 7 June 1985 (CRa).

Easton Pond, Middletown. A set of eggs was collected and 1 bird seen in the formerly massive cattail beds near the southeast corner of the pond 18 June 1893 (CED, HSH). Higher or fluctuating water levels in this reservoir, or the pumping of more acid waters from the Tiverton–Little Compton area, have slowly obliterated the formerly luxurious cattail stands in recent decades, to the detriment of all marsh breeders.

Sachuest marshes, Middletown. A pair was present during the summer of 1937 and a nest found in July 1938 (WHD). They were absent in 1940, and there were no further records until 1 was found here 18 June 1984 (CRa). Much of the cattail marsh was filled by the Middletown dump during the period 1950–1965, and the remaining marsh was opened to the sea, so the site is no longer suitable. The small marsh where the 1984 bird was seen slowly filled with *Phragmites* and recently was incorporated into a salt marsh restoration by USFWS, so is now no longer suitable.

Long Pond, Little Compton. A nest and 5 eggs were taken in June 1888 (Drowne, fide HSH); 3 adults and 2 young were shot in July during the 1890s (N. Dexter). A bird was present here 27 May to 1 July 2012, after a long absence (REm).

Briggs Marsh, Little Compton. One seen flying along west shore, 15 June 1962 (RLF). Thorough kayak survey by REm 6 July 2015 produced no evidence of Least Bitterns.

Block Island. There are several fine cattail marshes on Block Island, some of which are threatened by peat extraction. The only summer records are 1 seen near Dickens Point 13–24 June 1919 (EDi) and another heard calling 12 June 1984 (RBo). There is no more recent evidence of nesting there.

The only other summer records are of a wandering bird at Quicksand Pond, 2 July 1973 (DLE, REm), and another at Jerusalem 2 June 1989 (SSM). Two were at Trustom Pond 23 June–3 July 2012 (PAB et al.), and it is likely that 1 or a few pairs breed there (CRa, pers. comm.).

Nonbreeding

Rarely detected away from breeding localities. The earliest spring migrant was an individual standing on the flats at Jerusalem 25 April 1974 (DWa). One was at Easton Pond 5 May 2001 (CAv), and 1 at Jerusalem 9 May 1986 (DFi, DLK). Other than wintering birds, the latest fall dates are for 1 at Block Island, 5–6 October 1987 (DLK, RBo, DLE et al.); 1 at Jamestown, 7 October 1961 (RLC); and 1 at Warwick, 26 October 2003 (Doug Wilson). One injured bird was brought to the vet at URI 2 October 2012; it subsequently died.

Winter. One was found near the golf course at Sakonnet 19 December 1987 (RAC), and 1 was found freshly road-killed at Wesquage Pond, Bonnet Shores, on 20 December 1991 (DLK et al.). Another was reported on the Newport–Westport CBC in 1966. Howe and Sturtevant (1899) listed a record 28 February–1 March 1881 of an adult male, "killed by a boy with a bean snapper on the shore of the bay near the city," and brought to Southwick and Jencks (Jencks 1881). Examined a day later by William Brewster, who happened to be visiting the Natural History Store (Brewster 1881; Jencks 1881), it was found to be "thin in the flesh."

Extreme dates. (Apart from winter) 25 April 1974 to 20 December 1991.

Great Blue Heron (*Ardea herodias*)

Status

Common migrant and winter resident. Uncommon breeder, beginning in 1984.

Historical

From the late 1800s until the 1950s Great Blue Herons were much rarer than at present and did not occur during late fall and winter as they now do. Howe and Sturtevant (1899) considered them to be common migrants, but HSH saw his "first live one" in 1894 after 10 years of active experience in Rhode Island.

Breeding

Unlike other herons that select offshore islands for breeding, Great Blue Herons nest in beaver swamps at inland locations. The dead trees provide refuge from predators, especially raccoons. Breeding in Rhode Island was first recorded at the Buck Hill Management Area, Burrillville, in 1984. Other sites have gradually been discovered.

Buck Hill, Burrillville. Three adults and a single nest were discovered 12 June 1984 in the middle of a flooded 50-acre (20-ha) cedar swamp. A single chick was still present in the nest 24 July (REn). Three pairs and 9 young were noted 20 July 1985 (REn, CRa). Two birds were seen there 10 June 1988 (REn) and 2 pairs 22 April 1989 (JSJ). Then there were no reports until 1 reported on a nest 22 April 2001 (MLy). Fifteen nests were there in fall 2006 (C. Mostello, R. Veit).

Woonsocket Reservoir. Great Blue Herons were seen building at least 5 nests, Woonsocket Reservoir, Smithfield, 10 March 2002 (MLy)

Exeter. A nest with 3 young was seen at Metcalf Marsh, central Exeter, in July 1990 (CRa).

West Greenwich. A pair was found nesting in May 1991 (CRa). Ten nests were at Hazard Road 15 June 1995; there were 33 young and 5 adults seen (RLF). In May 1999 there was a colony of 2–3 nests in "Cardi Swamp" and about 10 nests in another swamp off New London Turnpike in West Greenwich; the second colony was first noted in 1998. This fits with the loss of the westernmost West Greenwich colony in 1998 (fide M. Tucker).

North Scituate. At the Scituate Reservoir northwest of the "horseshoe dam," several old nests were discovered in January 1993 (BCG), and 7 nests with 14 young were present 16 June 1993 (BCG). On 17 June 1995, this colony contained 16 nests with about 30 birds, including many young (BCG). There were 18 nests in 1995 (B. Gearheart), 21 nests in 1996, and 41 nests 14 June 2000 (RLF).

Foster. Two adults and 2 young 21 June 2002 (R. Graefe); 9 nests, 14 May 1994 (RLF); 10 nests, 8 June 1996, Dolly Cove Brook (RLC).

Tug Hollow, Richmond. There has been a colony for several years along the Beaver River that contained 48 nests in 2016.

Taney Brook, Richmond. A colony in a beaver impoundment contained 16 nests in 2016.

Route 6, Seekonk. There is a colony on the Massachusetts border, present for many years, that contained 29 nests in 2016.

Nonbreeding

Spring. Peak numbers of migrants appear in mid-April. Two checks of the nesting colony at Scituate Reservoir yielded no birds on 13 March and 11 birds on 22 March 1997 (BCG).

Spring maxima. 29, Watchaug Pond, 11 April 1930 (MSo); 25, Newport County, 1 April 1960 (SSD); 17 migrating over the ocean at Middletown, 13 April 1958 (JBa). Flocks are only occasionally encountered after mid-May (7 migrating at the East Providence Reservoir 29 May 1953; REW).

Fall maxima. 100, Block Island, 12 November 1910; 85, Seekonk River, 26 July 2017 (DFi, STa); 76, Dickens Point, Block Island, 11 October 1950 (EDi); 67, Block Island VDC, 6 November 2003 (SSM, CRa et al.). Counts away from Block Island or Point Judith are of fewer than 50 birds per day per locality. One was seen at Cox Ledge 25 November 2000 (SSM et al.).

Winter maxima. The largest numbers of wintering birds in recent years have been at Ninigret Pond and other South Shore coastal ponds: 74, South Kingstown CBC, 22 December 2001 (STs et al.); 70, Newport–Westport CBC, 15 December 2012; 46, Ninigret and Quonochontaug, 20 January 1974 (DFi, DLK); 40 flushed at Ninigret by a Golden Eagle, 29 January 1985 (DLK et al.); 35, Winnapaug Pond, 7 February 1992 (DLK).

Great White Heron (*Ardea occidentalis*)

Status
Vagrant. One record.

Nonbreeding
Great White Herons have occurred north to Massachusetts and Nova Scotia (Vennesland and Butler 2020). An all-white heron present intermittently 4–29 September 1983 at Weekapaug, which was the size of a Great Blue Heron and possessed a yellow bill and "bright yellow" legs (A. C. Anthony, A. Spicer), seems clearly to have been this form. We treat Great White Heron as a separate species, rather than as either a color morph or subspecies of Great Blue Heron, following Browning and Kushlan (2022), who describe multiple distinctive structural, behavioral, and genetic characteristics.

Great Egret (*Ardea alba*)

Status
Fairly common spring and common fall transient. Fairly common breeder. Substantial increase in abundance since 1980.

Historical
There were only 10 records of Great Egrets in Rhode Island during 1888–1914. The population was reduced by the plume trade in the late-nineteenth century, but there are only vague references to how numerous they were before this eradication. With protection, Great Egrets recovered more rapidly than did other herons and appeared regularly in small numbers after 1930. Counts steadily increased during the next 20 years, and a remarkable invasion occurred in the summers of 1948 and 1949. In 1948, *170* birds were found in Rhode Island during late August, 97 of these along the South Shore. Thirty-three were still present at Seapowet 3 October. In 1949, 52 were counted at Ninigret Pond on 3 September (DLK). Numbers did not again approach the magnitude of the 1948–1949 invasion until the 1990s.

Breeding
During the 1950s, Great Egrets began to appear annually at Seapowet during May and to stay for the summer, and some may have nested during that time. An unfledged juvenile picked up on the beach at Island Park, a half mile north of Little Gould Island in the Sakonnet Passage, and brought to NBS in 1964 (REW) was the first definitive nesting record. In 1965, a pair with 2 fledglings was found on Little Gould Island (REW, DLE, RLF). Seven pairs nested there in 1969, 27 in 1980, 48 in 1983, and 53 in 1984 (RLF, JEM).

Great Egrets colonized Hope Island in 1978 and Big Gould Island in 1983. There were 15 nests on Hope in 1980, and 3 nests on Big Gould in 1983 (RLF, JEM). In

2012, 170 nests were found statewide, with the biggest colonies at Little Gould (77 nests) and Hope Island (75 nests) (RLF, CRa). In 2016, 179 nests were found statewide, with the biggest colonies at Little Gould (24 nests) and Hope Island (131 nests) (CRa, JO). Great Egrets began to nest on Block Island in 1994 and have slowly increased. There were 20 nests on Block Island in 2016.

Nonbreeding

Spring. Arrival of first birds in spring shifted from early April during the 1960s and 1970s to late March in the 1980s; e.g., 15, Little Gould Island, 25 March 1984 (DLE); 12, Seapowet, 27 March 1985 (NPH); 11, Seapowet, 29 March 1992 (DLE).

Prior to their breeding in Rhode Island, Great Egrets rarely appeared in spring at all, and most occurred after mid-July, with greatest concentrations from mid-August to late September.

Fall maxima. 141, Seapowet, 27 August 1983 (HCa, JC); 50, Ninigret, 16 August 2001 (CRa); 10, Block Island, 23 September 1991 (MHC, SMC).

Winter. Great Egrets began lingering into December and early January during the latter part of the twentieth century (e.g., 1, Galilee, 1 January 1969 [HDa]; 1, Ten Mile River at Pawtucket, from December 1982 through 8 January 1983 [CRa, REn, RAC, DLE]; 1, Jamestown, 13 January 1992 [W. Gardner]), and the first successful overwintering occurred in 2001–2002 (*FNRIB*). There are now several CBC records across Rhode Island and additional late-winter records; e.g., 1, Weekapaug, 3–8 February 2002 (GW, SSM).

Winter maxima. 22, Newport–Westport CBC, 17 December 2011; 7, Napatree CBC, 23 December 2012.

Extreme dates. 4 March 2012, Easton Pond (1; Robert Weaver), to 21 January 1957, Block Island (1; fide EDi); 26 January 2002, Misquamicut (1; GLB et al.); and 11–28 January 1989, Potter Pond and Jerusalem (1; DFi, DLK). Individuals wintered at Seapowet 2011–2012.

Little Egret (*Egretta garzetta*)

Status

Vagrant. One record.

Nonbreeding

A juvenile Little Egret was found at Block Island 14 September 1998 (Simon Perkins et al.), 5 days after a similar bird was seen at Monomoy, Massachusetts (*BOEM* 27[1]: 33). Little Egrets colonized the West Indies from Europe and Africa, first nesting in Barbados in 1994 (Buckley et al. 2009) and subsequently Antigua (Kushlan and Prosper 2009). Since then, they have been widely recorded on the east coast of North America, north to Newfoundland.

Snowy Egret (*Egretta thula*)

Status

Very common breeder and transient.

Historical

Despite a few turn-of-the-twentieth-century vagrants to Massachusetts and an indefinite Rhode Island report by Talbot (1884b), Snowy Egrets were not certainly recorded in Rhode Island until one was found dead after an early spring snowstorm on the piazza of a house in Newport 2 April 1924 (Taylor 1924). Another was seen at Block Island, 26 August–3 September 1926 (EDi). The next was 28–31 August 1936 at Briggs Marsh (Horton, RBo, RCC), and probably a second one there after a storm, 5 September 1936 (RCC). Singles were at Jamestown, 3 August 1937 (Clarke, v.o.); Briggs Marsh, 21 August–5 September 1937 (RCC, WHD); and Quonochontaug, 9 September–8 October 1937 (perhaps the same bird; HSH). One was seen 14 May 1938 at Bonnet Shores after a violent southerly storm (WBD, DLK, CWo). Other early records include one at Seapowet on 22–28 August 1940 (RBo) and an exhausted bird that appeared in Warren on 16 September 1944 after a hurricane (RBo). In 1948, an unprecedented irruption of southern herons reached the Northeast and included at least 22 Snowy Egrets found in Rhode Island.

Breeding

In Rhode Island, Snowy Egrets nest invariably among Black-crowned Night-Herons on islands where they are relatively free from predation. Distribution of nesting colonies of herons and other colonial waterbirds through 1985, along the East Coast from Maine to Virginia are provided in Andrews (1990). In 1963 and 1964, adults were reported attending the Little Gould Island heronry in the Sakonnet Passage. Thirty nests were counted there 9 June 1965 (REW, DLE, RLF), and it seems likely that they had already been breeding there for a few years at that point. The Little Gould Island colony increased to 50 pairs in 1970 and 80 pairs in 1974 (RLF). The colony was abandoned in 1974 and was unoccupied until 1984, when there were 5 nests (RLF, JEM). The colony increased to 70 nests by 1991 and then declined again to no nests by 1996. In 1975, at least 43 adults were found at Hope Island, suggesting that the Little Gould Island birds shifted there. Between 1977 and 1983, the Hope Island colony contained from 109 to 330 (1979) pairs (RLF, JEM). The colony increased to 132 nests by 1993 but has declined again since that time. Some of the Hope Island herons apparently moved to Big Gould Island, where they were up to 76 pairs 1982–1984 (JEM, RLF). Snowy Egrets have not nested on Big Gould Island since 1991. They also began breeding on Dyer Island in 1983 (6 nests; RLF, JEM). Dyer Island was unoccupied between 1991 and 2004, but there were 4 nests in 2005. The colony at this site has fluctuated since that time and had 40 nests in 2016. Snowy Egrets were discovered nesting on Rose Island in 1991, when there were 4 nests. This colony increased to 45 nests by 2000 but then declined. There has been no nesting since 2009. The only other nesting records from Rhode Island are of pairs that bred at Block Island among night-herons in 1976 (HLa), 1981 (RBo), and 1994, and then almost annually since 1999, increasing to 10 pairs in 2016 (CRa).

Nonbreeding

Spring. Snowy Egrets arrive during late March, and the bulk of the nesting population appears in late April or early May. Larger early counts include 20 at Trustom Pond 3 April 1980 (RAC) and 35 at Seapowet 7 April 1974 (DLE, REm).

Fall. The largest concentrations occur when young from nearby colonies and wanderers from elsewhere join local adults in late summer and early fall.

Fall maxima. 175, Quonochontaug, 5 August 1975 (DWa); 147, Seapowet, 17 August 1974 (SSD); 95, Charlestown Breachway, 16 August 2001 (CRa). In general, they are relatively scarce at Block Island, despite occasional breeding. Until the 1990s, the largest reported count of transients there was 15 on 14 September 1960 (JBa), directly after Hurricane Donna. More recently, 100 were reported there on 23 September 1991 (MHC, SRC). There are only 8 records for inland locations, north to Pawtucket (5 on 5 September 1983; DLE), Smithfield, Lincoln, and the Cumberland Reservoir (3–4 there to 11 October 1980; REn, DLE).

Extreme dates. 20 March 1999, Weekapaug (2; CRa), and 21 March 1958, Quicksand Pond (1), to 22 December 1979, Newport (1; RTL), and 11 January 2006 at Briggs Marsh (1; GDe).

Little Blue Heron (*Egretta caerulea*)

Status

Uncommon breeder and transient.

Historical

There were 5 records of Little Blue Herons in Rhode Island between 1884 and 1899. They then increased and reached peak abundance during the 1970s. Perhaps because of reduced hunting pressure (relative to the white egrets), Little Blue Herons were more numerous than those species during the early twentieth century. There was a major irruption of southern herons in 1948 (Cottrell 1949), when about 67 Little Blue Herons were reported in Rhode Island.

Breeding

Beginning in 1960, adults foraged during summer in the Seapowet marshes, and in 1965, 2 territorial adults were present in the nearby heronry on Little Gould Island (RLF, REW). Estimates of nesting pairs at Little Gould Island rose from 5 pairs in 1967 and 9 pairs in 1971 to 18 pairs in 1974 (RLF, REW et al.). In 1975 the birds shifted to Hope Island. Seven adults were present there in 1975, and between 3 and

43 pairs (maximum in 1978) nested there from 1977 to 1983. The main Narragansett Bay-area nesting location then shifted again to Big Gould Island, where at least 2 pairs nested in 1982 (RLF, JEM). Little Blue Herons were discovered nesting on Rose Island in 1993, and by 1999 there were 9 pairs present (RLF, JEM). This site then declined, and nesting ceased by 2004. Three pairs were present at Hope Island 23 May 2007 (RLF, CRa). Elsewhere, 3 territorial adults were present at the Block Island night-heron colony on 7 June 1976 (RLF, HLa), but these apparently did not remain to breed (HLa). One pair nested on Block Island in 2001 (CRa). Since 2007, all nesting has been on Hope Island, with the statewide population 2 or 3 pairs annually.

Nonbreeding

Spring. The bulk of arrivals are from 15 April onward. In recent years, adults are present in spring.

Fall. Some birds, especially immatures, often remain into late summer (e.g., 37 at Tiverton 8 August 1987; JO) through October; e.g., 7 immatures at Seapowet 17 October 1971 (SSD, LBe), 5 immatures at Seapowet 23 October 1973 (FAP). There are only 5 inland records for Rhode Island: individuals at Lonsdale 6 September 1936 (AEE); Smithfield Reservoir 1–8 October 1951 (RCC); Wilson's Pond, Pascoag, 19 October 1988 (SMi); Great Swamp 13 July 2000 (SSM); 1, Diamond Hill Reservoir, 31 October 2002 (JSJ).

Extreme dates. 2 March 2002, Newport (RWa), to 23 January 1972, Cross Mills (DLK, ESa). There are a number of December records, including 1 on the Block Island CBC 20 December 1975.

Tricolored Heron (*Egretta tricolor*)

Status

Uncommon but increasing wanderer from the south.

Historical

The first Rhode Island Tricolored Heron was at Quonochontaug on 3 September 1949 (DLK) and the second at Galilee on 5–6 August 1950 (DLK). Others were at Seapowet on 12 September 1954 (RBo, NPH) and 7 June 1955 (RBo). There followed 12 records in the 1960s and 26 in the 1970s, and a minimum of 37 birds were seen between 1980 and 1984. Tricolored Herons have gradually increased and now appear annually in Rhode Island. They began nesting in New Jersey in 1948, western Long Island in 1955, and in Connecticut and Massachusetts during the 1970s. A pair possibly nested at Big Gould Island in 1989, as they were seen flying into the colony in May and June of that year (APo, NPo, DLE).

Nonbreeding

Despite breeding in neighboring states (including Massachusetts as early as 1976) and a supply of wanderers in May and June (to 2, 3, and 4 per day at Seapowet and Jerusalem), Tricolored Herons have not successfully nested in Rhode Island. They prefer extensive salt marshes with numerous salt pannes and extensive mudflats, environments that are limited in Rhode Island and only poorly approximated by the marshes at Quonochontaug, Galilee/Jerusalem, and Seapowet. Most records are from May–June and August–September. There are records northward to Riverside and Watchemoket Cove in East Providence, but none away from the immediate coast.

Extreme dates. 19 March 2010, East Beach, Charlestown (PLE), and 4–8 April 1970, Weekapaug (ESa, CWo, DLK), to 7 November 1991, Weekapaug (GTa, ATa).

Reddish Egret (*Egretta rufescens*)

Status

Vagrant. One record.

Nonbreeding

A 1st-year Reddish Egret was found at the Charlestown Breachway 5 July 2003 (STs, CPe, SMi et al.; see Photo 21). It was subsequently seen on Quicksand Pond, where it remained until 13 August (GDe et al.).

Cattle Egret (*Bubulcus ibis*)

Status

Former breeder (until 2003); now a rare migrant with 1 or 2 seen per year.

Historical

The conquest of the Americas by this Old World heron certainly ranks as one of the most remarkable natural range extensions of any bird species. Arriving in South America from Africa during the late 1800s (Telfair 2020), Cattle Egrets rapidly spread through that continent, the West Indies, and Florida during the 1940s, and vagrants reached the northeastern United States beginning in 1952 (Davis 1960). Like Glossy Ibises, Cattle Egrets feed on insects in upland fields and pastures, often in association with farm animals, and away from the salt marshes and open wetlands frequented by other waders.

The first Cattle Egret discovered in Rhode Island (and one of the first in the Northeast), appeared at West Kingston on 17 May 1955 (KKi). It remained in the same pasture for 2 weeks, was seen by over 80 people, and photographed by RCC. Subsequent records include one in Foster 16 April–15 May 1958 (Dexter et al.), one at Howard 14 May 1959 (JGe, RGe), and one at Cards Pond 11 July 1959 (DLK). There were six records of individuals in 1960 and three in 1961. An irruption of Cattle Egrets to the Northeast occurred during the spring of 1962, when flocks of 15 were seen at Newport and 21 at Block Island.

Breeding

There were no reports of summering birds until 1964, when small groups were observed feeding daily through May and June in Portsmouth (REW, RLF) and Seapowet (SSD). The same summer, several unidentified egret chicks were found in the night-heron colony at Little Gould Island (fide RLF). In 1965, 10 pairs of Cattle Egrets were found nesting on Little Gould Island, and a nestling was collected on 7 August (REW; MCZ #285848). Little Gould had 15 nesting pairs in 1968, 25 pairs in 1971, and 60 pairs in 1972 and 1974 (RLF et al.). Together with most other herons, Cattle Egrets then abandoned Little Gould and moved to Hope Island in 1975. In that year, 20 adults were present at Hope Island and between 33 and 120 nests were found there in the years 1977–1982 (RLF, JEM). Cattle Egrets abandoned Hope Island in 1983 and shifted to Big Gould Island. In that year, there were 20 nests at Big Gould (JEM). These declined to 3 pairs in 1988, the only ones in Rhode Island. The last Rhode Island nesting was 1 pair at Rose Island in 2003 (RLF, CRa).

It is possible that loss of foraging habitat was partly responsible for the decline of nesting Cattle Egrets in Rhode Island, as farm and pasture lands were certainly converted to suburbia during the 1980s. Nevertheless, the decline of Cattle Egrets closely paralleled that of other herons that use different habitats, so other factors were probably important as well.

Nonbreeding

Spring. Most transients occur in pastures of the south coastal area between mid-April and late May with a peak about 5 May (formerly up to 20 per day per locality, now much less common; e.g., 6, Jamestown, 5 May 2009 [JSJ]).

Fall. Migrants appear in late summer and early fall.

Fall maxima. 66, Little Compton, 13 August 1973 (JCB); 68, Tiverton–Little Compton, 6 September 1971 (DLE, REm); 63, Bristol, 6 October 1978 (fide CWo); 58, Tiverton, 29 August 1990 (DLE). Counts since 2005 have not exceeded 1 or 2 birds. Most are gone by late October, but there are nevertheless several November and early winter records.

Inland. There are a few records for the interior of western Rhode Island north and west to Greenville, mostly in spring, but none for northern Rhode Island.

Extreme dates. 18 March 1973, Beavertail (1; MDa), to 11 November–9 December 1970, Jamestown (1; MDa); 20–30 November 1960, Block Island, an emaciated bird (JBa, MCZ #261,400); and 9 November 2011 to 1 January 2012, Fort Getty–Jamestown (1; STa, DFi).

Green Heron (*Butorides virescens*)

Status
Common summer resident and transient.

Breeding
Green Herons nest solitarily or in small groups, sometimes in association with other species of herons. In Rhode Island, they are most common near salt marshes and to a lesser extent near the larger freshwater lakes and marshes inland. Nests have been found on small coastal islands close to the mainland (East Island off Sakonnet and Little Gould Island) but not on Big Gould, Hope, Dyer, or Rose Islands. Inland, they are present at the Great Swamp, along the Blackstone River, and locally about many of the ponds and streams north to Cumberland, but they are absent from large portions of the interior of western Rhode Island.

Nonbreeding
Spring. General arrival is during the last week of April, with migrants still moving in late May.

Spring maxima. 13, Summit, 24 April 1982 (JZy).

Fall. This heron reaches peak numbers during shorebird season along the coast in late July and most of August, when counts of 12–16 have been reported.

Fall maxima. 19, Winnapaug marshes, 27 July 1981 (CRa); 11, Lonsdale, 27 July 1982 (REn, CRa). Rare after September.

Extreme dates. 31 March 2000, Point Judith (JSJ); 9 April 1983, Sakonnet Point (RBo); and 11 April 1993, Lincoln Woods (DWa), to 3–9 January 1971, Ninigret Pond (FPe, CWo, DWa), and 8 January 2007, Lonsdale (DFi). A starving individual picked up in Portsmouth on the extraordinary date of 1 March 1965 and brought to the NBS (REW) could have been either a winter survivor or an exceptionally early migrant.

Black-crowned Night-Heron (*Nycticorax nycticorax*)

Status
Common summer resident and transient. Rare in winter, but some survive.

Historical
For most of the historical period this was the only colonially nesting heron in Rhode Island. Other herons and egrets have arrived since the 1970s and at times have outnumbered Black-crowned Night-Herons, with which they invariably form joint colonies. In the first half of the twentieth century night-herons often nested in wooded swamps on the mainland or on large islands, but in the second half of that century

most nesting was confined to small islands where raccoons and other predators were absent.

Breeding

Black-crowned Night-herons have been declining irregularly at Rhode Island colonies for most of the period 1950–2020 concurrently with the arrival and occupation of the islands by new breeding populations of at least 9 other species of gulls, herons, ibis, and cormorants. The possible competition for nest sites by these newer species, and the influence of their combined guano in modifying the vegetation on the islands, may have been detrimental to night-herons. Increases of predators such as raccoons and opossums may be part of the reason why the larger, inhabited Bay islands lost their breeding night-herons to the smaller Bay islands after about 1950. Further threats from pesticides and other contaminants (Custer et al.1983, 1985, 1986) have exacerbated the long-term decline of Black-crowned Night-Herons that continues into the twenty-first century. Rhode Island heron colonies are as follows:

Bristol. Howe and Sturtevant (1899) stated that there was a colony near Mount Hope in the late-nineteenth century. Another colony of about 100 pairs was found on both sides of the road through the swamp at the south end of the Colt estate from about 1930 to about 1954.

Prudence Island. A large colony was present in a swamp in the south-central part of this island before 1888. Doe visited this colony for eggs at least 7 times during the period 1887–1910. Hundreds of eggs were collected in May 1889. The colony was found deserted in 1891, but in 1893, 300 nests were counted and 1000 birds estimated (HSH, CED); 303 eggs were collected in 1894 (CED, HSH), and 1000 birds were present in 1904 (CED). This colony is last mentioned in literature for the year 1910, when they were "still increasing" (CED). This is probably the same colony visited by Talbot 7 June 1884, which numbered "into the thousands" of nests at an unstated location in what was presumably a maple swamp "in the southern part of Rhode Island" (Talbot 1884b). Note: In HSH's notebook #4 there is a map clearly showing the colony's location in 1893 to be in the *southwest* part of the island, perhaps where "crow swamp" is now.

Portsmouth. There was a large colony near St. Mary's Pond from at least 1937 to 1940. William H. Drury found 250–300 nests in 1937 and 300 nests in 1938. In 1940 this rookery had split into adjacent portions containing a combined 235 nests.

Conanicut Island. Several small colonies were at various points along the east shore of the island from 1936 or earlier through the mid-1950s; e.g., 50 pairs 1936–1940 (Clarke), 200+ pairs in 1946–1948 (RBo), 100 or fewer pairs at 4 places including the Wracket Road area through the mid-1950s (MDa).

Little Gould Island. A colony of about 100 pairs was discovered in 1963. This colony served as an initial magnet for several related species during the remarkable period in the 1960s and early 1970s when the 3 species of egrets, Little Blue Herons, and Glossy Ibises were all extending their ranges northward. The island's surprisingly diverse flora found itself host to 7 species of herons and Glossy Ibises during the period 1963–1974, when the number of night-herons remained relatively constant at 100–120 pairs. After a slump in breeding numbers there by most of the other species in 1974, numbers of night-herons dropped and remained in the range of 12–74 during the period 1975–1984 (RLF, JEM). The colony declined dramatically after 1995 but averaged about 50 pairs until 2008. Further declines then occurred, and by 2016 only 1 pair was observed on the site. By 1990, a skyrocketing nesting population of Double-crested Cormorants was already threatening to devegetate the island to the detriment of all nesting herons. This island is owned by ASRI.

Hope Island. This island was abandoned by the U.S. Navy about 1973. Casual observations suggest that both gulls and herons were increasing rapidly there during the early 1970s. In the period 1977–1982, all species of herons and ibis breeding on this island totaled around 1000 pairs, with 450–560 pairs of Black-crowned Night-Herons. These declined to 164 nests in 1984 and 233 nests were counted there 21 May 2012 (RLF, CRa). The island is now owned by the state and administered as a wildlife refuge by the RIDEM.

Big Gould Island. Occupied by the Navy until the mid-1970s, this island was colonized soon after by gulls and night-herons. There were no breeding herons in 1978–1979, but up to 72 night-heron nests were there 1980–1984 (JEM, RLF et al.). Forty-five nests were there in 2008, but the colony has since declined. Most of this island is now also administered by RIDEM.

Rose Island. Ten pairs of Black-crowned Night-Herons were discovered nesting on Rose Island in 1990. The colony increased to 97 nests by 1997 but then slowly declined. No night-herons have nested there since 2010.

Dyer Island. Colonized by gulls since the 1960s but probably insufficiently vegetated for herons until the late 1970s. There were up to 25 pairs breeding there with Snowy Egrets and Glossy Ibises during 1982–1984 (RLF, JEM). Birds abandoned the site after 1991 but recolonized in 2005. Twenty-five pairs were present in 2016 (CRa). Dyer Island is now protected as part of the NOAA Estuarine Reserve System.

Block Island. A colony of <10 pairs was north of Great Salt Pond 1941–1945 (EDi). Then none were reported until 1971, when 40 nesting pairs were found in the same location (ELa). This site contained 92 nests in 1976 (HLa), and 70–85 nests were regularly present during the next decade. In 1989 or 1990, the colony moved north to a hollow in the dunes south of the main gull colony. Eighteen nests were counted

23 June 1993 (CRa, REn). The colony moved again to Beane Point and has been present every year but with only about 10–15 nests per year.

Nonbreeding

Spring. Adults arrive in colonies during late March and early April; e.g., 40 in Warwick 4 April 1993 (DB).

Fall maxima. 200, Jamestown, 16 November 1957 (HDa).

Winter. One of the most regular wintering locations years ago was in the hemlocks of the Butler Hospital grounds along the Seekonk River in Providence, where 5–19 birds were tallied regularly on the old Providence-area CBC during the period 1922–1949. Also, a count of 12 made by flashlight before dawn at Green End Pond on 16 December 1984 (REn, CRa); 20 through December and January 1952–1953 at Jamestown (MDa); 30, Barrington, December 1952–January 1953; 24, Block Island CBC, December 1977; 11, Cross Mills, 6 February 2001 (CRa).

Inland. They are rare; SSM has recorded them just once at the Great Swamp, 28 April 1992.

Yellow-crowned Night-Heron (*Nyctanassa violacea*)

Status

Uncommon and irregular wanderer from the south. Uncommon breeder.

Historical

Formerly very rare, with only six Rhode Island records between 1886 and 1925, Yellow-crowned Night-Herons have gradually increased. They currently occur annually in small numbers and breed occasionally.

Breeding

At Block Island, 2 or more adults were present annually in summer and early fall from at least 1954 to the early 1970s. In 1958, 2 adults and 2 fully fledged young were seen feeding together. In June 1972, 2 adults were seen displaying beside a nest in a mid-island spruce grove (RLF). A pair was again present in 1973 at the same location, and a third bird was seen carrying sticks into a red pine plantation southeast of Sachem Pond (RLF), where a nest was finally located in 1974 (P. Rock). Both of these evergreen plantations are on high ground at a considerable distance from any of the sites along the west shore used by the island's shifting colony of Black-crowned Night-Herons. The next nest was not found until 1990, in a Japanese black pine in the Clay Head area (ELa, HLa, JEM; photo). Another pair was present through June 1992 (ELa). Three scattered nests were found on the island in 1994 (MHC, fide CRa), and 1 nested in a disjunct hammock known as Skipper's Island in the same area near the salt marsh but not in the Black-crowned Night-Heron colony on Gunners Hill.

They still nest on Block Island in low numbers. Away from Block Island, reports are somewhat clustered in the Jamestown Marsh and near the Charlestown Breachway, but there has been no confirmation of breeding from those areas.

In Tiverton, a pair of adults that fed daily all summer at Nannaquaket Pond through the years 1962–1964 aroused speculation about breeding. This contention was supported by a pair of birds at the nearby heronry on Little Gould Island 5 July 1965 (REW, RLF), by single adults seen at this colony in the 2 following summers (REW), and by 4 adults and 3 immatures at Seapowet 22 July 1963 (M. Gardler). Breeding may have occurred in this general area over a 6-year period, but no more substantive proof was obtained; there were no breeding-season reports from Newport County between the late 1960s and 1995, when an adult and a younger bird were found on Little Gould Island.

Nonbreeding

Most sight records of Yellow-crowned Night-Herons are of singles or small groups feeding in salt marshes during the summer months, with a peak in August.

Maxima. Five (3 adults, 2 immatures), Jamestown, 11 July 2014 (WM); 4 immatures, Charlestown Breachway, 19 August 2010 (DFi, ST).

Extreme dates. 6 April 2003, Middletown (Jay Manning), and 15 April 1964, Bristol (MHo), to 11 November 2010, Block Island (SSM et al.). One was found alive 16 April 1934 under the portico of the Paramount Theater on Broadway in Providence during a night of wind and rain (Taylor 1934).

IBISES AND SPOONBILLS (THRESKIORNITHIDAE)

White Ibis (*Eudocimus albus*)

Status

Vagrant. Increasing in frequency.

Nonbreeding

White Ibises are nomadic and have recently extended their breeding range north to Maryland and New Jersey, and this increase is associated with a general increase of vagrants to the Northeast (Heath et al. 2020). The first Rhode Island record is of an immature in a small pond at Dickens Point, Block Island, 13 October 1954 (EDi). An adult appeared at Wickford 9 March 1964 and remained until the 25th (T. Morton Curry, DLK et al.; photographed in color by SSD). An immature seen flying west at Sakonnet 14 September 1975 (SSD et al.) was possibly the same bird seen in Westport, Massachusetts, during the previous weeks. Immatures were in the Quonochontaug–Charlestown area, 12–16 August 1979 (DLK, MPe, ESa et al.), and Block Island, 23 September–6 October 1991 (MHC, SRC, CHa, JSJ et al.; photo by

RBo, *AB* 46:1). An adult was seen by CAv coming in to roost at Rose Island 16 May 1996, and another immature was at Block Island in August 1998 (CSM et al.). One adult, Jamestown, 9 May 2003 (WM et al.); possibly the same adult again on 1–2 June 2003, Jamestown (D. Cinotti), and again, 6 June 2003, Middletown (CPe et al.); 1, Cumberland, 2–8 September 2007 (MLy); 1, Weekapaug, 2–10 August 2011 (fide M. Tucker); 1 immature, Narragansett, 9–12 November 2012 (N. Anthes); 1 immature, 25 August–30 September 2014, Block Island (N. and C. Grindley).

Glossy Ibis (*Plegadis falcinellus*)

Status

Common summer resident and fairly common transient.

Historical

Glossy Ibises were first recorded in Rhode Island at Jamestown on 3 May 1931, where a flock of six appeared and two were shot by A. O. Taylor. The disposition of these specimens was not recorded, but one is probably a bird now at the RWPM labeled (probably erroneously) "Point Judith, 5/2/31." The next individual appeared at Block Island, 3–14 April 1947 (EDi). They appeared with increasing frequency thereafter, with 6 individuals seen during the years 1950–1955, 9 individuals during 1958–1966, and 31 birds found 1967–1970. The big push into Rhode Island occurred in the years 1971–1974, when annual totals rose to 64, 56, 80, and 221. The dizzying increase on the Rhode Island coast began about 8 years after a similar increase in 1959–1962 on Long Island (Bull 1964).

Breeding

Glossy Ibises forage in farm fields, especially those with cows present and also those with streams running through them; thus, maintenance of working farms is important to the conservation of Glossy Ibises in Rhode Island (Trocki and Paton 2006). Earliest summer records include 5 adults circling over the Little Gould Island heronry in late May 1972. A single territorial pair was seen among nests in the colony 13 June of that year (RLF), and the nest with 4 young was photographed 6–7 July (LOG, DLE, REm). None were present there in 1973, but 12 pairs were breeding in 1974, after which the birds transferred to Hope Island. Some may have already been breeding on Hope Island in 1974, and 38 adults were seen circling this colony 29 May 1975. Between 1977 and 1983, numbers of nests there ranged from 8 to 160 (1977) (RLF, JEM). Some of this colony apparently transferred to Big Gould Island beginning in 1981, where the number of nests for the period 1980–1984 reached 96 in 1984 (JEM, RLF; Meyers 1982). The colony subsequently moved to Dyer Island. Glossy Ibises were breeding on Rose Island by 1990. Nest counts there swelled to 97 in 1997 and 70 in 2000 (RLF, JEM). The colony then declined, and by 2010 was gone. Glossy Ibises also nested sporadically on Sandy Point Island, and 9 pairs nested there in 1990. The peak of the Rhode Island nesting population was during 1990, when 475 pairs were estimated statewide. In 2002, 377 pairs were present, but since that time there has been a reduction in birds and occupied sites. For the past few years, ibises have nested only on Dyer Island, which had 210 pairs in 2016 (CRa et al.). A steady stream of breeding adults has often been seen commuting across Narragansett Bay to and from the colonies. Counts of adults on the breeding islands include 90 over Hope Island 31 May 1978 (RLF, JEM) and 75+ over Big Gould Island 4 June 1984 (RLF, JEM). Mid- to late-summer feeding aggregations are often observed; e.g., 75, Seapowet, 5 August 1989 (APo, NPo). Glossy Ibises range widely while gathering food for their young in June and appear regularly inland at the Great Swamp, with records every month from April through August. There are usually 1–2 individuals, with a maximum of 10 on 15 April 1993 (SSM).

Nonbreeding

Glossy Ibises forage in wet agricultural fields as well as in salt marshes.

Spring. Ten, Narragansett, 23 March 2003 (RFa et al.); 16, Rose Island colony, 29 March 1991 (CAv); 60, Jamestown, 30 March 2002 (DFi).

Spring maxima. 200, Middletown, 12 April 1990 (JHo, REn); 194, Little Compton, 19 April 1999 (REm); 160, Jamestown, 29 April 2005 (JMa); 39, Napatree, 30 April 1999 (CRa).

Fall. Glossy Ibises leave rather abruptly in fall and are uncommon after early September. Nevertheless, there are a few records in November (e.g., 7, Block Island, 13 November 1996; MHC) and December.

Extreme dates. 18 March 1990, Sachuest Point (2; SHa), to 18–20 December 1990, singles at Block Island and Matunuck (RBo, DFi, SHa).

White-faced Ibis (*Plegadis chihi*)

Status

Vagrant. About eight records, all in spring. Increasing in recent years.

Nonbreeding

The first was 1, Middletown, 20–26 April 2001 (RTL, DFi), and another was at Jamestown, 1–3 May 2005 (J. Magill et al.). Others include 1, Jamestown, 2–3 May 2005 (DFi, STa); 1, South Kingstown, 22–26 April 2012 (L. Bostrom); 1, Jamestown, 1 May 2013 (CPe); 2, Middletown, 18–27 April 2015 (W. Munns, CPe); 1, Middletown, 18–27 April 2015 (C. Sidler, CPe); 1, Marsh Meadows, Jamestown, 17 April 2019 (T. Metcalf); and 1, Newport, 27 April 2019 (P. Copobianco). Most have been found among Glossy Ibises in the Middletown–Jamestown area.

Roseate Spoonbill (*Platalea ajaja*)

Status
Vagrant. One record.

Nonbreeding
An adult Roseate Spoonbill was well described and photographed "in the last week of July 1992" at Sachuest Point by N. Logothets. Although the photograph has apparently been lost, the credibility of this record is supported by the presence of 1 juvenile at Staten Island, New York, 20 August–5 September 1992 (Levine 1998), which suggests a northward irruption at the time (Veit 2008).

NEW WORLD VULTURES (CATHARTIDAE)

Black Vulture (*Coragyps atratus*)

Status
Uncommon resident and rare breeder.

Breeding
Black Vultures were confirmed nesting in Exeter during 2014 (CRa/RIARC). That site was again active in 2015, and a second pair nested in Charlestown during that year. During 2016, a third pair nested at a site in Exeter, different from that used in 2014 (CRa), and another pair nested in Hopkinton in 2016 and 2017 (CRa).

Nonbreeding
The first Black Vulture in Rhode Island, not previously published, was 1 collected in Smithfield 2 June 1928 by persons unknown and mounted by Angell and Cash. It was purchased by RWPM (#V4548) but somehow remained unknown to HSH and other chroniclers of Rhode Island's ornithology. The next record was 1 seen by MSo over Kimball Sanctuary, Charlestown, 30 April 1933. The third appeared over Apponaug 17 April 1948 and continued its flight in a southeasterly direction over Narragansett Bay (RBo). One was shot by a farmer in South Kingstown 5 March 1951, obtained by DLK, and donated to RWPM. Another was seen at Block Island 14–30 April 1951 by RCC and others. One was found in a weakened condition on the east side of Providence 11 April 1964 and taken to the zoo at Roger Williams Park, where it subsequently died (not preserved). One was seen in Exeter 17 June 1969 by KKi and MDa. Others appeared at Ell Pond, Hopkinton, 9 April 1978 (JAG), and West Greenwich 25 May 1979 (RAC). One or 2 individuals were seen in Cranston and North Kingstown, 16 and 21 July 1993 (both JGe, RGe). Prior to about 2000, the only fall record is for a weakened bird at Block Island, 12–19 October 1973 (ELa, HLa, MWh, MSl), and there was 1 midwinter record (during a record mild January) in Richmond 16 January 1990 (JGe, RGe). More recently, Black Vultures have been

found roosting with Turkey Vultures in Woonsocket; e.g., 7, 20 December 2011 (S. Wheelock); and 6, 6 February 2013 (J. Fecteau). There are increasing number of CBC records.

Maxima. 21, Newport–Westport CBC, 14 December 2014 (Westport Massachusetts roost); 15, Ashaway roost, 26 January 2002 (STs); 12, West Greenwich, 7 February 2013 (S. DelPozzo); 12, Napatree CBC 23 December 2012.

Turkey Vulture (*Cathartes aura*)

Status

Common summer resident and migrant; locally fairly common winter resident.

Historical

Turkey Vultures were rare vagrants from the south, with six records between 1861 and 1912 from Niantic, Conanicut Island, Newport, Point Judith, and Block Island. Reports did not increase until the 1930s. One was recorded in late March 1920 from Block Island (EDi), where they are still rare. They were otherwise unreported until 7 September 1935, when one was found in Tiverton (RCC). On 15 August 1936, three were seen at Jamestown (A. M. Clarke), and another was recorded at Kimball Sanctuary on the late date of 28 November 1936 (MSo). There were observations of 22 birds during 1939–1948, with a maximum count of four at Kimball Sanctuary 26 April 1943 (MSo). The spectacular increase noted in Massachusetts after 1948 (Bagg and Parker 1951) was evident in Rhode Island as well, as 26 birds were reported statewide in 1949, in 1 year doubling the total state records. A single high count of 10 was recorded at Wood River Junction 9 April 1949 (DLK).

Breeding

The first nesting was in May 1961, when an adult was shot and a nest with 2 eggs found among boulders off Sheffield Hill Road, Exeter (fide T. M. Curry Jr.; photo of adult examined). In 1969, 2 were noted near an inactive Westerly quarry, and 1 was twice seen emerging from a large pile of rocks among the woods (ESa). Birds were apparently in residence there in 1970. An investigation of this rock pile 19 June 1971 produced a hissing adult and 2 downy young from a deep chamber under the rocks (ESa, RLF; photographs). On 26 May 1980, the same chamber contained an adult and 2 eggs (ESa, CWo, RLF). Additional breeding records include a pair with an egg and downy young among rocks in central-western West Greenwich 1 June 1980 (REn, CRa); and a pair and nest in a rocky outcropping on Mt. Tom, Exeter, in August 1981 (Monahon, Gould, fide CRa, REn).

A dozen or more pairs (a conservative estimate) may now be breeding in the state. Two were seen emerging from rocks in a remote part of Parker Woodland, Coventry, in June 1980 (K. Dann), and they were thought to be nesting nearby along Sisson

Road, Greene, during the period 1965–1968 (fide RLF). A report of a nest in a cave in the Snake Den area of Johnston during the 1980s and persistent observations there strongly suggest breeding at a point very close to the Providence metropolitan area (REn, CRa). Turkey Vultures are now likely breeding in Glocester and Burrillville (e.g., Durfee Hill Management Area), judging by a sharp increase in records (6 in Glocester 24 April 1982 [JGe, RGe]; 5 there 5 May 1982 [CRa]; 5 at Wallum Lake 16 June 1983 [SMi]).

Eastward, a concerned pair was present in June 1984 over a rocky section of Coppermine Hill in Cumberland (RLF). The only suggestions of breeding east of the western border of Narragansett Bay are in the Tiverton highlands, where increasing June reports since the late 1970s have slowly translated into an apparent breeding population; e.g., 19 there 14 July 1988 (fide RBo) and 38 on 4 June 2001 (RLF). Wanderers are now found during the breeding season south into Little Compton; e.g., 10 there 18 May 1986 (DLE) and 6 at Warren's Point 11 June 1988 (RBo). Two were present on Conanicut Island 1–15 June 1993 (CPe).

Nonbreeding
Spring. With their gradual (re)occupation of southern New England, Turkey Vultures have arrived earlier in spring and departed later in fall.

Spring maxima. 14 together at Charlestown, 22 April 1977 (DWa).

Fall maxima. 30, Westerly, 9 October 1989 (GTa); 27 roosting in Hopkinton, 7 October 1978 (ASM); 20, India Point, 26 August 2001 (MLy), a Providence County maximum at the time; 20, Tiverton, 15 October 1993 (fide APo, NPo).

Winter. The first winter record was of 1 at Slocum 20 January 1966 (RLC). In 1976–1977, 2 birds wintered near Westerly (ASM, ESa, DWa). In early 1978, a bird was seen in the Westerly–Shannock area at least to 21 January (ESa, RAC). A late-fall roost of 30 in the North Stonington, Connecticut, area (4 mi [6 km] from the Rhode Island border), discovered on 29 November 1986, may have been the source of these birds. Wintering occurred later in Newport County: the first was in Middletown 25 January 1987 (LT). Four were seen at Quicksand Pond 3 December 1988 (RBo), and 3 were in Tiverton 7 January 1992 (RBo). They are occasional on Block Island; e.g., 2, Block Island VDC, 12 November 2005 (SSM et al.). They were rare or unreported there before the 1980s.

Winter maxima. 107, Newport–Westport CBC, 15 December 2001 (majority from Westport, Massachusetts roost); 76, Ashaway, 15 February 2003 (STs).

OSPREY (PANDIONIDAE)

Osprey (*Pandion haliaetus*)

Status

Common breeder and migrant. Increasing following major decline due to pesticides 1945–1970.

Historical

By virtue of their highly conspicuous nests and the general interest among the public in these birds, a very detailed history is available of the shifts in abundance of Ospreys during the past 125 years or so. An old clipping in HSH's notes states that in the 1890s, Rhode Island was "the only state that protect[ed] Fish Hawks and [laid] a heavy penalty on anyone who molest[ed] them." They "enjoy[ed] the immunity and ignore[d] the presence of humans [in Bristol County]." In the late 1800s and early 1900s, nests were concentrated along the Massachusetts–Rhode Island border near the towns of Rehoboth and Barrington. Bent (1937) counted 80 nests in this area in 1882. Between that time and the 1930s, the entire population shifted to the south, disappearing almost entirely from the Rehoboth–Barrington area and increasing in Bristol County. In 1936, 12 nests were counted in Tiverton and Little Compton, mostly around the Seapowet marshes, and in 1940 there were 56 nests in Swansea, Warren, and Bristol. In 1949, David Emerson counted 17 nests in Barrington, Warren, and Swansea and 48 nests in southern Bristol County. By 1954, the total number of pairs nesting in Rhode Island probably reached a maximum of 150–160 pairs.

Breeding

Robert E. Woodruff censused the whole state in 1963 and counted 42 nests; this drop in abundance continued precipitously, and by 1966–1967 the state population reached a low of 4 pairs. This dramatic decline has been conclusively attributed to high concentrations of the pesticide DDT and its metabolites, which cause the birds to lay very thin-shelled eggs that break easily (Spitzer et al. 1978, 1983). In 1968, DDT was outlawed in the United States (it is still widely used in Central and South America), and the Rhode Island Osprey population has steadily recovered since that time. There were 12 nests in 1977–1980, 19 in 1981, and 25 in 1986. There were 127 active Osprey nest sites in 2014, which produced 186 fledglings (1.46/active nest). The number of active Osprey nests from 1977–2014 shows an increasing trend; e.g., 31 (1990), 58 (2000), and 108 (2010), with a recent peak of 138 in 2013 (http://riosprey.info).

While most historical Rhode Island nests were in trees, Ospreys now use custom-built platforms. They have built nests on swiveling windmills, large bridges (e.g., the top of the old Jamestown Bridge), towers of every height and description, and utility poles, as well as the communication tower of a mothballed aircraft carrier at Newport Naval Base, a barge crane moored near Watchemoket Cove, and, more recently, on large channel markers within Narragansett Bay. Rhode Island Ospreys have regularly nested on duck blinds but, unlike those of Gardiner's Island and Plum Island, New York, have rarely nested on or close to ground level, even on predator-free islands in Narragansett Bay. A few pairs nested for years in the old Warren Reservoir on stumps about 4 ft (1 m) in height (Hathaway 1905). In 1990, a pair incubated eggs in a shallow (failed) nest on a floating boat dock in the Pawcatuck estuary (fide L. Suprock). A ground nest was discovered on the salt marsh on The Tongue, Hundred Acre Cove, in 1996–1997 (Suprock 1997). This nest produced young in its second year. Curiously, very few have nested on the smaller, well-vegetated Narragansett Bay islands. Nests have been present on Dutch, Patience, and Hog Islands but never on Dyer, Rose, or Fox. More recently, they have nested on an old U.S. Navy building at the north end of Big Gould Island, as well as on Hope and Little Gould Islands. While all but perhaps the last 2 lacked adequate nesting trees before and during the 1950s peak of abundance, all but Dyer now have trees. Block Island lacked nesting Ospreys until the late 1980s, after which a single pair began nesting on a tower near Old Harbor.

Nonbreeding

Spring. The vanguard of the breeding population usually appears in mid-March, with an average reported arrival of 20 March during the years 1969–1989 (fide DLE). In the record mild February of 1954, individuals appeared at Narragansett 12 February (F. Richards) and Warren 15 February (RBo). The breeding population is usually at nest sites by late March. Later, however, migration of apparently nonbreeding subadults occurs in May and even into early June; e.g., 7 at Block Island 26 May 1959 (MSl). Most of the 1-year-olds remain through their first summer in the tropics.

Fall. Ospreys disperse in mid- to late-summer after young are reared, and many probably leave the state before September, although a few sometimes remain later at nests. Locally reared fledglings depart about the same time. Migration usually peaks in late September, and Ospreys are rare after October. The largest counts are from all-day hawk watches at Watch Hill in the last half of September and the first week of October.

Fall maxima. 103, 19 September 1993 (JZy); 93, 5 October 1999 (JZy). The highest count away from Watch Hill is for 17 at Sakonnet 19 September 1993 (DLE). Smaller numbers of fall migrants are regular at Block Island and even over offshore waters (to 5 per day at Cox Ledge).

Extreme dates. 12 February 1954 (see above); 1 was reported at Block Island 26 December 1952 (EDi). One was in the Narrow River area 27 December 1987 (HWi). There is an increasing number of midwinter records, all of single birds: 3 January 2002, Frying Pan Road (CBl); 19 January 2002, Hope Valley (SRC); 20 December 2003, South Kingstown CBC (STs et al.); 2 January 2013, Providence (CPe).

MAP 1. Rhode Island state map showing counties, major municipalities, and highways. "South County" is a widely used local term referring to an area roughly but not exactly equivalent to Washington County. Map layer data is from the Rhode Island Geographic Information System's Geospatial Data Hub. The base map for the ocean is provided by the Environmental Systems Research Institute. All of the data is in the public domain.

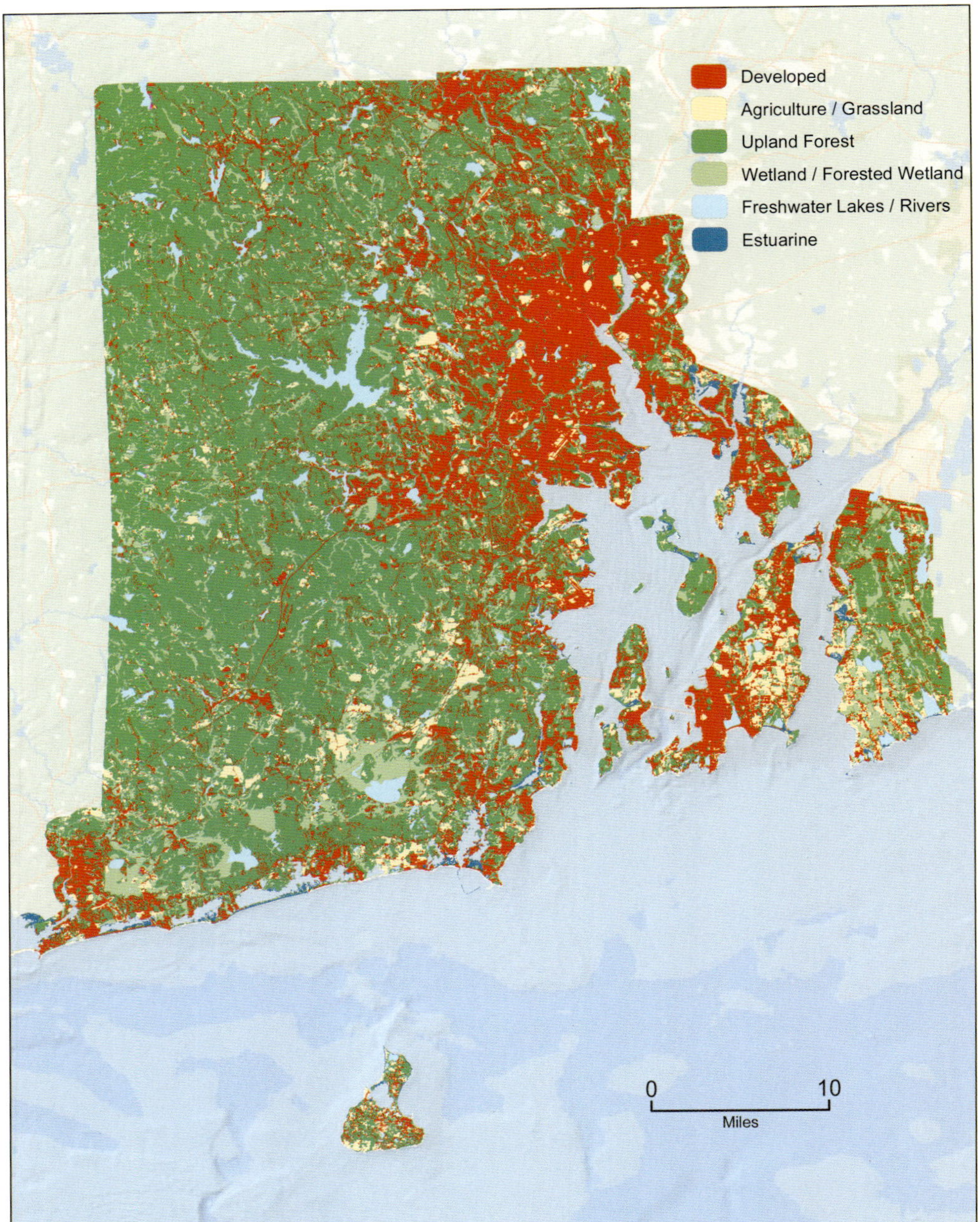

MAP 2. Rhode Island state map showing land cover as indicated. Developed areas (red) are mostly concentrated around Narragansett Bay.

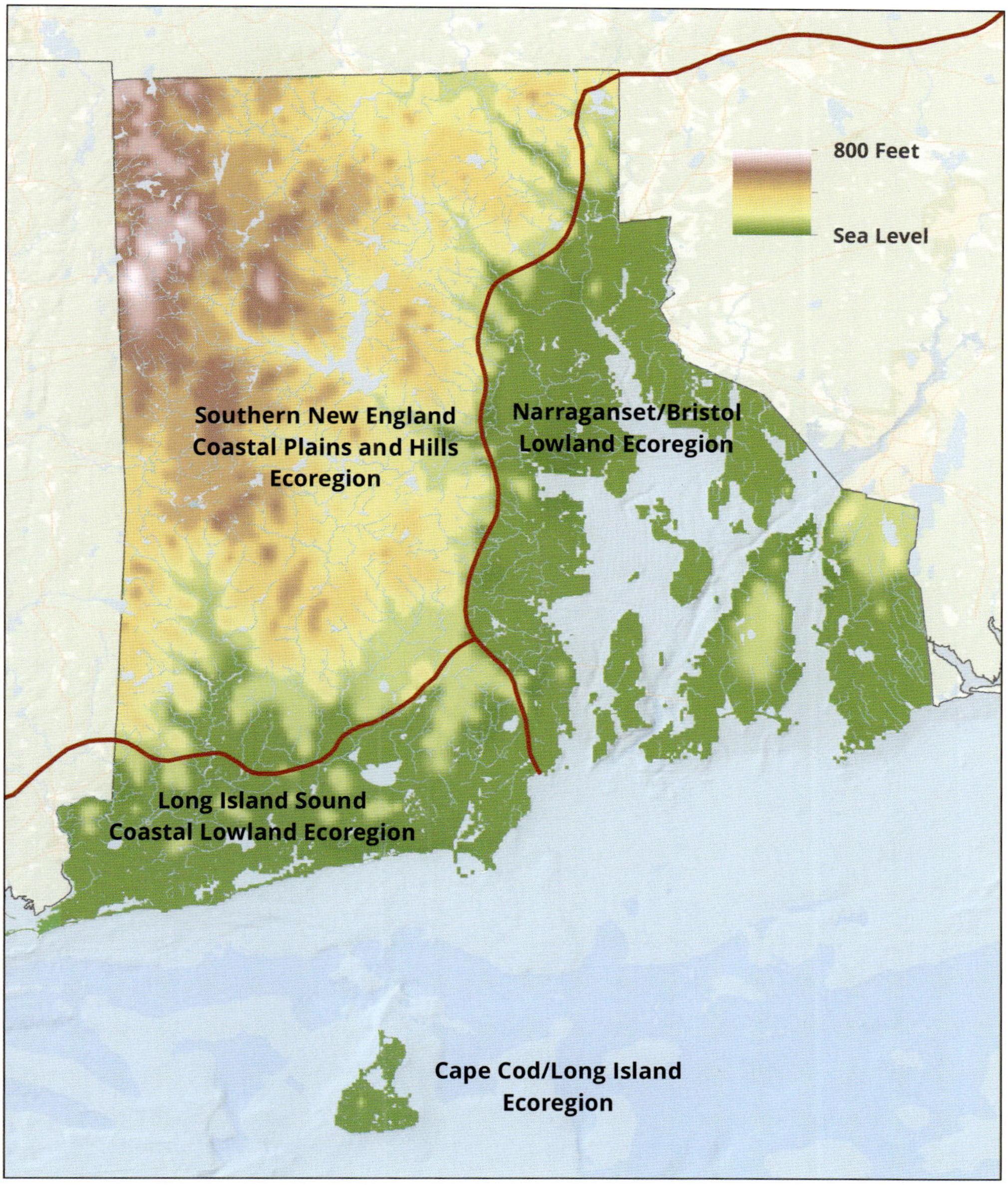

MAP 3. Elevation and ecoregions map of Rhode Island. Elevations shown by color in feet above mean sea level. Level IV ecoregions of Rhode Island as defined by the U.S. Environmental Protection Agency (https://www.epa.gov/eco-research/ecoregion-download-files-region#pane-01; accessed 8 July 2022).

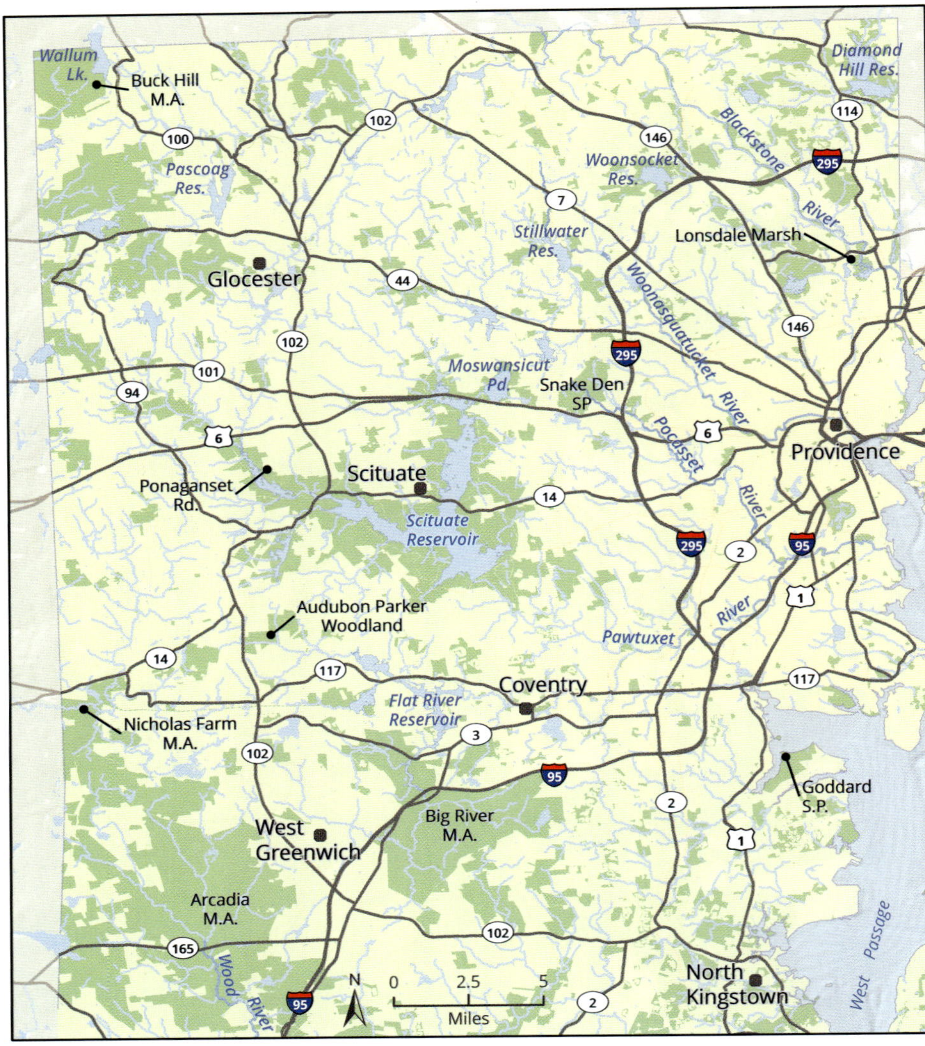

MAP 4. Northern and western Rhode Island. On this and the following maps: M.A. = state wildlife management area; NWR = national wildlife refuge; S.P. = state park.; green background = private and public conserved areas.

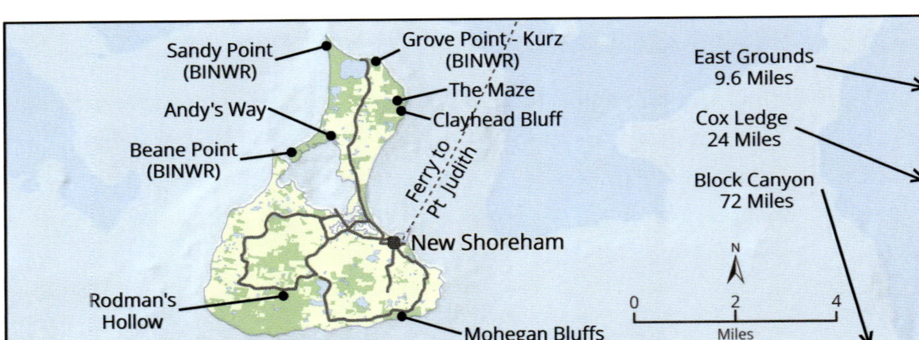

MAP 5. Block Island (BI), including distance and direction to notable pelagic areas.

MAP 6. Southwestern Rhode Island.

MAP 7. Narragansett Bay and southeastern Rhode Island.

PHOTO 1. Black-bellied Whistling-Duck, Apponaug Cove, 8 August 2021, © Carlos Pedro.

PHOTO 2. Graylag Goose with Canada Geese, Watchemoket Cove, East Providence, 22 December 2016, © Carlos Pedro.

PHOTO 3. Pink-footed Geese, Newport, 7 January 2007, © Paul L'Etoile.

PHOTO 4. Common Cuckoo, Snake Den Farm, 1 November 2020, © Carlos Pedro.

PHOTO 5. Calliope Hummingbird, Little Compton, 22 October 2013, © Geoff Dennis.

PHOTO 6. Snowy Plover, Goosewing Beach, 17 April 2017, © Geoff Dennis.

PHOTO 7. Sharp-tailed Sandpiper, Galilee Marsh, 13 November 2021, © Carlos Pedro.

PHOTO 8. Curlew Sandpiper, Napatree, 1 June 2018, © Carlos Pedro.

PHOTO 9. Red-necked Stint, Napatree Point, 5 July 2020, © Carlos Pedro.

PHOTO 10. Little Stint, Charlestown Breachway, 4 July 2012, © Carlos Pedro.

PHOTO 11. Terek Sandpiper, Napatree Point, 28 June 2020, © Carlos Pedro.

PHOTO 12. Wood Sandpiper, Jamestown, 13–30 October 2012, © Carlos Pedro.

PHOTO 13. Ivory Gull, Quicksand Pond, 23 January 2010, © Geoff Dennis.

PHOTO 14. Mew Gull, *Larus canus kamtschatschensis* (Dickinson and Remsen 2013), East Providence, 30 January 2006, © Carlos Pedro.

PHOTO 15. Bridled Tern, Little Compton, 28 August 2011, © Geoff Dennis.

PHOTO 16. Pacific Loon, Central Beach in Charlestown, 17 December 2017, © Carlos Pedro.

PHOTO 17. White-faced Storm-Petrel, near Block Canyon, 3 September 2018, © Carlos Pedro.

PHOTO 18. Band-rumped Storm-Petrel, near Block Canyon, 3 September 2015, © Paul L'Etoile.

PHOTO 19. Black-capped Petrel, near Block Canyon, 3 September 2015, © Paul L'Etoile.

PHOTO 20. Magnificent Frigatebird, adult male, Goosewing Beach, 31 October 2012, © Geoff Dennis.

PHOTO 21. Reddish Egret, Charlestown Breachway tidal flats, 12 July 2003, © Carlos Pedro.

PHOTO 22. Zone-tailed Hawk, Wood River Junction, 15 August 2015, © Carlos Pedro.

PHOTO 23. American Three-toed Woodpecker, Burlingame State Park, Charlestown, 24 February 1991, © Simon Perkins.

PHOTO 24. Cave Swallow, Scarborough, 14 November 2015, © Carlos Pedro.

PHOTO 25. Mountain Bluebird, Fort Getty, Jamestown, 10 November 2012, © Carlos Pedro.

PHOTO 26. LeConte's Sparrow in snowbank, Middletown, 25 January 1993, © Geoff Dennis.

PHOTO 27. Green-tailed Towhee, Second Beach, Middletown, 18 January 2011, © Carlos Pedro.

PHOTO 28. Townsend's Warbler, South Kingstown, 25 November 2017, © Paul L'Etoile.

HAWKS AND KITES (ACCIPITRIDAE)

Swallow-tailed Kite (*Elanoides forficatus*)

Status

Vagrant. Annual, or nearly annual, in spring.

Nonbreeding

On the morning of 7 June 1973, 2 probable 1-year-olds were discovered soaring over Clay Head, Block Island (RLF; photo, *AB* 27:1020). They appeared after several days of foggy southwest breezes and showed a strong inclination to continue their travels east-northeast. On 2 occasions, they completely disappeared eastward from the island, only to return, and they spent most of the day sailing along the east shore of the island between Sachem Pond and Mohegan Bluffs. On 27 May 1981, a single bird was seen over King Tom Pond, Charlestown, by DVT. One was found along Moonstone Beach Road in South Kingstown 18 April 1994 (DFi; photographs on file). One was on Block Island 22 April 1996 (Jim Stringer) and the same or another in Richmond, 1–14 May 1996 (Burkley Weilburg). There were 2 at Block Island in 2005: 1 on 28 April (C. Blane, fide RFa) and a second on 16 May (STs, JSJ et al.); others include Slocum, 12 May 2000 (P. Paton); Block Island, 18–20 July 2004 (M. Wagner). There is a total of 7 records since 2010, all during April and May, including 1 at Jerusalem, 7 April 2010 (v.o.; photo); 1, West Kingston, 25–29 April 2012 (I. Dewhurst); 1, Swan Point, 15 May 2013 (STa, DFi); and 2, Hopkinton, 22–24 April 2015 (L. Thompson, CRa et al.).

Golden Eagle (*Aquila chrysaetos*)

Status

Rare migrant and winter resident, appearing annually as a migrant in fall.

Historical

This spectacular species was collected several times during the shooting era (Howe and Sturtevant 1899), as follows: a 12-pound bird shot 17 February 1887 at Westerly (Chapman, fide J. M. Southwick) was the first taken in Rhode Island in 30 years (Dexter 1887). Another was shot at Sakonnet by Dexter in October 1893. John Hague, Providence taxidermist, had one in captivity for a few years prior to 1899, which he claimed was captured in Rhode Island. A female was shot somewhere in Rhode Island 13 or 15 December 1898 by a Mr. Grinnell.

Nonbreeding

From 1887–1989, there were about 30 Rhode Island records of Golden Eagles. More recent records include an immature found on the rocks off Briggs Beach 26 December 1937 (RCC, DLE, CGW) and a weakened immature caught alive on Prudence

Island 12 October 1945 and identified by HNG. The latter bird subsequently died, was mounted, and was displayed for years in the Bristol town office. Another was seen on Prudence Island 1 January 1946 (WBD), and the same or another reappeared there 15 May 1946 (WBD). One was found sitting on the ice at Quicksand Pond 11 January 1948 (RBo, HMP). An immature was found at Block Island 28 March 1970 (E. Dickerson, MWh). Another immature remained at Trustom Pond 25 December 1971 to 13 January 1972 (DLK, FPe et al.). Another immature was seen repeatedly at Block Island 11 May 1972 (ELa, MWh). A subadult was at Quicksand Pond 15 October 1972 (DLE, REm). Still later records are for individuals at Trustom Pond 26 January 1974 (DLK, FPe, ESa, CWo); in the Matunuck–Great Swamp area 26–30 November 1975 (GRa, STs et al.); Block Island 14 May 1977 (HLa et al.); South Kingstown 20 May 1978 (KKi, TAd); Weekapaug 16 November 1980 (v.o.); Jerusalem 2 October 1981 (DVT); Charlestown 3 October 1981 (RAC); 1 adult, Middletown, 28 December 2001 (REn); 1st-winter plumage, Little Compton, 14 February 2004 (GDe); 1 adult, Harmony, 1 March 2004 (DFi); 1 juvenile, North Kingstown, 31 October 2004 (JMa); and 1 immature, West Greenwich, 10 April 2005 (RFa, DVa). Since 2005 there are an additional 10–15 records, including 1, Providence, 6 February 2019 (J. Mitchell), and 1, Westerly, 23–24 November 2020 (R. Madison et al.); mostly single birds in late fall and early spring.

Northern Harrier (*Circus hudsonius*)

Status

Uncommon to rare summer breeder, common migrant and fairly common winter resident at the coast.

Historical

Northern Harriers were formerly far more widespread as breeders within the coastal zone and even nested in the interior. They disappeared from the interior in the early twentieth century and from the mainland shore before 1960. Howe and Sturtevant (1899) termed them "common summer residents" but said that they were declining, perhaps because of shooting. Hathaway found only one nest with four eggs, in Charlestown 26 May 1901. Doe, however, collected four sets of eggs in a cattail marsh near Point Judith between the dates 19 May and 17 June during the years 1915–1924, as well as two 5-day-old young from the same marsh in 1921. He also collected eggs from the south end of Prudence Island (then unwooded) in 1910. There is a record of a nest and five eggs from Kent County 23 May 1904. The only other inland nests were all discovered by CED, who took two sets of five eggs each from the same cranberry bog in Glocester, 15 May 1911 and 9 June 1911 (FlMNH #54526, #57089), and obtained two sets of four eggs each from another cranberry bog south of Buck Hill Road in Burrillville, 18 May 1924 and 26 May 1925 (FlMNH #53058, #53510).

FIGURE 8. Hawks migrating past Watch Hill, © Margaret La Farge.

Northern Harriers increased after 1900, perhaps because of reduced shooting. They bred in South County at Charlestown (nest and three eggs 6 July 1946; WBD), Quonochontaug (a nest and four young on the improbably early date of 20 April 1947; WBD), Bonnet Shores (a nest and three eggs in cattails 18 May 1947; DGa, RCC, WBD), and Trustom Pond (an adult and two barely fledged young on 11 August 1965; DLK). An inveterate hawk shooter named Snell told RCC of at least three locations in upland swampy thickets where he had found nests in the Sakonnet area during the early 1930s, and birds were suspected of breeding near Briggs Marsh in 1939 (DLE). They were also suspected of breeding at Seapowet Marsh in Tiverton in the 1940s, but no proof was ever obtained. William B. Dean calculated that there were three pairs breeding on Prudence Island in 1946. On Conanicut Island, this species was also a regular summer resident for many years, based on the regular presence of adult males in June, of pairs seen, or barely fledged young observed. They were present there in summer until the mid-1960s, after which they disappeared (MDa).

Breeding

Northern Harriers require sandplain grasslands and other open habitats for foraging, and much of their fluctuation in status likely reflects succession of grasslands and agricultural fields to forest. Harriers continue to nest at Block Island, likely owing, at least in part, to the lack of mammalian predators. The island population was conservatively estimated at 6 pairs in 1972 (RLF), and 6 nests were found there in 1976 (P. Serrentino). Richard Bowen estimated 10 pairs there in 1981. By 1984, however, there was some suggestion of a reduction (REn, CRa). Only 6 birds were present on the island 17–18 June 1987 (CRa, REn), and 5 pairs were estimated 25–26 April 1991 (CRa, REn). Eight birds were there 28 June 2000, 7 on 25 June 2001, 6 on 28 June 2003, and 10 on 30 June 2004, all BIBBS (SCo).

Nonbreeding

Spring. Away from Block Island, spring totals have not exceeded 8 per day. Their regular occurrence at the Great Swamp from late April through June stands in contrast to their rarity elsewhere inland at this season: 8 records of single birds 23 April–22 June, 1992–1998 (SSM).

Fall. Nonbreeders are regular throughout coastal areas in summer and they are joined by a few migrants from late July and early August onward (1 at Cox Ledge 30 July 1975; DLK, CWo).

Fall maxima. 34, Warren's Point, 24 September 1939 (DLE, DLK); 30+, Block Island, 11–13 October 1958 (ASRI); 25, Block Island, 8 November 1999 (CRa et al.); 15 in 3 hours, Napatree Point, 9 October 1983 (JZy, T. Swift).

Winter maxima. 42, Newport–Westport CBC 15 December 2007. At Block Island, December numbers since 1976 have occasionally exceeded 20 birds, but no more than 10 have been found there since 2009. Away from the coast, the only records are of

1 at Lonsdale Marsh, 27 January 1978 (REn), and 1, Cumberland, 22 January 2016 (REm).

Sharp-shinned Hawk (*Accipiter striatus*)

Status

Fairly common spring, common to very common fall migrant, and uncommon winter resident. Rare breeder.

Historical

Charles E. Doe collected eight clutches in the towns of Glocester and Smithfield during the period 1887–1911 and another clutch from Neutaconkanut Hill in Providence. Sprague collected sets in Chepachet in 1888, 1889, and 1890, and eggs were collected by HSH in Cranston (1903), Richmond (1906), and Kenyon (1909, 1914). This is an extraordinary number of nests by today's standards; there have not been more than one or two nests found in the entire state during the last 50 years. All of CED's and Sprague's nests were in white pines, usually in groves rather than isolated trees, and all in an era when overall forest cover was vastly less than today. Sharp-shinned Hawks disappeared as a summer resident by about 1915, judging by CED and HSH's inability to find it during the late 1910s and 1920s despite regular field work.

Breeding

The only records for the modern period are for a nest and 5 eggs at Warren 21 May 1939 (RBo) and an unconfirmed 1957 report from Hope. The reasons for the dramatic change in status are unclear. The decline from the early part to the latter half of the twentieth century coincides with a substantial reforestation of much of the state, which would seem to have produced more nesting habitat. Having a more northerly breeding distribution than Cooper's Hawks, it is possible that Sharp-shinneds have been moving northward in response to warming climate.

Nonbreeding

Spring. Sharp-shins have always been one of the most numerous hawks during migration. During peak migration in mid to late April it is not uncommon to see 1–5 birds per day per locality.

Spring maxima. Six, Fiskeville 28 April 1985 (JGe, RGe).

Summer. There are numerous records of summering nonbreeders.

Fall. They are occasionally abundant along the coast, particularly during northwest winds.

Fall maxima. 1221, Napatree Point, 25 September 1981 (CRa, GLB); 662 from 9 hawk watches, 28 September 1975 (v.o.); 483, Napatree Point, 27 September 1984 (CRa, REn); 451, Napatree Point, 8 October 1981 (CRa, REn). More recently, 129, Napatree Point, 7 October and 27 October 2001 (JZy); 129, Napatree Point, 12 September 2002 (JZy).

One of the largest recorded flights occurred 2 October 1982 when a coordinated effort recorded 600 at Sakonnet Point (DLE, KHo), 712 at Taylor's Lane in Little Compton (HCa), 64 at Sachuest (SRi), 366 at Brenton Point (RTL), and 530 at Napatree Point (CRa, REn, RLF). Also, 40, Block Island, 28–30 September 1984. Many of the above flights occurred during strong northwest winds accompanying the approach of high pressure into New England.

Winter. There are relatively few winter records from northern and interior western sections of the state. Sharp-shins have occasionally wintered at Block Island; several have been recorded on CBCs, and 1 banded there 21 November 1975 was recaptured 10 May 1976 (ELa, HLa), so survived the winter.

Winter maxima. 25, Newport–Westport CBC, 3 January 2009 (REm et al.). They are now less common in the winter than Cooper's Hawks; e.g., the 10-year mean on the Newport–Westport CBC (2007–2016) is 10.3, as compared with Cooper's Hawk's 10-year mean of 14.1.

Cooper's Hawk (*Accipiter cooperii*)

Status

Fairly common summer resident, common migrant, and uncommon winter resident. Generally increasing.

Historical

Cooper's Hawks were, before 1940 or so, regular summer residents. They were always more numerous than Sharp-shinned Hawks, perhaps due to their preference for swampy deciduous woodlands and less dependence on evergreen groves for nesting. Charles E. Doe's collection contains egg clutches from 32 Rhode Island locations from the years 1888–1930, including East Providence (four), Cranston (three), Lincoln (seven), Warren (two), East Greenwich (one), Warwick (one), and Providence (one), along with specimens of nestlings from Potowomut. Cooper's Hawks sharply declined beginning in the 1950s and almost entirely disappeared as breeding birds in the state. They were always rare in Newport County, with the possible exception of Tiverton and northern Little Compton, where persistent summer reports 1935–1941 suggested possible breeding (RCC, DLE). In the northern part of Conanicut Island, a territorial pair was found in Hull Swamp 1 May 1957, and a nest was found there in 1971 (MDa).

Breeding

The breeding population of Cooper's Hawks in Rhode Island began a slow recovery following the ban on DDT in the early 1970s. A pair was at Carolina 13 June 1972 (DWa), 3 were seen at Green Hill 13 June 1974 (DWa), and a pair with fledged young was in the Carolina State Forest in 1977 (NCl, RAC, DLK). A nest was found 50 ft (15 m) up in a white pine in the Arcadia Management Area in 1976 (REn). An adult was feeding 2 half-grown young in a nest 35 ft (11 m) up in a small hemlock in Weetamoo Woods, Tiverton, and another territorial pair was in the Sin and Flesh Brook drainage, Tiverton, 24–25 June 1991 (RLF). They have nested on Block Island (2 nests confirmed June 2004; M. Wagner), Aquidneck Island (Ocean Drive, Miantonomi Park), and even in Providence (Swan Point Cemetery).

Nonbreeding

Spring. Peak numbers occur from late April to mid-May. Cooper's Hawks have increased since their DDT-related decline and are now fairly common.

Fall. Cooper's Hawks occur as migrants from mid-August on, with peak numbers in late September and early October. During the 1960s and 1970s, they were practically unreported during migration, but hawk watches since the late 1970s have regularly reported 5–13 or more per day migrating at coastal points.

Fall maxima. 43, Napatree Point, 12 September 2002 (JZy, MPa); 41, Watch Hill, 8 October 1981 (CRa, REn); 27, Sakonnet, 25 September 1987 (DLE); 5, Block Island, 8 September 1993 (CRa, REn).

Winter. Cooper's Hawks have steadily increased since the 1970s.

Winter maxima. 22, Newport–Westport CBC, 15 December 2012; 17, South Kingstown CBC, 27 December 2008; 12, Block Island, 12 November 2012 (SSM et al.).

Northern Goshawk (*Accipiter gentilis*)

Status

Rare breeder; uncommon migrant and winter visitor.

Historical

Local taxidermists were indispensable in documenting the great flights that occurred during the shooting era. In 1899, F. T. Jencks wrote (in Howe and Sturtevant 1899) that "large numbers were killed in a winter about 25–30 years ago," referring to the great flight of 1870. During the winter of 1896–1897, 22 adults and 7 young birds were brought to J. W. Critchley, Providence taxidermist, before January 15. In the famous invasion of 1906–1907 (Deane 1907d), 65 specimens were received by Angell and Cash of Providence, all taken within 30 mi (48 km) of the city, and Harry Cash stated that this was the most remarkable flight since 1870, with numbers surpassing those of the latter year. There was an apparent flight in 1917–1918, when according to HSH 50 were brought to taxidermists in Providence in the early winter. In 1926 there was a very large flight; Angell and Cash received 11 during the month of October alone, and 85 had been mounted by them before early February. An echo flight occurred the following year, 37 being reported by HSH as brought to Angell and Cash during the month of December 1927. These were the last great flights documented for Rhode Island. No comparable recent flights have occurred since, with the exception of 1972–1973, when an extraordinary flight throughout the eastern half of the continent resulted in 13 Rhode Island reports, more than in any year since 1926–1927.

Breeding

A pair of territorial adults were observed throughout the spring of 1953 at Fiskeville, and a fledged young and empty nest were discovered on 1 June (RGe, JGe, RCC). These birds were apparently present at this location in several previous years but disappeared after 1953 when 1 was reportedly shot. In 1975 and again in 1977, aggressive adults were reported, and a nest and chick were discovered in the Hemlock Ledges area of extreme western Glocester (P. Bissell, confirmed by RLC). Another nest and an adult were discovered in a hemlock swamp in Foster, 11–19 June 1981 (REn). This old nest was vacant 12 June 1982, but adults were calling nearby and presumably breeding at a new site (REn, CRa). None was present in 1984, and only 1 adult was seen there 15 June 1985 (REn et al.). Investigation 30 June 1988 located another nest, this time only 20 ft (6 m) high in a deciduous tree (REn; photographs of calling adult and 3 fledged young). In 1984, an oak tree in the big pine woods of Eppley Sanctuary, Exeter, was found to contain a nest and 3 young. The same nest contained another brood in 1985 (HCa), and birds were seen there in subsequent years.

During the 1980s, adults were seen during summer in suitable breeding habitat in Richmond, Exeter, Greene, West Greenwich, Watchaug Pond, in the Great Swamp (South Kingstown), Lincoln, and Tiverton. A pair with a nest in a deciduous tree was seen near the north end of Briggs Marsh in Little Compton in June 1989 (DLE). There seems to be an inverse relationship between abundance of Cooper's Hawks and goshawks as breeders in Rhode Island. Goshawks were especially numerous during a period of scarcity of Cooper's Hawks (almost certainly pesticide related) through the 1980s but have since declined as Cooper's have dramatically increased.

Nonbreeding

Fall. Goshawks are late-fall migrants, often not seen until mid-October or later.

Fall maxima. In nonflight years, totals of 5 or fewer birds are reported statewide. In flight years they appear earlier; e.g., a total of 5 were reported from 3 locations 27 September 1984 (CRa; REn; DLE). They are uncommon at Block Island: 23–29 September 1982 (JCB, DLK, RBo) and 28 September 1974 (Natchaug Ornithological Society). The species is often less common in midwinter and again occurs as an occasional transient in early spring.

Winter. Despite a customary late arrival in fall, the species is scarcer in midwinter than in late fall, and it is particularly uncommon in late winter.

Winter maxima. Three, South Kingstown CBC, 20 December 2003 (STs et al.).

Spring. While also rare in spring, there is some apparent return migration in April. Records of presumed migrants (some reportedly actively migrating) exist for scattered locations (mostly South County) in March and through April. The latest bird unlikely to have been part of the new breeding population is 1 at Shannock 30 April 1955 (DLK).

Bald Eagle (*Haliaeetus leucocephalus*)

Status

Uncommon migrant. Occasional in summer and winter. Increasing. Four pairs currently nesting.

Historical

As winter residents, Bald Eagles were once fairly regular in Rhode Island, and they often remained long periods at some locations. They wintered along the Seekonk River in the 1880s and were common at Wordens Pond before 1900 (E. A. Mearns 1901). Eagles were present nearly annually in Barrington until the mid-1940s but were most regular and dependable during the period 1930–1965 at Watchaug Pond, where one or two (maxima of five in the winters of 1935 and 1940; MSo) could be found annually attending the considerable concentration of waterfowl that formerly spent daylight hours there on the ice. By the early 1960s, only one adult was present there, and this bird was last seen in March 1965 (DLK et al.). There are 13 CBC records 1950–1962, but only two during the period 1962–1988. The only extant specimen with data taken in Rhode Island is of an immature shot on 5 May 1900 at Sandy Point, Block Island (Capt. Sisson; in RWPM). This specimen seems to match the southern *H. l. leucocephalus*.

After the late 1950s, the number of eagles observed in Rhode Island dropped precipitously commensurate with their decline nearly everywhere in North America. The worst period in Rhode Island was between 1969 and 1976, when only two Bald Eagles were recorded. A slightly increased number was observed during the period 1976–1984, mostly during the colder months and largely from the Watchaug–Quonochontaug area. During the latter period, from one to three birds were present for parts of six winters in that area; they were noted in early fall, and there were four reports in late spring and summer. A further increase seems to have been in progress in the late 1980s and 1990s, based on annual reports.

Breeding
Adult Bald Eagles have been persistently present in the Scituate Reservoir watershed since the spring of 1994, and the first nest was found there on 30 May 2003, with 2 "sizable" young (RLF). There are now about 4 nesting pairs of Bald Eagles in Rhode Island. There are also 2 or more pairs nesting on the eastern shore of Narragansett Bay. There are records of nests for the historical period in western Massachusetts (Griscom and Snyder 1955) and for Gardiner's Island, New York, where they bred with some regularity until 1930 (Bull 1964).

Nonbreeding
Spring. Three immatures over Kingston 9 May 1982 (RAC).

Fall maxima. Five migrating over Jamestown, 16–25 August 1951 (MDa); 4 over Watch Hill, 24 September 1995 (JZy).

Winter maxima. Nine (all immatures), Seekonk River, 7 December 2014 (PP); 6 (1 adult feeding on deer carcass), Wordens Pond, 18 January 2016 (CRa); 5, East Providence Reservoir, 9 December 2012 (JSu); 5, South Kingstown CBC, 22 December 2012.

Mississippi Kite (*Ictinia mississippiensis*)

Status
Vagrant. Now occurs annually in spring.

Nonbreeding
This species from the river bottoms of the lower Mississippi River valley and southern plains has expanded its range into the Northeast; it appeared in Massachusetts a half-dozen times beginning in 1962 before being recorded in Rhode Island. The first Rhode Island record was an adult that appeared over a wooded area of Coventry 17 June 1986 (RLF). Then came several reports in rapid succession: an adult in Cranston 27 May 1987 (JGe); an immature flushed from the dunes at Quicksand Pond and seen at 10 ft (3 m) 7 April 1989 (AFr); and an adult in the Fogland area of Tiverton 21 May 1989 (RAC). More recently, singles were at Swan Point, 14 May 2003 (DFi, RFa et al.); Jamestown, 10 May 2005 (WM); East Matunuck, 4 July 2006 (STs); Coventry, 16 May 2009 (CPe); North Kingstown, 16 May 2011 (DFi, STa); 4, Richmond, 31 May 2011 (DFi, STa); and 1, Potowomut, 24 August 2011 (B. Glemboski). In 2019, Mississippi Kites were present at 4 sites west of the Bay 21–26 May, and presumed pairs were in East Greenwich and Richmond (v.o.). Two birds were in Richmond, 10 May–7 June 2020 (Tim Metcalf; photo). The only record from Block Island is 1, 14 June 2013 (Brendon Murtha).

Mississippi Kites have undergone a substantial range expansion in parts of the south (Parker and Ogden 1979) and have recently bred in New Hampshire and New

York. Birds have been present in East Greenwich during summer in recent years, but breeding has not been confirmed.

Red-shouldered Hawk (*Buteo lineatus*)

Status
Uncommon breeder and winter resident. Fairly common migrant.

Historical
Red-shouldered Hawks were common nesters in Rhode Island during the years 1880–1950. For example, three active nests were in Johnston within a few hundred feet of each other 18 April 1891 (CED, fide HSH). Hathaway (1943) considered it the most common hawk in Rhode Island. Doe's turn-of-the-twentieth-century collection contains 299 sets of eggs from Rhode Island, including 92 from Cranston, 42 from East Providence, 35 from Warwick, 23 from North Kingstown, 20 from Warren, 16 from Glocester, 14 from Bristol, and 5 from Potowomut in East Greenwich. Similar population levels continued into the mid-1950s and perhaps into the early 1960s, but thereafter the Red-shouldered Hawk declined substantially. For the first half of the twentieth century, Red-shouldered Hawks were far more common than the gradually increasing Red-tailed Hawks, but these positions have since been dramatically reversed.

Red-shouldered Hawks declined sharply in the early 1960s along with several other raptors. This decline is obvious from CBC data: 68 were recorded on the two major mainland counts during the period 1950–1963, but only 23 between 1964 and 1990. This decline is very likely due to pesticide contamination (Henny et al. 1973). By the mid-1970s, Red-shouldered Hawks were increasing again in Rhode Island. During the period 1970–1973, there were 40 records averaging 10 per year (up to 3 per day), including breeding-season reports from 10–15 locations west of Narragansett Bay.

Breeding
Before 1950, Red-shouldered Hawks were common throughout most northern and western districts west of Narragansett Bay and also many parts of South County away from the coast. They formerly nested in the outskirts of Providence (Blackmore Pond in Cranston, 1893 [HSH]; Saylesville, 1913 [CED]) and fairly commonly east through East Providence and Barrington. They nested regularly at the turn of the twentieth century close to Narragansett Bay at Potowomut and in the Mount Hope area of Bristol. They apparently never bred on islands in Narragansett Bay, perhaps because of the lack of forest there, and have generally been uncommon as breeders in the southeastern part of the state. Individuals, however, have been recorded 3 times on Conanicut Island in the breeding season (29 May 1946 [RBo]; 19 May 1975 [MDa]; and 17 June 1969 [MDa]). They bred at 2 places in southeastern Tiverton in the late 1930s (RCC), and 1 was seen there 6 June 1978 (RBo). They formerly

nested in Wilbour Woods, southern Little Compton, where a nest was found 11 May 1952 (DLE). With the gradual increase in recent years, others have been reported in Tiverton, 1–2 June 1983 (HCa); adults were seen near Simmons Pond 19 June 1988 (RBo), 26 June 1992 (RBo), and 1 August 1990 (DLE) and found breeding in Little Compton (GDe). A pair nested in East Providence in 2017 (fide RFa).

Nonbreeding

Red-shouldered Hawks favor inland mountain ridges during migration more than do accipiters and falcons, and for this reason are not readily detected as migrants in Rhode Island, where hawk-watching effort is focused along the coast.

Spring. Spring migrants appear in late March and April; they are uncommon and only rarely have more than 1 or 2 have been seen per day.

Fall. Red-shouldered Hawks are occasionally seen at Sakonnet, Brenton Point, or Watch Hill in September and October, rarely more than 2 per day.

Even though they never bred on the larger Narragansett Bay islands, they were formerly regular there in winter, particularly on Aquidneck Island. Even more than other buteos, however, Red-shouldered Hawks have always been scarce on Block Island. There are only 2 fall reports: 11 October 1959 (ASRI) and 10 October 1976 (ESa, CWo et al.), but 1 was seen on the Block Island CBC, 19 December 2013 (CRa, SCo).

Winter maxima. 15, Newport–Westport CBC, 15 December 2007 (REm et al.). Recorded every year since 2002 on Newport–Westport CBC, aggregate total of 97 birds in that period, current 10-year average (2007–2016) of 7.3. Also, 9, Napatree CBC, 18 December 2011 (SSM, GW).

Broad-winged Hawk (*Buteo platypterus*)

Status

Fairly common summer resident, especially in the heavily wooded sections of western Rhode Island, and common spring and early fall migrant.

Historical

Broad-winged Hawks were considered summer residents by Howe and Sturtevant (1899), as well as by Hathaway (1943). They were less common than either Red-shouldered or Cooper's Hawks. The gradual increase of Broad-winged Hawks during the historical period probably reflects the succession of farmland to forest. They were apparently absent from most southeastern sections of the state at the turn of the twentieth century, and the earliest report is for a nest in Bristol in 1915 (CED). They returned early to the Tiverton highlands, and three pairs were noted there in 1938 (RCC). Statewide, the species has slowly been increasing southeastward, in spite of accelerating urbanization of this area, and even began to breed during the 1960s in the Hull Swamp section of Conanicut Island (perhaps re-extirpated in 1985).

Breeding

Broad-winged Hawks are the commonest nesting buteo in wooded sections of western and northwestern Rhode Island. They are present in smaller numbers south to the vicinity of the South County moraine in Charlestown and South Kingstown. They breed to the very edge of the metropolitan area in Johnston and Cranston. In areas where they were historically absent, a territorial pair was seen and heard at the north end of Poppasquash Point, Bristol, 16 June 1982 (HCa), and they have nested in Little Compton for the past 5 years (fide RFa). They have only rarely been reported during the breeding season on the Narragansett Bay islands: on Prudence Island in 1946 (WBD) and at NBS on Aquidneck Island in 1961 (REW). Robbins et al. (1986) list Rhode Island among the states with the highest nesting densities in North America.

Nonbreeding

Spring. Migrants usually arrive in mid-April and peak numbers occur in late April. Numbers of nonbreeding 1-year-old birds often appear during May and June, especially at the coast.

Spring maxima. 13, Diamond Hill, 18 April 1981 (REn); 112, Coventry, 6 May 1977 (ASM).

Fall. The relatively narrow peak period for major flights in nearby states is in mid-September, particularly September 12–19 (Roberts 1996).

Fall maxima. 500, north of Quonochontaug Pond in Charlestown, 24–25 September 1981 (GLB). Between 0900–1030 hours the next morning, 196 were seen over Charlestown and Weekapaug (ESa), and these were probably part of 7 groups totaling 556 noted crossing the Pawcatuck River just north of Watch Hill in the same time period as observed from the Napatree dunes (CRa, GLB). Also, 30, Little Compton, 29 September 1951 (DLE, RCC); 25, Watch Hill, 25 September 1987 (CRa); 25 at Sakonnet and probably a different 25 at Newport, both 28 September 1957 (WBD; JGe, RGe).

Broad-winged Hawks are uncommon at Block Island at any season. There are 6 fall records, mostly of singles, 4 September–9 October. A flock of 7 appeared there 23 September 1935 (EDi). There are also several records for spring, including 2, 2 May 2005 (STs); 29 April 2014 (DFi, STs); and 3, 20 May 2014 (DFi, STs)

Extreme dates. 29 March 2002, Great Swamp (1; DFi, CNu), and 3 April 1982, Kingston (RAC), to 5 December 1991, Fiskeville (JGe, RGe), and 9 December 1961, Westerly (SSD, RMc).

Swainson's Hawk (*Buteo swainsoni*)

Status

Vagrant. Four records.

Nonbreeding

There is a record of a light-morph adult seen directly overhead at Matunuck 14 November 1964 (GRa). Another was at Matunuck 29 April 1976 (DWa). A dark-morph adult was well described from Tiverton on the unusual winter date of 28 February 1971 (DLE, REm). A small all-dark buteo that soared with a dihedral east across the Pawcatuck River at Avondale on 3 October 1981 was probably this species (RLF, RAC). This highly migratory prairie buteo has occurred annually in the East in recent years, especially at such coastal concentration points as Cape May, New Jersey (average 3 per year; Sibley 1993). The peak of eastern reports is in late October.

Zone-tailed Hawk (*Buteo albonotatus*)

Status

Vagrant. One record.

Nonbreeding

A Zone-tailed Hawk was seen and photographed 15 August 2015 soaring with Turkey Vultures over turf fields near Wood River Junction (WM, CPe; see Photo 22). A flurry of records between Delaware and Massachusetts in 2014–2015 may have pertained to 1 or more individual Zone-tailed Hawks. Apart from these records, there are only 2 occurrences in eastern North America.

Red-tailed Hawk (*Buteo jamaicensis*)

Status

Fairly common to common summer resident, common spring and fall migrant, and common winter resident.

Historical

Red-tailed Hawks were less numerous than Red-shouldered Hawks in Rhode Island during the last part of the nineteenth century. Howe and Sturtevant (1899) described them as "not uncommon" as a winter visitor and "uncommon" as a summer resident. They listed eight winter records and alluded to "numerous others," suggesting much lower abundance than today. Doe collected only 18 sets of eggs in his 50 years of egg collecting, almost entirely from the wilds of Glocester, or less than one-tenth of the 299 Red-shouldered Hawk clutches he found during the same time. Red-tailed

Hawks have substantially increased since 1950 and are now one of the most familiar sights of the state, especially along highways.

Breeding

Most breeding presently occurs in more remote districts in western and northern sections of the state. Birds are commonly present in summer throughout the state in an arc from Cumberland and Burrillville south to the South County moraine. They are relatively scarce as breeders along the shores of Narragansett Bay and in the Providence metropolitan area. However, a pair was noted mating in Providence's Swan Point Cemetery 22 March 1986 (fide RLF), and a pair nested on a *fire escape* of a Pawtucket factory in 1985 (REn et al.), as has been reported in New York City during the 1990s. On 22 June 1991, a pair was perched on the head and arm of the statue atop the capitol building in Providence, and another pair was resting the same day on a radio tower along Mineral Spring Avenue in southwestern Pawtucket (RLF). One much-ballyhooed adult spent several winters in the early 1980s catching pigeons in the vicinity of the Biltmore Hotel in Providence.

Nonbreeding

Spring. As in fall, spring migration is unspectacular, and few through-transients are noted. Seldom are more than 1 or 2 birds (maximum 6) seen actively moving per day in March and April, and these are probably incoming breeders in southeastern New England or outgoing wintering birds.

Spring maxima. 12, Scituate, 27 March 1993 (JGe, RGe).

Fall. Migrants have appeared where they do not breed from 13 August on, but little migration is noted until late September. Occasional small flights of from 5 to 7 per day are encountered a mile or 2 (1.6–3 km) north of the immediate coast.

Fall maxima. 17 passing north at Taylor's Lane, Little Compton, 10 October 1981 (DLE, KHo); 13, Watch Hill, 6 November 2001 (JZy).

Winter. Red-tailed Hawks are now one of the most common hawks in southern parts of the state during the winter months, particularly in more open areas (to 7 or more per day). They are comparatively less common in the northern and western parts of the state at this season, but they nevertheless seem to have benefited from the highway interchanges and powerline corridors there; e.g., 10 along Route 95 between Coventry and Ashaway 23 February 1974 (ASM). While they are among the most common winter hawks on the mainland, they are surprisingly uncommon at any season at Block Island. Through the end of the 1900s, there were only about 11 records for the island, mainly in fall. Two appeared there 24 May 1937 (EDi). They are now more regular there in small numbers, with 6 CBC records since 2005. However, they are not yet known to nest on island.

Winter maxima. 74, Newport–Westport CBC, 18 December 2010.

Rough-legged Hawk (*Buteo lagopus*)

Status

Uncommon migrant and winter resident; very rare in summer.

Nonbreeding

Rough-legged Hawks move south from their Arctic and sub-Arctic breeding grounds at irregular intervals, based on prey abundances (Bechard et al. 2020). While some are usually reported each winter, the total number seen varies considerably from year to year. They are sometimes very scarce, as in 1958–1959 (2 reports) and in 1956–1957 (1 report), but relatively common at other times, as in 1964–1965 (22 reports) and 1986–1987 (16 reports). They are seldom very numerous, to 5 per day; once 12 were seen at Trustom Pond 24 November 1949 (Coman).

Fall. The earliest fall records are 22 September 1941, Narragansett (Coman), and 22 September 2007, Moonstone (STs). Most usually arrive during November. In many winters, none appear until severely cold weather arrives during January and February. Once here, individuals may remain for long periods at such outer coastal locations as Sachuest Point through February and March.

Winter maxima. Nine, South Kingstown CBC, 22 December 2007.

Spring. Migrants are occasional in spring but are especially rare in the interior (recorded at Providence, 26 March 1942 [DLE]; Wallum Lake, 26 March 1982 [SMi]; Scituate, 16 April 1939 [WBD]). Migrants and departing winter residents occur irregularly to mid-April; there are 5 records for May.

Summer. A light-morph bird remained through May, June, and July 1962 at Middletown (REW et al.; photographed by RLF); a similar bird was seen at the same location 11 June 1966 (JBa, RLF, 25 NEBBA members); a dark-morph bird at Block Island 12 May 1972 (ELa) was seen again there 6 June 1972 (RLF); a dark-morph bird appeared at Sakonnet 2 June 1973 (DLE, REm). One seen 3 September 1960 at Middletown (REW, R. Weaver) may have summered.

BARN OWLS (TYTONIDAE)

Barn Owl (*Tyto alba*)

Status

Uncommon and local resident near the coast.

Historical

Only five occurrences were known to Howe and Sturtevant (1899) at a time when farmland, their preferred habitat, was at a maximum. Hathaway (1943) called them "very rare visitors." Barn Owls were first found nesting in Rhode Island in 1938, when workmen cleaning the Mount Hope Bridge discovered a nest and young (fide CGW). In 1939, DLE found five owlets in the possession of a farmer in Little Compton, who claimed that a pair had occupied his barn during the two previous years. They were found nesting in Warren in 1938 (RBo), and in 1947, three families were raised in Bristol, Matunuck, and Narragansett (RCC, DLK). Barn Owls have been present on Block Island since at least 1938.

Breeding

Barn Owls increased during most of the twentieth century, but in recent decades they have declined, particularly on the mainland. The most persistent breeding has been at Block Island. Pairs have been found breeding in abandoned hotels or on gravelly sea cliffs at Clay Head (1957 on [ELa, HLa]; 2 pairs in 1984 [HCa]) and apparently also at the Mohecan Bluffs and other bluffs (reports at 2 places in the 1980s). Three pairs were estimated on the island in 1981 (m.obs.). Chris Raithel walked the beaches under all of the bluffs on the island in 1991 and found 8 burrows active, based on signs of whitewash. Single pairs have been noted at other locations; e.g., Sakonnet (1949 on; DLE); Middletown (several places at least since the mid-1950s, especially NBS, into 1980s or later; JBa, REW, LOG et al.); Newport (1951); Jamestown (1951 on; 2 nests in 1959; MDa); Narragansett (1947–1950; DLK); Charlestown (1965 and other years); and again at Warren (1976; RBo). More recently, they bred in Middletown at 1 or 2 nest-box sites since about 2000; e.g., 6 young on 18 May 2005 (JZy). They have also been suspected of breeding in Portsmouth (adults and young at Glen Farm in 2016; m.obs.).

Nonbreeding

During the period 1950–1970, Barn Owls were regularly found roosting in evergreen groves at the coast during winter. Some were found starving or frozen in severe winters or were eaten by Great Horned Owls. After the mid-1960s, numbers found on CBCs declined, and at the same time breeding pairs disappeared from such regular locations as NBS and Block Island. The cause of this decline is unclear but was likely related to the same cold weather that devastated the Massachusetts population and also severely reduced Carolina Wren populations in the winter of 1960–1961. It is also possible that Barn Owls suffered from pesticides, as did many other raptors during the 1960s. After the mid-1970s, Barn Owls were again found at Block Island and at 2 or more locations in Middletown, now more often aging dairy barns than evergreen groves. The winter of 2014–2015 was very hard on Barn Owls, with several dead birds found at nest boxes and other areas. However, at least a few birds remained on Block Island; e.g., 1, 12 November 2015; 1, 14 November 2016 (both SSM et al.).

Recent maxima. Six, Newport–Westport CBC, 15 December 1979 (DLE et al.); 5, Block Island, 25 February 1978 (SRD); 7 there, 2 January 1986 (CBC).

OWLS (STRIGIDAE)

Eastern Screech-Owl (*Megascops asio*)

Status
Fairly common resident.

Breeding
Screech-owls are found wherever trees are large enough to provide nesting and roosting cavities, often in older suburban districts near the head of Narragansett Bay. While often absent from the most open portions of the coastal area, they are unquestionably most abundant in those milder southern and eastern sections of the state where there are patches of woodland, particularly on the larger Narragansett Bay islands, near the Sakonnet Passage, and north through Providence and the Blackstone Valley.

Recent evidence suggests screech-owls may be absent from extensive sections of the state, particularly the western and northern portions, where the Rhode Island Breeding Bird Atlas shows only a handful of records. Screech-owls have always been rare at Block Island, where there are only 4 reports of vagrants: 4 June 1915 (EDi), 22 October 1930 (EDi), December 1963 (CBC), and 28–30 September 1992 (RBo et al.).

Maxima. Most counts have not exceeded 7 per location or 10 per night. Affirming that the best habitat lies in the low, mesic habitats of the Narragansett Basin, 11 were found on Prudence Island 25 August 1989 (REn, CRa), and a partial search there 5–7 June 1990 produced 37 by means of taped vocalizations (REn, CRa); also, 9, Prudence Island, 18 June 2003 (REn). All recent high counts are from CBCs, especially Newport–Westport, which has best habitat in the state; e.g., 83, Newport–Westport CBC, 15 December 2012 (REm et al.); 67, Newport–Westport CBC, 17 December 2011 (REm et al.).

Great Horned Owl (*Bubo virginianus*)

Status
Uncommon to fairly common resident.

Historical
Great Horned Owls were much less common during the colonial period than at present, probably owing to the scarcity of woodland at that time as well as shooting.

Howe and Sturtevant (1899) considered them to be rare and irregular visitors and rare summer residents. Remington and Commissioners (1908) called them one of the rarest nesting birds in the state and claimed the first nesting record in 1904. However, Howe and Sturtevant knew of five nests from before that time and of eggs collected in Lincoln in 1880 (CED). Forbush (1927) called them "rare residents" in Rhode Island, doubtless following the opinion of HSH. Wright (1941), quoting RCC, stated that they were rare "permanent resident[s] and rare winter visitor[s]" and that they were somewhat less common than Barred Owls. During 1950–1970, however, Great Horned Owls were much more often reported than Barred Owls, especially in southern and eastern sections where Barreds are nearly or entirely absent.

Breeding

In recent years, particularly during work on the Rhode Island Breeding Bird Atlas (1980s), Great Horned Owls have been outnumbered by Barred Owls in many areas, particularly in the most heavily wooded sections of western Rhode Island where Barred Owls are common and widespread. The perceived densities of each species depend on the time and season of field work; Great Horned Owls are more secretive and less vocal during the day and are quietest during the summer when observers have been most active in the wooded western interior. In sum, Great Horned Owls are overwhelmingly more common than Barreds in eastern Rhode Island but a bit less common than Barreds in the westernmost parts of the state. Great Horned Owls breed in suburban areas such as Providence's Swan Point Cemetery. They had not been recorded on Block Island until recently, when 2 birds spent the winter of 1977–1978 in a mid-island pine grove; another was seen there 15 June 1981 (SRD, RBo), as well as during the fall of 1981 (ASRI).

Nonbreeding

Great Horned Owls breed very early in the season, usually in January. Therefore, they are often most vocal in December and January, including the CBC period. There are 24 records from Napatree Point. Most are in fall between 6 October and 31 December; e.g., 2–12 November 1983, 6 November 1986, and 3–8 December 1989 (all CRa). There are 3 records for spring: 3 March, 7 April, and 5 May in different years (CRa).

Maxima. 46, Newport–Westport CBC, 15 December 2012 (REm et al.).

Snowy Owl (*Bubo scandiacus*)

Status

Rare and highly irregular winter visitor. Recorded almost annually in recent years.

Nonbreeding

Generally found in coastal locations or near Narragansett Bay, but there are many records from inland locations. Snowy Owls are found with particular regularity at Block Island. The following is a chronological history of flights and notable counts based on Rhode Island data (Deane 1902, 1905, 1906):

1876–1877. More than 30 were shot (Southwick 1888).

1882–1883. Nineteen were received by Southwick and Jencks (1883), in contrast to only 1 in 1884–1885.

1886–1887. Seven were brought to taxidermists from Point Judith and Newport (HSH, old notes).

1887–1888. Eight or 10 were known to have been shot (Southwick 1888).

1890–1891. "Mr. Southwick had about 50 in the [Natural History] store, all taken within a short time of each other" (HSH, old notes).

1899–1900. One was killed in Warwick; 4 were seen on a beach near Point Judith on 10 December 1899 (HSH).

1901–1902. Eighteen specimens received by Harry Cash, Providence taxidermist, between 1 January–10 February; 3 were shot at Block Island, 4 on Aquidneck, and 1 in Pawtucket (Deane 1902).

1905–1906. A large, well-documented "flight" (Deane 1906, 1907c). Angell and Cash received 17 before 1 January and at least 12 others were known to have been shot. All were from coastal locations except Woonsocket, Warren, and Scituate.

1926–1927. Perhaps the largest incursion on record (Gross 1927). Angell and Cash received 156 to be mounted and C. B. Clarke mounted 22 shot near Newport. They were almost abundant at Block Island with up to 6 per day (EDi).

1930–1931. Three records are known from the mainland, but many others were reported to EDi on Block Island (to 3 per day).

1937–1938. Many reported shot in Little Compton, and others seen.

1941–1942. About 40 reported statewide. A major and apparently coincidental flight of both chickadees also occurred in the same year (Poor 1946). It should be noted that by this time, there had been a marked reduction in the shooting of nongame birds, and nearly all of the large taxidermy firms had disappeared. This dried up that traditional source of information and shifted the responsibility of documenting raptor incursions to a small but growing corps of active birders.

1945–1946. More than 45 noted statewide, mainly along the coast. One was reported from Smithfield.

1949–1950. A very large flight, perhaps equal in magnitude to that of 1926–1927. Most birds were along the coast, but many were at inland locations. No adequate idea was obtained as to how many were present in Rhode Island. Up to 8 per day at Block Island, 28 December 1949 (EDi et al.).

1960–1961. Fifteen were reported, including 2 in East Providence and 1 at Wayland Square, Providence.

1964–1965. Sixteen reported from coastal locations.

1980–1981. Seven records in late winter.

1986–1987. Sixteen birds in the state at 11 locations including 4 birds on the Block Island CBC.

1991–1992. About 15 birds, October to April.

2013–2014. About 30 birds found statewide; 12, Newport–Westport CBC, 14 December 2013 (REm et al.).

The first arrivals are generally in November or later with the last birds usually in early March.

Extreme dates. 25 October 1964, Napatree Point (ESa), to 7 May 1953, Bristol (fide RCC).

Northern Hawk Owl (*Surnia ulula*)

Status

Vagrant. Two records.

Nonbreeding

One is recorded in Howe and Sturtevant (1899) as taken by W. A. Aldrich, exact locality and date unknown. The collection of this bird was reported by Newton Dexter, but it is unclear whether he ever saw the specimen. Another was brought to Angell and Cash, Providence taxidermists; it had been shot 16 November 1906 at West Greenwich (Deane 1907a).

Burrowing Owl (*Athene cunicularia*)

Status

Vagrant. One record.

Nonbreeding

On 11 November 1963, a Burrowing Owl was found dead at Brenton Point, Ocean Drive, Newport, by RSt and brought to NBS. Robert E. Woodruff, who prepared the specimen and sent it to the MCZ, found it to be very fat with a gizzard full of earwigs. At MCZ, it was determined to be the western race *A. c. hypugaea* by Dr. Raymond L. Paynter Jr. (MCZ #285845). The fact that other Burrowing Owls appeared in the Northeast during this same season (Veit and Petersen 1993) supports this being an unassisted vagrant.

Barred Owl (*Strix varia*)

Status

Fairly common resident.

Historical

During the nineteenth century, substantial incursions of Barred Owls occurred that dwarf any more recent movements. During the winter of 1882–1883, 75 were received by Southwick and Jencks, and in 1884–1885, more than 50 were reported (Howe and Sturtevant 1899). Nothing remotely approaching these numbers has occurred since, though Barred Owls still occasionally wander to areas where they do not breed. There are old records from Newport County, such as at Aquidneck Island in 1900, 1902, and 1915 (Sturtevant); and one in Middletown 20 December 1986 (fide REn).

Breeding

Barred Owls primarily breed west or north of Narragansett Bay. They are not uniformly distributed through this area, however. They have been reported most often from a belt of swampy woodlands north of the Charlestown moraine, extending from the vicinity of Watchaug Pond, Charlestown, east through the Great Swamp to suburban Kingston. At such sites, it is not unusual to hear 3 or more individuals calling at once. East of Narragansett Bay, Barred Owls are resident in the Sin and Flesh Brook area of Tiverton, where breeding was confirmed by RCC in the 1940s and a pair seen 25 June 1991 (RLF) and where the species has been noted irregularly in winter. Judging by cast-off feathers, and a pair heard "dueting" there on 17 June 2017 (REm), they are also present southward in Weetamoo Woods, Tiverton. One was heard calling in Little Compton south of Simmons Mill Pond 8 July 1984 (RBo), and 2 were reported from Little Compton 9 June 1991 (KS, CS). Since about 2000, a few birds have been found throughout the year in wooded areas south of Watson Reservoir, west of Quicksand Pond, and occasionally in Wilbur Woods (GDe, REm).

An ailing bird was caught alive at NBS in lower Middletown 31 October–1 November 1992 (LP, CRa; AMNH #831221). There are only 3 occurrences each for Prudence and Conanicut Islands (once in summer at Jamestown; MDa); 1, Beavertail, Jamestown, 6 February 2012, found dead on road (Cathy Ginn); 1, Jamestown, 15 December 2012, found dead (fide CRa; photo); 1, 6 January 2013, dead on road (fide CRa). One was collected in East Providence 1 November 1909 (CED). There is a report for Block Island 27 December 1937 (fide EDi).

The scarcity of Barred Owls from eastern and coastal Rhode Island is consistent with their general absence from the coastal plain of southeastern Massachusetts and Long Island. Of particular interest, therefore, is their year-round occurrence at Watchaug Pond, only about 2 mi (3 km) from the ocean. There are also records of birds calling at Quonochontaug during July 1934 (HSH) and records of wanderers (mostly in the colder months) for such South Shore locations as Matunuck, Green

Hill, Perryville (2, 19 July 1991; MJM), Westerly, and Napatree Point (4 November 1989, and 1, 13 October 2012 [both CRa]).

Maxima. 29 reported, December 2012–January 2013.

Great Gray Owl (*Strix nebulosa*)

Status

Vagrant. About five records.

Nonbreeding

According to Newton Dexter, 1 was taken near Providence in 1870 and brought to John Hague, a Providence taxidermist. Another is extant from Fox Island, off Wickford, shot there on 25 March 1883. The mounted specimen was acquired by HSH and is in the RWPM collection. There was a major "flight" in 1890–1891, with numerous records for Massachusetts but none for Rhode Island. For years, the last record was of 1 shot 19 November 1906 at an unknown location within a mile of the Providence city limits and brought to Angell and Cash, Providence taxidermists (Deane 1907b). This bird was part of an incursion to New England in the winter of 1906–1907.

For a half century or more, no flights reached New England, and such incursions were thought to be a thing of the past. Then, in the winter of 1978–1979, a major incursion brought 60 birds to New York state, and 15 or more records accumulated in central and eastern Massachusetts, mainly in February and March (Vickery and Yunick 1979). As part of this flight, one of the only modern records of Great Gray Owl in Rhode Island was of a bird found perched in a tree in an open field in eastern Coventry on 19 February 1979 (RAC). At least 2 other unconfirmed reports of large owls by the public during the same winter also may belong here. Finally, a Great Gray Owl was photographed in Westerly near Dunn's Corners during the winter of 1996; the photograph was examined by DLE, who confirmed the identification.

Long-eared Owl (*Asio otus*)

Status

Rare, but likely overlooked, resident. Uncommon migrant.

Breeding

There are few breeding records, but Long-eared Owls are notoriously secretive, and they have likely been overlooked. On 21 June 1908, a nest was found "in Kent County" by T. L. Flanagan. It contained 3 young thought to be 3 weeks old (Remington and Commissioners 1908). Allen (1909) went so far as to give an egg date of

10 April, an undoubted extrapolation of the 1908 record. In late July 1950, adults were seen several times at twilight feeding young near Diamond Hill by DLE. On 28 August 1950, 1 was heard calling at Johnston by RCC. Long-eared Owls bred at Block Island during the 1970s and 1980s. A bird was found sitting on a nest there 8 May 1972 (ELa, MWh et al.), and an adult with 2 recently hatched young and an unhatched egg of an apparent second brood was photographed at the same nest 4 June 1972 (RLF). Others had been seen in April and May 1966 and during 1968. Elsewhere on Block Island, from pines along the west shore of Grace's Cove Road, 1 answered a tape 15 June 1981 (RBo). Other records in the breeding season include 1 calling at Shannock Hill 18–20 May 1977 (RAC), 3–4 seen and heard at North Kingstown 10 July 1979 (CWo), and 1 at Portsmouth 17 May 1980 (RBu, RAC, HWi et al.).

Nonbreeding

Most reports of Long-eared Owls are from November through March, when they often aggregate in communal roosts in conifers and other dense vegetation. One roost was present for several years during the mid-1950s near the Mount Hope Bridge in Bristol, where the highest count obtained was 15 in February 1957 (RLF). The highest counts since then are 6 on the Block Island CBC in 1982–1983 and 6–8 through February 1982 at Warwick (DLE, RBo). The maximum on the Newport–Westport CBC is 10, 27 December 1959 (DLE et al.).

Short-eared Owl (*Asio flammeus*)

Status

Uncommon transient and winter resident. Very rare in summer

Historical

Read (1887) stated that Short-eared Owls were found breeding 28 April 1879 at an unstated location in Bristol County in a "salt meadow . . . of large extent" (Hundred Acre Cove, Barrington?), where, in a corner of the marsh in a patch of rank grass and low bushes, a bird was flushed from a nest and five eggs on a "hummock" made from old reeds. There seems little doubt of the authenticity of this discovery by Read, whose other observations of various raptors and other species in the vicinity of Rehoboth are frequently and impeccably detailed in *Ornithologist and Oologist*. The only other suggestion of breeding involves one seen by HSH, WBD, and DLK in the marsh at Fresh Pond, Quonochontaug, 13 May 1933, and suspected of breeding.

Breeding

Short-eared Owls formerly (to about 2009) bred on Martha's Vineyard and Nantucket, and to a lesser extent on Long Island and Cape Cod, and these sources may explain the appearance of wandering birds in Rhode Island during summer and early fall. Short-eared Owls now (2017) do not nest anywhere in New England, although a few territorial birds have been present in Maine recently during spring and summer (Vickery et al. 2020). There are older records from summer; e.g., 10 June 1962 at Jamestown (MDa); 28 July 1946 at Quonochontaug (DLK); collected 14 August 1896 at Middletown (ESt; possibly the basis of the moot hypothetical breeding status given in Howe and Sturtevant 1899); 20 August 1946 at Galilee (DLK); 30 August 1959 at Trustom (CWo). The presence of birds in suitable habitat between mid-May and early August suggests breeding, and perhaps some nests of this inconspicuous bird were overlooked.

Nonbreeding

Winter. While single birds are usually reported, they have sometimes been found in small aggregations in favorable feeding or roosting habitat close to the coast. Such groups were more regular and even numerous at times (maximum 9 birds) in former years when large expanses of feeding habitat were still available. Seven have been found on Block Island CBCs 1962–1990. Since about 1960, there has been a pronounced decline in winter reports, and they have become very uncommon. The infrequent and scattered reports are usually of 6 or fewer birds per day; e.g., 6, South Kingstown CBC, 23 December 1995 (DLK et al.); 4, Trustom Pond, 10 February 2001 (JMu); 4, Matunuck, 27 February 2003 (DFi); 6, South Kingstown CBC, 27 December 2008, and they continue to decline.

Fall. There are records for 13 September 1941 at Tiverton (RCC) and 17 September 1941 at Middletown (WHD). Other early records run from 28 September and 9 October into November. Most reports are for the period November through March.

Spring. There are many reports for April (e.g., 1 migrating offshore at Matunuck 27 April 1975; DLK) and only a handful of records for May. The latest of these was at Middletown 21 May 1938 (WHD); perhaps a nesting or prospecting bird. While most often encountered near the outer coast, this species is occasionally reported along the shores of Narragansett Bay as far north as Warwick (collected, 19 October 1894), Green Island, Bristol, and Barrington; e.g., 1 at Sabin Point, 7 January 1995 (DFi). There are only 3 records from inland Rhode Island: a specimen from North Providence in the Doe collection, 1 January 1909 (FlMNH #2852); 1 at Fiskeville 3 April 1969 (JGe, RGe); and 1 at Lonsdale 15 April 1991 (REn). Unusual for the locality was 1 at upper Narragansett Bay, 24 November 2004 (GDe).

Boreal Owl (*Aegolius funereus*)

Status

Historically, a very rare winter visitor with several records prior to 1899 but none since.

Nonbreeding

Howe and Sturtevant (1899) listed all the following records: 1 shot near Providence during the winter of 1880–1881. One was shot in an old house in North Providence in December 1880 or January 1881 and brought to John Hague, Providence taxidermist; the specimen was examined by Newton Dexter. This bird was mounted and displayed in the Brown University collection (Jenks Museum) for many years, but its present whereabouts are unknown. One was shot in Scituate in the winter of 1882–1883 and mounted by Daniel Seamans. Another was brought to Southwick and Jencks, Providence taxidermists, on 18 December 1882 by a milkman but was probably taken just east of the Rhode Island line in Seekonk, Massachusetts. A bird thought to be this species was caught by a farmer in Bristol before 1899, and the remains were later examined by RHH. Even so, RHH's identification of the battered bird by means of its "white beak and reddish plumage" leaves something to be desired. Finally, 1 was shot several years before 1899 in the northern parts of the state and later acquired by HSH. This last bird was undoubtedly mounted and may have been placed at Roger Williams Park with HSH's other mounted specimens, but a search there by Ferren in 1971 failed to locate it.

In summary, there are at least 5 old records, all but 1 from 1880–1883, involving 2 specimens seen by competent ornithologists, but none of these specimens is known to be extant.

Veit and Petersen (1993) mention 18 specimens taken in Massachusetts between 1859 and 1905 and describe 4 flights during that period (none of which corresponds to any of the Rhode Island specimens above). None appeared in Rhode Island during the huge New England flight in the winter of 1922–1923, when more than 30 specimens were collected in Massachusetts, as close as Seekonk, or during the 1991–1992 flight, when 6 were recorded in Massachusetts.

Northern Saw-whet Owl (*Aegolius acadicus*)

Status

Rare summer resident; uncommon transient and winter resident.

Breeding

There are about 7 breeding records of Northern Saw-whet Owls in Rhode Island, though they probably breed more commonly than this, as the records below suggest. The first is for a female and 6 young discovered in a duck box in Foster and banded

16 May 1952 (given incorrectly as 9 May in *RNEB*) by RIDFW personnel, the nesting later confirmed by RCC. The second nesting record was coincidentally obtained 2 days later on 18 May 1952 when RCC and E. Southwick found another female and 3 young in an old Northern Flicker box at the Kimball Bird Sanctuary, Charlestown. A juvenile was seen at Buck Hill 8 July 1992 (BCG). Calling birds, repeatedly heard on ASRI property at Ell Pond and Long Pond, 1983–1985, were finally confirmed by the discovery of 2 adults with young 19 May 1986 (R. Petruska). A bird was discovered in a nesting box in Little Compton on 30 April 2011; 4 fledglings were photographed out of the box on 7–8 May 2011 (GDe). A pair with young was at Coventry in 2016 (fide RFa).

A number of other records of birds present in suitable habitat during the breeding season suggest more widespread and frequent nesting: 1 was calling at the edge of a cedar swamp in Pulaski Park in Glocester and Burrillville for 2 years, 1984–1985, through late April (REn, CRa). One was caught alive at Kingston 13 May 1978 (fide DWa), and others have been seen in South Kingstown 14 May 1977 (HWi, RBu, RAC), Charlestown 16 May 1982 (CRa, REn), Arcadia 17 May 1986 (CRa), and West Greenwich 18 May 1993 (CRa). Two were at the Blackstone River canal 26 May 1990 (LS). One was calling at Warren 9 June 1948 (RBo), and another was seen in the evergreens of Burlingame, Charlestown, 14 June 1986 (CRa). One was calling in Phantom Bog, Bradford, 14–15 June 1986, where it had been heard earlier in the year (CRa, REn). One was found near a Tiverton maple swamp south of Seapowet 25 July 1991 (REm).

In Massachusetts, Saw-whet Owls prefer Atlantic white cedar and red maple swamps, and groves of white and pitch pine for breeding. Searches of these habitats in Rhode Island would seem likely to reveal a more widespread nesting distribution.

Nonbreeding

Saw-whet Owls are uncommon fall transients and winter residents, when they are frequently caught in mist nets (Paton 1998). There have been a number of "flights" when they have been especially numerous: 1887; 1929–1930; 1936; 1946–1947; 1951–1952; 1965–1966; and late winter 1973. Twenty were recorded on the Napatree CBC 18 December 2005 (SSM, GW et al.)

The earliest fall records are 20 September 1965 and 22 September 1959 at Block Island, both banded by E. Dickerson. Most transients appear from mid-October through late November, with a peak in late October, but there have been occasional incursions in late winter. Overwintering there is confirmed by 1 banded there 21 November 1975 and recaptured 12 February 1976 (ELa, HLa). They are less numerous in spring than in fall, a pattern confirmed by mist-netting. One was banded at Block Island 24 April 1974 (ELa).

Maxima. 35 banded, Block Island, 8 November 1999 (S. Comings); 20, Napatree CBC, 18 December 2005 (SSM et al.).

KINGFISHERS (ALCEDINIDAE)

Belted Kingfisher (*Megaceryle alcyon*)

Status
Uncommon and local summer resident; uncommon spring, common fall transient. Fairly common winter resident along the coast.

Breeding
Like Bank Swallows, Belted Kingfishers are limited during the breeding season to the vicinity of sand and gravel banks, where they construct nesting burrows; they are therefore present locally throughout the interior where gravel mining has created suitable nest sites and along the outer coast where wave action maintains similar habitat in the form of sea cliffs; e.g., Block Island, Little Compton, and elsewhere. Their distribution, therefore, depends almost entirely on the state's glacial drift and sandplain geology and the natural mud banks along the Blackstone River in Lincoln and Cumberland.

Nonbreeding
Spring. Hathaway's older records indicate that kingfishers formerly arrived in the Cranston area, where they did not winter, in late March. Less common in spring than fall.

Spring maxima. Four, Swan Point Cemetery, 28 April 1983 (RBo).

Fall. Migration in fall is rarely evident except for occasional individuals seen moving westward at places such as Napatree Point. Fall counts, rarely more than 5 per day, are nevertheless often the highest of the year, suggesting both recruitment of young and a probable influx of northern migrants.

Fall maxima. 18, Block Island, 25–26 September 1971 (RBo, WBD); 12, Block Island, 27–28 September 1976 (RAC, DLK, HWi); 10 along the South Shore, 5 October 1947 (DLK)

Winter. In shore areas and close to Narragansett Bay, they are now found throughout the year, with totals of up to 11 birds on the CBCs. While there are no midwinter records for locations more than 5 mi (8 km) from Narragansett Bay or the outer coast, they often visit ponds a short distance from salt water (e.g., the East Providence Reservoir) in January and February, depending on the availability of open water.

Winter maxima. 20, South Kingstown CBC, 2 January 2011; 16, Napatree CBC, 22 December 2013 (SSM, GW et al.); 9, South Kingstown CBC, 26 December 2012.

WOODPECKERS (PICIDAE)

Lewis's Woodpecker (*Melanerpes lewis*)

Status
Vagrant. One record.

Nonbreeding
There is 1 record of a bird shot by a boy at Mount Pleasant, near Providence, 16 November 1928, and retrieved by the superintendent of Roger Williams Park (Bryant 1929). Another individual was said to have been seen flying around in the vicinity the same day. The specimen, reported at the time to be the first for any location east of the Mississippi, was mounted and placed in the RWPM collection; it was extant in the mid-1950s and seen by several ornithologists. In 1971, however, a thorough search of the museum's bird vault and exhibits by the author failed to locate the specimen, and it is presumed lost. They have since been recorded in Ontario, Massachusetts, New York, and Virginia (Vierling et al. 2020).

Red-headed Woodpecker (*Melanerpes erythrocephalus*)

Status
Rare but regular visitor, most numerous in fall. There are three breeding records.

Historical
This is one of the few Carolinian species whose range has not expanded north in recent decades. A general decline of Red-headed Woodpeckers after 1900 is roughly coincident with the introduction of European Starlings (which usurp nesting cavities) and the automobile (which kills low-flying birds crossing roads). They were formerly much more common in southwestern New England than at present (breeding commonly in the Connecticut River valley of Massachusetts and occasionally found there in flocks [see Bagg and Eliot 1937; Griscom and Snyder 1955; Veit and Petersen 1993]). There were many Rhode Island records and specimens known to Howe and Sturtevant (1899), suggesting that before 1900 Red-headed Woodpeckers were at least as common and likely more common here than at any time since. Since about 1980, there have been up to 10 records per year.

Breeding
There are 2 old instances of breeding in Rhode Island. A nest and chick were found in a small oak in Warwick 28 July 1882 (in Howe and Sturtevant 1899), and a nest with 4 eggs was found in a dead elm in "Providence County," 28 May 1902 (F. E. Newbury, in Remington and Commissioners 1908). The only breeding in modern times was for Cumberland in 1978 (CWo et al.; photographed). There have been

several other reports of adults present during the summer that may well have been breeding: 2 adults spent the summer of 1964 on Poppasquash Point, Bristol (Freeman, fide RLF); another pair was suspected of breeding in Barrington in April and May 1958 (RBo); another pair was thought to be nesting in Warwick in 1967 (DWa); 2 were seen in Smithfield 12 June 1975 (BMi); and 2, Barrington 9 June 2016 (fide RFa).

Nonbreeding

Red-headed Woodpeckers are rare migrants in May and again in September and October. At these seasons they are most often encountered along the outer coast or near Narragansett Bay, but they are also reported farther inland. They are recorded about twice as often in fall as in spring. There are numerous records for the winter months between November and April, including several birds that have remained all winter at feeders or other favorable sites with ample acorns. Most records of migrants are from Block Island, where as many as 3 per day have been seen in both spring and fall.

Maxima. Four immatures, Greenville, 12 November 1967 (DWa); 4, South Kingstown CBC, 27 December 1998 (DLK et al.); 3, Narragansett, 26 December 1982 (m.obs.); 3, Slocum, December 1998 through January 1999 (m.obs.).

Red-bellied Woodpecker (*Melanerpes carolinus*)

Status

Common resident and migrant; greatly increased since about 1965.

Historical

Two old specimens of Red-bellied Woodpeckers were known to Howe and Sturtevant (1899) as having been taken some years before 1880, but without other data (formerly in BSNH and BMS, they have apparently disappeared; C. Kirdahy, BMS, pers. comm.). Otherwise, Red-bellied Woodpeckers were unrecorded in Rhode Island until the 1950s, when they began a substantial northward expansion of their range.

The increase of Red-bellied Woodpeckers noted in New Jersey in the 1940s and southern New York after 1955 began to reach Rhode Island in 1959 and became a full-blown invasion during the next 5 to 6 years. The first Rhode Island individual was at a feeder in Jamestown 14 October 1959, where it remained through the winter and spring to 7 May (Mrs. W. Miner; photographed by RBo). This well-studied bird was soon followed by other individuals: 11 November 1960 at Ocean Drive, Newport (CWo et al.); 5 December 1960 on a wire at Third Beach, Middletown (REW); 23 April 1962 at a feeder in Westerly (Pendleton); 1 May 1962 at Middletown (REW); 30 November–13 December 1962 at a feeder in Lincoln (Bokelund); 9 April 1963 at Johnston (M. Gardler); one banded at Block Island 18 December

1964 (MSl); 29 April–6 May 1969 at Westerly (ESa); and 6 May 1970 at Kimball Sanctuary (E. Southwick). One attended a feeder in Tiverton from the fall of 1969 to early spring 1971 or later (DLE, REm et al.).

Breeding

The preferred habitat for breeding in Rhode Island seems to be mature oak forest. A breeding pair of Red-bellied Woodpeckers discovered during 1977 in Adamsdale (North Attleboro), Massachusetts, was just 100 yd (91 m) from the Rhode Island state line, and the pair's foraging territory was partly in Cumberland, Rhode Island (R. A. Forster et al.). Birds appeared after 1982 at two or more locations in Saunderstown, including a pair visiting a feeder year-round (Gardiner). Individuals were seen after 1974 many times in Swan Point Cemetery in May, including pairs in 1979 and 1981. One was carrying food there 18 June 1983 (DFi). Breeding was established 27–28 June 1985, when DLK saw a pair with 2 fledged young at the Hempstead feeder near the Exeter/Kingston town line. Adults with 1 young were at the same feeder in 1986; 2 pairs with 2 young were there in 1988, and 2 pairs with 3 young in 1989. In 1990, 2 pairs and 6 young were seen at the same location. Other records of birds probably breeding include a pair in Coventry through March and April 1984; 2 at Watchaug Pond, 1–13 May 1986, and pairs there again in 1990 and 1991 (CHa). Pairs were present in Cumberland after 1986. In 1988, at least 1 pair was present at the Gilbert Stuart birthplace in North Kingstown, and 2 pairs were resident in Exeter. In 1989–1990, 2 pairs were in Kingston including 1 that produced 2 young on DLK's farm. A pair was in Narragansett 25 June 1991 (fide REn). A pair with 1 young attended a feeder at Wallum Lake through June and July 1992 (SMi). Others continued to be seen in northern Tiverton, including 3 pairs there 23–25 June 1991 (RLF).

Nonbreeding

Migrants occasionally appear at Block Island and elsewhere along the coast, mainly in May and October.

Maxima. 62, Newport–Westport CBC, 18 December 2010 (REm et al.); 60, South Kingstown CBC, 2 January 2011 (STs); 60, South Kingstown CBC, 27 December 2014 (STs et al.); 17, South Kingstown CBC, 2 January 1993.

Yellow-bellied Sapsucker (*Sphyrapicus varius*)

Status

Uncommon spring and common fall transient. One recent nesting record. Occasional but increasing in winter.

Breeding
A pair of Yellow-bellied Sapsuckers was observed courting and later feeding a chick in Johnston during April through June of 2008 (JMu). This is the only nesting record, despite their breeding fairly commonly in western Massachusetts (Veit and Petersen 1993; Kamm et al. 2013).

Nonbreeding
Spring. Uncommon in southeastern sections and more regular west and north of Narragansett Bay, even though they are occasionally numerous at Block Island. Peak numbers of transients occur in late April (to 6 per day at Swan Point Cemetery) and the first week of May. Late-spring records include 1 at Kimball Sanctuary 24 May 1938 (MSo), 1 at Swan Point Cemetery 24 May 1983 (CRa), and another there 28 May 1977 (REn).

Spring maxima. 30, Block Island, 15 April 1961 (MSl).

Fall. The earliest record in fall is for 1 banded at Block Island 12 September 1959 (G. Cant). Peak numbers appear in late September and early October.

Fall maxima. 75, Block Island, 4–6 October 1968 (ASRI).

Winter. There are now more than 60 records for the period December through March, primarily from locations near the coast, north to Warwick, Providence, and even Woonsocket (16 December 1990; LS). The following records are of individuals that survived the winter in Rhode Island: 1 from October 1973 through the last week of February 1974 at Jamestown (MDa); 1 from November through 14 March 1959 at Warwick (KKi); 1 at Westerly 9 January–13 March 1969 and also seen at the same location from December through February the following year (ESa); 1 remained at a Charlestown feeder from 3 January 1971 through the rest of the winter (CWo, FPe, ESa et al.); 1 all winter and through March 1987 in Warwick (KKi); 1 at Swan Point Cemetery, 10 January–23 March 1971 (LBe, SSD, BRi). Eight were found statewide in December 1989. There are at least 9 additional March records.

Winter maxima. 12, Napatree CBC, 22 December 2013 (SSM, GW et al.); 11, Newport–Westport CBC, 15 December 2012 (REm et al.); 11, South Kingstown CBC, 21 December 2013 (STs et al.).

American Three-toed Woodpecker (*Picoides dorsalis*)

Status
Vagrant. Three records.

Nonbreeding
An American Three-toed Woodpecker was seen at Block Island by Edward and Mabel Dodge 21 November–3 December 1943. Presumably, the same bird was seen

again by EDi near her home at Dickens Point 16 December 1943 and then rediscovered by the Dodges 15 March 1944. These observations were not accepted by HSH, who omitted them from his *Bird Notes*, and no descriptions survive. There is another unconfirmed report of an American Three-toed Woodpecker from Barrington for 1946. Despite somewhat unsatisfactory documentation of these records, the occurrence of 3 birds in Massachusetts during the period 1940–1942 (including 2 in Westport, just across the border from Rhode Island) suggests an irruption of these notoriously irruptive woodpeckers at that time and lends credence to the Rhode Island records. A cooperative and unambiguous female appeared in Burlingame State Park, Charlestown, where it remained for months to feed on dying red pines. First discovered by Texas birder Richard Armstrong 6 January 1991, it eluded local observers until February, after which it was thoroughly examined and photographed 8 February–15 April 1991 (DVa, JO, DLK et al.; see Photo 23).

Black-backed Woodpecker (*Picoides arcticus*)

Status

Very rare visitor, subject to episodic invasions, with such invasions sometimes protracted over a period of several years.

Nonbreeding

The first record of a three-toed woodpecker, not certainly this species, was a bird shot by a boy in Barrington 25 December 1905 and purchased by Angell and Cash, present whereabouts unknown (fide HSH). Another from the turn-of-the-twentieth-century era, date unknown, was collected in North Smithfield and acquired by CED (FlMNH #89888; recently re-examined and determined to be *P. arcticus*). Another *P. arcticus* was seen in a grove of pitch pines on Potowomut Neck, Warwick, 22 February 1911, where HSH fired twice at close range but failed to secure the specimen.

According to Forbush (1927), another flight to southern New England occurred in 1923–1924 when "some entered Rhode Island and Connecticut." Veit and Petersen (1993) say that about 31 were reported in Massachusetts during the 1923–1924 flight, but no Rhode Island reports were known to HSH or his contemporaries.

During southern New England's largest and most extended invasion (1956–1957, continuing in varying numbers to 1966), there were 4 Rhode Island records: 1 at Bristol 8 December 1956 (G. Low), a female at Perryville 20 January 1957 (Dr. W. Gee), 1 in Coventry 30 October 1957 (Mrs. R. E. Snow, fide SSD), and 1 at Carolina 1 January 1959 (CWo). Massachusetts recorded a minimum of 124 birds during this long invasion (Bull 1964; Griscom and Snyder 1955; Veit and Petersen 1993).

Veit and Petersen (1993) summarized 13 Massachusetts reports in 1974–1975, but none appeared in Rhode Island at that time. Despite an increase in observers, more than 55 years have now passed without further records. Van Tyne (1926) and others have connected the multiyear incursions of this species with Spruce Budworm (*Choristoneura fumiferana*) and subsequent bark-beetle infestations in northern forests that in turn fuel southward movements by a greatly enhanced output of young birds.

Downy Woodpecker (*Dryobates pubescens*)

Status

Common and ubiquitous resident except on treeless portions of islands and in the most heavily urbanized areas. Irruptive transient in highly variable numbers from year to year.

Breeding

Downy Woodpeckers are many times more numerous as breeders than are Hairy Woodpeckers in most parts of the state. They are also several times more numerous than the latter in fall and winter, and this is particularly true in coastal areas due to their greater acceptance of secondary growth habitats.

Nonbreeding

While still rare in summer at Block Island, they are regular transients and winter residents there.

Fall. The earliest fall transient at Block Island was 11 August 1959 (EDi). One was seen about 100 miles south of Block Island 13 July 1997 (PAB, RRV).

Fall maxima. 30, Block Island, 5–7 October 1973 (ASRI); 25, Block Island, 25 November 1960 (JBa).

Winter maxima. 154, South Kingstown CBC, 27 December 2014 (STs et al.); 132, Newport–Westport CBC, 14 December 2014 (REm et al.); 99, Newport–Westport CBC, 15 December 2012 (REm et al.).

Hairy Woodpecker (*Dryobates villosus*)

Status

Common resident in more forested areas, scarce or absent as breeder from the coast. Irruptive transient in highly variable numbers from year to year.

Breeding

During the late-nineteenth century, Hairy Woodpeckers were relatively rare, and as late as 1899 only 2 sets of eggs and 2 nests with young were known to HSH and other early oologists. In the Cranston–Warwick area, HSH considered them to be uncommon winter residents. With regrowth of the Rhode Island woodlands, Hairy Woodpeckers are now more widespread but rather uncommon. They are rare and local breeders on Conanicut and Prudence Islands (possibly 1 pair each on these islands in the 1980s; RLF; REn, CRa) and are unreported on Aquidneck Island. Lacking in the lowland portions of Tiverton and Little Compton near the Sakonnet Passage, they breed sparingly in the higher eastern parts of those towns. They are absent from Block Island. Within the interior, they prefer rich or mesic woodlands. They are therefore more common in Lincoln and Cumberland.

Nonbreeding

Hairy Woodpeckers seem to become more abundant during the winter in some years, perhaps owing to influxes from outside the state. They are moderately common winter residents nearly throughout the state, including the more forested coastal regions, but are uncommon at that season in brushier districts near the coast. They are rare in winter at Block Island, but may appear there annually (MWh, SRD, RBo et al.). On CBCs between 1962 and 1990, the ratio of Hairy to Downy Woodpeckers was about 14% on Block Island and 15 % on the mainland. Transient Hairy Woodpeckers are fairly regular at Block Island during fall with increased abundance in some years, suggesting occasionally modest incursions; e.g., 25 there 11–13 October 1957 (ASRI). Migrants or wanderers have been noted at Block Island on dates ranging from 12 August 1971 (banded; ELa) to 3 May 1961 (EDi).

Maxima. Notable flight in fall and winter of 2014–2015: 40, South Kingstown CBC, 27 December 2014 (STs et al.); 21, Newport–Westport CBC, 14 December 2014 (REm et al.); 3, Block Island CBC, 18 December 2014 (SCo et al.).

Northern Flicker (*Colaptes auratus*)

Status

Common summer resident throughout the state except at Block Island and very common transient; declined since 1990s. Fairly common winter resident, more numerous near the coast. Has declined considerably since 1950s and 1960s.

Breeding

Northern Flickers breed throughout the state, wherever woodlands are sufficiently mature to provide natural cavities for nesting. There is some recent suggestion that flickers have declined, perhaps because of fewer nest cavities being available as mature forests are replaced by secondary growth.

Nonbreeding

Spring. Under mild conditions in spring, a few migrants reach inland locations in late February, doubtless from nearby wintering areas. Otherwise, most birds appear in March, and peak numbers occur in early to mid-April.

Spring maxima. 60, Block Island, 17 April 1960 (MSl); 38, Block Island, 19 April 1959 (MSl); and a notable flight the same day at Sakonnet (SSD).

Fall. In fall, individuals may begin wandering in August; e.g., 1 at Cox Ledge 13 August 1966 (Trimble). Peak numbers occur at the coast in late September and early October.

Fall maxima. 400 seen and 3000 estimated at Block Island, 25 September 1954 (RCC, DLK); 1000, Block Island, 30 September–2 October 1994 (ASRI); 642 leaving Block Island from Sandy Point, 13 October 1958 (Ian Nisbet, RLF). One was seen at Cox Ledge 2 October 2013 (m.obs.).

Winter. Owing probably to the abundance of open country that once existed in coastal districts in the nineteenth century, this rather terrestrial woodpecker has been a common winter resident in these mild (if raw) and usually snowless areas throughout the historical period. Historical data suggests that flickers can survive the winter in the most exposed and treeless sections of the coastal uplands.

Winter maxima. 186, Block Island CBC, 18 December 1995 (CRa et al.); 164, Block Island CBC, 19 December 1989; 153, Newport–Westport CBC, December 1952. Farther inland, they are now rare but were formerly more frequent (especially in western Rhode Island) during the agricultural era; e.g., regularly to 5 per winter day in Cranston in the 1880s (HSH). While wintering birds are now very scarce in northern Rhode Island, they have been seen in winter in recent decades north to Lonsdale and rarely to Woonsocket. Two wintered at Wallum Lake in 1983–1984 (SMi).

Pileated Woodpecker (*Dryocopus pileatus*)

Status

Uncommon resident.

Historical

The first observation in Rhode Island came from a most unlikely place, treeless Dickens Point at the southwest corner of Block Island, where EDi saw one on a telephone pole 22 September 1925. Her records show that its appearance coincided with a 10 °F [–12 °C] drop in temperature. The next Pileated Woodpecker was found 14 June 1933 near Long Pond, Hopkinton, where HSH investigated a report and found recent tree workings but no birds. Roland C. Clement and other members of the RIOC endeavored to confirm the breeding of this species in parts of the northwest corner without success but managed to locate two individuals in westernmost Glocester during April and May 1949. Pileated Woodpeckers are mainly found in the western third of Rhode Island from Arcadia northward.

Breeding

Pileated Woodpeckers are permanent residents in parts of western Kent and Providence Counties, northwesternmost Washington County, and the northern and western fringes of the greater Providence metropolitan area. They were rare east of Narragansett Bay until recently, when 1 was reported in December 1981 at a feeder southeast of Pachet Brook Reservoir, Little Compton (fide LOG), and another appeared in eastern Portsmouth, 20–28 January 1982. Excavations in these areas were confirmed by RBo and DLK. Pileated Woodpeckers have been found at one location along Elbow Rock Road, Glocester. In 1965, an injured young bird barely able to fly was discovered there by a local resident. An adult was also seen there 29 June 1972 (RLC, MDa, KKi, FPe), and the author was shown a recently constructed but abandoned nest hole in June 1973. A male was noted drumming 1 mi (1.6 km) north of this site 4 June 1989 (REn). Elsewhere on Snake Hill Road, southern Glocester, a nest hole was discovered; a female or immature briefly protruded its head out of the opening 22 June 1983 (REn, CRa). There is another record of a bird seen in Glocester 28 April 1966 (RLC). One bird was seen at Buck Hill, Burrillville, 3 June 1963 (RLC), and 2 were calling and drumming in Pulaski Park, Burrillville, 7 June 1993 (REn). One was seen at Wallum Lake 27 May 1979 and again 1 July 1979 (CWo). A few other observations, mostly in spring, suggest the possibility of breeding at locations south of the northwest corner of the state, although some may represent wanderers; 2 "probable" breeders were recorded in Cumberland in 2016 (BCG). Singles were seen near a pond in the Arcadia Management Area, Exeter, 25 April 1973 (KKi), 11 May 1978 (KKi, FPe), 1 May 1979 (RAC), and 18 June 1992 (BCG). One was seen at Wyoming 17 May 1977 (DWa). Others have been seen at Hopkinton 29 April 1980 (fide CWo) and 21 June 1991 (DRi, HR), Carolina 9 May 1981 (FPe, KKi et al.), Kingston 1 May 1977 (STs, JC), and Charlestown 12 April 1983 (v.o.).

Nonbreeding

Fall. One was seen at treeless Dickens Point at the southwest corner of Block Island on 22 September 1925 (see above). Pileated Woodpeckers do occasionally disperse large distances, and there are 1 or 2 records from Nantucket in the 1950s (Veit and Petersen 1993).

Winter. Since the early 1940s, there have been winter reports of characteristic oval excavations from Westerly, Charlestown, Hopkinton, East Providence, Wordens Pond, Great Swamp, Coventry, eastern Exeter, Warwick, Cranston, Pawtucket, Saylesville, Cumberland, and Lincoln. Average tree sizes over large areas of western Rhode Island seem adequate and comparable to those in many parts of Massachusetts where Pileated Woodpeckers are common, but the hard, dry, and slow-rotting oaks may prevent them from finding adequate food or nest sites.

FALCONS (FALCONIDAE)

American Kestrel (*Falco sparverius*)

Status
Rare breeder and winter resident; uncommon migrant.

Historical
American Kestrels were rare summer residents in the nineteenth century (Howe and Sturtevant 1899; Forbush 1927). Mearns saw only three birds in 3 years of observation at Newport, 1900–1902. Arnold saw his first one on 9 October 1912 after 17 years of daily observations at Woonsocket and his next one on 14 December 1921, when he commented at length on its rarity (Arnold, personal diary). Doe failed to collect any eggs during the period 1882–1908 but then obtained more than 19 sets of eggs and specimens of nine nestlings in the vicinity of Providence during the period 1909–1923. Hathaway (1913) stated that there had been a decided increase in the previous 10 years and noted several nests and two winter occurrences. Prior to this, the only known winter records were for one collected in 1882–1883 (Southwick and Jencks) and another 25 January 1896 in West Warwick (Southwick and Jencks).

Breeding
American Kestrels are now uncommon to rare as breeders in open country. They nest in natural cavities, nest boxes, and in the eaves of buildings and factories in urban areas. Surveys in the 1980s showed that they are more common in open habitats in the eastern half of Rhode Island and in coastal districts but rare or absent in the heavily wooded western and northwestern sections. Some of the higher densities were in the Blackstone Valley and in the Providence–Pawtucket metropolitan area, especially near abandoned textile mills. But with successional growth of the state's remaining grasslands, there is little suitable habitat left, and the decline of American Kestrels reflects this.

Nonbreeding
Spring. In spring, residents often appear in late February and March, earlier than migrants. Peak numbers of migrants occur in mid-April.

Fall. In fall, a few migrants have been noted moving along the outer coast in early August, and the sprinkling of August transients gradually becomes more substantial by the first week in September.

Fall maxima. 1287 total for the state, 28 September 1975; 564, Watch Hill, 8 October 1979 (RLF); 394 in 4.5 hours, Napatree Point, 30 September 1994 (CRa); 269, Charlestown Beach, 3 October 1977 (Slack); 144 in 6 hours, Napatree Point, 28 September 2002 (CRa). Kestrels have declined in abundance as migrants; there has been no daily total over 100 since 2002 and none over 30 since 2009. Unlike Merlins and

Peregrine Falcons, substantial numbers of kestrels do not appear in fall at Block Island, where counts have never exceeded 20 per day.

Winter. Kestrels are most numerous in open coastal regions (including Block Island).

Winter maxima. 68 on the two mainland CBCs. Occasionally survives the winter at inland sites. Since 2005, the highest CBC total is 8, Newport–Westport, 3 January 2009.

Merlin (*Falco columbarius*)

Status

Common in fall; uncommon spring migrant and winter resident. Rare inland. Recent nesting records on Nantucket and Martha's Vineyard, combined with increased winter abundance, suggest that nesting in Rhode Island may occur soon.

Nonbreeding

Fall. Unlike American Kestrels, Merlins are commonest in fall at Block Island, where numbers peak in late September and October.

Fall maxima. 50, Block Island, 1–3 October 1990 (RBo et al.); 42, Quonochontaug, 28 September 1975 (NCl); 39, Watch Hill, 8 October 1979 (RLF).

Winter. Since 1950, Merlins have been regular winter residents in coastal Rhode Island and now appear annually on CBCs: 9, Newport–Westport CBC, 14 December 2014; 7, South Kingstown, 2 January 2011. There are now many reports for the months of January through March, suggesting ample survival through the winter. Most winter records have come from near the coast or lower Narragansett Bay, although they have occasionally been reported north to Barrington and Pawtucket, and inland observations have accumulated from Carolina, Fiskeville, North Providence, Lonsdale, and Woonsocket.

Spring. In spring, Merlins are rare. Dates range from 5 April 1980 (a migrant passing off Point Judith; RLF) to 23 May 1965 at Galilee (HWi) and 26 May 1951 at Quicksand Pond (RBo). One was seen at Block Island 20 June 2000 (REn); recent nesting records on Martha's Vineyard and Nantucket (Kamm et al. 2013) suggest likely future nesting at Block Island.

Inland maxima. Four, Lonsdale, 21 September 1981 (REn). There are 5 April records for East Providence, Providence, Pawtucket, Cumberland, and Wallum Lake.

Gyrfalcon (*Falco rusticolus*)

Status
Rare fall and winter visitor.

Nonbreeding
Howe and Sturtevant (1899) summarized earlier records as follows: a gray bird shot in the winter of 1864–1865 near Providence by Newton Dexter was said to be in the MCZ but now lost. A black bird was shot at Point Judith 11 November 1883 (formerly BMS #9806, now apparently lost; Jencks, 1884a). Another black bird was shot on Conanicut Island 22 November 1891 (Taylor) and is now on display at NBS; a second bird was reportedly seen at the same time. Still another black bird was shot at Newport 28 October 1896 and is now in the RWPM collection, recently examined by the author. A fourth black bird was shot in Tiverton 26 December 1896 by A. Scudder, for years in the collection of A.C. Bent (MCZ #252710; recently examined). Another record known to HSH was of a black bird shot at Matunuck 28 January 1915 by E. Parker.

On 16 October 1950, a dark bird was seen at Block Island (RCC). Another dark bird was seen resting on the cliffs from a boat off the southwest side of West Island, Sakonnet, 23 December 1950 (DLE, DLK). A gray bird was seen flying over NBS 11 November 1958 (JBa). A black bird was seen over Clay Head, Block Island, 11 October 1959 (JBa, SSD, RLF). A gray bird appeared at Block Island 5 October 1963 (HWi, RMa, O. S. Pettingill). A dark bird was found sitting on the ice at Long Pond, Sakonnet, 16 October 1966 (RBo, RKl, HMP). A dark bird was reported at Block Island 5 October 1979 (GRa et al.). On 17–18 December 1983, observers were able to contrast a gray-morph bird with a Peregrine Falcon off Sakonnet Point (RAC, GLB, DLK). This individual remained in the vicinity of the Sakonnet islands all that winter, where it reportedly fed mainly on Herring Gulls; it was seen pursuing a Peregrine there 22 February 1984 (RBo) and was reportedly still present 10 April. A gray bird was seen at Dickens Point, Block Island, 8 October 1985 (DLE, DLK); it disappeared toward Long Island. A gray bird was present at Sakonnet 20 December 1986 (RAC), and probably the same gray-morph bird was seen over West Island 10 January 1987 (DLE). A series of observations of very dark birds in 1993, thought to involve the same individual, were at several widely separated locations: Block Island 1 October (RBo, DLK); Quicksand Pond 30 October (BCG, GGa); Fresh Pond, Quonochontaug, 7 November (DLK); and Sakonnet Point 25 November (JO). A gray bird was seen in Charlestown 5 November 1995 (STs). Two birds (1 dark, 1 white) were observed on Block Island, 25 October 1996 (ELa, HLa).

Extreme dates. 5 October–10 April. A large falcon that captured an American Avocet at Napatree Point 22 August 1994 (CRa) may have been a falconer's Gyrfalcon × Saker hybrid.

Peregrine Falcon (*Falco peregrinus*)

Status

About 10 nesting pairs as of 2015; fairly common fall and uncommon spring migrant; uncommon winter resident.

Breeding

Peregrine Falcons had been wintering in downtown Providence from 1933 or so, roosting on the Industrial Trust Building and feeding on pigeons. They were present year-round from 1947 on (WBD) and suspected of nesting. In 1951, WBD talked to a person in charge of maintenance at the Industrial Trust Building, who said that he had taken 2 young hawks from a nest high on the building. The adults were not seen thereafter in summer, but at least 1 bird wintered regularly in the downtown area until 1957. It seems likely that Peregrines did in fact colonize downtown Providence "on their own" in the late 1940s or early 1950s.

At the time of the Peregrine's endangered listing, the eastern subspecies was actually extinct and the migratory Arctic population was severely reduced. By the 1980s, there were still a few Arctic migrants in Rhode Island; these appeared in a regular pattern during fall (October) and occasionally in spring (March or April). The re-establishment of nesting Peregrine Falcons in Rhode Island followed the typical track for a colonizing species. Even though there were no releases in Rhode Island, birds produced by successful programs in other northeastern states dispersed into the state. By the 1980s, immature birds were being seen in Rhode Island at unusual dates, that is, outside of the normal migratory window. By 1990, adult birds were consistently seen perched on large structures, especially the bridges over Narragansett Bay. Initially, these observations occurred mostly during winter and did not attract undue attention because Arctic birds could also winter in Rhode Island. However, as adults began to establish a more consistent presence on the bridges and lingered into the nesting season, nesting seemed imminent. In the mid-1990s, JZy thought it would be a good idea to put up nesting boxes at likely locations in the state. On his own initiative, he constructed and deployed 2 nesting boxes in Providence: 1 on the Industrial National Bank Building, and the other on a warehouse in the Port of Providence. The bank box was almost immediately occupied by Peregrines, and nesting has continued at that site nearly continuously since that time.

After this success, attention then turned to the bridges, where the Rhode Island Turnpike and Bridge Authority and the Rhode Island Department of Transportation were hugely supportive in monitoring birds and watching for nesting activity. Eventually, nesting was confirmed on the Newport (Pell) Bridge in about 2001 and the Mount Hope Bridge in 2004 (CRa, RLF). Nesting boxes were placed on these sites, they were used, the population continued to expand, and new birds continue to appear and establish territories elsewhere in Rhode Island. In 2011 there were 5 active Peregrine Falcon nests in Rhode Island.

Nonbreeding

Fall. Few people in North America and none in New England have seen more Peregrines than EDi, whose farm at Dickens Point on the southwest corner of Block Island provided Rhode Island's best location for observing them as they circled over the fields or soared along the nearby bluffs before departing westward toward Montauk Point. In more than 50 years of daily observations made in the process of carrying out her daily tasks, she recorded a total of 1057 birds in fall (see also Slack and Slack 1981).

Fall maxima. 44, Block Island, 4 October 2003 (RFa et al.); 25+, Block Island, 6–12 October 1959 (JBa, RLF); 20, Block Island, 3–5 October 1986 (RFa).

Winter. While Peregrines were always uncommon during winter in the past, there are a great many older reports for that season, especially near Narragansett Bay (Tiverton, Portsmouth) or along the coast. Undoubtedly, some of these winter birds were the same as 1 or more that roosted annually on buildings in downtown Fall River, especially in the late 1940s (RBo), or others that roosted on the Mount Hope Bridge during the early 1950s. In addition to scattered individuals seen along the outer coast in the early 1990s (e.g., Block Island and Sakonnet), occasional individuals have taken up residence on the old Jamestown Bridge and Sakonnet River Bridge between Portsmouth and Tiverton.

Spring. Peregrines are always rare in spring; dates range from 21 March and 3 April to 31 May. There were only 6 spring sight records during the period 1957–1991.

Prairie Falcon (*Falco mexicanus*)

Status

Vagrant. One record.

Nonbreeding

One was seen and photographed at Block Island 7–9 October 1989 by JDi, Ann Shaw, John Walsh, AFr et al. This bird showed no sign of having been captive, and the date is appropriate for a wild vagrant.

PARROTS (PSITTACIDAE)

Monk Parakeet (*Myiopsitta monachus*)

Status
Introduced. Resident in the Providence area.

Breeding
At the height of the initial spread of Monk Parakeets in the late 1960s and early 1970s, 1 appeared at Block Island 11 March 1973 (MWh) and 2 others were apparently seen elsewhere in Rhode Island. The Block Island record is of special interest as it suggests "wild" rather than human-assisted dispersal. At that time, there was a substantial breeding population in the New York City area, with other reported breeding outposts ranging north to New Hampshire and into the Midwest.

With concern for the potential destruction of crops with further spread of this South American temperate parrot, the USFWS conducted a "retrieval" program in the years 1973–1975. This took place in the species' principal centers of abundance and succeeded in effecting a substantial reduction in some areas such as New York and California (Neidermyer and Hickey 1977). Inasmuch as there were no known outposts in Rhode Island in the early 1970s, none of the eradication was carried out here or in eastern Massachusetts (fide CRa).

Monk Parakeets were reported in Rhode Island during the early 1970s, and by the early 1980s they were breeding in the Warwick Avenue area of Warwick. A total of 14 birds was present there in 1982. The Warwick birds were thought to be sustained by bird feeders, and numerous stick nests became very conspicuous in oak trees in the vicinity of Warwick Avenue. During the 1980s, this colony was the main or only one known in Rhode Island, and by 25 April 1989 had grown to 81 pairs (B. Cassie). By 2008, the colony had declined to about 15 birds.

Other individuals and small flocks were noted in Cumberland in 1980–1981 and Barrington in 1983. Another group of 30 birds and 5 nests was located on Bullocks Point, Riverside, 1 October 1989 (v.o.), and 35 pairs were estimated there 7 May 1993 (DLE et al.); there were 22 birds and 14 nests there 18 October 2002 (RFa), but the colony declined to 9 birds by 2006 (MLy). A small group of 3 birds and 1 nest was located in Lincoln 27 March 1993 (BCG), and 8 birds were present in 1995. Six birds and 2 nests were there 18 October 2002 (RFa et al.).

Another colony was in Bristol, where at least 2 nests were present in 1994 (DLE). Three wanderers were noted in Westerly in the fall of 1988 (CRa), and another was present on Block Island through June 1988 (REm, REn). Four were seen at Conimicut 7 May 2009 (PAB et al.), 3 at Sachuest Point 12 November 2000 (CaP), and 1 at India Point 15 July 2001 (MLy). There have been few reports of this species in recent years and only 2 reports since 2010.

TYRANT FLYCATCHERS (TYRANNIDAE)

Elaenia sp. (*Elaenia* sp.)

Status

Uncertain; a flycatcher seen and photographed at Moonstone Beach 10 November 2008 (N. Conway) was thought likely to be some species of elaenia, perhaps most likely either of the South American species White-crested (*E. albiceps*) or Small-billed (*E. parvirostris*) Elaenia (Howell et al. 2014). A White-crested Elaenia was found at Padre Island, Texas, 9–10 February 2008, and another elaenia, thought to be Small-billed, was seen in Cook County, Illinois, 17–22 April 2012. At least 4 species of elanias, including the 2 mentioned above, perform substantial northward migrations in South America during the austral autumn, and that behavior may enhance the likelihood that they appear in North America. The record is listed here in anticipation of potential (climate-related?) increase of vagrancy of South American species to North America.

Ash-throated Flycatcher (*Myiarchus cinerascens*)

Status

Vagrant. Now occurring annually, mainly during late fall.

Nonbreeding

An immature female, the first record for the Northeast, was collected at Block Island on 15 September 1960 by JBa (Baird 1962; see Murphy 1982). This individual was determined by Lanyon to be *M. c. cinerascens* of the Rocky Mountains and is now MCZ #261402.

There are a number of other sight records, including 1 seen at close range at Block Island's north end 25 September 1976 by Frazier, Niosi, Koebel, Vogel, and other members of the Urner Ornithological Club. Another was found near Watchemoket Cove, East Providence, 12–30 November 1979 by HWi and RBu, seen by numerous observers, and photographed by RBo. Another was photographed in Acoaxet near Quicksand Pond (about 300 yd [274 m] east of the Rhode Island line) 15 December 1979 (DLE, REm et al.). One was present on the Lewis-Dickens Farm, Block Island, 1–3 October 1993 (F. Gallo et al.; photos). One was present in Barrington 23–29 October 1995 (DVa, RFa, BCG, MJM, DFi et al.; photographed); 1 was in Middletown 28 October 2001 (HWi et al.). One was at Middletown, 4–12 December 2011 (P. Champlin); 1, Sachuest Point, 5–13 November 2014 (v.o.); 1, Matunuck, 12–13 November 2016 (STs, v.o.). Zawadzki et al. (2019) have shown how the occurrence of Ash-throated Flycatchers on the East Coast is driven by population growth within the core of the breeding range.

FIGURE 9. Migrant landbirds at the North End of Block Island, © Margaret La Farge.

Great Crested Flycatcher (*Myiarchus crinitus*)

Status

Common summer resident; common spring and fairly common fall transient.

Historical

Hathaway considered this woodland flycatcher to be very uncommon during the 1880s near Providence at a time when few woods of any size existed there or elsewhere. Even so, CED managed to collect 12 sets of eggs in Cranston, 1892–1911.

Breeding

Great Crested Flycatchers are common summer residents in the western parts of Rhode Island, decreasing to uncommon in some eastern parts of the state, particularly in Cumberland and Lincoln (only 5 in these towns during an intensive survey 15–23 June 1984; RLF) and most of the Providence metropolitan area. On the islands in Narragansett Bay, however, they are surprisingly numerous. They are sparsely distributed in the more open parts of Aquidneck Island (3 in West Aquidneck 15–16 June 1990; RLF) but apparently breed regularly in several places on Newport Neck and in greater numbers on Conanicut Island (25 there, 22–25 June 1985 [RLF]; 17 on 22 May 1988 [ChP et al.]), including several locations on the Beaver Neck and on Prudence Island (8, 6–8 June 1990; REn, CRa). A thriving population is present in the Tiverton highlands; e.g., 15 in Weetamoo Woods 13 June 1992 (REm). Smaller numbers inhabit wooded districts of eastern Little Compton. They do not breed on Block Island, but wanderers and migrants occur there in June, and a pair was noted courting in a north-end locust grove there 18 June 1981, suggesting that they may soon colonize the island. In general, data from the author's June surveys on the mainland suggest that there is a gradual increase from north to south within western Rhode Island. The greatest density is probably in deciduous forests of southern Rhode Island. Thirty-nine individuals were found from 59 point counts surveyed in Burlingame Management Area, Charlestown, in late May and June 2006 (CRa).

Nonbreeding

Spring. The earliest record in spring is for 1 at Green Hill on 22 April 1975 after a southerly storm (DWa). General arrival is in early May, and migration beyond late May is only detectable in such coastal locations as Block Island, where late migrants or wanderers are common into June; e.g., 1 banded 13 June 1972 (ELa).

Spring maxima. Seven, Miantonomi Park, 5 May 2012 (J. Sweeney).

Fall. In fall, this tropical winterer apparently makes a very early westward departure, seldom with any numbers appearing in coastal areas.

Fall maxima. Six, NBS, Middletown, 10 August 1967 (DWa). One was photographed at Little Compton 30–31 October 1999 (GDe), and 1 was banded at Block Island 5 November 1980 (ELa, HLa).

Extreme dates. 22 April 1975, Green Hill (DWa), to 16 November 1969, Sakonnet (SSD), and 16 November 1975, Matunuck (JC).

Western Kingbird (*Tyrannus verticalis*)

Status
Uncommon to rare but regular fall transient or vagrant.

Historical
The first Rhode Island Western Kingbird was found 24 September 1913 at Watch Hill (Griscom 1914b). None were then reported for many years, although EDi often saw "Crested Flycatchers" at Block Island on many unlikely dates from fall 1913 onward, during an era when the present species was not suspected. Western Kingbirds fairly burst into the Rhode Island records in 1939 with five state reports and then continued to be found more or less annually. A simultaneous increase was noted in Massachusetts (Griscom and Snyder 1955). Veit (2000) and Zawadzki et al. (2019) have shown how the occurrence of Western Kingbirds on the East Coast has fluctuated in accordance with population growth at the eastern limit of their breeding range.

Nonbreeding
Western Kingbirds are now more or less regular transients along the coast, variable in numbers from year to year and less frequent during 1980–2015 than in 1950–1980. Reports have ranged from 0–10 per year since 1957, except in 1960 when *21* were seen at Block Island (JBa). Reports are about equally distributed between the months of September, October, and November, and almost all are from the immediate coast. The earliest record in fall also constitutes one of the only inland occurrences: 19 August 1960 in East Providence (RTL). Other inland records include 23 August 1940, Barrington (HNG); 5 December 1984 at the Scituate Reservoir (fide RLF); and 10 November 1984 along the Blackstone River (REn).

There are remarkable midsummer records of singles at South Kingstown 11–12 July 1980 (DRu, DLK, KKi, FPe) and Napatree, 20 June 2014 (T. Auer; photo).

Extreme dates. 19 August 1960 (note year), East Providence (RTL), to 9 December 1961, Watch Hill (SSD, RMa), and 17 December 2006, Napatree Point (RLF).

Eastern Kingbird (*Tyrannus tyrannus*)

Status

Common summer resident, and common to abundant transient, especially in late summer–early fall.

Historical

For a passerine, this species is a notable diurnal migrant. They are very common in settled parts of northern New England and eastern Canada and often move through southern New England in sizable flocks, especially in fall. They often collect along the South Shore, where HSH would sometimes encounter numbers perched on wires along the old Post Road in the Charlestown–Westerly area. The fact that most of the best counts were made before 1935, however, and that these older records have not been seriously challenged in the last 50 years despite a geometric increase in the amount of observation in coastal areas, strongly suggests that a general decline has occurred in populations to the north that supply Rhode Island's coastal migrants.

Breeding

Eastern Kingbirds are one of Rhode Island's more conspicuous birds. They breed throughout the state (including Block Island), forage in open country, and build rather messy and easily identified nests high in tall trees. This is a widespread but somewhat thinly distributed species, with largest counts along shores of ponds and rivers; e.g., 10 territories in Great Swamp Management Area 5–11 June 2009 (CRa).

Nonbreeding

Spring. The earliest spring record is for 1 that appeared at the Great Swamp on the incredible date of 28 March 1986 after a period of southerly winds (P. Swanson). Others have been seen at Block Island 4 April 1920 (EDi) and 5 April 1973 (MWh). General arrival is during early May.

Spring maxima. 34, Conanicut Island, 20 May 1990 (ChP); 27, Napatree Point, 16 May 2005 (CRa); 25, Block Island, 12–14 May 1999 (CRa, REn); 22, Point Judith (where they invariably pass westward along the shore or southwestward over the ocean toward Long Island), 27 May 1984 (RLF).

Fall. Fall migration reaches a well-defined peak in the last week of August and first few days of September, at which time large aggregations of kingbirds have often been seen on telephone wires and dead branches.

Fall maxima. 134, Tiverton, 2 September 1940 (RCC). There are at least 9 counts of 75–100 between the narrow dates of 23 August–2 September. Late flights of 20 were reported from Block Island 21 September 1985 (HWi).

Extreme dates. 28 March 1986 (above) to 22 October 1977, Barrington (DFi); to 27 October 1989, Little Compton (KS, CS).

Scissor-tailed Flycatcher (*Tyrannus forficatus*)

Status
Vagrant. About 12 records.

Nonbreeding
Based on information from neighboring states, Scissor-tailed Flycatchers seem more likely to occur in spring than in fall, unlike most vagrants from western North America. Veit and Petersen (1993) list 25 spring records but only 8 for fall.

Spring. One was seen along Tower Hill Road, South Kingstown, 13–14 June 1969 (v.o., fide DLK, SSD). One was at Richmond 18–20 June 2003 (SRC, m.obs.), and (perhaps the same bird) 29–30 June at Wood River Junction (C. Thompson, REn et al.). One was at Jamestown 18–19 May 2005 (MJM et al.), and others at Block Island, 23 May 2005 (fide RFa), and Carter Preserve, Charlestown, 3 June 2014 (S. DelPozzo, DFi et al.).

Fall. An immature was present 16–19 October 1965 at Matunuck, discovered by RGe and JGe and later seen by numerous observers including DLK and CWo and photographed in color by SSD. An immature female remained at a farm in South Kingstown 3 August–2 September 1995 (PAB; DLK, RBo et al.), and shortly after, a young male appeared less than a mile away in South Kingstown 12–16 September 1995 (BCG, DLK, PAB). On 16 October 1999, PAB saw a third individual, this time a pink-sided adult (almost certainly a male by its extremely long tail), cross the road head high at the Narrow River heading south toward Narragansett.

Fork-tailed Flycatcher (*Tyrannus savana*)

Status
Vagrant. At least four records.

Nonbreeding
The first Rhode Island occurrence is of 1 simultaneously discovered by DLK, DLE, and RBo near Middle Pond, Block Island, 9 October 1985. Perched on an electric wire, it was photographed by RBo. The second record was of 1 at Central Beach, Quonochontaug, that came to the attention of the birding community 27 October 1994. Reported by residents to have been present as early as 20 October, it remained through 29 October and was seen and photographed by innumerable observers (JO, DLE, RBo, DLK, RGe, JGe, RLF et al.). A third bird was present at Block Island 9 May 2000 (Megan Whitman, SCo) for 1 day only. One was at Richmond 7–10 September 2003 (DFi, m.obs.). Many Fork-tailed Flycatchers in the Northeast occur during August–November, the season at which migratory *T. s. circumdatus* "should be" migrating south (Monroe and Barron 1980). It may be that both migratory seasons, spring and fall, are times when long-distance dispersal in any direction is likely to occur.

Olive-sided Flycatcher (*Contopus cooperi*)

Status

Uncommon transient.

Historical

Hathaway saw only two in 50 years of Rhode Island field experience, both of them in spring. Even with the great increase in observers in recent years, they have gone unrecorded in some years during recent decades. They are most frequently observed in Swan Point Cemetery, Providence, and are rare in the southeastern part of the state, as is true of most Canadian-zone land transients. There are, however, a number of records for Block Island. There is no evidence of past breeding in Rhode Island despite widespread breeding in Massachusetts, where the species bred around Boston and even in the Cape Cod pine barrens before 1900 (Griscom and Snyder 1955).

Nonbreeding

Spring. Most occur in the last week of May and first week of June; e.g., 1, Block Island, 10 June 1973 (RLF). One was seen migrating northeast over the ocean off Block Island 16 June 1987 (DLK, DLE, DFi).

Summer. There are several summer records, including individuals at Foster 25 June 1981 (JGe, RGe), Breakheart Pond 8 July 1979 (JGe, RGe), and Jamestown 22 July 1963 (MDa).

Fall. The majority of fall transient records are in late August and early September. The earliest of these include 1 at Kingston 6 August 1969 (DLK) and another in Exeter 13 August 1979 (MPe). There are a few later records for September and 4 for October.

Extreme dates. 5 May 2012, Swan Point Cemetery (JSJ), to 21–22 November 1953, Tiverton (Smith; RCC, RBo, DLE, WBD et al.). There were other November records during the 1950s in Massachusetts but none since.

Eastern Wood-Pewee (*Contopus virens*)

Status

Fairly common resident and migrant.

Breeding

Fairly common to common summer resident in more mature woods throughout most of the state but local in most parts of the coastal area and islands and absent from Block Island.

Nonbreeding

Spring. Eastern Wood-Pewees are normally one of the latest of spring land transients, with peak numbers being found in late May. Apart from the 16 April record cited below, the earliest arrival is for Hopkinton, 1 May 1983 (DWa), and there are 2 records for 4 May. Breeding birds first arrive on territory during the third week of May, while migrants peak in the last week of May and the first week of June. Transients are always conspicuous on unforested Block Island and are seen there through mid-June

Fall. In fall, transients are encountered most frequently in late August and September.

Extreme dates. 16 April 1961, Hopkinton (MDa, HDa), to 18 October 1995, 1 banded at Block Island (ELa, HLa); 22 October 1949, Sakonnet (1; DLK); 22 October 1966, Tuckertown (1; DLK); 22 October 2001, Block Island (1; STs, JSJ et al.); 25 October 1971, 1 calling at Sakonnet (REm, LBe); and 15 November 2014, 1 photographed, Charlestown Breachway (DFi, STa).

Yellow-bellied Flycatcher (*Empidonax flaviventris*)

Status

Uncommon spring and fairly common fall transient.

Nonbreeding

Spring. Yellow-bellied Flycatchers are remarkably uncommon in spring away from Swan Point Cemetery in Providence and are particularly rare in southern and southeastern sections of the state. Indeed, there are no spring records whatsoever for Newport County, and DLK never saw one during more than 60 years of spring coverage in South County. At Block Island, they have been banded only a few times in spring. Even at Swan Point Cemetery they are rare, with seasonal totals ranging from 1 to 3. The latest records of spring migrants are of singles banded 9 and 10 June 1972 at Block Island (ELa).

Spring maxima. Two, 16 May 1990, Swan Point Cemetery (DFi, RBo et al.); several other records of 2 birds per day.

Fall. Yellow-bellieds are more numerous in fall, particularly at Block Island and at some other coastal locations. The best data on this secretive species comes from the mist-netting of JBa, Dickerson, DLK, and ELa. The earliest fall dates are for 1 seen 15 August 1982 at Little Compton (RBo) and others banded at Kingston 16 August 1971 (DLK, RLF) and 16 August 1972 (DLK). There are 4 single-day banding totals of 3 birds, ranging from 23 August 1968 to 18 September 1970, all at Block Island (Dickerson, ELa).

Fall maxima. A total of 15 were banded during the fall of 1970 at Block Island by ELa.

Extreme dates. 2 May 2001, Coventry (DWa), to 10 October 1988, 1 banded at Kingston (DLK), and 14 October 1991, 1 banded at Block Island (ELa).

Acadian Flycatcher (*Empidonax virescens*)

Status

Uncommon summer resident and transient; increased since recolonization in about 1970.

Historical

This southern *Empidonax* has had a curiously checkered career as a summer resident in New England and elsewhere in the Northeast. They formerly bred in parts of Long Island and elsewhere in southern New York and southwestern Connecticut, and once in Massachusetts (Bull 1964, Griscom and Snyder 1955, Veit and Petersen 1993), but then disappeared from these areas after about 1900. During the 1960s, they recolonized some of this region and bred north to southeastern Connecticut, and by the mid-1970s they bred north to western Connecticut and Massachusetts. Interestingly, these northernmost breeders often use hemlock gorges, a habitat with interior and northern affinities that is quite distinct from southern floodplain forests.

Breeding

Before the 1970s, there was only 1 unconfirmed report of breeding in Rhode Island: in 1939, 2 pairs raised young in a wooded swamp with large red maples and sphagnum moss in Hull Swamp of north-central Conanicut Island (Clark 1939, and in. litt. to Griscom and RCC). A native of Philadelphia, Clark was previously familiar with the species. The RIOC was not convinced about this record, however, and members were unable to locate territorial birds in succeeding years. However, the species' later history in Rhode Island suggests that Clark's report was valid.

Beginning in about 1970, Acadian Flycatchers colonized Rhode Island, as documented below:

Parker Woodland, north-central Coventry. A pair was discovered along a brook among sizable woods 27 June 1976, and a nest found 2 days later in a witch-hazel directly over the stream (FPe, KKi, DLK). Three young fledged by 26 July, and birds were present at the site until at least 5 August. In 1977, a nest was found and 4–5 adults were present. (KKi, FPe, JGe, RGe). In 1978, a single bird was noted 19 May, 2 on 27 May, and 2 nests were found 30 June, with 4 singing males found in the general area (KKi, FPe, CWo). In 1979, a pair was on territory on the remarkably early date of 30 April (GRa), and 4 males and 2 nests were found 5 July (v.o.). In 1980, birds were present at 2 places and young seen (REn). As many as 7 pairs have nested at this locality since then.

Burlingame State Park. On the "Clawson Farm" north of Watchaug Pond, Charlestown, birds have been found in a damp area of richer woods near some rocky outcrops lacking any stream, as well as near a stream farther east. Two to 3 males have been present here since 1983 (DRu, DLK).

South Kingstown. Along the upper Pawcatuck River (the western outlet to Wordens Pond). This is a maple swamp environment with green ash and pin-oak. A pair was seen, and a nest was in a low tree over the stream examined by canoe 3 July 1984; another bird was seen singing elsewhere along the same stretch of water the same day (CRa, REn). Numerous records have come from this part of the Great Swamp more recently.

Stepstone Falls. West Greenwich. Seven birds present 12 June 1989 (CRa). Seven there 5 June 1990 (DLK, MJM).

Tiverton and Little Compton. An agitated pair was seen along the Sin and Flesh Brook in Tiverton, 11 June 1983 (REn), and 2–6 territorial birds have been resident in June in the Weetamoo Woods/Fort Barton woodland during the breeding season in most years since about 1997 (REm). Also, a female was feeding 3 young south of Watson Reservoir in Little Compton, early August 2005 (GDe), and a bird was photographed on a nest there in 2016 (GDe). Farther south in Wilbour Woods, a bird has been singing on territory almost every year since about 1997.

Since 1977, singing males have also been noted in the following locations: Westerly, 12 June 1977 (ASM, ESa); Wallum Lake, 28 June 1979 (REn); Alton Jones Campus, West Greenwich, 20 May 1982 (FPe, KKi); Pulaski State Park, Burrillville, 17–23 June 1982, with an old nest suspected to have been built by this species found over a rushing stream among hemlocks (RLF); Escoheag, 28 May 1984 (R. Noss); Foster, in a hemlock swamp, 29 June 1984 (REn, CRa); Carolina, 11 July 1984 (DRu, DLK).

All nesting birds were in wet areas among large trees away from the immediate coast, often along rocky streams among hemlocks or in red maple swamps. Special efforts to find this flycatcher by following watercourses on foot or by boat would probably yield many additional nesting pairs.

Nonbreeding

The first modern occurrences of Acadian Flycatchers in Rhode Island came in a rapid-fire sequence. These included single individuals banded at Block Island 12 September 1969, 25 September 1970, 21 May 1974, 28 May 1975, 21 and 23 May 1976, 20 and 21 May 1977, and 20 September 1981 (ELa et al.). A singing transient was found at Green Hill 12 May 1974 (DWa), and 1 was seen and heard in Swan Point Cemetery 23 May 1979 (RBu). *Twenty* individuals were found throughout the state 6 May–7 July 2012.

Extreme dates. 30 April 1979, Parker Woodland, 30 April 2009, and 30 April 2011 at Little Compton (GDe), to 25 September 1970, Block Island (see above), and 1 October 2003 at Little Compton (GDe).

Alder Flycatcher (*Empidonax alnorum*)

Status

Rare to uncommon migrant.

Nonbreeding

Until recently, the status of *E. alnorum* has been uncertain due to confusion with *E. traillii*.

Alder Flycatchers are common breeders from Alaska to Newfoundland and south to central New York, western and occasionally central Massachusetts, and northwestern Connecticut. Recent observations of singing Alder Flycatchers show that they are at least irregular visitors in summer and suggest that they could breed. Recent summer records of apparent migrants or wandering individuals singing the "fee-bee-o" song include 1 in the Wood River area 4 June 1976 (DWa); 1 at Burlingame 9 June 1992 (CRa), 1, Exeter, 23 June 1977 (RAC); 1 singing, Beavertail 27 June–13 July 1980 (REn, CRa); 1, Warwick 21–24 June 1982 (TAd, KKi, FPe); and 1 in a stretch of alders in the Sneech Brook area of Cumberland 15–16 June 1982 (CRa; RAC et al.). On 30 May 1995 at the Great Swamp, a male sang persistently in appropriate breeding habitat but was not observed again (SSM). A singing Alder was seen in North Smithfield, 16 June 1996 through July 1996 (BCG; m.obs.), then in the same location 24 May through July 1997 (BCG).

Migrant "Traill's Flycatchers" occur from mid-May to early June and from mid-August through mid-September, with a peak in early September.

Extreme date (Traill's). 2 October 1974, Block Island (ELa).

Willow Flycatcher (*Empidonax traillii*)

Status

Fairly common breeder and migrant. Has increased substantially since the 1960s.

Breeding

Willow Flycatchers, those with the "fitz-bew" song type (Stein 1963), are now far better known than *E. alnorum* in Rhode Island because they now breed fairly commonly. Since the mid-twentieth century, they have dramatically extended their range southward to the New Jersey coast and Long Island and eastward to Rhode Island and southeastern Massachusetts. Most recently, they have pushed northeastward to occupy the coasts of New Hampshire, southern Maine, and parts of eastern Canada. Birds singing "fitz-bew" have bred in the New York City area (north of the New Jersey and Long Island coastal plains), since at least 1888 (Bull 1964) and in several eastern Massachusetts locations since 1902–1909 (Griscom and Snyder 1955). Their invasion of the Rhode Island coast occurred more or less simultaneously with their appearance in the mid to late 1950s on the coastal plain of Long Island (Bull 1964).

Territorial birds were first noted by DLK 27 May 1950 along the Saugatucket River south of Wakefield and in 1951 along Moonstone Brook, South Kingstown. In the Moonstone area, 1 to 3 singing birds were present annually in the early 1950s, and 4 singing males and 2 nests (1 with eggs) were located in 1956 (DLK). In the same year a bird was present through June at Seapowet (RBo), and in 1957, 1 occupied a territory near Briggs Marsh (RCC). They appeared in Little Compton by 1963, with counts there and in the Moonstone area of 7 singing males each in the 1970s (DLE, REm, v.o.). Also, in the 1960s, they expanded their range, with additional territorial birds found at Watch Hill, Weekapaug, Misquamicut, Quonochontaug, Green Hill, near Point Judith, in the Ocean Drive area of Newport, and at Melville.

During the 1980s, the number of Rhode Island breeding locations nearly doubled from those of the 1970s, and the population at these sites increased as well; e.g., 13 singing from Ninigret to Jerusalem 15 June 1980 (RAC), 6 at Galilee 30 May 1982 (RAC), 8 in lower Middletown 14 June 1984 (CRa, REn), and 6 at Beavertail 23 June 1985 (RLF). They are rare or absent at Block Island, where a pair was at last found north of Great Salt Pond in 1982 and 1984 (RLF; RBo), and singing males were found 19 May 2012 and 24 June 2013 (CRa). Most of the early locations were near the coastal ponds along the outer coast, with few near Narragansett Bay and fewer at any distance from salt water. From initial Narragansett Bay nesting locations at Melville and Warren, they have spread widely in the maritime scrub along the western slope of Aquidneck Island (14 males in western Portsmouth and Middletown 15–16 June 1990; RLF) as well as to other locations close to the Bay. They breed in beaked willow–arrowwood habitat throughout the interior of the larger Narragansett Bay islands (16 in Middletown, 28 May 1990; CRa).

Along the coast the optimal habitat is extensive thickets of arrowwood, multiflora-rose, small black cherry, and other mesic coastal brush within a half mile of salt water, often close to coastal ponds that contain beaked willow, narrow-leaved cattail, or *Phragmites*. These habitats now contain the largest numbers of Willow Flycatchers and were the first Rhode Island habitats to be occupied during the 1950s. Subsequently, many arrowwood thickets in the drier coastal uplands began to attract territorial birds.

Now, inland concentrations occur in the Lonsdale marshes, the Great Swamp, and West Greenwich. At Lonsdale marshes, a single pair as early as 7 June 1974 (RLF) had gradually increased to 6 males by June 1984 (CRa, REn). At the Great Swamp, as many as 6 territories were occupied in the early 1990s, and multiple nests were found (SSM).

Nonbreeding

Spring. General arrival is in mid to late May, and transient "Traill's" occur into mid-June.

Fall. Territorial birds are present throughout June and July but most cease singing in August and quietly depart, with only occasional birds vocalizing into early September. Peak numbers of "Traill's" Flycatchers occur in late August and early September; perhaps most of these are *E. alnorum*.

Extreme dates ("Traill's"). 11 May 2002, Jamestown (CPe et al.), and 14 May 1982, Jerusalem, to 2 October 1974 (see Alder Flycatcher account), and 1 banded, 15 October 1995, Block Island (ELa, HLa). Rare before mid-May and after mid-September.

Least Flycatcher (*Empidonax minimus*)

Status

Uncommon and local summer resident and fairly common transient.

Historical

Least Flycatchers were formerly much more widespread breeders in shade trees and orchards throughout Rhode Island. They were common breeders up to the 1930s and 1940s in the outskirts of Providence and throughout the state's farmlands. Hathaway's egg collection contains sets for 1896–1897 from East Providence, Westerly, and two from Quonochontaug. Charles E. Doe's collection contains 10 clutches from Cranston as well as others from Bristol and Wickford. Sir Charles Blagden (1903), writing during the American Revolution in Newport, described this species as "common in gardens." Sturtevant wrote, "These birds are common in the spring in the city of Newport [undoubtedly indicating breeding] but I have never seen one in the adjoining countryside," where the landscape was then nearly treeless. They were probably absent from much of the similarly barren parts of Newport County except perhaps in the villages of Little Compton and at Tiverton Four Corners, where they persisted into the early 1950s (RBo). They bred annually at Kimball Sanctuary on the Charlestown moraine at least until 1932 (M. Southwick). There are eggs from Peace Dale for 1883 (Hazard). They bred in the village of Kingston until 1958 and remained at West Kingston until 1971 (DLK). As many as 10 or more territorial birds were present at Jamestown 29 May 1948 (DLK), and they remained at Bonnet Shores to 1959 (DLK).

The decline of Least Flycatchers was apparent to HSH by the 1930s and 1940s. However, Least Flycatchers were still common in some areas during the late 1940s; e.g., 29 in northern Rhode Island 8 June 1946 (RCC, DLK, WBD), so the steepest decline seems to have occurred in the late 1940s and 1950s. A slight increase from a nadir during the 1970s to the twenty-first century is apparent. The causes of the decline are not entirely clear, as there seems to be plenty of unoccupied habitat available. Factors that may have been important include the loss of pear and apple orchards (from which many egg sets were obtained) in the early twentieth century, pesticides, and competition from House Sparrows.

Breeding

Least Flycatchers are now mainly confined to the western halves of the western counties from the Arcadia area northward but also eastward into northeastern Rhode Island, especially Cumberland. Recent counts include 7, Breakheart Pond, 21 May 1988 (CRa); 10, Arcadia throughout June 1990 (CRa); 6, northern Rhode Island, 23 June 1976 (KKi, FPe); 4, Foster, 2 June 1983 (v.o.). During the 1980s, there were 2–4 pairs breeding in Little Compton (e.g., 4 singing, 13 June 1987; RBo), but they have been absent as breeders from Little Compton since about 1992 (REm). On the western mainland, there have been few recent indications of breeding south of Arcadia. Examples include males singing persistently at Burlingame Reservation through June and July 1982–1984 (CRa) and at the Great Swamp on 15 dates from 7 May through 5 June 1994 (SSM), as well as 2 other records from the vicinity of the Great Swamp: along the Pawcatuck River northwest of Wordens Pond 3 July 1984 (REn, CRa) and along Ministerial Highway, South Kingstown, 1 June 1983 (CRa).

Griscom and Snyder (1955) stated that this species had never been found breeding in Massachusetts in "natural habitat," that is, away from cultivated or otherwise disturbed situations such as about farmyards and along roads. This statement is generally true for Rhode Island, even though territorial birds have been found on a number of occasions in open second-growth woodland (near Breakheart Pond, for example) well away from orchards, dwellings, or recent human disturbance. Like Warbling Vireos, Yellow-throated Vireos, and Red-shouldered Hawks, Least Flycatchers seem partial to wooded habitats in the vicinity of water, a habitat that is present at Arcadia, Scituate, Durfee Hill, and Watchaug Pond.

Nonbreeding

Spring. Least Flycatchers arrive earlier than other *Empidonax* species, usually by late April or early May. Once a common transient, it is now decidedly uncommon, parallel with its slide as a summer resident. The earliest arrival record is for 22 April, year and locality unstated (Howe and Sturtevant 1899), and migrants are found in nonbreeding areas to early June.

Spring maxima. Six, Breakheart Pond, 29 April 1992 (CRa); 3, Durfee Hill, 30 April 1993 (JMu).

Fall. The earliest transient date is for 1 banded at Kingston 16 August 1971 (DLK, RLF); also, 1 on 16 August 2006, Napatree Point (CRa). The peak abundance of fall migrants is in late August and early September (DLK, ELa et al.).

Extreme dates. 20 April 2002, Durfee Hill (JSJ) and Ponagansett Road (RFa et al.), and 22 April (see above), to 13 October 1977, banded at Block Island (ELa); 15 October 1995, 1 banded, Block Island (ELa, HLa); and 4 November 2012, Little Compton (m.obs.). Also, *Empidonax* sp. banded at Block Island 18 October 1972 (ELa).

Hammond's Flycatcher (*Empidonax hammondii*)

Status

Vagrant. One or two records.

Nonbreeding

One *Empidonax* sp., very probably this species, was seen and photographed 5–6 November 1995 north of Trustom Pond by STs, DLK, RBo et al. A second individual, clearly *hammondii*, was discovered on the Walker Farm near Hundred Acre Cove, Barrington, 23 November 1995. This bird remained near a brush and mulch pile and was seen by many through 6 December (RFa, DVa; DFi, PAB, RBo, DLE).

Eastern Phoebe (*Sayornis phoebe*)

Status

Common breeder and common to abundant migrant. Occasional in winter.

Breeding

Eastern Phoebes breed most commonly in interior western sections of the state. In this state, they are almost wholly dependent on artificial structures for nest sites. They usually nest in such situations as under bridges (perhaps a majority) or in dilapidated sheds or other old buildings, usually near water. The essential lack of mountainous terrain in Rhode Island allows little opportunity for nest placement along the rocky faces of stream gorges (probably their original New England habitat), as is still common in adjacent states. Nevertheless, there are records of nesting on the face of a rocky ledge at the "cascades" on the Pocasset River in 1918 (HSH), as well as on rocky ledges at Limerock and Glocester in the 1980s (REn, CRa). In addition, phoebes sometimes place their nests in the root masses of large fallen trees, including once at NBS in Middletown (JBa) and regularly in the swampy woods along the Chipuxet and upper Pawcatuck Rivers (SSM).

Summer maxima. 44, Hopkinton, 4 June 1973 (DWa).

Nonbreeding

Spring. Arrival dates closely follow the first thaw, generally in early March but sometimes earlier or later, depending on the severity of the winter. The earliest dates in spring are 26 February 2002 at Trustom Pond (S. Keeble). Singles 15 February 2012 at Block Island (SSM et al.) and 14 February 2011, Great Swamp (C. Brown), were likely new arrivals. A late transient was banded at Block Island 24 May 1974 (ELa); another migrant was at Point Judith 5 June 1995 (RLF).

Spring maxima. 21, South County, 3 April 1953 (LRBC).

Fall. Phoebes are particularly common at the immediate coast during fall. A few individuals may appear where they do not breed by late August, and migration picks up through September with a peak in early to mid-October.

Fall maxima. 150+, Block Island, 28–30 September 1975 (GRa); 80, Sakonnet, 28 September 1957 (WBD, DLE). Some migration continues through late October and November.

Winter. Beginning in 1945, numerous records have accumulated for the early winter period, almost all from coastal areas (1 inland at Lonsdale 24 December 1987; SDz). As of 1995, there were more than 30 reports on CBCs in late December, 7 records for January, and a remarkable 12 for February, including several for February 1954, a very mild season. Some individuals survive the winter in Rhode Island; e.g., 1 in Little Compton, October 2006 to March 2007 (RFa).

Winter maxima. Five, Newport–Westport CBC, 16 December 1995 (DLE et al.).

Say's Phoebe (*Sayornis saya*)

Status

Vagrant. About five records.

Nonbreeding

The first Say's Phoebe from Rhode Island was 1 found along Ocean Drive, Newport, 18 November 1962 by RSt. This individual remained in the same area until at least 25 February 1963, thus apparently surviving the winter, and was seen by numerous observers and photographed in color by SSD. A second 1 was seen at the north end of Block Island 1–2 October 1966 by MacLachlan, Sanborn, Kenneally, and numerous others (Kenneally 1966). Still another was seen near the north end of Block Island 25 September 1972 by KPA and ELa. Another bird seen at Cox Ledge 30 September 1977 (R. Forster, P. Vickery, N. Proctor) may have been the same as 1 banded later at Block Island 10 October 1977 (ELa, HLa, CWo). The most recent record is of 1 at Ninigret Park 9 September 2011 (P. Blake).

VIREOS (VIREONIDAE)

White-eyed Vireo (*Vireo griseus*)

Status
Fairly common summer resident and uncommon transient.

Historical
Forbush (1929), Griscom and Snyder (1955), and Veit and Petersen (1993) all quote Brewster who stated that White-eyed Vireos were common in Massachusetts before about 1880, including regular breeding near Boston and nests in the inland river valleys, but then declined during 1880–1930 (even in Bristol County, where they are now numerous). Talbot (1884b) stated [they were] "very common [at Warwick Neck] and have taken over a dozen nests in the last two years." Hathaway thought it rare in his haunts in Cranston in the 1890s but found them commonly on Prudence Island during the same years. Doe found a nest at Rocky Point in 1883, five nests in Cranston 1896–1911, and one as far inland as Lincoln 7 June 1904. There were no other breeding records from the interior, although there are turn-of-the-twentieth-century spring specimens from Johnston and Smithfield.

Breeding
White-eyed Vireos are currently most numerous near Narragansett Bay in Newport County; e.g., 12, Wilbour Woods, 14 July 1954 (DLE); 25, Little Compton, 16 July 1954 (DLE); 51, Conanicut Island, 22–25 June 1985 (RLF); 10, Prudence Island, in 1966 (WBD); 34, Prudence Island, 6–8 June 1990 (REn, CRa). Smaller numbers inhabit southern South County west of Narragansett Bay (19, Narragansett–South Kingstown, 23–28 May 1984; RLF). On the immediate South County coastal plain, however, numbers are comparable to, or greater than, those in Newport County. At least a dozen pairs nest regularly on the Trustom Pond NWR, and 12 were found on Ninigret NWR 10 June 1983 (DVT). They were absent from Block Island until the early 1970s but have increased in pace with the succession of woody habitats: 6 nesting territories were identified there in June 1981 (RBo); 17 birds were reported on BIBBS, 28 June 2003; and 15 on 30 June 2004 (SCo).

Farther inland, greatest densities have been found in southern South County, north of the moraine; e.g., a surprising 14 singing males at Hopkinton 19–22 June 1987 (RLF) and at least 8 territories at the Great Swamp in June 1994 (SSM). Farther north, territorial males have been found in western Exeter (Arcadia), northeastern West Greenwich, central-northern Coventry, along the Moosup River in northwestern Coventry, West Greenwich, Foster (JGe, RGe), Lonsdale, Smithfield, and near Sneech Brook, Cumberland, where 2 males were singing in extensive multiflora-rose 11 June 1982 (CRa, REn).

Nonbreeding

Spring. White-eyed Vireos arrive during late April and early May, the exact dates depending on weather conditions; there are at least 12 records from late April. Most spring reports are from coastal areas; e.g., 23, Jamestown, 11 May 2002 (ChP, CaP et al.). They are close to the northern limit of their range in Massachusetts, and few migrants occur in interior Rhode Island, as illustrated by their rarity at Providence's Swan Point Cemetery.

Fall. Moderate numbers regularly remain into September.

Fall maxima. 16, Prudence Island, 6 September 1972 (RS); 14, Ninigret NWR, 8 September 1988 (LAH); 8, Little Compton, 9 October 1955 (SSD). Occasional birds have been heard singing on territory through mid-September. Rare after October.

Extreme dates. 5 April 2010, Little Compton (GDe), and 7 April 2010, Napatree (CRa), to 18 December 1993, Fogland Point, Tiverton (FAP, STs). There are 3 recent CBC records, including 1 that remained in Little Compton until 3 January 2016 (GDe).

Bell's Vireo (*Vireo bellii*)

Status

Vagrant. One record.

A Bell's Vireo was found at Sachuest Point NWR, Middletown, 24 November 2007 (POn), where it remained until 1 December and was clearly photographed. There is a recent increase of records of this midwestern species to the Northeast (Kus et al. 2020).

Yellow-throated Vireo (*Vireo flavifrons*)

Status

Fairly common summer resident, including east to Tiverton and Little Compton, and rare on any Narragansett Bay islands. Uncommon to rare transient.

Historical

Yellow-throated Vireos were formerly much more common and widespread near human habitation, breeding commonly at the turn of the twentieth century in village shade trees (especially American chestnuts) in Woonsocket, Kingston, Cranston, and East Providence. They bred in Roger Williams Park, and nested annually in a chestnut tree in HSH's yard at South Auburn in Cranston. Doe, continually in the field in northwestern Rhode Island, found only seven nests between 1885 and 1913, and these were all in the Warwick/Cranston/Johnston area. A sharp decline began about 1917 throughout the Northeast (Griscom 1949), and they disappeared from Woonsocket in 1919 (Arnold). Hathaway (1943) suggested they were gone from Rhode Island by 1922 and attributed the vireo's disappearance to the chestnut blight, an Asian fungus that decimated American chestnuts throughout the east.

Breeding

There has been a substantial increase of Yellow-throated Vireos since the 1950s in Rhode Island, especially the western sections, at least partly related to reforestation of these now well-wooded regions. They are moderately common in the isolated western halves of the towns from Burrillville south to Hopkinton and locally farther east around the Great Swamp, where at least 7 pairs nested annually in a relatively small area during the early 1990s. They nest along boundaries where forests meet fields or open water, and are widespread, but rarely numerous at any one locality. There is relatively high density in the Great Swamp Management Area, where 9 territories were counted in 2009. They are also fairly common at edges of clear-cuts, as in DeCoppett Management Area, Richmond, where 5 Yellow-throated Vireo territories were counted in 10 point counts, 15 June 2014 (CRa).

Summer maxima. 27, Foster to Coventry, 16–24 June 1986; 21, Richmond to Hopkinton, 15–21 June 1987 (RLF); 15, Glocester to Burrillville, 17–23 June 1982 (all RLF). The southernmost birds in South County are from Burlingame State Park north of Watchaug Pond, Charlestown (6 in Watchaug area 3 June 1992; CRa). Breeding birds, however, have been regularly noted in the area of Gilbert Stuart Road in North Kingstown (especially in the 1950s and 1960s). The most isolated breeders are in Tiverton and Little Compton, where they were first noted in Tiverton in June 1945 (Williston, exact location unknown), and where 4 singing males were in the Sin and Flesh Brook area of Tiverton in June 1974 (RLF). South of there in northeasternmost Little Compton, adults were feeding young south of Adamsville in July 1983, and another was seen at a nearby location 23 June 1984 (RBo). Robert Emerson has generally found between 2–8 pairs or sites of actual or presumed breeding Yellow-throated Vireos in Weetamoo Woods and Fort Barton (including Pocasset Ridge in 2015). These are usually singing birds on territory or agitated birds, occasionally attended fledglings; e.g., 20–30 June 2015: 8 pairs or sites in the Weetamoo area. Yellow-throated Vireos are firmly entrenched as breeders in the Tiverton woods, and breed to the south at Simmons Mill Management Area near Adamsville (REm).

Nonbreeding

Spring. Most arrival records are in the second week of May or later, though there are at least 6 records from the last few days of April. Rarely are more than 3 birds per day per locality reported.

Fall. Occasional birds may still be found singing on their territories into mid-September (latest, 22 Sept), but transients or wanderers have appeared earlier at nonbreeding locations (29 August 1957 at NBS, Middletown; JBa). As fall transients in coastal areas, they are by far the least frequent of the vireos (rarely to 2 per day; only 6 banded at Block Island in 25 years).

Extreme dates. 5 April 2009, Block Island (DFi), and 8 April 2009, Napatree (bird found dead; RTL), to 13 October 1958 (ASRI), and 16 October 2007, Miantonomi Park (2; RTL).

Blue-headed Vireo (*Vireo solitarius*)

Status

Uncommon to fairly common summer resident and fairly common transient, perhaps decreasing since about 1990. Rare in early winter.

Historical

Near the turn of the twentieth century, Blue-headed Vireos were uncommon summer residents. They were recorded from Washington before 1899 (Howe and Sturtevant 1899), West Greenwich in 1902 (W. Angell), Summit in 1909 (HSH), and in the Great Swamp, Richmond, where HSH found a nest in a holly tree in 1907. Doe obtained 20 sets of eggs from Glocester and vicinity between 1911 and 1926, and a nest was located in "heavy hemlock and pine woods" in Burrillville in 1925 (Havemeyer collection). Territorial birds have been found in hemlocks near Beach Pond, Exeter, since the early 1930s (fide HSH).

Breeding

With the regrowth of woodland during the twentieth century, Blue-headed Vireos have increased and are now uncommon to fairly common as breeders south to the Carolina State Forest near Richmond.

Summer maxima. Four, Burrillville, 8 June 1946 (RCC, WBD, DLK); 4 singing males in hemlocks east of Bowdish Reservoir, 13 June 1974 (RLF); 23, Glocester to Burrillville, 17–23 June 1982 (RLF); 10, Exeter to West Greenwich, 29 May–5 June 1983 (RLF); 6, George Washington Management Area, 20 May 2006 (MLy).

Territorial birds favor groves of white pine or mixed hemlock–oak or white pine–oak. A small number were present in planted Norway spruce around the Scituate Reservoir until Hurricane Gloria in 1985. At Greene in 1954, a nest was located in an apple tree not far from mature white pines (FPe). Atlas data from the 1980s indicate that a few pairs inhabit white pines east to the eastern sections of West Greenwich and Exeter, and they are present in mature white pines at Eppley Sanctuary, South Kingstown (HCa). Blue-headed Vireos are localized and uncommon in Rhode Island, with rarely more than 1 bird within earshot. There are occasional reports of individuals during breeding season in the east Narragansett Bay area; e.g., 1, Weetamoo Woods, 18 June 2000 and again 17 June 2001 (REm); but no evidence of breeding.

Nonbreeding

Spring. Most move through during late April.

Spring maxima. 12, Block Island, 12 May 2002 (DJA, STs); 9, Little Compton, 6 May 2000 (RFa). They are rare away from nesting locations after mid-May; 1 was at Block Island 29 May 1972 (LBe, REm).

Fall. There are occasional records of birds wandering where they do not breed in late summer; e.g., 5 August 1971 at Watchaug Pond (CWo) and 6 August 1961 in Middletown (JMu), but most transients do not appear before September. The peak of migration is during late September–early October.

Fall maxima. 30, Block Island, 9 October 1995 (RLF et al.); 15, Block Island, 18 October 2015 (REm, LBe).

Extreme dates. 11 April 1910, Beach Pond (HSH), and 12 April 1970, Block Island (ELa), to 2 January 2012, Providence (1; DFi). There are several December records and 12+ November records since 2005.

Philadelphia Vireo (*Vireo philadelphicus*)

Status

Uncommon transient in fall, rare in spring.

Historical

Only one record was known to Howe and Sturtevant (1899), a spring bird (surprisingly) collected 31 May 1891 at Drownville. The specimen was acquired by HSH and later placed in the RWPM collection, but it is no longer extant. Another was shot 21 May 1902 at West Greenwich by Walter Angell and identified by William Brewster. Philadelphia Vireos were then unrecorded until 29 August 1930, when one was shot in West Barrington (fide HSH); the species was next seen at Middletown 10 October 1937 (RCC).

Nonbreeding

Spring. There are about 20 records during spring, all occurring between 7 and 31 May. Most appear about the third week of May.

Spring maxima. Three banded, Block Island, 19 May 1996 (MHC, SMC).

Fall. Recent decades have seen a great increase in reports, at least in part due to increased effort; banding efforts have shown that they are regular transients along the coast in fall, especially at Block Island.

Fall maxima. Seven, Block Island, 15 September 1972 (RLF); 8 banded, 21 September 1970 among an exceptional 52 banded at Block Island during the fall of 1970 (ELa). Since 2005, there have been 2 spring and 41 fall reports between 4 September and 15 October.

Extreme dates. Early May to 11–12 November 1953, Fiskeville (JGe, RGe).

Warbling Vireo (*Vireo gilvus*)

Status

Common summer resident and uncommon transient.

Historical

Warbling Vireos were moderately common summer residents of village elms, pear orchards, and other planted trees in many upstate areas of the geological Narragansett Basin, especially the outlying streets of most of the cities and towns in the arc from Warwick to Bristol. About 1920, HSH (1943) noted a decrease; he attributed this to "the spraying of elm trees in cities and towns for the suppression of the elm leaf beetle (*Xanthogaleruca luteola*) and Gypsy [Spongy] Moth." Warbling Vireos formerly occurred in town at Bristol (to about 1940) and Tiverton (to 1945) and were formerly found along streets in Rumford, Barrington, Providence, Pawtucket, Cranston, Warwick, East Providence, Woonsocket, and even Wickford. Neither Sturtevant nor Mearns ever encountered territorial Warbling Vireos in the Newport area, despite an abundance (after 1900) of adequately sized shade trees. At Jamestown, a pair, nest, and chicks were found in July 1939 (A. M. Clark), and another sang there through early summer 1960 (MDa). One sang all summer 1932, and again in 1934, on the shores of Watchaug Pond at Kimball Sanctuary, Charlestown (MSo), and another was seen at Charlestown 19 June 1970 (ESa). One remained on territory at Wakefield through early June 1946 (DLK).

Breeding

There has been a continuing increase of Warbling Vireos breeding in Rhode Island since the 1980s, and they are now found throughout the interior. At 37 locations where Warbling Vireos were singing in 1984, the common tree species were American elms, willows, ashes, red and ash-leaved maples, black cherry, sycamore, and black and white oaks (RLF, unpublished).

During the "atlas period" in the 1980s, REn (Enser 1992) found "many along the banks of the Pawtuxet River from the malls in Warwick all of the way to the [Narraganset t] Bay; especially in areas along the border of Cranston and Warwick where wide wooded strips are still found along the river, as along the Blackstone." One was found in Cranston 5 June 1987, perhaps in the same area (KKi). Birds have been found regularly since 1980 in the Cowesett Road area of Warwick (KKi) and near the ponds at Roger Williams Park; e.g., 17 July 1980 (KKi). Rachel Farrell noted an increase from 0 to 10 singing males in Swan Point Cemetery between 2004 and 2015, and CRa found 14 males along Blackstone River near the Valley Falls Marshes, 12 June 2015.

Nonbreeding

Spring. As transients in spring, Warbling Vireos are uncommon away from established breeding locations

Fall. Warbling Vireos are less common in fall than spring in Rhode Island. Residents are often heard singing into mid-September, and numbers gradually decline from then until early October. Fewer than 30 were banded at Block Island 1959–1999, with a maximum of 6 per year, and counts by birders at that season have not exceeded 2 per day per locality. They are rare by early October, but singles were banded at Block Island 13 October 1977 and 14 October 1976 (ELa).

Extreme dates. 20 April 1959, Jamestown (MDa, KKi), to 20 October 1949, Charlestown (DLK), and 21 October 2013, Block Island (CNu).

Red-eyed Vireo (*Vireo olivaceus*)

Status

Very common summer resident and common transient.

Breeding

As with a number of other common New England woodland birds such as Wood Thrushes and Rose-breasted Grosbeaks, Red-eyed Vireos are less numerous in Rhode Island than they are farther north and west, probably because of the dry and acidic nature of Rhode Island oak habitats. Greatest densities in the uplands are found where neutral soils provide habitat for white ash and sugar maple or along streams where better-watered soils allow the oaks to be relieved by red maples. While often cited as the most common New England landbird, Red-eyed Vireo ranked 10th most common in a statewide survey of Rhode Island breeders during the 1980s by the author.

Summer maxima. 526, Richmond to Hopkinton, 15–21 June 1987 (RLF); 477, Foster to Coventry, 16–24 June 1986 (RLF); 42 males in Burlingame Management Area, 29 May to 6 June 2006 (CRa); 26, Conanicut, 22–26 June 1985 (RLF).

After a record drought in 1980 and unprecedented defoliation by Spongy Moths

(*Lymantria dispar*) in 1981, an all-day tour of Exeter and West Greenwich 20 June 1981 produced only 3 singing males (RLF). At Block Island, they have always been rare, being first recorded in 1919 (EDi) but remaining rare and local up to the twenty-first century (1–2 pairs in 1972–1973 [RLF]; lacking again in 1981 [RBo]). Also absent on any of the smaller islands in Narragansett Bay.

Nonbreeding

Red-eyed Vireos declined significantly over the years 1969–2001 as fall migrants at both Kingston and Block Island banding stations (Osenkowski et al. 2012).

Spring. Red-eyed Vireos appear in peak numbers in mid-May and often do not reach maximum abundance in marginal habitats such as dry oak moraines until the second week of June. Latest spring transient at Napatree 2 June 1986 (CRa).

Spring maxima. Counts of 30–40 birds per day per locality are routine during May.

Fall. In fall, migrants occur in mid-August, and Red-eyed Vireos are common as transients in late August and September.

Fall maxima. 50, Little Compton, 1 September 1953 (NPH); 50, Warren, 11 September 1940 (RBo); 35+, Warren, 29 August 1941 (RBo). Peak migration at Block Island is in September. Migration continues on a smaller scale through mid-October; there are numerous records from November, including 4 since 2005, all from Block Island.

Extreme dates. 11 April 1993, Jamestown (ChP), and 14 April 1987, Swan Point Cemetery (RBo), to 17 November 1956, Little Compton (SSD), and 19 November 1972, Block Island (ELa).

Black-whiskered Vireo (*Vireo altiloquus*)

Status

Vagrant. One record.

A Black-whiskered Vireo was found at Little Compton 21 May 2018 (Lipton, Waters); this was the first record for eastern North America north of Virginia. Since that observation, additional birds have been seen at Martha's Vineyard (21 April 2019) and Kiptopeke, Virginia (15–16 September 2020). This remarkable cluster of records suggests enhanced, perhaps climate-related, northward dispersal.

SHRIKES (LANIIDAE)

Loggerhead Shrike (*Lanius ludovicianus*)

Status
Rare fall migrant or vagrant.

Historical
Loggerhead Shrikes have fluctuated in abundance in Rhode Island and since about 1980 have been as rare as at any time in the historical period. They were considered to be rare by Howe and Sturtevant (1899) and by Howe (1903), who knew of only three fall records (all specimens). They were also considered to be rare by Hathaway (1913), who listed eight additional fall records including *five* seen in August 1912 on Block Island (EDi), which suggests that they were fairly common then, by today's standards. Hathaway shot one 17 September 1916, his first specimen in a quarter century of collecting, and he saw only two others in 22 years of regular field work at Quonochontaug. They were seen regularly by EDi and others at Block Island from 1912 until about 1970. Elsewhere in the state, they were uncommon but regular fall, and rare spring, migrants, and occasional birds lingered into early winter.

Nonbreeding
Beginning in about 1970, Loggerhead Shrikes rather suddenly abandoned their northeastern breeding range; this was roughly coincident with the disappearance of Henslow's Sparrows and Sedge Wrens from the same areas. They declined as migrants in Rhode Island during this same time frame.

Spring. Loggerhead Shrikes have always been scarcer in spring than fall. There are numerous older records for March along the coast that probably represent successful overwinterers. Most occur during April. Two were in Narragansett in April 1972 (DWa); otherwise, all reports are of single birds. One was in Narragansett 31 March 2002 (DWa). Always rare away from the coast at any season, they have been recorded upstate in fall at Cranston, Barrington, Warren, and Warwick. The only spring record from an inland location is for 1 in full song at the East Providence Reservoir, Pawtucket, 7 April 1956 (RLF, REW, JSu).

Summer. There is a remarkable recent record of 1 at Charlestown 15 June 1981 (DVT), a year when this shrike was otherwise unreported in the state.

Fall. The earliest fall transient date is for 1 on 9 August 1973 at Warwick (DWa). The bulk of fall reports are concentrated in September through November.

Winter. Between 1949 and 1973, a total of 73 individuals were reported on the Newport–Westport CBC (9 and 8 in 1959 and 1968, respectively); the last Newport–Westport CBC record was in 1973. This is a surprisingly high number of individuals relative to the rest of the Northeast. Two were banded at Middletown 19 January 1957 (JBa). One at Charlestown Breachway 19 November 1995 (DLK, STs) was said by DLE to be the second Rhode Island report in a decade, and there have been no records since then.

Northern Shrike (*Lanius borealis*)

Status

Rare winter resident; irruptive, so erratic in occurrence from year to year.

Nonbreeding

It is remarkable how much less numerous Northern Shrikes are in Rhode Island than in adjacent Massachusetts. For example, in 1978–1979, when 100 were found just in eastern Massachusetts, only a single bird was found in Rhode Island. The largest flight in recent years was during 1995–1996, when 15 were found on the Block Island CBC. The irruption of Northern Shrikes in December 1995 was "utterly unlike anything ever recorded in Rhode Island in the last 60 years. A total of 36 birds appeared throughout Rhode Island in December, and this was about four times all the birds recorded in Rhode Island in December for the prior 26 years" (*FNRIB*). The first bird was reported at Matunuck 10 December (BG, WG), and high counts included: 15, Block Island CBC, 18 December (CR et al.); 9, South Kingstown CBC, 23 December (DLK et al.); and 4, Newport–Westport CBC, 16 December (DLE et al.). Approximately 8 birds were reported in January 1996, primarily single birds along the coast, and single birds were reported from at least 10 locations in February, along with 5 on Block Island 19 February 1996 (CR et al.). There were 4 reports in March, the last occurring at Middletown 22 March (BG, WG), and 1 report from Sachuest 15 April (N. Logothets).

Other than in 1995–1996, counts have rarely exceeded 2 per day per locality.

Maxima. 15, Block Island CBC, 18 December 1995; 4, Newport–Westport CBC, 16 December 1995 (DLE et al.). Reported in 7 years between 2006 and 2016, with most reports during 2007.

Extreme dates. 19 October 1888, Middletown (Sturtevant), to 8 May 1996, Little Compton (photographed; GDe).

CROWS AND JAYS (CORVIDAE)

Blue Jay (*Cyanocitta cristata*)

Status
Very common summer resident, transient, and winter resident.

Breeding
In summer, Blue Jays are one of the most numerous birds in Rhode Island's oak woodland and residential areas. They are ubiquitous as breeders on the mainland and common on the 3 larger Narragansett Bay islands. At Block Island, however, they are only fairly common as summer residents and sporadic as transients. Despite being present year-round in Rhode Island, Blue Jays are in fact highly migratory.

Nonbreeding
Spring. Sizable flocks of Blue Jays arrive at Point Judith from the north and northwest almost daily in late spring, and these flocks often attempt to depart southward or southeastward over Block Island Sound from near the Point Judith lighthouse.

Spring maxima. 366, Point Judith, 5 June 1981, and 271, Point Judith, 24 May 1980 (RLF); 300, Watchaug Pond, 8 May 1986 (CHa).

Summer. A flock of 9 was seen in a northwest wind venturing out Napatree Point and flying north across the bay to Pawcatuck, Connecticut, 22 June 1992 (CRa). A single flock of 30 was seen moving northwest across Block Island 19 July 1983 (CRa, REn); another flock of 66 was noted moving west over Westerly 2 August 1983 (REn). These midsummer movements, as well as the late-spring maxima noted above, suggest the wanderings of immature birds and nonbreeding adults.

Fall. In fall, Blue Jays are sometimes abundant as diurnal migrants in late September and early October. Loose flocks of up to 200 or more have commonly been observed after cold fronts in fall, usually moving southwestward. Like hawks, they are best observed along the outer coast where their flight paths are channeled or deflected by the ocean and other water barriers. As is the case for migrating hawks, the best locations to observe migrating Blue Jays are in the Sakonnet and Watch Hill areas, where the "timid" flocks of jays are forced to deal with the prospect of crossing large expanses of water.

Fall maxima. 2925, Watch Hill, 4 October 1981 (RLF, REn, CRa); 1800, Watch Hill, 25 September 1987 (CRa); 1575, Watch Hill, 14 October 1979 (RLF); 1061, Napatree/Watch Hill, 1 October 1990 in 6 hours (CRa). During the fall of 1981, a few months after an extensive defoliation of most of southeastern New England's oak woods by Spongy Moths (*Lymantria dispar*), extremely large numbers of Blue Jays were recorded migrating southward. Coastal counts that fall included 950 at Brenton Point, Newport, 26 September (RTL); 1000 at Sakonnet (RBo) and 2750 at Watch

Hill on 3 October (RLF, REn, CRa); and 2370 at Watch Hill on 10 October (RLF, REn, CRa). See Veit and Petersen (1993) for other notable migrations in Massachusetts, some of which may have been related to freezes of oak flowers in spring and consequent failures of acorn crops.

Winter. In most years, Blue Jays are common throughout the state in winter. The combined totals of the 2 mainland CBCs have ranged from approximately 200 to more than 800 in recent decades and probably reflect abundance of food resources such as acorns.

Winter maxima. 195, Burrillville, 29 December 1991 (SDz); 81, Shannock Hill, 9 February 1978 (RAC). Winter numbers at Block Island have always been sharply lower than for any comparably sized mainland area.

Eurasian Jackdaw (*Corvus monedula*)

Status
Vagrant. One record.

Nonbreeding
The first specimen of Eurasian Jackdaw for the Americas was shot at Block Island on or about 15 April 1984. The bird was shot by Christopher Blane, a Block Island resident, who assumed it was a crow but thought it was a little unusual in appearance. It was shown to MSl, who tentatively identified it as a jackdaw. Frozen, it was later given to RBo, who carried it off island and put it in his freezer. The news of other jackdaws in the Canadian Maritimes during 1984 caused the crow to be removed from the freezer, where W. Smith and RBo determined it to be a jackdaw. It is now MCZ #331728. The specimen, an immature male, was compared with a series at the AMNH and proved to be the western European race *C. m. spermologus*. The North American records of jackdaws, including the birds mentioned above (plus 2 in Massachusetts during 1982–1985), were part of a larger westward irruption from western Europe (Smith 1985).

American Crow (*Corvus brachyrhynchos*)

Status
Common resident.

Breeding
Once wary and confined to rural areas, crows now breed and forage regularly close to the center of Providence and within other urban centers. Block Island now has an apparently more or less stable and resident population of about 30 pairs, but crows

are relatively recent recolonizers there. They were numerous on the island in 1693, when over 90 were killed for bounty (Livermore 1877). None, however, had been present for decades prior to 1875 because of virtually complete deforestation, and crows did not return as breeders much before 1930 (EDi). A census of Block Island on 18–22 June 1990 produced 63 (REn, CRa), and censuses on 28 June 2003 and 30 June 2004 each produced 97 birds (S. Comings).

Nonbreeding

Roosts. Upper Narragansett Bay has been the most important rendezvous for the crows of eastern Rhode Island and nearby Massachusetts for a century or longer, and it seems possible that during most of that time a single roost has formed at changing locations. Since the 1950s and perhaps earlier, wintering crows present during the day in Little Compton and the rest of Newport County have coalesced in late afternoon across Aquidneck and Prudence Islands to a roost on the west shore of Narragansett Bay; e.g., 250 across Portsmouth 20 February 1993 (RLF, CRa). This roost was located in the Buttonwoods area of Warwick from at least the early 1980s through the mid-1990s and on several occasions was estimated to contain 7000–8000 crows (REn, CRa et al.). The numbers using this roost at seasons other than midwinter are poorly known, particularly in fall and spring; 800 were estimated there 3 April 1994 (DWa).

Elsewhere, winter crow roosts were unknown until recently and the crows of the southwest quadrant of Rhode Island were thought to aggregate toward either the Warwick roost or toward southeastern Connecticut. A new and previously undiscovered roost, however, was located in 1970 close to the most urbanized area of Westerly. This roost has continued to attract birds from southwestern Rhode Island and parts of southeastern Connecticut. A total of *9114* entered this roost 15 January 1982 (CRa). The Block Island roost numbered 450 on 20 December 1993 (CBC).

Spring maxima. 500, Seapowet, 19 March 1955 (v.o.).

Fall. A flock of 200 moving west past Watch Hill 6 November 1982 was likely migrating (CRa), as were 150 moving from Watch Hill to Stonington 13 November 1995 (CRa); also 247, Exeter, 6 November 1985 (HCa).

Fall maxima. 500, Matunuck, 10 November 1982 (KKi, TAd, FPe).

Fish Crow (*Corvus ossifragus*)

Status

Uncommon and local resident.

Historical

Fish Crows were unrecorded by Howe and Sturtevant (1899), despite a "colony" found in 1901 in Connecticut "within sight" of Watch Hill (*NRIORN* 2:24; five nests found) and another group eastward along the shores of Buzzards Bay in 1905 (Forbush 1929). Sturtevant himself unknowingly obtained the first specimen for the state, shot by G. Shattuck 11 February 1899 in Middletown and later discovered in ESt's collection by JBa. For years, the supposed first record for the state was one randomly shot in Riverside 3 February 1927 and brought to Angell and Cash to be mounted. They were then unrecorded but probably overlooked until 1936–1940, when small numbers were found on several occasions in Middletown, Tiverton, and Little Compton, and a flock of 20 was seen (RCC, WHD). A possible source for these eastern Newport County birds, previously unpublished, comes from Conanicut Island, where A. M. Clark noted them regularly in summer 1936–1939, with counts of up to 18 birds. He stated (in litt. to RCC) that he "could always see them near Conanicut Point [the north end of the island]; they were very evidently breeding there." He could find none in 1940, however, and observations by MDa and other observers after 1949 suggest that none were present in later years on the island.

On 28 March 1941, RBo saw and heard nine birds arriving in Warren from the south in a tight group, exhibiting unusually excited behavior with much calling and circling. These remained in the vicinity throughout the spring but decreased locally to four in early May and to a single pair through July. Again, on 14 March 1942, another group of four arrived at Warren, remained throughout the spring, and increased to eight on 29 July. This group included some obvious immatures, probably locally reared. Some of these birds remained into late September, and one was seen that December and in February 1943. Richard Bowen again saw a flock of 10 on 14 March 1943. These were likewise reduced to a single pair by 18 April, and on 30 May, a nest was located in a black oak about 60 ft (18 m) up, with the bird still incubating on 9 June 1943 (shown to A. C. Bent). Birds again appeared in March 1944, 1945, and 1946, generally disappearing in fall with occasional observations in winter. On 9 May 1946, RBo found another pair completing a nest near the previous nest location, and on the same date found a flock of 30 in a pigpen near Bristol, with the latter flock still present in late May.

Thereafter, Fish Crows increased in the Warren and Bristol area and soon spread to Seekonk, Rehoboth, and Swansea in Massachusetts. They also increased during winter and in the early 1950s appeared in small numbers in the Tiverton to Portsmouth area on CBCs. They frequented the East Providence Reservoir after the late 1940s, particularly in winter and spring, and were suspected of breeding in 1959

(HEC, RLF, REW). Records increased greatly for a time in the Rumford–Seekonk area, and Fish Crows were conspicuous during fall and winter among American Crows going to roost near Hundred Acre Cove (maximum 40 in Rumford on 21 March 1953; HEC).

Breeding

Fish Crows then declined, possibly because of the decline in pig farming in the area. In the 1960s, the center of abundance apparently spread eastward into Massachusetts; e.g., 50 or more daily through the summers of 1962–1965 at a large piggery north of Taunton, including many young birds accompanying adults (RLF). They also spread to the Boston area. Fish Crows bred annually in or near Swan Point Cemetery through the late 1960s (CWo), and another Rhode Island nest was located in a 30-ft (9-m) oak along the west shore of the East Providence Reservoir in 1985 (REn, CRa).

Numbers of Fish Crows found in the Westerly–Watch Hill area during the 1980s suggested that they might be breeding there, and occupation of this area continued into the 1990s (CRa). Twenty Fish Crows attended a roost and fed at the nearby Westerly dump through May, June, and July in the early 1980s, but no nests were found. Birds in the 1980s visited Napatree Point (where they were previously unknown) at all seasons (e.g., 8 there 22 June 1992; CRa), and others now occasionally visit the South Shore; e.g., 10 at Misquamicut 29 September 1990 (DLK). At the same time that they have increased near Westerly, they have declined to the north of Providence and Pawtucket, with scattered reports of small numbers in the Blackstone Valley north to Lonsdale, Berkeley, Arnold Mills, Lincoln, Cumberland, and Woonsocket (DMK; Veit and Petersen 1993).

Fish Crows colonized Block Island in the late 1990s, and this tiny colony has gradually grown; they are now regular around the jetties at Old Harbor. Fish Crows were first recorded on the Block Island CBC 19 December 1994, and by 2016, this CBC total was 97. There has been a large increase of Fish Crows during the last 15 years, and they are now found virtually throughout the state during summer, even in the forested interior. This increase seems to coincide with a decrease of American Crows (CRa). Fish Crows have been confirmed nesting along the Connecticut border at Beach Pond (CRa).

Winter maxima. Many of the large counts have been obtained at nocturnal roosts: 350+, Providence, 4 April 2012 (REm); 237, Newport–Westport CBC, 19 December 2015 (REm et al.); 200, Wakefield, 29 September 2008 (STs); 97, Block Island CBC, 22 December 2016; single flock of 46, Kingston, 26 December 2015 (SSM).

Common Raven (*Corvus corax*)

Status

Rare resident and winter visitor; increasing rapidly.

Historical

The only record known to HSH was of one shot in Bristol in the fall of 1925 and brought to Angell and Cash to be mounted (present whereabouts of specimen unknown). Common Ravens invaded western Massachusetts and Connecticut during the 1980s, and counts of more than 25 birds were made at hawk-watching sites in interior Massachusetts. Ravens have bred in the Quabbin and Wachusetts regions of central Massachusetts and even in the vicinity of Bigelow Hollow State Park in northeastern Connecticut (only a few miles from northwestern Rhode Island) beginning in the late 1980s. In Rhode Island, a bird was seen soaring over route I-95 in Warwick 10 May 1991 (RBo). Another was seen and heard while being mobbed by crows in Cumberland 28 March 1992 (WBo). Still another was seen and heard at Wickford 25 April 1992 (RAC). Two others were seen and heard along the Blackstone River at Manville 29 March 1994 (JO). Another was reported from Beach Pond 12 March 1997 (D. DeForest).

Recent records. There has been a substantial increase since 2000, coincident with the decline of American Crows. In 2012, about 30 birds were reported; the maximum was 4, Howard Conservation Area, 23 November (DFW). Ravens have nested in Coventry, Providence, Hopkinton, Cumberland, and South Kingstown and have been suspected of nesting in several other locations. Nests have been located on utility towers, in bridges, within buildings, and beneath overhanging ledges. First proven nesting occurred on a cell tower in Cumberland 23 May 2005, 2 adults and nest with 2 large chicks (G. Valade).

Maxima. 18, Durfee Hill Management Area, 12 March 2008 (RFa et al.); 13, Great Swamp, 28 January 2016 (CRa).

CHICKADEES AND TITMICE (PARIDAE)

Black-capped Chickadee (*Poecile atricapillus*)

Status

Common resident; southward incursions occur in some years.

Breeding

Black-capped Chickadees are common and widespread through most of Rhode Island's woodlands. Local abundance fluctuates from year to year, with increases associated with large incursions during the fall. However, they were absent from Block

Island before about 1970. Even as transients they were historically scarce except during the largest invasions. After such an invasion during the fall of 1980, "family groups" were found at 4 places 9–20 June 1981 (RBo), and individuals were noted on such isolated outposts as Hope Island (1, 30 May 1981; RLF, JEM) and Dutch Island (1, 21 June 1981; RLF).

Nonbreeding

There are sometimes significant incursions during fall and winter, confirmed by banding. Particularly noteworthy was the incursion of 1913–1914, when large counts were noted at Watch Hill and "numbers" reached Block Island. In 1941–1942, DLK noted more than 100 along the South Shore 21 September 1941, and they were "abundant" during the following month. This flight was noted from Minnesota eastward and was particularly noticeable in early fall around New York City (Poor 1946). Other flights were noted in 1952 and 1957, when in the latter year exceptional numbers were noted in coastal areas as late as 16 November (SSD). There were minor fall flights in 1973 and 1975; in the latter year, birds reached Block Island in moderate numbers. Larger flights occurred in the fall of 1971, again in 1980 (9 banded at Block Island), and 1990 (62 at Ninigret NWR 25 October 1990 [LAM]; 13 at Napatree 15 November [CRa]).

Maxima. 862, South Kingstown CBC, 26 December 2005 (DLK et al.); 611, Napatree CBC, 27 December 2015 (SSM, GW et al.); 539, Newport–Westport CBC, 15 December 2012 (REm et al.); 165, Block Island VDC, 12 November 2005 (SSM et al.).

Boreal Chickadee (*Poecile hudsonicus*)

Status

Rare and irregular winter visitor. Almost always appears during "flight years" of Black-capped Chickadees.

Historical

The first Rhode Island Boreal Chickadee was collected 1 November 1880 at Smithfield (Jencks 1880a). There was a tremendous flight of these birds in the fall of 1913 in the Boston region (Wright 1914, 1918), and some were noted in Rhode Island. On the morning of 30 October 1913, Ludlow Griscom, spending the fall at Watch Hill, stepped out of his house and noticed an obvious movement of Black-capped Chickadees and later saw one Boreal Chickadee. The next day he saw four Boreals together at Watch Hill, and, on 5 November, found three singles at Napatree Point (Griscom 1914a). Another was seen 25 December 1913 at Warwick and collected 2 days later by HSH. Again, there was a large "flight" to the Boston region in 1916–1917, and one was collected by HSH 18 March 1917 in East Providence (Wright 1914, 1918).

Nonbreeding

During the chickadee flight of 1941–1942 (Poor 1946), the only Boreal Chickadees found in Rhode Island were at Block Island, 30 October and 28 December 1941 (EDi). These are among the very few records ever from southern New England's offshore islands. Another large flight occurred during 1954–1955, when birds attended feeders at Hope, Hope Valley, Scituate, East Greenwich, and Bristol, with the latest date reported from Bristol 19 March 1955 (fide FPe). Minor flights occurred 1959–1963, and the largest recent flight occurred in 1969–1970, when at least 15 birds were reported statewide, primarily west of Narragansett Bay but including Portsmouth, Middletown, and Bristol. They arrived exceptionally early in this flight (10 October 1969, 2 in Bristol; CWo), and some were reported throughout the winter. Another flight produced 4 birds from Riverside, Saunderstown, Pawtucket, and Green Hill from 10 November 1975 to 28 March 1976. There were no further records until the early 1980s, when a single bird remained among the Burlingame evergreens 27 December 1981 to 18 March 1982 (RBo, DLE et al.).

Tufted Titmouse (*Baeolophus bicolor*)

Status

Common and widespread resident on the mainland; 1 record for Block Island.

Historical

Tufted Titmice have gradually extended their range northwards starting in the early twentieth century, at least partly owing to warming climate (Post 1979; Wilson 1999) Tufted Titmice were not recorded in Rhode Island until 24 October 1935, when one appeared at a feeder in Rumford and stayed 3 days (Hussey, confirmed by A. H. Walter). Another was at the same feeder during the fall and winter of 1942. In 1944 and 1945, a male was reported singing at Westerly from April through the summers by Carrie L. Saunders, who stated that bird(s) had been there since 1941. Three birds appeared at a feeder in Apponaug 10 January 1945 (Berkander), and another visited a feeder at Wakefield 20 November 1952 and remained through March 1953 (Coman, DLK). In 1954–1955, there were birds at Green Hill and at two locations in Kingston. Individuals were reported from Apponaug and Saunderstown and others at Kimball Sanctuary through the summer and fall of 1957; six were at Watchaug Pond 16 November 1957 (Wiley).

Breeding

The rate of increase in Rhode Island accelerated during the period 1960–1964. Nearly all of the reports before the late 1950s were from fall and winter but during 1960–1964, titmice were reported throughout the year and broadly distributed in the state, including Barrington and Riverside in 1960 and Providence, Pawtucket, and Rumford in 1961. Aquidneck Island, separated from Tiverton by a quarter mile of water, began receiving vagrants after 1969. Twelve were at a feeder in southern Middletown during December 1974, and the first 2 on Newport Neck were in 1979. Nine were in western parts of Portsmouth and Middletown 15–16 June 1990 (RLF). A single bird appeared at Jamestown 19 December 1971, a pair was there in 1973 (MDa), and a slow increase then ensued. An island-wide population of 35 was found on Conanicut 22–25 June 1985 (RLF), and 52 were there 31 December 1989 (ChP et al.). A single colonist was noted on Prudence Island in 1971 (RS), and a resident population is now present (20 there, 6–8 June 1990; REn, CRa).

A roving flock of 13 Tufted Titmice spent the weekend 30 April–1 May 1983 roaming the Point Judith barrens and repeatedly attempting to leave southward over the water toward Block Island (RLF). The first titmice on Block Island were on the CBC, 19 December 1994, when 2 birds appeared on Corn Neck (SRC et al.), and 1 was banded there on 20 October 1995 (ELa, HLa), but they have not been seen on Block Island since then. They were first seen on Dutch Island in 2016 (CRa et al.).

Maxima. 454, South Kingstown CBC, 22 December 2007; 217, Newport–Westport CBC, 15 December 2007; 201, Napatree CBC, 28 December 2014 (SSM, GW et al.).

LARKS (ALAUDIDAE)

Horned Lark (*Eremophila alpestris*)

Status

Uncommon and local breeder, mainly near the coast (most or all *E. a. praticola*). Common to abundant migrant and winter resident (mainly *E. a. alpestris*).

Historical

Horned Larks from midwestern North America ("Prairie" Horned Larks, *E. a. praticola*) colonized and began breeding in New England during historical times. They reached New York in the 1870s, Connecticut in 1891, and eastern Massachusetts between 1900 and 1905 (Forbush 1927). Hathaway saw what must have been "Prairie" Horned Larks in Cranston in late summer and early fall 1892, and they could have been breeding in the area at that time. On 7 April 1906, a pair was found in a field near Woonsocket, where a later check found the field plowed and planted (HSH). The first confirmation of nesting was on 22 April 1913, when an unfledged juvenile was collected at Glocester (CED; FlMNH #59152). A nest with four eggs was then found at Dickens Point, Block Island, 6 June 1914 (EDi). A pair was present and probably breeding in a Johnston pasture 18 June 1914 (HSH), where a nest had been reported the previous year (fide HSH). A pair was seen near Watch Hill, 25 June

1914 (Griscom 1915). A nest and chicks were photographed in Cranston 5 June 1918 (H. Madison). In the years following this rather sudden and substantial arrival, Horned Larks continued to breed annually in the pastures at Dickens Point, Block Island, and became more widespread as summer residents, being found in Cranston, Slocum, Weekapaug, and Narragansett.

Breeding

During the early 1980s, Horned Larks were revealed to be among the rarest breeding birds in the state. There are no birds breeding at inland locations, and the only remaining breeders are at 1 or 2 coastal airports.

Recent maxima. 15 territorial birds, Quonset/Davisville, 20 June 1981 (RLF). Much of the former habitat at Quonset is now developed into parking lots and businesses, and the airfield cannot be as easily surveyed.

Nonbreeding

Spring. During the 1940s and 1950s, when they were relatively widespread, locally breeding Horned Larks arrived after the earliest thaw in late winter. There are many records of arrival in February, with singing heard as early as 28 January.

Fall. In the past, Horned Larks (presumably *E. a. praticola*) were occasionally numerous in late summer and sometimes in fall, but maxima never approached fall and winter counts of *E. a. alpestris* ("Northern" Horned Lark).

Fall maxima. 62, Wakefield, 29 August 1941 (Coman); 20 *E. a. praticola*, Moonstone, 2 November 1946 (DLK).

Winter. General arrival of *E. a. alpestris* in fall is usually during late October or early November. The largest flocks of "Northern" Horned Larks (500–750 birds) have historically been seen in early winter in places where farmers spread manure on coastal fields, but with suburbanization of the coast these rich situations have gradually been disappearing.

Winter maxima. 510, South Kingstown CBC, 27 December 2014 (STs et al.); 400, Newport–Westport CBC, 14 December 2014 (REm et al.); 300, Slocum, 28 November 2012 (CRa); 300, Richmond, 3 February 2013 (PP); 200, Kingston, 25 January 1999 (SSM). Smaller numbers are still found regularly in other open coastal habitats (such as the wrack zone above the beach), particularly along the immediate coast. A small flock of 3–11 *E. a. praticola* was present year-round on the closely cropped fields of Dickens Point (close to where they bred), from at least 1917 to about 1940 (EDi, in *Bird-Lore* for those years). By late March, few remain from the wintering population; the latest *alpestris* are 4 at Matunuck on 14 April 1940 (DLK).

SWALLOWS (HIRUNDINIDAE)

Bank Swallow (*Riparia riparia*)

Status
Fairly common but local summer resident; fairly common to common spring, but uncommon fall, transient.

Breeding
Bank Swallows nest in sea cliffs of glacial sand and clay. Coastal colonies have been documented from Warren's Point, Seapowet, Conanicut Island, Aquidneck Island, Napatree Point, and most particularly Block Island, where eroding cliffs on the northeast and southwest sides of the island have created excellent nesting habitat and where Bank Swallows have been more numerous than elsewhere (300 birds estimated at Block Island in June 1981 [RBo]; 875 active burrows island-wide, June 1991 [CRa]). Three hundred pairs bred at Warren's Point, Little Compton, in the early 1940s (DLE, RBo et al.), and these decreased to 50–75 pairs along the coastal embankments between Warren's Point and Goosewing Beach since about 2000 (REm). They also nest in banks created by excavations for sand and gravel pits at inland locations. For example, 100 pairs were found nesting at the URI campus, Kingston, 26 May 1983 (GLB). They have not been found nesting in riverbanks in Rhode Island.

Nonbreeding
Spring. General arrival is in late April and early May.

Spring maxima. 100, Napatree Point, 16 May 1983 (CRa).

Fall. In fall, most Bank Swallows migrate in late July and August, somewhat later than Northern Rough-winged Swallows but much earlier than the majority of Tree Swallows. Most have departed by late August or early September.

Fall maxima. 200, Napatree Point, 10 and 16 July 2002 (CRa); 150, Napatree Point, 14 July 2012 (CRa); 80+, Cross Mills, 5 August 1974 (DWa); 30, Napatree, 25 July 1980 (REn); 25, Napatree, 29 August 1998 (CRa, REn).

Extreme dates. 7 April 1989, Little Compton (AFr), to 2 November 1918, Block Island (4; EDi), and 12 November 1961, Napatree Point (2; DLK).

Tree Swallow (*Tachycineta bicolor*)

Status

Common spring and abundant fall transient. Fairly common breeder; rare in winter.

Breeding

Tree Swallow distribution is surprisingly sparse in summer compared with the species' abundance as transients. They are absent from large sections of countryside in the interior and from much of the greater Providence area. Their original breeding habitat was very likely in wooded swamps, and such habitats are still among the most important breeding habitats. They also breed in artificial nest boxes throughout the state and even occupy unused Wood Duck boxes. A maximum concentration of 32 pairs bred in nest boxes provided at the Kimball Sanctuary, Charlestown, in 1933 (E. Southwick).

Nonbreeding

Spring. General arrival is usually in late March or early April, but a few often appear in late February; e.g., 2, Block Island, 20 February 2017 (BCG, STa, DFi, REm).

Spring maxima. 1300, South County, 3 May 1953 (LRBC); 500, Scituate, 3 April 1960 (JGe, RG); 500, Little Compton, 10 May 1986 (DLE).

Fall. Tremendous numbers of Tree Swallows aggregate along the outer coast in late summer and early fall to feed on flying insects and bayberries. Like Purple Martins, Tree Swallows often spend the night in very large roosts at this season.

Fall maxima. 150,000 at Galilee, 7 September 1981 (STs); 120,000+ at dusk entering *Phragmites* at Galilee, 6 September 1975 (RLF); 100,000, Ninigret, 2 September 1981 (KKi, FPe, TAd). A count of 7500 29 September–1 October 1989 (RFa et al.) may be a maximum for Block Island, where Tree Swallows are less numerous than along the mainland coast.

Winter. Tree Swallows survive mild winters in Rhode Island, although usually in very small numbers. In 1931–1932, 100+ appeared in Charlestown on 13 February, and at least 10 of these survived until 5 March (MSo). In 1932–1933, 40 remained through December in Charlestown, and 16 of these were still present 24 February (MSo). Eighteen birds, possibly including overlap with the Charlestown birds, were at Point Judith 19 February (WBD). In 1971–1972, 12 birds survived at Charlestown until 4 March (v.o.). There are several December records since 2005, including a remarkable 87 over Quicksand Pond on 23 December 2015 (GDe); 65 were at Ninigret Pond 14 February 1999 (RLF); and 40 were at the Charlestown Breachway 2–16 February 1999 (STs, SSM et al.).

Northern Rough-winged Swallow (*Stelgidopteryx serripennis*)

Status

Fairly common spring transient and summer resident, especially near water. Rare fall transient.

Historical

This southern and midwestern swallow invaded the Northeast during the twentieth century by way of southwestern New England. They bred throughout central and northern New York and in western Massachusetts during the late-nineteenth century and nested at Saybrook, Connecticut, in 1891, well before they were recorded on Long Island or in Rhode Island. Howe and Sturtevant (1899) noted that they "should occur" but gave no records.

According to Hathaway's *Bird Notes* (1943) and his diaries, he collected the first state record 13 May 1917 at Chapman Pond, Westerly. These sources also indicate that he found the first nests along the South Shore at Westerly in June and July of the same year. Two purported occurrences for Charlestown in 1905 listed in *Bird Notes*, as well as a 1908 clutch listed for Weekapaug in HSH's egg catalog, were not mentioned in Hathaway (1913).

The speed with which HSH found numerous Northern Rough-winged Swallows, together with the paucity of early field work in South County, suggest that they could have been quietly breeding along the western South Shore earlier than 1917. During the period 1917–1920, HSH found nests annually at Westerly and at several other locations along that coast east to Quonochontaug. Doe found a breeding pair at Matunuck in 1919. At Block Island, EDi saw two near Old Harbor on the suggestive date of 19 June 1924, but breeding was not established there for many decades. On 10 June 1933, a pair was found breeding in a drainpipe emptying into tidewater at Pawtucket, and in May 1935, a pair bred in a gravel bank along the Blackstone River in Lonsdale (Eynon 1936). Since few observers had been active in these areas prior to this period, in all probability these inconspicuous birds had invaded these locations somewhat earlier. Between 1936 and 1938, HSH found 15 nests along the South Shore. In the Sakonnet, Middletown, and Newport areas, where they had probably been overlooked previously, nesting probably occurred from at least 1934 on (RCC, DLE et al.). They may have reached saturation densities at coastal locations by 1940 and have probably only marginally increased since. However, in the Sakonnet area, they seem to have declined during the period circa 1950–1990 (DLE). In South County, a modest increase has been noted during the same time period (v.o.).

Breeding

Northern Rough-winged Swallows are locally common near the coast and Narragansett Bay and at Block Island (2–3 places; 4 pairs in 1981; RBo) and Conanicut Island (18 at Jamestown 15 May 1993; ChP et al.). On the smaller Narragansett Bay islands, they have been reported only on Big Gould Island (pairs entering drainpipes in 1993 and 1995; RLF, JEM).

The center of Northern Rough-winged Swallow abundance in Rhode Island seems to have gradually shifted from the coast to interior fresh water. Like Bank Swallows, Rough-wingeds sometimes nest (usually singly) in gravel and mud banks. The majority, however, breed in the vicinity of rivers, larger streams, and ponds near inhabited areas, especially where the faces of dams, power stations, and old textile mills provide drainpipe openings in walls that channelize the waters. Such formerly well-harnessed rivers as the Pawcatuck, Pawtuxet, and particularly the Blackstone and its tributaries attract the largest numbers today (57 in Lincoln–Cumberland, 16–22 June 1984; RLF), far more than along the coast. The species even tolerates the fetid environment of the lower Moshassuck and Woonasquatucket Rivers as they flow through downtown Providence (10 there, 22 June 1991; RLF)

Nonbreeding

Spring. Northern Rough-winged Swallows are fairly common in spring and have been arriving earlier in spring (regularly by mid-April) in recent years

Spring maxima. 25, White Brook, 25 April 1987 (LRBC); 23, South County, 30 April 1940 (DLK); 20, Westerly, 29 April 1990 (v.o.).

Fall. Midsummer concentrations of small numbers are occasionally reported (maximum 30 in the Sakonnet area 17 July 1954; DLE), and the annual departure of local breeders and their young is in late July or slightly later.

Fall maxima. An aggregation of 47 was feeding over the ponds at Roger Williams Park 7 August 1995 (RLF). Unusually numerous and late were 40 at East Beach 21 August 1990 (CHa). A late flock of 43 was in Cranston 21 September 2003 (STs), and 20–25 were counted at the East Providence Reservoir 21 September 2004 (RFa). One late bird was seen at Napatree Point 27 September 1986 (CRa, REn). Remarkably late were a flock of 10 leisurely observed at close range in Lonsdale, 3 October 1989 (REn), and 60 at the East Providence Reservoir 3 October 2004 (MLy). A single bird was at Quonochontaug 5 October 1958 (DLK); 15 were on the Seekonk River 19 October 2003 (MLy), and 1 was at Trustom Pond 27 October 2001 (JMu, DFi).

Extreme dates. 27 March 2009, Exeter (v.o.), to 17 November 2012, Sakonnet Point (REm; photos).

Purple Martin (*Progne subis*)

Status

Uncommon to fairly common summer resident; uncommon and irregular as a transient. Rare away from colonies.

Historical

By 1880, Purple Martins seem to have been confined as breeders to South County, the traditional stronghold that still holds the majority of the state population. Hathaway thought it exceptional that CED had collected eggs from "a large martin house" 15 June 1889 in the Mount Pleasant section of Providence. Doe had also taken three sets of eggs from a box between Clifford and Friendship Streets in 1883. These eggs represent the only known breeding records from the city of Providence. Hathaway's 1887 diary comments on how they were "formerly quite common . . . the English Sparrow seems to have driven it away." By 1899, Howe and Sturtevant were able to call them "uncommon summer resident[s] and common migrant[s], formerly much more numerous."

Hathaway (1913) indicated that the persistent June rains of 1903 (given mistakenly as 1904) "nearly exterminated" martins from Rhode Island, as well as the rest of southern New England. He noted that even after a decade of recovery, the Rhode Island population in 1913 consisted of only two colonies, located at Shannock and Charlestown. A slow expansion was evident, however, in the next two decades, and by about 1930, martins were common in their limited haunts along the South Shore; e.g., 50 at Charlestown 23 May 1931 (WBD). Elsewhere, a single pair bred in an ordinary birdhouse near the Wenscott Reservoir about 1938 (WBD), a nesting at what remained the most northerly known site for Rhode Island until recent Glocester nestings.

Hathaway noted that in 1889, they were "very common in large flocks of *thousands* around the outskirts of Providence . . . a large number of them appear to be young birds." Again, in September 1892, he noted large numbers roosting with Tree Swallows in Roger Williams Park and at Blackmore Pond, Cranston, and also said they were very common in Cranston in the fall of 1894. Sturtevant found 100 transients and collected one at Middletown 6 September 1898.

Breeding

Purple Martins require artificial nest boxes with numerous compartments for breeding. In 1946, all of the state's 8 colonies were on the western South Shore: at Watch Hill (1), Westerly (2), Charlestown (2), Bradford (1), Shannock (1), and Carolina (1) (Clement 1956). In addition, martins bred at Galilee in 1942 and again in 1953 (DLK). In the decade 1946–1956, RCC located several new colonies, mostly away from the South Shore, at Hope Valley (1942 on), near Point Judith Pond, near Waterman Reservoir in Greenville (1955–1958), South Kingstown, West Greenwich, Barrington, the Conimicut Point area of Warwick, Garden City, and Oakland Beach.

In general, this history indicates that a slow but accelerating increase in the Rhode Island breeding population occurred after the nadir in the years following 1903. By 1974, 374 pairs were estimated in Rhode Island (MDa), extending north to western Coventry, Warwick, and Barrington. In Newport County, despite a number of available boxes, there have been only 2 nesting records for Aquidneck Island, and the first breeding for Little Compton occurred in 2017 when a pair fledged 5 young from a martin box at Little Pond Cove that was also occupied by House Sparrows (REm). The increase after 1903 seems likely to have been slower that the original colonization during the period circa 1765–1870. The original colonization occurred at the height of the farming era, when House Sparrows and European Starlings were absent and when more open space was available. Introduction of "gourd" type nest boxes has resulted in further increase of the population (R. Marr et al.).

Nonbreeding

Spring. Purple Martins have often been the first swallows to arrive in spring after Tree Swallows, but average arrival is usually during the third week of April. Late transients are known through late May; e.g., 1 at Little Compton 21 May 1938 (DLE), 1 at Block Island 7 June 1976 and perhaps the same bird at Cox Ledge the next day (RLF; NHBC); 1, Block Island on 15 June 1926 (EDi).

Spring and summer maxima. These are mostly from established breeding colonies; e.g., 100, West Kingston, June 1974 (MDa); 50, Westerly, 24 May 1960 (DLK); 50, Woodville, 29 May 1983 (MDa); 50, Quonochontaug, 5 June 1977 (RLF); 25 migrating, Napatree Point, 16 May 2005 (CRa).

Fall. The only sizable fall counts in the twentieth century away from breeding areas are 125 at Block Island 2 September 1931 (EDi), 85 at Lonsdale 7 September 1936 (AEE), and 46 at Sakonnet Point 5 September 1988 (DLE). According to MDa, the major exodus from local colonies during the 1970s, and general departure out of Rhode Island for most of the local population, was during the third week of August. Postbreeding roosts: Purple Martins often aggregate in late summer at roosts, which are separate from any breeding colony. Counts of birds made in the vicinity of these roosts include 150–200 at Jerusalem 29 July–9 August 1977 (Munson, fide MDa); 250+ at Saunderstown 3 August 1977 (v.o.); "hundreds" at Coventry through the first week of August 1977 (fide MDa); 200 at East Matunuck 10 August 1981 (RAC et al.); 290 at Wordens Pond 20 August 1981 (KKi); 35, Great Swamp, 7 August 2012 (PJL, SSM). Some birds apparently begin migrating in July; e.g., 5 moving west at Napatree Point, 16 July 1985 (CRa).

Extreme dates. 22 March 2016, Providence and East Providence (DFi), to 26 September 1967, Buttonwoods (2; DWa), and 30 September 2009, Goosewing Beach (GDe). Ten were migrating over West Greenwich 23 September 1989 (CRa).

Barn Swallow (*Hirundo rustica*)

Status

Common summer resident in rural areas; common spring and very common fall transient.

Breeding

Barn Swallows nest within barns, abandoned buildings, rural garages, and occasionally under bridges in open country. They are therefore scarce in the greater Providence area and the heavily wooded northwestern part of the state. They are currently most numerous in open country near the coast, especially on Aquidneck Island, at Block Island, and along the South County coastal plain. Several pairs still breed in Purgatory Chasm, Middletown. They nest in a cleft in the rocks within the chasm, midway up the south side, where they have bred continuously for over 100 years. This location is one of the few places in the Northeast where Barn Swallows still nest in their ancestral habitat, although they apparently did so at Newport in the last century (Howe and Sturtevant 1899). They breed at many isolated places such as Hope, Big Gould, and Dutch Islands, as well as Napatree Point, where they use abandoned military structures. Like Cliff Swallows, they have been declining as breeders coincident with the disappearance of farm buildings. Only 12 could be found at Block Island 18–22 June 1990 (REn, CRa).

Nonbreeding

Spring. There are several records of singles in late March; early records of multiples include 10, Block Island, 11 April 2000 (CRa); 20, Cranston, 13 April 1905 (Mason). Most move through in late April and May.

Spring maxima. 500 moving west at Napatree Point, 16 May 1983 (CRa); 500, Charlestown, 25 May 1940 (DLK). Late migrants or wanderers occur into June over offshore waters; e.g., 3 at Cox Ledge 7 June 1975 (RLF); 2, Cox Ledge 8–12 June 1983 (RBo, DFi); 1, 25 mi (40 km) south of Block Island, 12 June 1974 (RLF, RS).

Fall. Fall migration begins in July and is heaviest in August, when a continuous westward movement is evident over the outer beaches.

Fall maxima. 1000, Trustom, 28 August 1947 (DLK); 525, Napatree, 6 August 2014 (CRa); 452, Napatree, 24 July 2016 (SSM, PJL); 400, Napatree Point, 27 August 1990 (CRa). While most of the breeding population departs by mid-August, transients are common at the coast through early September, with occasional individuals to mid-October. Much later in fall, numbers of Barn Swallows sometimes appear again, often after mild weather and southwesterly winds. For example, there are 20 records of Barn Swallows during November:

Late-fall maxima. 30, Little Compton, 3 November 1956 (SSD, RLF); 20, South County, 7 November 1948 (DLK). A total of 25 were reported from 6 locations

4–17 November 1981, including 10 at Newport on 15 November (RTL). There was a large flight of Barn Swallows in November 2005. Eighty-eight were seen in Rhode Island, including 41 on the Block Island VDC 12 November, along with 1 Cave Swallow and 2 unidentified *Petrochelidon*.

Extreme dates. 19 March 2011, Point Judith (SSM), and 20 March 1986, Napatree Point (CRa), to 16 December 1937, Block Island (EDi) (at about the same time that another appeared at Jones Beach, Long Island [Bull 1964]), and 15 January 2013, Moonstone (B. Shamgochian).

Cliff Swallow (*Petrochelidon pyrrhonota*)

Status

Rare summer resident and uncommon transient.

Historical

Howe and Sturtevant (1899) considered Cliff Swallows to be uncommon summer residents, "becoming rarer each year," but unfortunately did not elaborate on what they might have known of their breeding distribution. In general, the decline and now the likely disappearance of Cliff Swallows seems to have resulted from the disappearance of unpainted barns or other buildings located in relatively open country. Another declining requirement is mud suitable for nest-building purposes, traditionally available where cattle and horses are fed and watered or where animals or equipment have churned the soil. In recent years, the few remaining barns are more frequently painted and decreasingly used for housing animals, while at the same time open country has been rapidly disappearing through suburbanization and agricultural abandonment. Hedgerows and ornamental trees have come to surround agricultural buildings that were once starkly naked on an open and agrarian landscape. Persecution by House Sparrows has been a serious problem at some colonies in the past but has not likely been a factor in the Cliff Swallow's long-term decline since peak populations of both species occurred almost simultaneously during the twentieth century.

Breeding

During the twentieth century, Cliff Swallows were rare and very local summer residents and have declined since the 1960s. They are also scarce as transients in Rhode Island. Birds once bred as close to the ocean as Westerly, Quonochontaug, Charlestown, and Point Judith. There is no historical record, however, of their breeding in any Rhode Island locality east of Narragansett Bay or on any of the Narragansett Bay islands or at Block Island, although they formerly bred on outer Cape Cod (Hill 1965). In Rhode Island, Cliff Swallows most recently bred at Westerley (1960s), Quonochontaug (1965), Charlestown (1936), South Kingstown and Narragansett

(1959), Hopkinton (1988), Richmond (1975), Exeter (1960), North Kingstown (1950s), West Greenwich (1984), Coventry (1989), Foster (1971), Glocester (1956), Cranston (1894), Lincoln (1936), Woonsocket (1908), and Cumberland (1952). Nesting was suspected in Tiverton in 2000, 2001, and 2003, when at least 1 bird was found with a colony of Barn Swallows nesting under a bridge at the southern end of Nonquit Pond (REm, RTL). Birds have been present in summer recently, including 3, Moonstone, 28 July 2012 (B. Sherman), and 1, Richmond, 30 July 2012 (PAB). Two were on territory, Hopkinton, 1 June 2002 (GW)

Nonbreeding
Spring. In spring, Cliff Swallows are uncommon away from breeding locations and appear later in the season than other species of swallows. Whether transient or otherwise, the species has always been notably late for a swallow, and most dates for migrants in recent decades have been in May.

Spring maxima. 30, Fosters Cove, Charlestown, 25 May 1941 (DLK); 23, Middletown, 19 May 1900 (ESt). A count of 30 in the Great Swamp 20 May 1990 was unique for more recent years (DLK). There are numerous transient records for late May into early June. The latest transients or wanderers were 2 sitting on wires in heavy fog at Moonstone 12 June 1985 (DLK) and 1 at Newport 21 June 1970 (RTL).

Fall. In fall, migrants are scarcer than in spring. There are relatively few records for August and September, and the species has sometimes been unrecorded as a migrant in fall. Transients appear at the coast in July: 2, Napatree Point, 16 July 1986 (CRa); 2, Charlestown Breachway, 16 July 2016 (CPe et al.); and 1, Cox Ledge, 20 July 1984 (fide RLF).

Fall maxima. 15, Block Island, 15–20 September 1959 (ASRI); 11, Napatree Point, 9 October 1989 (CRa); 9, Napatree Point, 24 September 1989 (CRa).

Extreme dates. 2 April 2013, Great Swamp (BCG), to 15 November 2004, Napatree Point (CRa).

Cave Swallow (*Petrochelidon fulva*)

Status
Vagrant. Now regular most years in small numbers, mainly during early to mid-November.

Nonbreeding
Three Cave Swallows were seen on the morning of 7 November 1999 during a major irruption to the Northeast: 2 passing east to west at Napatree Point (SSM) and 1 at Point Judith (JSJ, D. Abbott). The species was not recorded in Rhode Island again until 2003, but since then it has been reported in all but 4 years 2003–2016. Major

incursions occurred in 2005, 2007, 2008, 2010, and 2012. A total of 23 birds were seen 1 November–8 December 2012, including 5, Napatree Point, 15 November (CRa); 4, Quicksand Pond, 1 November (M. Resch); 4, Beavertail, 6 December (K. Mueller); 3, Point Judith, 8 December (J. Magill). It seems worth enumerating the recent occurrence of this rather extraordinary vagrant:

2003: 1 report of 1 bird
2004: 9 birds reported between 4–27 November
2005: 39 birds reported between 8–19 November, plus 11 unidentified *Petrochelidon*
2007: 24 birds reported between 8 November and 6 December
2008: 55 birds reported between 8–19 November
2010: 21 birds reported between 24–27 November
2013: 1 report involving 3 birds
2015: 10 birds between 14–17 November (see Photo 24)
2016: 1 Moonstone Beach 24 October (SSM), plus 1 unidentified *Petrochelidon*
2020: 1 Trustom Pond on 9 May 2020 (Bill Thompson)

KINGLETS (REGULIDAE)

Ruby-crowned Kinglet (*Corthylio calendula*)

Status
Common spring and fall transient, uncommon in winter.

Nonbreeding
Spring. Peak numbers move through in late April and early May; Ruby-crowned Kinglets are rare after mid-May. Late migrants include 1 at Green Hill 5 June 1975 (DWa) and another at Block Island 8 June 1973 (RLF).

Spring maxima. 35, Block Island, 12 May 2002 (DAt, STs); 24, Swan Point Cemetery, 5 April 2000 (RFa).

Fall. Ruby-crowned Kinglets first appear in late September (earliest: 1, Kingston, 6 September 1995 [DLK]), and they are most numerous during the first half of October. There are counts along the mainland coast to only 30 per day, but largest counts are from Block Island.

Fall maxima. 365, Block Island, 3–4 October 1981 (RFa et al.); 350, Block Island, 9 October 1995 (RLF, DLE).

Winter. These kinglets were practically unknown in early winter until 1949, but intensive coverage on CBCs in recent years has shown them to be fairly regular in coastal woods and thickets in late December.

Early winter maxima. Nine in the Glen, Portsmouth, 20 December 1975 (RLF et al.); up to 20 on the Newport–Westport CBC; 24, Newport–Westport CBC, 16 December 2006; and 6, Cumberland, 1 December 2001 (MLy).

Midwinter maxima. Two, Warwick, 22 January 1994 (DWa). There are an increasing number of reports of singles in midwinter throughout the state. One survived at Jamestown from December 1971 to at least 20 March 1972 but was found dead a week later (MDa). One bird attended a suet feeder in Jamestown 5 November 1988 to 16 April 1989 and perhaps the same bird again 2 November 1989 to 4 April 1990 (ChP, CaP). There are many January and February records, as well as 3 from March, since 2005.

There is a summer record of 2 at Scituate, 22 July 1999 (fide RFa); Ruby-crowned Kinglets have bred in western Massachusetts and a singing male was present in Ipswich, Massachusetts, in a planted spruce grove during summer 1977 (Veit and Petersen 1993; Petersen and Meservey 2003), so the possibility of a Rhode Island nesting is not outlandish.

Golden-crowned Kinglet (*Regulus satrapa*)

Status

Common to very common fall and fairly common spring, transient; fairly common winter resident. Irruptive. Rare and local breeder.

Breeding

Beginning in the 1960s, Golden-crowned Kinglets were found during late spring and summer in planted spruce groves in Rhode Island. Douglas L. Kraus and RLF investigated such a grove near Watchaug Pond, Charlestown, on 1 June 1973, and found a pair of territorial birds but no further evidence of nesting. At Charlestown, these adults were seen feeding a young cowbird 26 June (KKi). Stimulated by this find, RLF examined another similar stand of planted spruce east of the Scituate Reservoir's main dam 2 June 1973 and found 2 more pairs there. Since then, territorial birds were present at both locations through 1985. In sum, 5 or more pairs were present at 4 sites almost every year 1973–1985. Additionally, 1 was at Ell Pond, 17 June 1999 (JGe, RGe).

These Rhode Island nestings were part of a broader pattern of invasion of spruce plantations in Massachusetts, New York, and Pennsylvania. Compared with other sites in the Northeast, the Charlestown location is unusual for its elevation (50 ft [15 m] above sea level) and proximity to the ocean (4 mi [6 km]). The plantations used in Rhode Island are primarily composed of Norway spruces; balsam firs were also planted in the WPA era of the 1930s (cf. Petersen and Meservey 2003). Hurricane Gloria devastated some of the groves at the Scituate Reservoir 27 September 1985, and no Golden-crowned Kinglets have been found in summer since then, except for

1 seen in a grove of white pines on Frosty Hollow Road, Arcadia, 3 July 1988 (DLK, DRu). Hurricane Bob destroyed stands of conifers in the Burlingame area in 1991, and Golden-crowned Kinglets have not been seen there during summer since.

Nonbreeding

Spring. Golden-crowned Kinglets migrate north in late March and early April, the exact date depending on weather. The latest date away from breeding habitat is for 1 at Green Hill 14 May 1976 (DWa).

Spring maxima. 200+ in the brush on Sachuest Point during a record heat wave, 6 April 1991 (RRo); 25, Block Island, 12 April 1953 (A. M. Bagg).

Fall. The earliest fall record at Napatree is 16 September 2001 (CRa). The numbers of migrants vary from year to year; e.g., at Block Island the number banded in 1977 was 0, while in 1972, 157 were banded, including 39 on 14 October (ELa, HLa). It is a frequent sight in October to find exhausted and pathetically tame Golden-crowned Kinglets hopping about on lawns and low shrubs searching for food, usually after a sharp cold snap.

Fall maxima. 445, Block Island, 9 October 1995 (RLF et al.); 250 there, 17 October 1976 (RLF); 200 there, 3–5 October 1986 (RFa et al.).

Winter. Golden-crowned Kinglet abundance fluctuates from year to year, and totals for the Newport–Westport CBC have varied from 0 to 84.

Winter maxima. 222, South Kingstown CBC, 21 December 2013 (STs et al.); 124, South Kingstown CBC, 22 December 2012.

WAXWINGS (BOMBYCILLIDAE)

Bohemian Waxwing (*Bombycilla garrulus*)

Status

Vagrant. Highly episodic.

Historical

A record of six Bohemian Waxwings near Blackstone Boulevard, Providence, 24 January 1912 (Mrs. Gladdings, listed in HSH's notes), is the only record listed prior to the 1973 occurrence below. Reportedly she "clearly saw the white tips of the wing quills and that they were larger than a cedarbird." Wright (1921a), however, described a large flight into New England in the winter of 1918–1919 and quoted HSH and Alice Hall Walter as both stating that they knew of no authentic record for Rhode Island in that year or in any previous year. The absence of records for Rhode Island is puzzling, given the long history of occurrence in Massachusetts and New York (Veit and Petersen 1993; Forster 1994; Levine 1998).

Recent records. A single Bohemian Waxwing was seen with Cedar Waxwings 7 March 1973 at Fiskeville by RGe and JGe, both previously familiar with the species. A flock of 24 Bohemian Waxwings appeared in RLC's yard in Jamestown 23 November 1995. One was at West Kingstown 22 December 2007 (SSM). Two were at Block Island 30 March 2008 (DFi, STa), and another was at Genesee Swamp, South Kingstown, 11–12 April 2008 (PLE, CS, CPe).

Cedar Waxwing (*Bombycilla cedrorum*)

Status

Common transient and resident. Varies widely in abundance according to supply of fruits and berries.

Breeding

In some years, waxwings are absent in early summer but then arrive in flocks by midsummer to breed. Many build nests during June. Generally, they are more numerous near the coast, where there is an abundance of fruiting trees and shrubs.

Nonbreeding

Cedar Waxwings depend for food on fruits of wild and ornamental shrubs, and their erratic movements seem to track the fruiting patterns of these plants. Flocks may remain all winter, and small groups are often found in early spring, but there is little sign of major spring migration until the last week of May. Migration is often very noticeable in early June.

Spring. Cedar Waxwings are especially erratic in their occurrence at all seasons, both in terms of abundance and dates when peak numbers appear. Usually, they arrive rather late in spring (mid-May or so) and start breeding later than most other passerines.

Spring maxima. 100, Point Judith, and 100, Bradford, both 27 May 1984 (RLF, CRa); 100, Lincoln Woods, and 100, Providence, both 25–27 May 1941 (v.o.).

Fall. They are most common and widespread in late summer and early fall, when they are often seen feeding fledged young well into September. Migrant flocks are highly visible in fall and common throughout the state, and the largest groups are seen near the coast.

Fall maxima. 800, Block Island, 29 September–1 October 1989 (RFa et al.); 500, Block Island, 26 September 1981 (DWF). Migrants are often seen through November; e.g., 50, Watch Hill, 5 November 1983 (CRa).

Winter. Cedar Waxwings are highly mobile and opportunistic in winter, and during the colder months nomadic flocks are commonly encountered in suburban areas of the interior, where fruiting shrubs are abundant. They were particularly numerous during the mild winter of 1952–1953, and again during the winters of 1956–1957 and 1958–1959.

Winter maxima. 600, Watch Hill, 11 December 2005 (GW); 363, Newport–Westport CBC, 14 December 2013 (REm et al.); 350, Warren, 23 November 1952 (RBo); 150, Block Island CBC, 28 December 1992. Large flocks are occasionally seen late in winter; e.g., 210, Beavertail, 13 March 1995 (RLF); 130, Lincoln, 18 March 1989 (SDz).

SILKY FLYCATCHERS (PTILIOGONATIDAE)

Phainopepla (*Phainopepla nitens*)

Status
Vagrant. One record.

Nonbreeding
An immature female of this southwestern species was caught in a mist net 14 November 1975 at Block Island by ELa; it was first identified by HLa and examined in the hand the next day by DLK, CWo, and ESa. It was subsequently taken to AMNH, where it was compared with museum specimens and later placed in the captive collection of the New York Zoological Society. This individual, and another at Nantucket, Massachusetts, in October of the same year (Veit and Petersen 1993), are the second and third records of Phainopepla east of Oklahoma.

NUTHATCHES (SITTIDAE)

Red-breasted Nuthatch (*Sitta canadensis*)

Status
Common and irruptive transient, fairly common winter resident, and rare breeder. Abundance fluctuates from year to year as a consequence of southward irruptions.

Historical
Charles Doe found the state's first Red-breasted Nuthatch nest in a "hemlock and laurel swamp" in Glocester 26 May 1913, and collected the female, nest, and eggs (eggs extant; FlMNH #55427). Doe's notes acknowledge this find as a "rarity for the state" (though he never reported it) and indicated that he knew of "two or three other sets being secured," finds apparently unknown to HSH. The only other suggestion of breeding before modern times was 1 July 1932 in "Arcade Woods," east of the

East Providence Reservoir's South Pond, where five birds including several young were seen in large white pines and eastern hemlocks (WBD). This location is on the Massachusetts/Rhode Island boundary, and the nest may not have been in Rhode Island. Bull (1974) stated that a considerable range extension into lower elevations in New York had been under way in the 1960s, due primarily to the maturation of planted spruce groves and increased occupancy of eastern hemlocks and pitch pines, and the same applies to Rhode Island. There are several records from Greene of birds in white pines during three summers 1958–1964, including one excavating a nest hole (FPe). On several other occasions 1966–1975, birds were seen in pitch pines during June near Wood River in Exeter, but no breeding evidence was secured (MDa, KKi). Three were displaying in 1971 and 1972 in mixed pines along Molasses Road, West Greenwich (KKi, MDa). Following a heavy flight in fall 1973, Red-breasted Nuthatches were reported in June from about 10 locations, mainly in western Rhode Island, including birds in a spruce stand north of Watchaug Pond and as many as 12 others in planted spruces around the Scituate Reservoir.

Breeding

There are the following records of confirmed breeding in Rhode Island: a pair was feeding 2 young at Greene 25–28 June 1976 (FPe); a pair fed a brood of 3 near Stepstone Falls, West Greenwich, 24 June 1984 (RLF); a pair attended a nest in a red maple in Arcadia 20 April 1985 (CRa); 2 pairs fed young in pitch and red pines, respectively, at Carolina 12–13 June 1987 (RLF); a pair at Greene had 2 young 20 June 1990 (FPe); a pair with 2 young was near a spruce grove in Fiskeville 11 June 1993 (JGe, RGe). There are many more breeding-season records from woods containing white, red, or pitch pines or mixtures of these conifers with deciduous trees. Two birds were present in a dense white and red pine grove at the Diamond Hill Reservoir 12 June 1988 (DMK) suggested breeding there. There is 1 breeding record for Block Island: 3 birds and a nest were located 12–14 May 1999, and birds were still present at the nest site through 22 June (CRa, REn).

Nonbreeding

Spring. Spring migration is diffuse and lacks the big aggregations seen during fall.

Fall. Fall migration is protracted, and its timing varies from year to year. In flight years (Bock and Lepthien 1972) migrants appear in August and sometimes even July, but peak numbers almost always occur in October and November.

Fall maxima. 1200, Block Island, 3–4 October 1981 (RFa, WRP et al.); 800, Block Island, 29 September–1 October 1989 (RFa et al.); 500, Block Island, 16 September 2012 (DFi, STa).

Winter maxima. 137, Block Island CBC, 18 December 1995 (CRa et al.), and 61 still present there 19 February 1996 (CRa et al.); 127, Napatree CBC, 23 December 2012 (SSM, GW et al.); 121, Block Island CBC, 22 December 1997 (CRa et al.).

White-breasted Nuthatch (*Sitta carolinensis*)

Status

Fairly common to common resident.

Breeding

Howe and Sturtevant (1899) spoke about White-breasted Nuthatches "breeding locally," no doubt because of the scarcity of their preferred deciduous woodlands at that time. Now greatly increased, they are widespread and common breeders in northern and western Rhode Island and, in smaller numbers, south on the east side of Narragansett Bay to Tiverton and parts of eastern Little Compton. They are rare and local in most of western Little Compton and on the 3 larger Narragansett Bay islands. On Block Island, there is a summer record of 1 at Gunners Hill 16 June 1992 (REn). Sample breeding-season counts by the author include 39, Glocester and Burrillville, 17–23 June 1982; 77, Foster to Coventry, 16–24 June 1986.

Nonbreeding

One was seen 30 mi (48 km) south of Sakonnet 12 September 1929 (HNG). White-breasted Nuthatches occasionally irrupt southward and to coastal areas where they do not breed. At treeless Dickens Point, Block Island, EDi saw 29 in 51 years, 19 in fall (13 September–21 October), 8 in spring (8 April–26 and 29 May), and only 2 in winter. Incursions to Block Island are evident from yearly banding totals of 39 in 1965 and 15 in 1968; a very large daily total of 65 was counted there 3–4 October 1981 (RFa, WRP et al.). Fifteen left Grove Point, Block Island, for the mainland on 9 October 1995 (RLF et al.).

Winter. Along with Black-capped Chickadees, these nuthatches are ubiquitous winter residents at feeders.

Winter maxima. 154, South Kingstown CBC, 22 December 2007; 152, South Kingstown CBC, 2 January 2011 (STs et al.); 42, Burrillville, 29 December 1991 (SDz); 25, Hopkinton, 31 January 1982 (DWa); 5, Block Island CBC 18 December 1995.

CREEPERS (CERTHIIDAE)

Brown Creeper (*Certhia americana*)

Status
Fairly common spring and common fall transient. Uncommon breeder and winter resident.

Breeding
Hathaway "exhausted every habitat where it might occur" but was unable to find Brown Creepers nesting in Rhode Island. Perhaps even more compelling negative evidence is in the records of CED, who never encountered the species in the northwestern part of the state during over a half century ending in the 1920s. In the mid-1930s, however, creepers were heard singing regularly, and probably bred, in white pines and hemlocks at the East Providence Reservoir (DLK et al.). Those groves were destroyed by the 1938 hurricane or were flooded in the subsequent expansion of the reservoir. In 1959, a nest with 6 eggs was found along Fish Road, Tiverton (Baird 1959a), and a pair was seen in Greene (FPe). Since then, they have been observed singing in summer or been found nesting at numerous locations throughout the state south to Wilbour Woods in Little Compton, the shores of the Sakonnet Passage in Portsmouth, and the South County moraine in Westerly. Brown Creepers currently breed in white pine, hemlock, and planted spruce in northwestern sections; in oak woodland without conifers (Tiverton and South County); the pure pitch pine barrens of western Coventry; in swampy hardwoods (Great Swamp); and in a coastal grove of American holly (Little Compton).

Summer maxima. Eight, Frosty Hollow Road, Coventry, 20 June 1989 (CRa, REn); 6 (5 singing) in spruces at the Scituate Reservoir, 10 June 1982 (REn); 6, Curran State Park, Cranston, 26 June 1989 (CRa, REn); 5, Breakheart Pond, 18 June 1983 (ASM, JZy).

Nonbreeding
Spring. Brown Creepers are uncommon spring transients, with peak numbers occurring in mid to late April, decreasing in early May.

Spring maxima. Six, Richmond, 29 March 1975 (RAC); 6, Arcadia, 13 April 1983 (JGe, RGe); 6, Scituate Reservoir, 7 April 1991 (JGe, RGe); 8, Block Island, 26 April 1961 (JMu). The latest recent transients include birds banded at Block Island 18 May 1977 and 19 May 1974 (ELa). Another was seen there 5 June 1973 (fide RLF).

Fall. Brown Creepers are conspicuous migrants at the coast, especially Block Island, where they often alight on observers. For example, on 13 October 1958, 6 creepers alighted on Nisbet and RLF during observations of diurnal migrants at the tip of Sandy Point, Block Island; 55 were counted leaving for the mainland, and birds were virtually everywhere else on the island (ASRI). Other maxima have risen to 50 per day at Block Island in the second week of October; 24 were banded there on 18 October 1995 (HLa, ELa).

Fall maxima. 284 banded, Block Island, fall 1965 (E. Dickerson).

Winter. Brown Creepers are erratic and sometimes rare in winter. They are more numerous in wooded inland districts than in more open coastal areas and are sparingly present on the larger and more wooded Narragansett Bay islands.

Winter maxima. 23, South Kingstown CBC, 21 December 2013 (STs et al.); 6, Block Island CBC, 17 December 2001.

GNATCATCHERS (POLIOPTILIDAE)

Blue-gray Gnatcatcher (*Polioptila caerulea*)

Status

Formerly rare, now uncommon, spring and fall transient. Fairly common breeder. Very rare in early winter.

Historical

Howe and Sturtevant (1899) listed only three records, including three or four seen by F. T. Jencks 23 May 1875 in Providence and another shot in East Providence on the unusual date of 24 June 1875, during one of the coldest and most backward springs in history. Hathaway (1943) listed two additional records: a late-fall bird in 1916 (see below) and one seen at Block Island 11 September 1919 (EDi). At least three were seen in 1947, during a marked flight to Massachusetts and Connecticut. From 1950–1955, from one to five were reported annually in spring, then a large increase occurred in 1956, when 16 were reported, and similar numbers were found in 1958, 1959, and 1961.

"Several" gnatcatchers in the American holly grove at Wilbour Woods, Little Compton, 1 August 1956, suggested the possibility of nesting, and in 1961, a gnatcatcher nest was found in Warwick on 10 May; birds were still carrying food to nestlings on 4 June (KKi). In 1962, birds arrived at the same location but apparently did not nest, and in 1965, a second nest was found at a nearby location (KKi). During the early 1970s, gnatcatchers were still rare; single pairs nested in Charlestown and Westerly.

Breeding

During the period 1982–1986, greatest numbers during the breeding season were present in northwestern Rhode Island. Lowest numbers are in Newport County and elsewhere close to the coast. Blue-gray Gnatcatchers breed in the Tiverton highlands and in central Little Compton, but there are no breeding records or summer reports on any of the Narragansett Bay islands, despite the presence of the species in June close to the western edge of the Bay at such locations as Saunderstown and Warwick.

Nonbreeding

Spring. The earliest records in spring are for 1 at the Great Swamp, 4 April 2002 (CRa) and 1 at Kingston on 5 April 1978 (DWa). Others were at Kingston 7 April 1991 (PAB, FBu), Burdickville 8 April 1989 (CAv), and Coventry 8 April 2000 (MT). The peak of spring migration is in late April, with occasional lingerers into June; e.g., 4 at Block Island 5 June 1972 (RLF).

Spring maxima. 50, Barden Reservoir, 4 May 1996 (Jan Hoftun); 14, Great Swamp, 1 May 2004 (GW).

Fall. Before they bred in Rhode Island, gnatcatchers were more frequent in fall than spring, but this pattern reversed beginning in the 1950s, and they are now much more common in spring. Migration is occasionally evident by early August; e.g., 2 on 6 August 2014, Napatree (CRa). Most are seen in September.

Fall maxima. Five, Sakonnet, 3 September 1995 (REm); 5, Block Island, 17–23 September 1972 (RLF); 3, Point Judith, 3 October 1990 (PAB). There are a number of later records of fall lingerers, including 12+ records for November and 10+ for December: 15 December 2012, Sakonnet Point (REm et al.); 26 December 2015, South Kingstown CBC (STs et al.); 1, Newport–Westport CBC (Briggs Marsh), 17 December 2016 continuing through 9 January 2017 (GDe). One not mentioned by Veit and Petersen (1993) was seen in nearby Acoaxet, Massachusetts, 16 December 1984 (DLE, REm, JCB).

WRENS (TROGLODYTIDAE)

House Wren (*Troglodytes aedon*)

Status

Common summer resident and common transient. Rare in early winter.

Historical

House Wrens were once common everywhere in southern New England, but they declined before 1880, perhaps because of nest-site competition with House Sparrows, introduced during the 1860s. Hathaway (1943) wrote that he saw two pairs on Prudence Island 9 June 1889 but that he had "never seen any around Providence, although they are reported as being found at Johnston and C. E. Doe has found eggs there." He later stated that on 4 June 1893 in Cranston he "heard the song . . . for the first time in this region." These facts, supplemented by a lack of summer records in

HSH's other early notes, suggest a certain rarity. Doe, however, collected 28 sets of eggs in Cranston during the period 1887–1923. Howe and Sturtevant (1899) called them "locally common," but they were unknown as breeders to ESt in Middletown until 1913. Since then, House Wrens have gradually increased in Rhode Island, with the decrease of House Sparrows, reforestation, and climate among the probable contributing factors.

Breeding
House Wrens are now common in woodlands and wooded suburbs and are more numerous inland than in more open coastal areas. They have decreased in heavily suburban locations, where they were more common in the 1930s. On relatively treeless Block Island, House Wrens did not breed until 1971 (ELa), and at least 3 pairs nested there in the early 1980s (RBo).

Summer maxima. 58, Conanicut Island, 20 May 2008 (ChP et al.); 31 pairs in artificial boxes on the Kimball Sanctuary, Charlestown in 1931 (MSo); 21, Burrillville, 8 June 1973 (DWa).

Nonbreeding
House Wrens declined significantly as fall migrants at Kingston, but not Block Island, 1969–2001 (Osenkowski et al. 2012); however, they seem to have substantially increased both as fall migrants and as attempted winterers on CBCs since then

Spring. Arriving summer residents are common and vociferous in spring. General arrival is in early May

Fall. House Wrens are regular migrants, never in any numbers (to only 6 per day per locality), with a peak in early to mid-October. There are now more than 35 records for late December and the first week of January, nearly all on coastal CBCs, and most from Newport County. Multiflora-rose is abundant in this area and is thought to be an important component of wintering habitat for House Wrens.

Winter. One, Charlestown, 10 February 2007 (DFi); 1, Portsmouth, 25 February 2007 (RLF).

Extreme dates. 13 April 1984, Fiskeville (JGe, RGe), and 14 April 1973, Wakefield (DWa), to 14 January 1991, Narragansett (2; Karlson), and February (see above).

Winter Wren (*Troglodytes hiemalis*)

Status

Uncommon spring and fairly common fall transient and winter resident. Rare breeder.

Breeding

The first breeding record for Rhode Island was of a pair with a nest containing 6 young 24 May 1908 in the roots of an overturned tree in the Richmond portion of the Great Swamp (HSH, A. C. Bent, J. Flanagan; a nestling preserved). A newly fledged juvenile was found in a woodpile on the Kraus farm, Kingston, 28 July 1950 (DLK). Since then, Winter Wrens were found to be widespread breeders in cool swamps and along fern-, rock-, and hemlock-studded streams. During the 1980s and 1990s, singing birds were present during summer at many sites throughout the western mainland in Glocester, Chepachet, Coventry, West Greenwich, Exeter, Arcadia, Hopkinton, South Kingstown, Fisherville Brook, Johnston, and Charlestown. Rick Enser wrote David Emerson that he found Winter Wrens on all his 1993 BBS routes in western Rhode Island. There were no late-spring or summer records anywhere east of Narragansett Bay until 29 May 1989, when 3 singing males were located in the rocky bed of Sin and Flesh Brook in the Fort Barton area of Tiverton (DLE). One male was still there 22 June 1989 (DLE), and others were present at the same location in 1992 and 1993. Also, there was 1 male in Tiverton, 12 July 2000 (RFa), but generally absent as a breeder in Tiverton for the last 10–15 years based on Robert Emerson's June BBS routes.

Since about 2000, reports of potential breeders in appropriate habitat and on dates later than 13 May have been much reduced and more concentrated in the northwestern part of the state at sites such as Beach Pond (Exeter), Pulaski SP (Burrillville), George Washington Management Area (Glocester), and Big River Management Area (West Greenwich).

Nonbreeding

Spring. Birds in March could be overwinterers or early migrants, while increased reports during April likely represent northward movement of transients. Late-spring migrants in places where they do not breed include 1 in West Kingston 7 May 1975 (RAC); 1, Swan Point Cemetery, 11 May 2011 (RFa); 1 at Block Island 16 May 1979 (ELa); and 1 singing in Wilbour Woods, Little Compton, 24 May 1975 (DLE).

Fall. Earliest fall record at Napatree is 19 September 2009 (CRa). Fall migrants are most numerous from mid-October to late November, when up to 5 per day per locality are regularly found.

Fall maxima. 24, Block Island VDC, 12 November 2005 (SSM et al.).

Winter. Winter Wrens have probably increased as winterers as forests have regenerated. Wintering birds are most concentrated in wooded habitats not far from the outer coast and lower Narragansett Bay. Winter records are occasional north along Narragansett Bay to Barrington and Swan Point Cemetery, but the species has only rarely been reported at this season farther inland; e.g., at Cranston, Rumford, Fiskeville, North Scituate, Foster, Wallum Lake, and Woonsocket.

Winter maxima. 25, South Kingstown CBC, 21 December 2013 (STs et al.); 21, Block Island CBC, 22 December 2005 (CRa et al.).

Sedge Wren (*Cistothorus stellaris*)

Status

Rare fall transient and very rare winter lingerer. Formerly (until 1940s) a rare breeder.

Historical

Both diminutive and exceedingly secretive, Sedge Wrens have fluctuated greatly in abundance over the years and since about 1970 have been at an all-time low in the Northeast. There were 10 records between 1969 and 1981, singles in 1989 and 1993. There is only one specimen from the state: a mislabeled bird found in the Block Island Collection and reidentified by Ferren, collected at Block Island 16 October 1936.

Breeding

Sedge Wrens wander considerably during the breeding season and are known to begin new "colonies" quite late in the season. Two pairs of Sedge Wrens were found feeding young on private property in Newport County, 23 June 2005 through July (RFa et al.), but did not return the following year. Another was acting in a territorial fashion in Durfee Hill Management Area, Glocester, 18–21 August 2007 (CBr, CRa).

Older breeding records of Sedge Wrens in Rhode Island are open to question, as they are from cattail marshes. This is a habitat in which Sedge Wrens do not breed but Marsh Wrens do. Hathaway's old notes record a trip to Newport with CED 18 June 1893 when the latter found a Sedge Wren nest in cattails at the east end of Easton Pond in a colony of several dozen Marsh Wrens. Doe's notes report collecting this set, but it is no longer present with most of his collection at FlMNH (T. Webber, pers. comm.) and was apparently given to HSH, whose egg collection contains a set for the same date and place labeled "Short-billed Marsh Wren." Hathaway's notes, as well as CED's, indicate that CED later returned alone 2 July 1893 and took another nest and 5 eggs at the same location, perhaps from the same pair. These items were formerly in CED's collection (FlMNH #53650, #55400) but can no longer be located

(T. Webber, pers. comm.). Howe and Sturtevant (1899) did not mention either of these records, and in 1936, HSH wrote RCC, "I have always felt the eggs were a set of albino Long-bills, as they sometimes do lay white eggs." Such an explanation may also suffice for alleged breeding at Newport mentioned in Howe and Sturtevant (1899) as discovered by La Farge (egg date 6 June, year unstated). The same applies to a clutch mentioned by Hathaway (1913) as taken in cattails at Easton Pond 7 June 1879 by Howland, said to be in AMNH. Allen (1909) repeated records given in Howe and Sturtevant (1899). In 1937, a pair was reported on territory in the Third Beach marsh, Middletown, but the details are now lost.

Nonbreeding

Spring. The only spring record is of a singing male in a grassy field on Durfee Hill, Glocester, 27–30 May 1970 (RRi, DLK et al.).

Fall. In previous decades, postbreeding transients had the well-known habit of appearing in late summer along the immediate coast, near salt marshes, in areas of switchgrass. The fact that these birds frequently sang caused further speculation as to the breeding of the species in Rhode Island. The earliest fall transient is 1 at Briggs Marsh 21 July 1941 (DLE). Maxima at Briggs include 6 each 6 August–4 September 1939 and 8 there 8 August 1941 (DLE). A similar record is of 3 at Sakonnet 28 August 1975 (DLE). These records contrast with the seasonal occurrence of migrants now, which is in late September and October, and suggests that the Briggs Marsh birds were breeding there. It seems worth listing all the recent fall records: 1, Avondale, 2–4 October 2005 (GWi, JSJ et al.); 1, Jerusalem, 1 October 2009 (B. Harris); 1, 16 September 2014, Trustom Pond NWR (T. Auer, m.obs.); 1, Pawtuxet River, 22 October 2011 (DFW).

Winter. There are many December records (e.g., 4 Newport–Westport CBC records including 3 on 21 December 1952; DLE et al.), but none since 1967. There are 3 records in January including 1 for 11 January.

Marsh Wren (*Cistothorus palustris*)

Status

A locally uncommon to fairly common summer resident and spring transient; fairly common fall transient. Uncommon in early winter; a few have apparently overwintered.

Breeding

Marsh Wrens nest in the following habitats in Rhode Island:

Common cattail. Common cattail (*Typha latifolia*) is the dominant species of cattail in interior marshes. Rhode Island's acid soils and waters, however, prevent the develop-

FIGURE 10. Marsh Wrens in cattails, © Margaret La Farge.

ment of extensive stands of this plant in most areas, and Marsh Wrens are therefore rare in the interior. Sites with apparently large enough stands are

1. Junction of Pawcatuck River and Buckeye Brook (to 3 pairs in 1961; DLK);
2. Lonsdale marshes along the Blackstone River, the most extensive such habitat in the state (to 15 singing males through the 1980s but none there in 2015–2017; CRa);
3. North end of the East Providence Reservoir in Pawtucket (probably most numerous prior to the building of a higher dam in 1938; to 5 pairs in the 1950s, essentially absent after about 1965 due to the replacement of the cattails by pickerelweed and other plants; 1 pair in 1989 [RLF]); and
4. Hammond's Pond in Pawtucket (wrens formerly "common" until the pond was drained by the WPA in 1937). Most of the stands of this plant in the state have been shrinking, possibly because of acid rain. The introduced purple loosestrife has so far had only a minor impact on Marsh Wren habitat in Rhode Island, in marked contrast to the situation in eastern Massachusetts, where it is a major competitor with *T. latifolia*.

Narrow-leaved cattail. Narrow-leaved cattail (*T. angustifolia*) is largely confined to freshwater ponds near the coast and the brackish margins of salt marshes. The extent of Rhode Island's narrow-leaved cattail habitat has been slowly eroding since the 1950s or earlier because of replacement by *Phragmites australis*. While that introduced European plant is sometimes used by Marsh Wrens for nesting, it is clearly suboptimal.

1. A fringe around the northern and western parts of Briggs Marsh, once a salt marsh but now a brackish coastal pond (maxima 10 pairs in 1973 [DWa]; 20 pairs in 2015 [REm])
2. A small pond southwest of Tunipus Pond, Little Compton (1 pair in 2000)
3. The Long Pond marshes, at least formerly (maximum 30 pairs in 1938 [RBo]; down to 7 pairs in 1985, in almost pure *Phragmites* [CRa]; consistently seen through 2014, but then *Phragmites* control was initiated and cover temporarily gone [REm])
4. Peckham's Marsh south of Nonquit Pond, Tiverton/Little Compton (a few noted in mid-1930s; RCC)
5. Almy's Pond and another small pond along Ocean Drive, Newport (eggs collected by CED; 7 territorial birds in 1985 [CRa])
6. Boyd's Marsh, Portsmouth (to 5 pairs)
7. Easton Pond in Newport/Middletown (formerly abundant; e.g., 25 sets of eggs collected 18 June 1893 [HSH, CED]; 150 seen 5 June 1897 [ESt]; this colony now absent)
8. Two separate marshes behind Second (not present in 2015–2017; CRa) and Third Beaches in Middletown (to 5 pairs)

9. Conanicut Island (1919–1938 at least, no counts or exact location)
10. Wesquage Pond; Moonstone Brook, Matunuck (to 5 pairs through 2016; CRa)
11. Fresh Pond, Quonochontaug (to 10 pairs early in the century and 7 pairs in the 1950s; 6 in 1984 but none in 2015–2017 [CRa])

Salt-water cordgrass. Salt-water cordgrass (*Spartina alterniflora*) is the primary grass in the wettest parts of salt marshes. The best stands have developed where salt marsh peat meets protected estuarine waters, along well-flushed tidal creeks, and around some small ponds and salt pannes that fill at the highest tides. Since the average daily tidal fluctuations in upper Narragansett Bay are about 2 ft (0.6 m) greater than in the ponds along the outer coast (due to a "Bay of Fundy" effect), some of the best stands of *Spartina* are located northward in the state, and most of the few instances of breeding in this habitat by wrens have been in the northern Narragansett Bay area. Locations include the edges of Runnins River; the northwestern inlet of Hundred Acre Cove (in 1950s; RCC); northeastern portions of Hundred Acre Cove (4 males in 1981; SRi); the Palmer River marshes, Warren (3 nests with eggs, 1981; SRi); and Galilee (south of the Escape Road), where 1 or a few pairs bred to about 1958, when the part of this marsh occupied by the wrens was destroyed by state-authorized dumping of dredging spoil taken from Galilee Harbor.

Phragmites australis. Formerly much more uncommon and local in the 1930s, this plant has spread widely in coastal areas and has supplanted many beds of narrow-leaved cattail. While *Phragmites* did not harbor nesting Marsh Wrens until the 1950s, this habitat is now commonly occupied but of unknown quality. Known locations are along the Runnins River (in 2000, 4 pairs; present in 2016 [CRa]); Briggs Marsh (2–3 pairs); Long Pond (7 pairs, 1985; CRa); Galilee (5+ pairs, 1962 on); Green Hill (to 10 pairs); Charlestown Beach (to 10 pairs).

Mixed aquatic vegetation. In addition to the above-mentioned relatively homogeneous plant communities, wrens have been found on territory in mixed aquatic vegetation including buttonbush, various grasses, and sedges; with scattered cattails at Moonstone Brook; in the more open parts of the Westerly Cedar Swamp; in Newton's Marsh, Westerly; in the Great Swamp impoundment; and in central Middletown. There are no river meadows in Rhode Island such as those that have held numerous wrens in parts of Massachusetts, such as those along the Sudbury River (Griscom 1949, Walton 1984).

Recent surveys in 2015 and 2016 revealed disappearances from some former nesting areas, including Lonsdale marshes, Second Beach, Belleville Pond, Fresh Pond, and Long Pond (see above). Marsh Wrens still breed in Runnins River, Moonstone Brook, Trustom Pond, Briggs Marsh, and Round Pond (Little Compton), possibly elsewhere.

Nonbreeding

Spring. The first migrants arrive in mid-April but maximum numbers in "colonies" do not occur until mid-May. Latest spring transient at Napatree Point 15 May 2004 (CRa). Much less common in spring than fall.

Fall. Earliest fall record at Napatree 20 August 2014 (CRa). Peak of migrants at the coast during early October, and some occur into late December along the coast and more rarely in interior marshes. One was seen at Cox Ledge 22 September 2012 (CPe et al.).

Winter maxima. 16, Napatree CBC, 18 December 2005; 12, Little Compton, 23 December 1950 (DLK). There are numerous coastal records for January and February, and some for inland sites: 5, East Providence Reservoir, 19 January 1955 (RLF); 1, Lonsdale, 18 January 1986 (REn); 1, Runnins River, 6 February 2000 (CBr). There are several records of apparent winter survivors near the coast: 2, 20 March 1992, Great Swamp (CRa, ChP); 1, 30 March 1941 at Middletown (WHD); 1 through February and singing 30 March 1991 at Potter Pond (DLK); 1, 1 April 1951 at Tiverton (DLE); and 1, 4 April 1981 at Bonnet Shores (DWa).

Subspecies

The race *C. p. dissaeptus* of the interior has been collected as a fall transient. Breeding series collected by the author and others appear to be intermediate between this race and *C. p. palustris* of the Long Island and New Jersey estuarine marshes, but these specimens could bear more study and may prove closer to *dissaeptus*.

Carolina Wren (*Thryothorus ludovicianus*)

Status

Common to very common resident

Historical

The "Great Carolina Wren" was very rare in Rhode Island during the nineteenth century and for at least the first third of the twentieth century (Jencks 1880a; Howe 1919). The tireless CED, collecting eggs between 1880 and 1925, never encountered this species in Rhode Island, much less its once-coveted nest or eggs. There are, nevertheless, nine general notes in the *Bulletin of the Nuttall Ornithological Club* and *The Auk* between 1880 and 1919 attesting to their presence or nesting in Rhode Island. Sturtevant first found them breeding at Middletown in 1899. A marked increase occurred throughout southern New England in 1907–1911 (Hazard 1907, 1908; Townsend 1909; Forbush 1929). The population then disappeared, likely because of the severe winter of 1911–1912. After almost no records for 20 years, Carolina Wrens reappeared and very gradually increased through the 1930s, and by the summer of 1947 there was a substantial population. This crashed badly in the winter of 1947–

1948 when snowfall accumulated to a record 75.6 in (192 cm) at Providence and 122 in (310 cm) fell in Foster. Douglas L. Kraus counted 10 pairs in Kingston in 1947, and only 1 pair remained in 1948.

Breeding

Prior to the end of the twentieth century, Carolina Wrens were most numerous on the islands of the southeastern part of the state, though they now nest throughout. Lower wren numbers in the western mainland may have reflected a more continental winter climate and (especially in Kent and Providence Counties) drier, more porous, and acidic soils, as well as the oak-dominated vegetation. Like Eastern Screech-Owls, Carolina Wrens seem to thrive best where somewhat richer, swampy woodlands of mature red maple, black cherry, and black tupelo are found close to tidewater, such as on the larger Narragansett Bay islands or in the lowlands of Tiverton and Little Compton. While almost abundant at all seasons in most coastal regions in the 1980s and 1990s, and thoroughly confirmed as a breeder, the number of nesting records in these same areas in the decades prior to 1970 was relatively few, despite a significant winter presence. During the period 1940–1970, Carolina Wrens were much more often reported as fall and winter residents in shore districts than during the breeding season, and coastal areas that were well populated with wrens in early winter were often found to lack them in spring and summer. Despite sharp fluctuations during the last century, Carolina Wrens in recent years have been vastly more common overall than in former decades and in the 1980s and 1990s thoroughly surpassed previous highs in the population; e.g., circa 1916, 1946, 1960. Each succeeding high has exceeded previous numbers. The increase is similar to that of other Carolinian breeders and likely reflects warming climate, increased feeding, and succession in vegetation.

Block Island was colonized as recently as the mid-1970s but has come to host an exceptionally dense population (Mitra and Raithel 2001)—though still subject to periodic reductions due to severe winters. It is interesting to note that Frank Chapman (1908) found a large population of Carolina Wrens on (then as now) well-wooded Gardiner's Island, near Montauk Point, Long Island. There, during the exact period (as early as 1900–1907) when the species was regarded as a "straggler" on the largely deforested main portions of Long Island as well as in coastal Rhode Island and Connecticut, he found numbers that sharply contrasted with those in surrounding areas. Not surprisingly, Forbes and Brewster established the first breeding record for Massachusetts on Naushon Island in the Elizabeth chain in 1901 (Veit and Petersen 1993), a location that then contained the only significantly wooded area remaining in coastal southeastern Massachusetts.

Summer maxima. 126, Block Island, 18–22 June 1990 (REn, CRa); 67, Conanicut Island, 22–25 June 1985 (RLF); 47, Ocean Drive, Newport, 9 June 1986 (RLF); 20, eastern Middletown and Portsmouth, 25 June 1985 (RLF); 25, Prudence Island, 8 June 1990 (REn, CRa).

Nonbreeding

There are few data on migrant numbers for Carolina Wrens. Banding data and observations at locations where they do not breed suggest that peak migration of Carolina Wrens occurs in late April–early May and then again in the first half of October.

Fall maxima. 84, Block Island, 12 November 2005 (VDC).

Winter maxima. December numbers on the Newport–Westport CBC ranged between 62 and 95 for the years 1983–1990 and then jumped to 250 in 1991; 169, Block Island CBC, 21 December 1999; 311, Newport–Westport CBC, 15 December 2012; 247, Block Island, 12 November 2012 (SSM et al.). There was a severe reduction on Block Island after the winter of 1994; the CBC total went from 140 in 1994–1995 to 27 in 1995–1996. There was another reduction after the winter of 2014–2015, when the Block Island CBC total dropped from 121 in 2014 to 26 in 2015, and the Newport–Westport CBC totals plummeted from 302 in December 2014 to 17 in December 2015.

Bewick's Wren (*Thryomanes bewickii*)

Status

Vagrant. One or two records.

Nonbreeding

On 4 October 1969, 1 was caught in a mist net near Clay Head, Block Island, by ELa and shown to 200 participants of ASRI's Block Island Campout (AMNH #793543) (Lapham 1971). An earlier report, suppressed by HSH because of a lack of a state specimen, is of 1 described by J. F. Mehner in a residential section of Newport 13 October 1845. The date and location seem appropriate for a vagrant Bewick's Wren, and Carolina Wrens would have been scarce or absent in 1845.

MIMIC THRUSHES (MIMIDAE)

Gray Catbird (*Dumetella carolinensis*)

Status

Abundant summer resident and common to abundant transient. Fairly common in winter at the coast; less numerous inland.

Breeding

In deforested Newport County, as well as at Block Island, catbirds were uncommon and local around 1900. They are now astonishingly abundant in the coastal shrubbery of these same areas, where the densest populations presently occur. They are ubiquitous, except in the most urban areas, where they nevertheless breed in the smallest patches of habitat. Their preference for thickets and woodland understory makes them most common in the eastern and southern portions of the state, where they often outnumber all other landbirds. There can be few places in North America with a greater density of catbirds than parts of the larger Narragansett Bay islands such as Conanicut and Prudence. Even in the dry oak woods from Burrillville to Hopkinton they outnumber all other birds. Sample counts by the author include 409, Glocester to Burrillville, 17–23 June 1982; 565, Cumberland to Lincoln, 16–22 June 1984; 823, Foster to Coventry, 16–24 June 1986; 370, Exeter to West Greenwich, 29 May–5 June 1983; 711, Narragansett to South Kingstown, 23–28 May 1984; 1530, Conanicut Island, 22–25 June 1985; 279, Newport Neck, 9 June 1986.

Nonbreeding

Gray Catbirds declined significantly as fall migrants at Kingston but not Block Island, 1969–2001 (Osenkowski et al. 2012).

Spring. The 51-year record of EDi at Block Island includes 14 instances of catbird appearance at treeless Dickens Point 13–29 April with a median arrival date of 23 April; 4 were seen there 19 April 1934. In the interior, the earliest records for arrivals in areas where they did not winter are for 1 at Fiskeville 17 April 1989 (JGe, RGe) and 2, Rumford, 21 April 1951 (BRi). Most spring migrants pass through in early May, when counts of 150 per day per locality have been made. The latest unassailable record of a transient or wanderer is for 1 at Cox Ledge 8 June 1983 (RBo).

Fall. In late summer, a gradual exodus of resident birds begins in August, and by mid-September only a few are present in the interior. Catbirds are among the most abundant migrants in September in most parts of the outer coast such as Block Island, where peak migration is in mid to late September (to 200 per day per locality; most numerous at the coast). According to Kim Gaffett, banding studies on Block Island suggest that most of the island catbirds and their offspring leave the island in late summer before off-island catbirds arrive on fall migration or for wintering.

Winter. Gray Catbirds were first found during winter in 1902 (Purdum 1902) and again in 1910, 1924, and 1926. They have dramatically increased as winterers since that time, though abundance varies from year to year depending on the mildness of the weather and the availability of fruits and berries.

Winter maxima. 98, Napatree CBC, 18 December 2005 (SSM et al.); 96, Block Island CBC, 22 December 2005 (CRa et al.); 90, Newport–Westport CBC, 17 December 2005 (REm et al.); 31, Conanicut Island, 27 December 1957 (JBa). Gray Catbirds regularly survive through February at coastal sites, as demonstrated by the Block Island PDCs, with a 21-year average of 14 birds and a maximum of 48 recorded on 20 February 2006.

Brown Thrasher (*Toxostoma rufum*)

Status
Fairly common summer resident and transient. Uncommon winter resident.

Breeding
Brown Thrashers inhabit brushy habitats statewide but are nowhere particularly numerous. They are uncommon in many districts that are now dominated by maturing oak woodland and are also scarce in heavily suburban districts. They seem unexpectedly scarce on brushy Block Island; e.g., 6 there, 18–22 June 1990 (CRa, REn).

Nonbreeding
Spring. Birds seen during early April (e.g., 7 April 1980 at Cowesett; KKi), may be local winterers, and general arrival of migrants is in mid- to late April (4 at Block Island on 18 April 1973; ELa). Spring counts have not exceeded 10 per day.

Fall. Most migrants are found from late September to mid-October and rarely are more than 30 seen in a day; numbers have declined since the 1980s.

Fall maxima. 150, Block Island, 25–26 September 1971 (RBo, WBD).

Winter. A Brown Thrasher was found in winter in 1886, but there were few records from then until the 1930s. Since that time, they increased to reach peak numbers in the 1970s and have since declined. This pattern of abundance probably reflects to a certain extent the growth and subsequent disappearance of secondary growth.

Winter maxima. 44, Newport–Westport CBC, 26 December 1970. There are many records of birds surviving through February and March in coastal sections north to Warwick and Barrington. They have been recorded on 11 of 21 Block Island PDCs, with a maximum of 5 on 20 February 2006. Four survived to mid-April 1975 on the Kraus farm, Kingston, by eating cracked corn (DLK). Farther inland, there are fewer records, and there are no late-winter records for the northern half of the state. One was at Eppley Wildlife Refuge 4 April 1988 (HCa).

Northern Mockingbird (*Mimus polyglottos*)

Status
Common resident.

Historical
The first mockingbird for Rhode Island was found in 1877, and five occurrences were known to Howe and Sturtevant by 1899. They were rare for most of the first half of the twentieth century. Wright (1921b) listed 28 Rhode Island records 1900–1921, many of them seen at Block Island by EDi. They were found nesting in Barrington in

1911; near Fields Point, Providence in 1922; and at Charlestown in 1924. A pair bred near the city hospital grounds in Providence, 1949–1952.

Before about 1950, most mockingbirds appeared as vagrants in fall, and they often lingered into, and through, the winter. At Block Island, they often arrived in August and early September and remained through the winter to depart in early spring (EDi).

An invasion occurred during the fall of 1956, when mockingbirds were widely reported, and many remained to be counted on CBCs. The following summer (1957), pairs were found breeding in Chepachet, Greenwood, and Hope; then at Jamestown in 1958, Bristol in 1960, Portsmouth and Tiverton in 1962–1963, and Barrington and Wickford in 1964. By the early 1970s, mockingbirds had become common and conspicuous residents. The population seemed to level off during the late 1970s.

Breeding

During the period 1950–1970, mockingbirds were most common in coastal districts, particularly in Newport County. After that, they increased inland (e.g., 24, Woonsocket area, 8 June 1973; DWa), and this pattern of higher abundance inland during summer has persisted. Sample counts include 23, Glocester to Burrillville, 17–23 June 1982; 72, Foster to Coventry, 16–24 June 1986; 62, Exeter to West Greenwich, 29 May–5 June 1983; 59, Narragansett to S. Kingstown, 23–28 May 1984; 90, Cumberland to Lincoln, 16–22 June 1984 (RLF et al.).

A significant factor in the expansion of the mockingbird population has been the proliferation of multiflora-rose (Stiles 1984). This nonnative species was recommended for introduction by the U.S. Soil Conservation Service after 1938 as an ideal erosion-control agent, a "living fence" for livestock, and an excellent provider of food and cover for wildlife. In the decade after World War II, a vast number of seedlings were distributed and planted by landowners throughout rural areas. Most recently, mockingbird abundance appears to have declined. Counts on the Block Island PDC averaged 9.7 per year from 2007 through 2016, compared with 23.9 per year 1997 through 2006.

Nonbreeding

Mockingbirds are highly mobile in nonbreeding seasons, particularly in early fall and, to a lesser extent, in early spring. The mobility of the supposedly nonmigratory mockingbird is most apparent at Block Island.

Migratory maxima. 30, Block Island, 28 September 1981 and 17 October 1981 (DWF; DLE, REm).

Winter maxima. 222, Newport–Westport CBC, 17 December 1994 (DLE et al.); 187, Newport–Westport CBC, 18 December 2010.

STARLINGS (STURNIDAE)

European Starling (*Sturnus vulgaris*)

Status

Abundant resident and transient. Native to Eurasia.

Historical

European Starlings appeared as natural vagrants in Greenland in 1851 and Labrador in 1878 (Bent 1950); four other specimens were also taken in Greenland during the period 1925–1927 (Bent 1950). Starlings were also deliberately introduced to New York City in 1890 and 1891. Thus, the origins of European Starlings in North America are not so unambiguously linked to the New York releases as has generally been assumed (Cabe 2020). They apparently dispersed on their own to New London, Connecticut, in 1906–1907 (Allen 1909; Zeranski and Baptist 1990) and were first noted in Rhode Island in 1908 at East Providence, where I. R. Sheldon saw eight to nine birds and told HSH that they nested at Silver Spring near Watchemoket Cove (Hathaway 1913). Hathaway himself saw his first starlings (a flock of 32) in Warwick 17 December 1911. Prof. H. E. Walter saw seven in Rumford 2 April 1912, including two entering holes in trees. One was found dead at Bristol in February 1912, and another was collected on Prudence Island 20 April 1912 (CED). At Block Island, Lovell Dickens saw his first starling 2 October 1912, which was followed a month later by a flock of 20. A flock of 200 was seen in Warwick and 1 collected 23 November 1912 (HSH). During December 1912, about a dozen roosted in the steeple of a church in Rumford. Hathaway saw the first individual in his own yard 29 May 1913, and another the following October. An observation of a flock of 40 at Swan Point Cemetery 23 November–6 December 1913 was published as a general note in *The Auk* (Caduc 1914). A pair that nested in an Elmwood Avenue building in 1914 (HSH) was probably representative of an increased scattering of breeders in the metropolitan area. By this time, they also seem to have become more generally distributed in the Providence area at all seasons, and in December 1914, HSH saw a flock of 1000 in Warwick. Hathaway noted them nesting in his yard in South Auburn for the first time in 1915. Doe collected his first starling eggs in 1915, in Matunuck, and obtained four other sets in 1916. At Woonsocket, Arnold saw his first starling 4 March 1916 and noted that they had become "regular" there by 1919. Hathaway began to encounter fall transient flocks along the South Shore in 1916. They were breeding in Burrillville at least by 1924 (CED). Douglas L. Kraus told the author in the late 1990s that by 1924, he regularly watched evening flights of tens of thousands going to their winter roosts in Providence.

Breeding

Flocks of hundreds of starlings are now ubiquitous in metropolitan and suburban areas. Starlings seem to nest in just about any conceivable crevice or cavity in trees, buildings, and bridges. Some have built nests in crevices in the rocky cliffs that have been created by highway construction (e.g., along routes I-295 in Smithfield and 138 in Portsmouth), and others have nested in natural cliffs and rock crevices along the coast (e.g., Little Gooseberry Island, 1997; RLF). Starlings aggressively displace other cavity-nesting species such as Purple Martins, bluebirds, and Red-headed Woodpeckers from birdhouses and natural cavities. Starlings begin nesting early in the year and often raise 2 or even 3 broods.

Nonbreeding

Starlings are often abundant, especially in fall, when many thousands may appear along the coast, particularly after cold fronts.

Fall maxima. 24,000, Little Compton, 24 October 1981 (DLE, REm). The acrobatics of starling flocks are often one of the best indications of hawk movements through coastal areas, and wheeling flocks of attendant starlings can sometimes be seen as far away as 10 mi (16 km). Roosting flocks of several to many thousands are common during fall in patches of *Phragmites* and pine groves.

Winter. Numbers decline somewhat in some coastal areas as winter approaches, and this withdrawal is even more noticeable in many parts of rural western and northern Rhode Island. Most of the starlings of Aquidneck Island have roosted in Fall River in recent years and necessarily drain through Portsmouth, where flocks have totaled approximately 15,000 in several winters (RLF).

NEW WORLD THRUSHES (TURDIDAE)

Eastern Bluebird (*Sialia sialis*)

Status

Uncommon to fairly common transient and summer and winter resident.

Historical

Eastern Bluebirds were common until the 1940s, and after about 1950 they substantially declined and reached a low but relatively stable population in Rhode Island, with only modest decline through the 1960s (Sauer and Droege 1990). The main reasons for the decline of bluebirds in the Northeast seem to be both direct and indirect effects of pesticides, usurpation of nest sites by European Starlings and House Sparrows, and the decline of fruit orchards.

Breeding

Between 1950 and 1977, there were only about 35 reports of nesting bluebirds statewide, most of which were in western and northern sections from Westerly and Green Hill north to Burrillville and Cumberland. Most reports were from Westerly, Charlestown, and Coventry, but there were also reports of nests or birds seen in summer in Cranston, Lincoln, Johnston, and East Providence. They bred on Prudence Island (3–4 pairs up to 1946; WBD), Conanicut Island (to about 1950; MDa), and to 1958 in Portsmouth on Aquidneck Island and in eastern Little Compton (JBa). They bred in eastern Tiverton in 1983–1984 (DLE; CRa, REn). There were 2 pairs (1 with young) in central Tiverton 23–24 June 1991 (RLF) and 2 in Little Compton 4 July 1990 and again on 21 June 1991 (RBo). They have never bred at Block Island.

Breeding bird atlas work in the early 1980s determined that Eastern Bluebirds were still uncommon breeders, being most numerous on sand plains with pitch pine and oak forest. Numbers of nesting bluebirds were also found where Spongy Moth (*Lymantria dispar*) infestations had created clearings in the forest. Some of the densest populations of bluebirds in Rhode Island presently appear to be in the Richmond–Hopkinton area; e.g., 15 territorial adults in these towns, 15–21 June 1987 (RLF).

In addition to oak–pine forests, bluebirds nest in flood-killed trees within extensive wooded swamps. Two nests were found in 1984 in South Kingstown and another in Cumberland in such swamps (RLF). Another was at its nest in a beaver pond within a colony of Great Blue Herons in West Greenwich 15 June 1995 (RLF). Burning and clearing by Native Americans may have provided suitable bluebird habitat before the historical period by increasing the number of pines and dead trees (Bromley 1935).

Nonbreeding

Spring. Migration peaks during late March and April, and late migrants have been recorded into June; e.g., 1, Point Judith, 7 June 1981 (RLF); 1, Block Island, 13 June 1919 (EDi).

Spring maxima. 32, Cumberland, 3 April 2004 (MLy); 25, Kingston, 12 March 1951 (DLK).

Fall. Only the oldest observers remember that bluebirds were once common transients in fall. Sizable flocks of diurnal migrants were seen near the coast or in roving flocks over agricultural areas, with lesser numbers reported from the interior.

Fall maxima. "Hundreds" at Block Island 29 October 1939 (EDi) and 1–2 November 1949 (EDi); 100 migrating over Kimball Sanctuary, Charlestown, 27 September 1932 (MSo); 69, Block Island, 8 November 1999 (CRa et al.).

More recently, it has been unusual to see more than 50 in a day, even at Block Island. This is particularly true on the Newport–Westport count, where in the period 1949–1966, 120 bluebirds were recorded, in contrast to only 1 seen in the period 1967–1990. Lacking for years in winter at Block Island, 1 was noted there 19 December 1989 (CRa et al.). In a bygone era, they were frequently encountered in winter near Providence; e.g., 6 during a train ride between Providence and Wickford Junction 5 January 1900 (R. Hoffmann); 5 wintering near HSH's home in South Auburn, Cranston, in 1899–1900. Numbers have increased steadily since the 1990s, as illustrated by CBC totals at Napatree and South Kingstown.

Winter maxima. 85, South Kingstown CBC, 22 December 2001 (STs et al.); 65, Newport–Westport CBC, 15 December 2012 (REm et al.); 65, South Kingstown CBC, 27 December 2014 (STs et al.); 22, Westerly, 12 February 1945 (ESa).

Mountain Bluebird (*Sialia currucoides*)

Status

Vagrant. Two records.

Nonbreeding

A female Mountain Bluebird was salvaged, along with a number of more common birds, from a highway in southern Rhode Island by JZy during the winter or early spring of 1994–1995. A year or more later, CRa examined the frozen birds and picked out the Mountain Bluebird. The identification was later confirmed by A. Andors at AMNH, where the specimen agreed with a series of female *S. currucoides* "in all particulars." Another was in Fort Getty Campground 10–22 November 2012 (Marcie Lindsay et al.; see Photo 25), and a female was found at the Big River Sandpits, West Greenwich, 10 April 2018 (Dylan Pedro; photos). A number of records for the Northeast have accumulated, including a group of 3 from Cape Cod in 1995.

Townsend's Solitaire (*Myadestes townsendi*)

Status

Vagrant. Three records, all in late fall.

Nonbreeding

The first record of Townsend's Solitaire for Rhode Island was of an individual discovered along the east side of the Kickemuit River in Warren 19 November 1978 by RBo during his 39th annual Thanksgiving census of "Point Farm" (Bowen 1979). This bird remained through most of the winter, feeding on fruits in the rich thickets of the same area and accommodated many who were able to observe it at close range (numerous photographs by RBo, Darling, RLF, and others). It was last reported 11 February 1979 (RBo). A second was at Montaindale Pond, Smithfield, 23–29 March 2002 (Deb and Dan Berard Sr.), and a third at Burrillville 31 December 2011 (SDz et al.).

Veery (*Catharus fuscescens*)

Status

Common summer resident and transient.

Breeding

Veeries are one of the most common forest nesting birds in Rhode Island, especially in moist deciduous forests. A relatively high density of 109 was recorded on 60 point counts in Burlingame Management Area 6–14 June 2006 (CRa). Sample counts by RLF include 214, Glocester to Burrillville, 17–23 June 1982; 366, Foster to Coventry, 16–24 June 1986; 159, Exeter to West Greenwich, 29 May–5 June 1983; 129, Narragansett to South Kingstown, 23–28 May 1984. They are much less common in the Narragansett Basin and south of the South County moraine. They are common on some of the larger Narragansett Bay islands (e.g., 38 on Prudence, 6–8 June 1990; REn, CRa), and on nearby Patience Island (9 pairs, 4 June 1985; RLF, CRa, REn). In northeastern Rhode Island, they are far less common than they are a few miles to the west (only 16 from Cumberland to Lincoln, 16–22 June 1984; RLF). Veeries are common breeders in parts of Tiverton and Little Compton but do not breed at Block Island.

Nonbreeding

Veeries declined significantly over the years 1969–2001 as fall migrants at both Kingston and Block Island banding stations (Osenkowski et al. 2012).

Spring. Veeries are common transients in spring and early fall. The earliest spring record involves 1 at Swan Point Cemetery 21 April 1986 after passage of a warm front (RBo). There were a number of records in 1983 following the passage of a storm: 2 at Sakonnet 25 April 1983 (RBo, HMP, RKl); 4 banded at Block Island 26–28 April 1983 (ELa). Late transients have been banded at Block Island to 2 June; a late transient was noted on Gooseberry Island, Newport, 10 June 1982 (RLF).

Fall. Veeries are one of the earliest transient thrushes. Night calls of migrants are often heard from mid-August on, and peak numbers are at coastal points during the first half of September. Sixty were heard migrating over Narragansett Bay 24 August 2008 (GDe). Later records include birds banded at Kingston 14 October 1974 (DLK), Middletown 15 October 1959 (JBa), and Block Island 19 October 1980 (ELa).

Fall maxima. 14 banded, Block Island, 9 September 1970 (ELa).

Gray-cheeked Thrush (*Catharus minimus*)

Status

Uncommon migrant.

Nonbreeding

Gray-cheeked Thrushes were formerly rare transients but perhaps overlooked owing to their shy nature and the difficulty of distinguishing them from other thrushes; only 1 record was known to Howe and Sturtevant (1899). They are uncommon now, and numbers of migrants recorded are always smaller than for Swainson's Thrushes. Gray-cheeked Thrushes generally pass through later than other thrushes in both spring and fall.

Spring. The earliest records in spring and the only April occurrences were associated with a storm: 4 banded, Block Island, 26–29 April 1983 (ELa). Early May dates include 1 well seen at Swan Point Cemetery 3 May 1979 (SDz, CWo) and 1 each at Kingston and Swan Point Cemetery 5 May 1983 (DLK; RBo). Most other records are after 10 May, and peak migration is during the last half of May. Counts (e.g., at Swan Point Cemetery) rarely exceed 3 per day.

Spring maxima. 25, Swan Point Cemetery, 27 May 1996 (RFa, DFi); 7, South County, 18 May 1947 (RCC, WBD).

Fall. The earliest fall record is for 1 at Point Judith 29 August 1990 (PAB).

Fall maxima. 33 banded, Block Island, 9 October 1968 (EDi, ELa), a remarkable number for anywhere on the East Coast; 4 banded, Block Island, 9 September 1983 (ELa). Peak numbers have occurred in early October in recent years. Late individuals include 1, 3 November 1974, Warren's Point (DLE); another on 6 November 1974 at Fiskeville (JGe, RGe), and 2 banded 6–7 November 1976 at Block Island (ELa). Totals banded in fall on Block Island in 1968 and 1970 were 173 and 146, respectively.

Bicknell's Thrush (*Catharus bicknelli*)

Status

Rare migrant.

Nonbreeding

Rhode Island specimens of *bicknelli* were said to have been collected in the past but none are now extant. Bicknell's Thrushes are certainly less numerous as migrants through Rhode Island than are Gray-cheeked Thrushes, as confirmed by examination of mist-netted birds, but they occur during approximately the same dates in both spring and fall as do Gray-cheekeds. The status of Bicknell's will undoubtedly become clearer as records of this recently split species accumulate. Five recent banding records include 3 birds carefully measured by DLK at Kingston 7 October 1991, 3 October 1992, and 25 September 1993, and 2 at Block Island 24 September 2016 (SCo) and 23 October 2012 (KGa). Two Bicknell's were identified among 25 Gray-cheeked Thrushes, by song and call note, in Swan Point Cemetery 27 May 1996 (RFa, DFi). Pyle (1997) details the complexity of distinguishing Bicknell's Thrushes from Gray-cheeked Thrushes, as well as from other species of *Catharus*, especially Veery. Many of the records listed above may best be regarded as tentative, given the close similarity of Bicknell's and Gray-cheeked Thrushes.

Swainson's Thrush (*Catharus ustulatus*)

Status

Uncommon spring, and fairly common fall, transient; formerly more numerous. All large counts are at least 30 years old, despite an increase in overall effort in recent years.

Nonbreeding

Swainson's Thrushes declined significantly over the years 1969–2001 as fall migrants at both Kingston and Block Island banding stations (Osenkowski et al. 2012).

Spring. In spring, Swainson's Thrushes are particularly erratic; they are sometimes numerous at Swan Point Cemetery but absent in other years and are particularly uncommon as spring transients in the southeastern sections. The earliest record is of 1 banded, Block Island, 26 April 1983 after a storm (ELa). First migrants usually arrive in early May with peak migration 17–25 May.

Spring maxima. 150, Block Island, 18 May 1957 (WBD, DLE); 20, Swan Point Cemetery, 25 May 1974 (DLE, REm); 15, Block Island, 25 May 1981 (STs); 9, Swan Point Cemetery, 16 May 2005 (RFa et al.), a very large count by recent standards. Transients occur to the end of May (1, Swan Point Cemetery, 1 June 2000; RFa, STs).

Fall. In fall, Swainson's Thrushes are common at the coast, particularly at Block Island, where they are probably the most common thrushes after American Robins and much more common than either Wood Thrushes or Veeries. They occur earlier in fall than Gray-cheeked Thrushes, with peak numbers in late September to early October. The earliest records are for 1 at Little Compton 16 August 1953 (DLE) and for birds banded at Kingston 18 August 1972 (DLK) and Middletown 20 August 1955 (JBa).

Fall maxima. 40, Block Island, 2–5 October 1987 (RFa et al.); 25, Block Island, 25 September 1954 (RIOC). They remain fairly common in coastal thickets through the first third of October. There are reliable reports through late October (banded, 31 October) and others into November to 18 and 23 November. One was well seen at Block Island on 14 December 1976 (SRD).

Hermit Thrush (*Catharus guttatus*)

Status

Fairly common summer resident; fairly common spring and common fall transient. Uncommon in winter.

Historical

Hermit Thrushes were first found nesting in Cranston in 1900, and others were in Washington in 1905. Hathaway located three singing males in Washington in June 1907. The general rarity of this species before the turn of the twentieth century is underscored by the fact that CED failed to find nesting birds in his first 35 years of tramping northwestern Rhode Island, and only managed to find four nests in Glocester during the period 1914–1925. A total of 12 birds in Burrillville 8 June 1946 was for that time a remarkable number (RCC, WBD, DLK). As with many other woodland species, Hermit Thrushes benefited from post-agricultural reforestation. During the last half of the twentieth century, they increased and are now locally common summer residents.

Breeding

Hermit Thrushes are most abundant in dry oak and white pine forests of western and northwestern Rhode Island, where in some regions they outnumber Wood Thrushes. Sample summer counts give some idea of their abundance: 49, Glocester to Burrillville, 17–23 June 1982; 59, Foster to Coventry, 16–24 June 1986; 46, Exeter to West Greenwich, 29 May–5 June 1983; 25, Exeter, 20 June 1981 (RLF); 27, George Washington Management Area, 26 June 2004 (MLy). They are present in smaller numbers south through most of the interior of South County, particularly in the South Kingstown and Charlestown pine barrens, on parts of the South County moraine, and in rocky oak woods to within a few miles of the mainland coast and Narragansett Bay. They are local in drier parts of Cumberland and Lincoln, unreported in the stretch from Pawtucket to Bristol, but regular in the granitic Tiverton hills south to the vicinity of Lafayette Road. Still farther south, singing birds were found in Wilbour Woods, Little Compton, in the summers of 1952 and 1954 (DLE). A "good colony" was found in deep woods north of the center of Conanicut Island in the summer of 1938 (H. Russell Jr.), where a limited amount of oak–pitch pine habitat is still present.

Hermit Thrushes are most common among white pines, eastern hemlocks, or pitch pines. On the other hand, territorial birds are also common in places as widely separated as Burrillville and Charlestown at sites where no evergreens of any kind are found. Usually, such situations are rocky, oak-laden, sterile hillsides with underlying granite and silicaceous soils, often with a dense cover of huckleberries. Thus, it seems that evergreens enhance the habitat quality but are probably secondary in importance to the presence of acidic, dry, or rocky slopes covered with ericaceous plants. Curiously, these same modest-looking Rhode Island woodlands are also well tolerated by several other species elsewhere regarded as relative specialists, such as Veery, Pine Warbler, and Black-throated Green Warbler.

Nonbreeding

Suspected migrants have appeared as early as 25 March; 31 March 1919 at treeless Dickens Point (EDi); and 31 March 1989, Swan Point Cemetery (DLE). There are numerous records from unlikely wintering areas 1 April on. The peak of migration is in mid to late April, while diminishing numbers occur through the first week of May, and late transients occur to about mid-May. One was at Block Island 18 May 1958 (WBD), and another was banded there 19 May 1977 (ELa). Hermit Thrushes declined as fall migrants at Block Island but not Kingston from 1969 to 2001 (Osenkowski et al. 2012).

Spring maxima. High counts are on the order of 10 birds per day per locality.

Fall. Fall migrants may appear where they do not breed by mid-September (1, 5 September 1958 at NBS, Middletown; JBa), and they are regular on the coast from 20 September on, with the largest coastal flights in late October.

Fall maxima. 100, Block Island, 27 October 1928 (EDi); 67, Block Island VDC, 12 November 2005 (SSM et al.); 64, Swan Point Cemetery, 3 November 2003 (RFa); 50, Block Island, 28 October 1922 (EDi); 40 along the South Shore, 12 October 1955 (DLK); 26 banded, Middletown, 12 October 1955 (JBa).

Winter. This thrush is hardier than others and is a common lingerer in protected coastal thickets in winter. Numbers in late December may fluctuate widely with the severity of the weather and availability of food. They are more difficult to find after mid-January, and by February become relatively rare. There are, nevertheless, numerous instances of proven winter survival. Late-winter records are almost entirely confined to districts not far from the outer coast or Narragansett Bay, occasionally north to Providence. They are scarcer in the interior but are probably overlooked in northern Rhode Island during winter. One was seen in Burrillville 3 February 2000 (JMu), another survived at Summit until at least 15 February 1983 (JZy), and another remained at Fiskeville to at least 8 March 1985 (JGe, RGe).

Winter maxima. 80, Block Island CBC, 21 December 2000; 51, Newport–Westport CBC, 16 December 2000 (DLE et al.); 50, Napatree CBC, 18 December 2005 (SSM et al.); 24, Block Island PDC, 19 February 2001 (SSM et al.)

Wood Thrush (*Hylocichla mustelina*)

Status

Common summer resident. Common spring, fairly common fall, transient.

Breeding

Wood Thrushes are found in greatest densities in rich, leafy woods, a relatively scarce habitat in Rhode Island. They are scarce in tracts of well-drained forest dominated by oak and pine, such as are found in southern South County and in western Coventry, Glocester, and Burrillville, and also in acidic swamps with interspersed drier areas of oak. Wood Thrushes are usually outnumbered by Veeries in most of western Rhode Island. Ferren censused thrushes in 8 towns in western Rhode Island during the 1990s and found 868 Veeries and 515 Wood Thrushes. In more mesic or richer vegetative conditions in the towns of Cumberland and Lincoln, corresponding counts obtained from a week's field work in June 1984 were 16 Veeries and 160 Wood Thrushes (RLF). The highest Wood Thrush densities found by RLF were 43 during a 5-mi (8 m) hike, Limerock, 20 June 1984 (RLF).

Wood Thrushes are common in interior wooded sections of the 3 larger Narragansett Bay islands: 66, Conanicut Island, 22–25 June 1985 (RLF); 26, Prudence Island, 6–8 June 1990 (REn, CRa); and they are even present on tiny Patience Island (2 pairs, 4 June 1985; REn, CRa, RLF). Offshore at Block Island, singing birds have been sporadically present at only 2 locations since the early 1970s but breeding has not been confirmed.

Nonbreeding

Spring. The earliest spring date is for 1 at Westerly 21 April 1951 (ESa). There are at least 31 other April records from 23 April on, including 4 at Jamestown 29–30 April 1957 (MDa) and 3 each at Kingston, 30 April 1968 and 30 April 1984 (DLK). The bulk of migrants appears during early May.

Fall. At Block Island, Wood Thrushes are banded far less commonly than any other thrush, despite their abundance in summer on the adjacent mainland. Among 3233 thrushes banded there in fall during the period 1959–1986, only 2.7% were Wood Thrushes, while corresponding percentages for other species were Veery, 16.3%; Gray-cheeked, 18.7%; Swainson's, 36.7%; and Hermit, 25.3%. Most residents disappear by early September, at a time when most of the other thrushes are increasing. Some increase has been apparent since about 1960 with the species' gradual spread as a breeder from southern New Hampshire into northern New England, and there has also been a corresponding increase in the number of late-fall occurrences. Occasional lingerers have been noted into early October and more rarely to mid-October. Exceptional were 3 at Block Island 17 October 1976 (RLF; HLa; SRD) and 3 banded there 18 October 1972 (ELa). One was at Lincoln Woods 21 October 1978 (REn), 1 was at Kingston 23 October 1966 (DLK), and individuals were banded at Block Island 30 October 1979 and 4 November 1993 (both ELa).

Redwing (*Turdus iliacus*)

Status
Vagrant. One record.

Nonbreeding
A Redwing was found in South Kingstown 26 December 2005 (CNu), where it was carefully studied and heard singing. This is the only record for Rhode Island, though there are single records from New York in 1959 and New Hampshire in 2016 and multiple records from Newfoundland (1950 on).

American Robin (*Turdus migratorius*)

Status
Abundant summer resident. Abundant transient and common winter resident. One of the most familiar of Rhode Island birds, they are most noticeable in spring and early summer as ubiquitous birds of lawns and shade trees.

Breeding
Robins nest virtually everywhere in the state, except in the densest forests. They are particularly abundant in open coastal areas of the mainland and have always nested abundantly at Block Island, even during the early decades of the twentieth century when they nested on stone walls because of the absence of trees (EDi). They are still abundant at Block Island (123 on BBS route, 28 June 1989; REn), but they are rare on such moderately sized Narragansett Bay islands as Hog, Patience, and Dutch, and absent from all of the smaller islands: Hope, Dyer, Big Gould, Little Gould, Big Gooseberry, and Rose.

Nonbreeding
American Robins declined significantly as fall migrants at both Block Island and Kingston 1969–2001 (Osenkowski et al. 2012), which is consistent with their longer-term decrease in New England 1966–2015 recorded by BBS.

Spring. Robins move about a good deal during the winter, and in mild seasons, singing males may appear during February. Ordinarily, spring migrants arrive during mid to late March, and peak numbers occur in mid to late April. While substantial movements occur through most of April, the tremendous flights of northbound migrants sometimes encountered in eastern Massachusetts in April (Veit and Petersen 1993) apparently pass substantially to the north of Rhode Island.

Spring maxima. 500, Beavertail, 13 March 1995 (RLF); 300, Perryville, 22 March 1975 (MDa); 300, Charlestown, 10 April 1973 (DWa).

Fall. Robins can be abundant during late summer and fall both in nocturnal roosts and as diurnal migrants, the latter especially at the coast. The largest roost so far detected has been at the East Providence Reservoir, where 5000 were counted entering the pines at dusk during August–September 1952–1956 (RLF, REW, HEC et al.). Numbers of diurnal migrants at the coast peak during late October to early November.

Fall maxima. 8500, Tucker Pond, 24 November 2000 (SSM); 3500 at dawn over Watch Hill, 8 November 2015 (CRa); 2400+ in 1 hour at dawn over Watch Hill, 14 October 1979 (RLF); 1250, Richmond, 2 November 1988 (v.o.).

Winter. The distribution of robins in winter is mercurial, with Newport–Westport CBC totals ranging from 2 to 1863 (in 1994) and South Kingston CBC totals from 6 to 11,500 (in 1994). Wintering birds feed on fruits such as bullbrier and winterberry that are most abundant near the coast; birds move about as supplies of these foods become depleted.

Winter maxima. 22,046, South Kingstown CBC, 26 December 2011; 18,876, South Kingstown CBC, 21 December 2002 (STs); 18,544, Newport–Westport CBC, 16 December 2000 (DLE et al.); 9970 at roost, Tucker Pond, 26 December 1994 (SSM). Flocks of hundreds of birds have been seen with increasing frequency through January and February.

Subspecies

There are 2 sight records of *T. m. nigrideus*, the "Newfoundland Robin": 2 November 1958 at Sakonnet (JBa) and an adult male 21–28 April 1985 at Point Judith (RLF).

Varied Thrush (*Ixoreus naevius*)

Status

Vagrant. Three records.

Nonbreeding

One was found in the yard of Mrs. S. Engdahl in Johnston and seen by many observers 31 January–10 February 1966. Despite its retiring habits, it was identifiably photographed in color by SSD. Another was seen on Doug Kraus's farm in South Kingstown 20 November 1997 to 21 January 1998 (DLK, m.obs.). A third record is of 1 banded and photographed on Block Island, 1 November 2018 (Clara Cooper-Mullin).

These are the only occurrences for Rhode Island, in contrast to 16 or more records for Connecticut (Zeranski and Baptist 1990) and 43 or more for Massachusetts, where they have sometimes occurred in marked "flights" (Veit and Petersen 1993).

OLD WORLD THRUSHES (MUSCICAPIDAE)

Northern Wheatear (*Oenanthe oenanthe*)

Status
Vagrant. About 20 records.

Nonbreeding
There are 2 old observations from Block Island (details lost), and 1 was reported 25 October 1964 near Jerusalem. More recent records include 1 seen north of Great Salt Pond, Block Island, 27 September 1975 (J. Kenneally, I. Black, R. Langly, S. Knox, and many other participants on a Massachusetts Audubon Society excursion to the island); 1 well seen at Charlestown 14 September 1976 by David Roach; 1 at Block Island 1–3 October 1982 (R. Forster et al.); 1 photographed at Block Island 8–11 October 1989 (JDi, AFr, DLE, DFi et al.); and 1 at Block Island 25–29 September 1992 (DSp, BCG, DFi, RBo, DLK, DLE et al.). Singles were at Fogland Point 16 September 2003 (E. Oldmixon); West Greenwich 28 May 2006 (D. Cooper); Middletown 31 May 2007 (CAv); Slocum, North Kingstown, 15–29 September 2008 (PLE et al.); Block Island 5 October 2012 (DFi, STa); Peckham Farm 19–25 October 2013 (CPe, RTL); and Bristol 20–24 September 2016 (M. McBrien, m.obs.).

Subspecies
The subspecies of Rhode Island wheatears is unknown, but 1 found dead in Massachusetts was *O. o. leucorhoa* of Europe, Greenland, and northeastern Canada (Veit and Petersen 1993).

WEAVERS (PASSERIDAE)

House Sparrow (*Passer domesticus*)

Status
Introduced from Europe. Abundant resident and common migrant.

Historical
House Sparrows were "successfully" liberated in Green-Wood Cemetery, Brooklyn, in 1853 (Barrows 1889) and then introduced to Rhode Island as well; e.g., from the *Narragansett Historical Register* of 1884–1885: "The first importation of this bird into America of which we have knowledge is that of Mr. Joseph P. Hazard, about the year 1858, who sent the birds from Liverpool, England, to his nephew (the Hon. Rowland Hazard at Peace Dale, R.I.), who released them here on their arrival. The shade trees at this time were sadly infested with canker worms and various species of caterpillar,

and these birds were thought beneficial in arresting their depredations. The birds in this country have increased wonderfully, and being naturally of the English spirit of pugnacity, they have been pronounced a nuisance. We think, however, there is room for all, and certainly these little fellows make our streets lively at all seasons."

Further redundant introductions were made at Providence in 1866 or before and at Newport in 1874. At least by 1882–1886, they were widespread and common throughout Rhode Island, and by 1888 were referred to as a "perennial nuisance." They reached peak abundance during the late 1800s, when horses provided the major means of transportation, and House Sparrows fed on seeds in the manure. They were soon recognized as pests. During a 2-month period in 1884, the sexton of St. John's Church on North Main Street, Providence, removed two cartloads of nesting materials from the vines around the church and destroyed 970 sparrow eggs.

Breeding

Summer maxima. 319, Richmond to Hopkinton, 15–21 June 1987 (RLF); 165, Cumberland to Lincoln, 16–22 June 1984 (RLF).

Nonbreeding

House Sparrows are, contrary to many statements, migratory, as evidenced by flocks seen passing coastal vantage points; e.g., 75 migrating at Napatree 15 October 2015 (CRa); 50 attempting to leave Block Island's north end after a cold front 8 October 1995 (RLF et al.); 30 migrating, Napatree Point, 5 October 2012 (CRa). At Napatree, House Sparrows (like House Finches) aggregate during late summer, peaking in August and then again in October. At the latter season, flocks gather at the end of Napatree and leave the point flying to the west.

Winter maxima. 1136, Newport–Westport CBC, 18 December 2010 (REm et al.); 973, Newport–Westport CBC, 15 December 2012.

PIPITS AND WAGTAILS (MOTACILLIDAE)

American Pipit (*Anthus rubescens*)

Status

Common fall and uncommon spring transient. Uncommon but increasing in winter.

Nonbreeding

American Pipits are found in the most open of habitats, including agricultural fields, turf farms, and beaches such as Napatree Point.

Fall. The earliest fall arrival dates are for 1 at Napatree Point 21 August 2000 (CRa), 1 at Quonochontaug 24 August 1969 (KKi), and 2 at Matunuck 28 August 1966 (SSD). Peak migration occurs in late October and early November.

Fall maxima. 360, Matunuck, 4 November 1990 (SHa); 250, Little Compton, 1 November 1953 (RBo); 200+ at a turf farm in Kingston, 31 October 1980 (DFi); 203, Wood River Junction turf farm, 21 November 2001 (JMa); 200, Portsmouth–Middletown, 4 November 1952 (RBo). There has been a tendency toward smaller flocks being recorded in fall during recent years, perhaps as a consequence of declining agriculture.

Winter. Birds have only rarely been reported in the northern Narragansett Bay area in midwinter; e.g., 3 in Bristol 3 January 1994 (RBo).

Winter maxima. 186, Matunuck, 22 December 2001 (STs; CBC); 180, South County, 28 December 1975 (DLK et al.). Smaller numbers have often remained into January: 50, Matunuck, 30 January 1983 (STs); 30, Moonstone, 5 February 2000 (WBo). There are only a few instances of proven winter survival, including 10 or so records of a few birds remaining through the season at Napatree Point. A flock of 15 wintered through January and February 1989 at Point Judith (JMa et al.). One wintered on Ocean Drive through February 1995 (JO, RLF).

Spring. Moderate flocks of suspected nonwinterers occasionally appear from early April on.

Spring maxima. 42, Westerly, 13 April 1975 (ESa, CWo); 30, Moonstone, 6 April 1975 (DLK); 30, Kingston, 2 May 1954 (DLK). Rare after early May: 2, Napatree, 10 May 1999 (CRa); 1, Hopkinton, 12 May 1986 (DRi, HR); 1, Block Island, 12–14 May 1999 (CRa, REn); 1, Napatree, 20 May 1996 (CRa); 7, Long Pond, Little Compton, 30 May 1936, and a single bird the next day at Tiverton (RCC, AEE); 1, Ninigret, 31 May 2002 (DFi).

FRINGILLID FINCHES (FRINGILLIDAE)

Evening Grosbeak (*Coccothraustes vespertinus*)

Status

Formerly common winter resident and migrant. Two breeding records.

Historical

Evening Grosbeaks expanded their range eastward from the Rocky Mountains during a series of large irruptions between 1850 and 1900. They first reached New England and New York in 1889–1890 and Rhode Island in 1910–1911 (Gillihan and Byers 2020). The first Rhode Island birds occurred during a major irruption to New England during 1910–1911, when one was seen in Cranston 7 January (Cobb) and seven were in Woonsocket 7–13 January (Arnold; Hathaway 1913). They appeared again at Woonsocket in late winter 1914 and at two locations during late winter 1916.

The first specimens were collected in the widespread flight of 1916–1917, when birds remained into May (HSH). They were first reported at Block Island in 1918 (EDi). They occurred nearly annually after that, with notable flights in 1918–1919, 1933–1934, 1941–1942, 1943–1944, 1945–1946 (a very large flight), and 1951–1952.

Breeding

On 15 June 1973, DLK was shown a pair, reportedly present all spring, that were bringing 2 barely fledged chicks to a feeder 1 mi (1.6 km) west of Wakefield (G. Scott). A pair fledged 4 chicks in Exeter, 31 July–20 September 2005 (RFa et al.).

Nonbreeding

Following their first appearance in the early 1900s, Evening Grosbeaks appeared in Rhode Island almost every winter, usually in substantial numbers. There were some winters during which they were scarce, such as 1952–1953, 1956–1957, 1961–1962, and 1970–1971. After this extended period of abundance, they began a decline in the early 1980s. During the winter of 1988–1989, not a single Evening Grosbeak was seen, and this scarcity persisted through the 1990s. In summary, Evening Grosbeaks were restricted to the Rocky Mountains before 1850, expanded into New England and Atlantic Canada in the late 1800s, reached a period of abundance in the Northeast from 1930 to 1980, then declined. They have been uncommon to rare in the Northeast, including Rhode Island, since about 1980. Both when common and when rare, Evening Grosbeaks have been especially erratic, even for a "winter finch," with surprising appearances at odd times of year. Migrants occur rather frequently throughout May, and 1 was at Block Island 28 May 1972 (REm, LBe).

Extreme dates. 7 September 1981, Warwick (DWa), to 6 June 1975, Green Hill (DWa). Adult males were seen at Hopkinton 12 July 1980 (DWa) and at Escoheag 17 August 1993 (CRa). Two males and 2 females came to a Saunderstown hummingbird feeder on 4 July 1995 (PAB).

Recent maxima. 43, South Kingstown CBC, 23 December 1995 (DLK et al.). Since 2005, there have been only about 15 reports, mostly at feeders and not exceeding 4 birds per day per locality.

Pine Grosbeak (*Pinicola enucleator*)

Status

Rare and irregular winter visitor.

Nonbreeding

Pine Grosbeaks stage periodic southward irruptions from boreal forests. Rhode Island is usually toward the southern limit of these incursions; they are noticeably less numerous here than farther north in New England, even Massachusetts. Pine

Grosbeaks feed primarily on conifer seeds, but during southward incursions into deciduous zones they often feed on berries of ornamental plantings, crabapples, and ash seeds (Veit and Petersen 1993). In general, the largest numbers of Pine Grosbeaks in New England have occurred far inland; however, as with Bohemian Waxwings, there are a surprising number of coastal (including Block Island) records during fall migration. Howe and Sturtevant (1899) and Brewster (1895) described irruptions prior to 1899, notably 1853, 1863–1864, 1872–1873, 1883–1884, and 1892–1893. There have been additional irruptions since 1899–1900, perhaps 1906–1907, 1910–1911, 1919–1920, 1929–1930, 1946–1947, 1951–1952, 1961–1962, 1968–1969, and 1977–1978. The largest irruptions were in 1892–1893, 1951–1952, 1961–1962, and 1977–1978. In nonflight years, often none at all are found in Rhode Island.

Maxima. 60, Diamond Hill Reservoir, 1 February 1947 (Bense); 50, Arcadia, February 1969 (fide RLF). The 2 recent records are 1, Diamond Hill Reservoir, 2 December 2007 (MLy); and 9, Burrillville, 24 December 2012 to 11 January 2013 (JSJ, m.obs.).

Extreme dates. 20 October 1919, Block Island (1; EDi), and 25 October 1954, Kingston (6; DLK), to 28 April 1911, Providence (14; Murphy 1911), and 7 May 1972, Charlestown (6; ESa).

House Finch (*Haemorhous mexicanus*)

Status

Introduced. Very common resident and migrant.

Historical

House Finches appeared in eastern North America following the release of birds caged for the pet trade in the early 1940s near what is now John F. Kennedy Airport in New York City (Elliott and Arbib 1953). They quickly began breeding, and the population started to spread in the late 1940s. House Finches reached southwestern Connecticut in the early 1950s, southwestern Massachusetts in 1955, southeastern Connecticut in the early 1960s, and eastern Massachusetts in 1963 (Hamilton 1992; Veit and Lewis 1996; Badyaev et al. 2020). They reached Rhode Island in late fall of 1964, during a region-wide invasion. The first Rhode Island records were 3, Warwick, 9 December 1964 (KKi); 1, Cumberland, 10 December 1964 (Mercer); 25, Jamestown, 15 December 1964 (MDa); and 5, Middletown, 27 December 1964 (CBC). Many birds remained through the winter at scattered locations (including 30 at Watch Hill), and at least as early as the following spring, the species reached Block Island, where two were banded 24 March 1965 (MSl). Nesting probably began in mainland coastal areas during the summer of 1965; breeding was first confirmed in 1969.

Breeding

House Finches are highly gregarious, even in the breeding season, and the nesting cycle is earlier (late April and May; young fledged by early June) than for most landbirds native to New England. House Finches increased dramatically following first breeding; by the mid-1970s, they outnumbered House Sparrows at many shore and suburban locations. By the mid-1980s, they had colonized throughout the state, generally in close association with human development. They are most numerous along the coastal strip, near beach developments, feeders, scrubby vegetation, scattered conifer groves, and exposed sandy soil.

Summer maxima. 287, Conanicut Island, 22–25 June 1985; 192, Narragansett to South Kingstown, 23–28 May 1984 (all RLF); 38, Block Island, 18–22 June 1990 (CRa, REn).

Nonbreeding

Despite statements to the contrary, House Finches migrate substantial distances (Veit and Lewis 1996), and this migratory behavior has evolved to be more pronounced in eastern birds since their release (Able and Belthoff 1998). Flocks of House Finches are commonly seen migrating at the outer coast during May and July–October.

Maxima. 809, Napatree CBC, 27 December 2015 (SSM, GW et al.); 727, Newport–Westport CBC, 14 December 2014 (REm et al.); 309, Tiverton, 1–31 May 1992 (KS, CS); 185, Matunuck, 30 July 1983 (RAC); 159, Middletown, 25 July 1981 (RTL); 155, Napatree Point, 19 October 1988 (CRa, REn).

Purple Finch (*Haemorhous purpureus*)

Status

Irruptive; uncommon summer resident, occasionally common transient and winter resident.

Breeding

Purple Finches are boreal, coniferous forest breeders, near the southern limit of their breeding range in Rhode Island. Numbers, both of residents and migrants, fluctuate considerably from year to year. They nest in northern red cedars, large ornamental evergreens, spruce plantations, or native white pines and eastern hemlocks. While most common in the interior, they are also fairly common in coastal and Narragansett Bay locations where groves of evergreens and red cedars are available. Some coastal areas that had numbers of Purple Finches nesting during the 1960s now seem to have only House Finches. However, there are recent records of Purple Finches breeding near the coast; e.g., 4, Patience Island, 4 June 1985 (REn, CRa, RLF); 1, Prudence Island, 6–8 June 1990 (REn, CRa).

Summer maxima. 29, Glocester to Burrillville, 17–23 June 1982; 18, Foster to Coventry, 16–24 June 1986 (RLF). They are less numerous to the south and east of these areas.

Nonbreeding

Purple Finch populations fluctuate considerably, and numbers of migrants in Rhode Island vary, as do their dates of occurrence.

Spring. General arrival is during mid-April, occasionally earlier, and birds are territorial by early May.

Fall. During flight years, migrants appear in August or early September, and hundreds may be counted at coastal vantage points. In other years, few if any are seen and arrival is not until October or November.

Maxima. Major invasions have occurred, as in 1938–1939, when 700 were counted at the Kimball Sanctuary in Charlestown 12 March 1939 (M. Southwick; Weaver 1940). Other counts include 200 at the same location 14 February 1941 (MSo) and 200 at Westerly through March 1963 (C. Saunders). More recently: 73, Block Island, 16 February 2008 (SSM et al.), and 60, Point Judith, 27 October 2014 (RFa et al.).

Common Redpoll (*Acanthis flammea*)

Status

Irruptive; uncommon to occasionally common winter visitor.

Nonbreeding

Common Redpolls frequent open weedy fields, overgrown vegetable gardens, brushy fields, and birch and alder thickets near water, where they feed on seeds, buds, and catkins. They are easily attracted to bird feeders. In most years, redpolls are unreported until late December or January.

Maxima. 1000, 25 March 1890 (HSH); 200, Rocky Point, 21 January 2005 (Mason); 200, Matunuck, 13 January 1994 (BCG).

Extreme dates. 22 October 1977, Little Compton (CWo), to 12 May 1972, Providence (at a feeder; ASi). There are only 3 counts of 100+ birds since 2005.

Subspecies

The majority of Rhode Island specimens are of *A. f. flammea*. There is 1 extant specimen of the "Greater" Redpoll, *A. f. rostrata*, taken by HSH in East Providence 14 March 1896 and identified by William Brewster (Brinkley et al. 2011).

Hoary Redpoll (*Acanthis hornemanni*)

Status

Vagrant. Most records have been during incursions of Common Redpolls.

Nonbreeding

Two specimens of Hoary Redpolls were collected in Middletown by JBa on 29 March 1956 (MAS #6855) and 31 January 1960 (MCZ #260596) (Brinkley et al. 2011). Other records, almost certainly all of *A. h. exilipes* (Brinkley et al. 2011), are as follows:

LOCATION	DATE	OBSERVER(S)
Little Compton	16 January 1944	RBo
Quicksand Pond	23 January 1949	RCC, REm
Matunuck (1 or more)	20 February 1949	DLK
Fiskeville	6 March 1953	JGe, RGe
Sachuest Point	28 January 1962	CWo, DLK
Westerly (2)	February, March 1969	ESa, DLK
Providence (1 or 2)	27 February–29 March 1974	FPe, CWo
Swan Point Cemetery	5 February 1978	CWo
Manville	28 February 1978	fide CWo
North Scituate	early March 1978	ALH
Little Compton	24 January 1994	RBo
East Providence	19 February–18 March 2008	Seitz, Davies
Goosewing Beach	3 January 2013	GDe
Block Island	21 February 2013 (*A. h. exilipes*)	RRV
Little Compton (2)	4 January 2014	GDe

Red Crossbill (*Loxia curvirostra*)

Status

Irruptive; occasionally common in winter.

Nonbreeding

Howe and Sturtevant (1899) considered Red Crossbills to be common, if irregular, winter visitors, but they appear to have generally decreased in the twentieth century.

While usually rare, they are occasionally numerous and more widespread in marked flights, as in 1896–1897, 1919–1920, 1952–1953, in late winter of 1964, in late fall of 1969, 1973–1974, and 1975–1976. Recent flocks have only occasionally exceeded 25 birds.

Maxima. "Hundreds" at Block Island in flocks to 75 through November 1919 (EDi); 200+, Middletown, 22 January 1920 (ESt); 200, Sakonnet, 2 November–30 December 1973 (v.o.); 166, Napatree CBC, 23 December 2012 (SSM, GW et al.); 140, Ninigret Beach, 17 November 2012 (m.obs.); 100, Misquamicut, 2 November 1969 (ESa); 98, Napatree CBC, 23 December 2007 (SSM et al.); 50, Sakonnet, through November 1969 (SSD).

Summer or early fall. Male and female collected 31 July 1911 and 23 August 1911 (CED; FlMNH #59396, #59397); 3 at Westerly 12 August 1972 (ASM); 1 at Greene 17–21 August 1969 (FPe). These suggest the possibility of nesting in Rhode Island, but such possible nesting has not occurred more recently. Single birds on EDi's roof at Dickens Point, Block Island, 15 September 1920 (EDi); 5 at Block Island 24 September 1944 (EDi); and 1 at Block Island 27 September 1976 (RAC, DLK) were early migrants. Red Crossbills have bred in Massachusetts and on Long Island on a number of occasions, usually in pitch pines or ornamental evergreens.

Subspecies

The small Rhode Island series of specimens are mostly *L. c. minor*. In the notable flight of 1919, HSH collected 4 specimens of the Newfoundland race *L. c. percna* (formerly *L. c. pusilla*); there are extant males from Westerly and Davisville collected 16 and 22 November, respectively, and examined by Bent and Griscom. An accidentally killed specimen from a very late flock of 20 present at a Barrington feeder 28–31 May 1973 (LBe, RLF) was confirmed at AMNH as *L. c. pusilla* (=*L. c. percna*) by J. Bull. Red Crossbills are weakly segregated into 8 different "types" based on flight calls; the 8 types also differ in size and bill size (Benkman 1989, 1993; Groth 1993). Call types 2, 3, and 4 breed in northeastern North America, and type 8 is the Newfoundland-breeding form *L. c. percna* (Groth 1993; Benkman and Young 2020). Parchman et al. (2006) found genetic differences among types but not sufficient differentiation to warrant specific status for any of the types.

White-winged Crossbill (*Loxia leucoptera*)

Status

Irruptive; less frequent than Red Crossbills in Rhode Island, but large numbers have appeared at very irregular intervals.

Nonbreeding

White-winged Crossbills appeared in the major irruption of 1899–1900, and a flock of 40 was found in Johnston 20 November 1906 feeding on hemlock cones; several specimens were collected by HSH. They appeared again in 1919, 1924, and in February 1953, when small flocks of up to 15 birds each were recorded statewide beginning on 5 February, with lingerers at Westerly to 19 March 1953. Other flights occurred in 1959–1960 (flocks of up to 15 birds), late fall 1963, and April 1964. In late fall and winter 1965–1966, 50 were at Little Compton 9 January 1966 (CWo). Finally, 27 were in Middletown 20 December 1975 (CBC).

Early fall dates include 1 accompanied by a Red Crossbill at Block Island 14 September 1920 (EDi) and another there 15 September 1924 (EDi). Four were seen in Arcadia 14 May 1988 (CRa). Other late records include a flock of 11 that remained at a feeder in Providence through April and most of May 1972; 1 of these stayed until 30 May (ASi; photographs examined). That same year, a female was photographed in a spruce tree on Block Island 8 June 1972 (RLF). Again in 1974, 9 remained at the Silverman feeder in Providence through 9 May, with 2 of these lingering until 31 May 1974.

Maxima. 650, East Beach, Charlestown, 19 November 2012 (D. Gochfeld); 455, Napatree CBC, 23 December 2012 (SSM, GW et al.); 300, Ninigret Beach, 27 November 2012 (m.obs.); 199, Block Island CBC, 22 December 1997 (CRa et al.); 140, Newport–Westport CBC, 3 January 2009 (REm et al.).

European Goldfinch (*Carduelis carduelis*)

Status

Introduced from Europe; now probably extirpated in North America. One Rhode Island record may have been of a vagrant rather than an escaped cage bird.

Nonbreeding

European Goldfinches were introduced about 1865 to the Boston area and about 1870 near New York City. In both places, they bred sparingly for a time and even spread to the Connecticut River valley of Massachusetts and to Long Island. The replacement of semi-agricultural country by housing developments apparently caused their decline, at least on Long Island (Bull 1964). They disappeared from Massachusetts by the 1930s and on Long Island by the late 1950s. A record of a specimen, here published for the first time, is of a male in the collection of Henry O. Havemeyer Jr., shot 8 September 1892 at Newport (Yale Peabody Museum #10517, specimen discovered by D. W. Finch).

Pine Siskin (*Spinus pinus*)

Status

Irruptive; uncommon to abundant fall migrant and winter visitor, less numerous in spring. Rare in summer, with 3 recent breeding records.

Breeding

Following the last major irruption year in 2008–2009, 2 pairs nested in Rhode Island: 1 in Swan Point Cemetery (RFa) and another in South Kingstown (BG), both during April. In 2015, another nesting pair was found in Coventry; the pair produced 2 young in June (M. St. Sauveur).

Nonbreeding

Pine Siskins occur in distinct irruptions, during which hundreds or thousands of birds appear, with peak numbers usually in late October or early November. In nonflight years, far fewer birds appear, sometimes just singles or none at all. Following some flight years, a few individuals remain into May and June, but no specific evidence has been obtained in Rhode Island.

Major irruptions include the following:

1913–1914: Hathaway saw numbers passing Quonochontaug on 13 October, and flocks to 2000 were noted at Block Island in late October and early November.

1952–1953: The first birds appeared on 23 October at Westerly, and in the next few weeks many thousands passed through the state; e.g., 1000+ along the South Shore 9 November 1952 (DLK) and 400 at Newport and 500 in Little Compton 11 November 1952 (DLK, RBo). They were much reduced after 20 November but remained common throughout the winter; e.g., 350 in Tiverton 19 February 1953 (Proctor). They were still common at Tiverton in April (150, 3 April 1953), and some remained through 19 May 1953 at Swan Point Cemetery (CWo).

1957–1958: In 1957, numbers were seen from 28 September through early October, decreasing thereafter.

1968–1969: 500, Charlestown Beach, and 150, Quonochontaug; both 16 October 1968 (DLK).

1973–1974: 150, Block Island, 6–7 October 1973; 1500, Charlestown, 4 November 1973 (DLK); 250, Trustom, 4 November 1973 (DLK); 300, Misquamicut, 5 November 1973 (ESa).

1975–1976: 400, Matunuck, 5 November 1975; 300, Green Hill, 6 November 1975 (both DWa).

1977–1978: 300, North Kingstown, 3 November 1977 (DWa), and numbers all winter, with birds present to 20 May 1978.
1990–1991: 509, Perryville, 18 November 1990 (MJM).
2008–2009: 1000 reported statewide; 220 on 11 thistle feeders, Rumford (RFa); 4400 passed Napatree Point in 3.5 hours 10 November 2008 (CRa).
2012–2013: 591 passed Napatree Point in 3.5 hours 21 October 2012 (CRa).

Extreme dates. Six remained at Wallum Lake to 28 May 1982 and 2 to 7 June 1982 (SMi). Six remained at West Kingston to 10 June 1984 (GTa, ATa). There were a number of late records after a flight in the fall of 1986, and late birds included 2 in West Kingston to 3 June 1987 (GTa, ATa), 5 and later 2 in West Greenwich to 9 June 1987 (CRa), and 6 at Wallum Lake through 20 June 1987 (SMi). A flock of 16 was present at Wallum Lake to 1 June 1988, with 2 still present 4 June 1988 (SMi); 2 others remained in Kingston to 1 June 1988 (v.o.).

American Goldfinch (*Spinus tristis*)

Status

Common summer resident, migrant, and winter resident.

Breeding

There is wide variation in the numbers of prospective breeding American Goldfinches in various parts of the state during early summer, with relatively small numbers in wooded western sections and more birds to the east and south. Even in more open sections of western Rhode Island, they are not common, and by far the largest numbers appear to be in the relatively open parts of the lower Narragansett Basin, such as on Aquidneck Island.

Summer maxima. 137, Newport Neck, 9 June 1986; 94, Conanicut Island, 22–25 June 1985; 60, Foster to Coventry, 16–24 June 1986 (all RLF).

Nonbreeding

Spring. American Goldfinches nest rather late in the season, so migration extends from mid-April through May. One was seen at Cox Ledge 28 May 1984 (DLE). In line with its late-breeding habits, flocks to 30–40 are even present in early June in some open-country areas, with nesting beginning in midsummer and young being fed through late summer, occasionally as late as 10 October.

Fall. Goldfinches are occasionally numerous in marked flights.

Fall maxima. 1795 migrating past Napatree Point in 4 hours, 15 November 2015 (CRa); 1600, Napatree Point, 4 October 1981 (RLF, CRa, REn); 1000 migrating past Napatree Point in 3.25 hours, 15 November 2003 (CRa); 804, Napatree Point, 15 October 1983 (CRa, RLF); 799, Block Island PDC, 8 November 1999 (CRa et al.); 528, Napatree Point, 19 October 1988 (CRa); 200, Westerly, 20 November 1958 (ESa).

Winter. Goldfinches are commonest near the coast in winter but weed fields inland can attract substantial flocks.

Winter maxima. 346, South Kingstown CBC, 22 December 2007 (DLK et al.); 325, Newport–Westport CBC, 14 December 2013 (REm et al.).

LONGSPURS (CALCARIIDAE)

Lapland Longspur (*Calcarius lapponicus*)

Status

Uncommon migrant and winter resident. Usually found with Snow Buntings or Horned Larks.

Nonbreeding

Lapland Longspurs frequent areas close to the coast, rarely north to Fields Point, Providence. Elsewhere in inland areas, there are known occurrences only from Warwick, the East Providence Reservoir, and Slocum. They are usually found in flocks with Snow Buntings or Horned Larks in very open habitats, including bare dirt fields. They usually appear in mid-October and are rare after March. The earliest record is for 1 at Little Compton 19 September 1939 (DLE), and another appeared in Little Compton on 24 September 2001 (REm). The latest records are 8 April 1967 at Kingston (DLK) and 13 April 2017, Napatree Point (v.o.).

Maxima. 75, Portsmouth, 14 January 1962 (SCr); 40, Newport, 28 September 1957 (JGe, RGe).

Maxima since 2000. 14, Block Island, 7 November 2003 (PJL, SSM); 4, Napatree CBC, 18 December 2005 (SSM, GW et al.); 3, Goosewing Beach, 15 November 2015 (GDe); 3, Moonstone Beach, 3 November 2014 (T. Auer).

These Arctic breeders seem to be disproportionately rare in Rhode Island during fall and winter, perhaps due to the lack of extensive open areas. They were more regular in past decades, at which time a few individuals or small flocks frequently joined flocks of Horned Larks and Snow Buntings to feed in manured fields and along beaches.

Smith's Longspur (*Calcarius pictus*)

Status

Vagrant. One record.

Nonbreeding

On 30 October 1965, after a strong cold front with 20 mph (32 kph) northwest winds, a Smith's Longspur was found at Moonstone Beach accompanied by 100 Snow Buntings and 20 Lapland Longspurs (DLK, FPe, KKi). The bird was observed leisurely at close range.

Snow Bunting (*Plectrophenax nivalis*)

Status

Fairly common migrant and winter resident.

Nonbreeding

Like many birds of open country, it is now almost wholly confined to locations near the coast and Narragansett Bay. Even near the coast, sizable flocks of Snow Buntings swirling over freshly manured fields have gradually become rare, and most flocks are now found in parking lots and along beaches near the outer coast. They were presumably more numerous in the interior before the decline of agriculture. In recent years, the largest inland counts have come from fields around the URI campus in Kingston: 40, 20 December 2003 (SSM); 25, 26 December 2013 (SSM).

Fall. The migration has a well-marked peak in late October and early November.

Fall maxima. 511, Block Island, 8 November 1999 (SSM et al.); 500, Green Hill, 30 October 1960 (CWo).

Winter. During winter, they are almost always found in smaller numbers, rarely exceeding 200.

Winter maxima. 500, Matunuck, 12 January 1977 (DLK); 300, Portsmouth, 13 March 1960 (JBa); 300, Misquamicut, 26 November 2013 (CRa); 31, Greene, 3 February 1994 (JZy).

Extreme dates. 1 October 1983 (DLK) to 20 May 1996, male in full breeding plumage at Napatree (CRa).

NEW WORLD SPARROWS (PASSERELLIDAE)

Grasshopper Sparrow (*Ammodramus savannarum*)

Status

Rare and local summer resident, formerly more numerous.

Breeding

Grasshopper Sparrows have always been commonest in South County and on Block Island. They prefer shortgrass fields with scattered shrubs, so are often found on the sandplain grasslands of glacial outwash plains of the immediate coast. While formerly more common farther inland, they have never been reported in Rhode Island more than 10 mi (16 km) from salt water. Piecing together the history of land use in Rhode Island, it seems likely that Grasshopper Sparrows were most numerous during the height of the agricultural era from about 1750 to 1870. The data on specific breeding localities are as follows:

Cranston. Hathaway occasionally saw 1 in a field near the corner of Park and Pontiac Avenues, and 3–6 males were regular nearby on "Coal Mine Road," 1893–1894.

Warwick. Bred near Warwick Pond in 1909 (HSH) and 1932 (WBD) and present there to 1935. One to 3 birds present at T. F. Green Airport 1974–1976 (GRa).

Quonset. Singing males there 10 June 1975 (JGe, RGe) and 1 June 1982 (RLF); 5 males were there 13 June 1985 (CRa), and 1–2 males in 1995–1996 (BCG et al.). Suitable but inaccessible habitat has existed for years along the southern and eastern edges of the naval airfield, mostly on fill dumped into Narragansett Bay.

West Greenwich. 12 males singing 14 May 1982 and 6 males, 7 May 1983 (CRa); 4 males, 18 May 1989 (AFr). Absent in recent years as habitat was converted to an industrial park.

Warren. Singing males east of the Kickamuit River in 1937–1938 (RBo).

Bristol. Two males on the Colt Estate in 1941 and 4 in 1951 (RBo).

Tiverton. Regularly up to 6 males along the Sakonnet Passage near Seapowet Marsh, last found on territory in 1957 (v.o.).

Little Compton. Known primarily from a high hill east of Peckham's Marsh on the Tiverton/Little Compton line (up to 1950, maximum 4 males) and from the area between Briggs Marsh and Tunipus Pond (to 9 males). They remained near Briggs through the 1960s (2–3 males) and the early 1970s (1 male) (DLE, REm). A pair was present and confirmed nesting east of Quicksand Pond during the period 1977–1984 (RBo).

Aquidneck Island. One or 2 in Middletown, 1937 (WHD).

Prudence Island. Present near Potter's Cove in 1894 (HSH). Ten males found island-wide in 1946 (WBD). The habitat had probably succeeded to unsuitable conditions before 1960.

Conanicut Island. Always regular in southern sections, especially near Dutch Harbor; also in fields north and northwest of the center of Jamestown. To 15 in 1948 (DLK, RBo); 8 in 1950 (DLK); 8 in 1958 (MDa); decreasing through the 1960s (1–2 pairs). Last reported on Beavertail (antenna fields) in 1972; persisting on the "Carr Farm" into the 1980s and still present on the Watson Farm, west side, in 1985.

South County. Douglas L. Kraus counted 22 males between Bonnet Shores and Quonochontaug in 1942. They were still present in the 1970s at Charlestown (to 7 males); also present at Green Hill (to 3 males), and Matunuck–Trustom (to 3 males). One was noted at Woodville 8 July 1981 (CRa), and single birds or pairs have been near Wood River Junction 1996–2004 (REm, CRa et al.). Single birds were at Charlestown and Trustom in 1982 and a sporadic male was in Westerly (ESa, CRa, v.o.). A pair was present near Hundred Acre Pond, Kingston, in 1980 (v.o.). A pair bred at the Carter Preserve in 2012 (PAB, BCG), and a single male was there in 2015.

Block Island. Always common near Dickens Point (to 10 or more males audible daily from the Dickens farmstead, 1913 on; EDi diaries). Much decreased by 1945 but still numerous to the east on the "Lewis Farm with resumption of cattle grazing"; e.g., 12 males there and 1 near the southeast lighthouse 8 June 1973 (RLF); 8 males in June 1981 and 1982 (RBo; REn, CRa); 7 males 19 July 1983 (REn, CRa), but only 3 males in 1984 and 1988; 5, 11 June 1985 (REn, CRa); 4 in 1989 (REn, CRa); 2 birds in 1990 and 1991 (REn); 1 bird in 1992 (CRa, REn); and 4 pairs in 1993 (CRa, REn). One pair was present on the Lewis Farm in 2012 and 2015 (CRa). According to Veit and Petersen (1993), the best areas, at least on the Elizabeth Islands, are where sheep are regularly grazed. Block Island habitat was once grazed by sheep, but despite annual mowing and intermittent grazing by cattle, the Lewis-Dickens property where most Grasshopper Sparrows have been present (now owned by ASRI) is now covered with bayberry scrub, making it less suitable for Grasshopper and Savannah Sparrows or Upland Sandpipers. In 2014, there were 2 breeding sites known for this species: 1–2 males on Dickens Farm, Block Island, and 1–2 males on a large dry field in Charlestown (Carter Preserve). The Quonset Naval Air Station used to have them but access is now limited, and their current status unknown.

Nonbreeding

Away from nesting areas, there have been only 9 reports since 2005, all in fall. Grasshopper Sparrows are secretive and rarely reported during migration. In spring, breeders arrive in late April and early May; in fall there is a concentration of records around 20 October, plus a handful of records in November.

Winter. One was well studied in a weedy garden at Portsmouth 19 December 1987 (E. Neumuth, RLF); 1 was closely observed at Block Island 2 January 1986 (CRa, REn), and another was found dead in Portsmouth in early January 1957 (JBa). One was at Middletown, 14 December 2013 (Jay Manning, Jay Wright), and another at Fogland 15 December 2007 (RFa et al.).

Extreme dates. 12 April 1927, Kimball Sanctuary, to early January (above).

Lark Sparrow (*Chondestes grammacus*)

Status

Rare fall transient; also, a rare or very rare winter resident and spring migrant.

Historical

The first Rhode Island Lark Sparrow was one observed by members of an Audubon bird class in Roger Williams Park on the unusual (by recent standards) date of 3 April 1937 (Hathaway 1943). Another was seen at Bonnet Shores 21 August 1949 (Nash). There were no further records until 1954, when one attended a feeder in Cumberland during August 1954 (Arnold, CWo, DLE et al.). This pioneer was followed by *eight* others seen during September and October of the same year, and one wintered in Rumford. In 1955, six birds were seen during fall, including a specimen collected (JBa), and there were two seen during winter 1955–1956 in East Greenwich. Five were reported during the fall of 1956. Thereafter, reports gradually declined, averaging two per year after 1960. The fluctuating numbers of this western vagrant are reminiscent of the similar historical pattern of Dickcissels as transients to the Northeast; Dickcissels also increased to a peak during the mid-1950s and then declined (Veit 1990, 2000).

Nonbreeding

Fall. Lark Sparrows are now regular transients during fall, with the majority of reports coming from coastal areas. The earliest fall records are 2 August 1956 at Block Island (BRi, RRi) and 3 August 1965 at Moonstone (DLK). Most appear in September and October, with fewer occurrences during November. There have been 43 reports since 2005, all in fall or winter.

Fall maxima. Three, Moonstone, 15 September 1954 (Murdock); 3, Matunuck, 22 October 1973 (GRa, STs).

Winter. There are about 18 December records. One remained at Galilee 1–12 January 1958 (JBa, DLK); another appeared at Little Compton 20 January 1979 (fide CWo). One wintered at the East Providence Reservoir 9 December 1954–26 March 1955 (RLF, JSu). An apparent winterer was seen at Block Island 26 March 1990 (v.o.).

Winter maxima. Two in 1 bush at Portsmouth 16 December 1984 (RLF, E. Neumuth).

Spring. Lark Sparrows appear much less frequently during spring. One was at Point Judith 19 May 2005 (JSJ). There are 3 specimens collected by JBa, all of which have been determined by Aldrich and Burleigh as the eastern race, *C. g. grammacus.*

Lark Bunting (*Calamospiza melanocorys*)

Status

Vagrant. About 10 records.

A male spent most of the summer of 1936 in the fields at Dickens Point, Block Island, where EDi observed it closely on 29 July and stated that it had been present "all season." All other records are from fall. One was near the Galilee–Great Island causeway 7 September 1969 (RDe). There is an excellent photograph of 1 that momentarily alighted on a stone wall at Block Island, 5 October 1970 (SSD). The most recent records involved 1 found feeding on a lawn at Block Island 30 September 1996 (RLF, BCG, JSJ, RFa, DVa); 1, Middletown, 15 December 2001 (Newport–Westport CBC; CRa); and 1, Watch Hill, 15 August 2015 (at feeder, photo; K. McCormick).

Chipping Sparrow (*Spizella passerina*)

Status

Very common summer resident and common to abundant migrant; rare in winter.

Breeding

Chipping Sparrows are the most common breeding sparrow of the interior uplands of western Rhode Island, where, excluding towhee, they equal or exceed the populations of all other native sparrows combined. They are numerous in suburban habitats and often nest in planted conifers, and they are also common in more natural woodlands interspersed with open areas, especially where white or pitch pines are abundant. Chipping Sparrows bred at Block Island in 1916, and birds were present through 1924 (EDi), but several careful searches in the 1970s, 1981, and 1990 (RLF, RBo, CRa, REm) failed to detect any.

Summer maxima. 298, Richmond to Hopkinton, 15–21 June 1987; 296, Foster to Coventry, 16–24 June 1986 (both RLF). They are less numerous in northeastern Rhode Island (59, Cumberland to Lincoln, 16–22 June 1984; RLF) and on the Narragansett Bay islands (15, Conanicut Island, 22–25 June 1985; RLF).

Nonbreeding

Spring. There is usually an influx of migrants in early April, and most birds arrive in early May. Six appeared at Wallum Lake 1 April 1986 (SMi), and 16 were seen at Green Hill 9 April 1977 (DWa). Late migrants or wanderers were recorded at Block Island 26 May 1949 (EDi), 3 June 1954 (EDi), and 8 June 1981 (RBo).

Spring maxima. 96, statewide, 8 May 1993 (RIOC); 30+, South County, 11 May 1968 (DLK).

Fall. Gatherings of young from local breeders are often found in late August and new transients appear at Block Island by mid-September. Peak migration occurs during the second week of October.

Fall maxima. 750, Block Island, 8 October 1960 (JBa); 500, Block Island, 13 October 1958 (ASRI); 350, Block Island, 11–13 October 1957 (ASRI).

Winter maxima. 42, Hope Valley, 1 January 2017 (CRa); 20, South Kingstown CBC, 26 December 2011; 8, Richmond, 27 December 1997 (MHC, SMC); 5, Perryville, 27 December 1987 (DLK). They have overwintered with increasing frequency: 1, Kimball Sanctuary, 22 January–9 April 1961 (ESo); 2, Cowesett, 1–31 March 1965 (KKi); 3, Cowesett, 1 March–15 April 1975 (KKi).

Clay-colored Sparrow (*Spizella pallida*)

Status

Uncommon but regular fall transient. Very rare in winter and spring. One probable nesting record.

Breeding

A pair was present all summer at the Francis Carter Preserve in Charlestown 2010–2013, and they probably nested in 2013 (RFa).

Nonbreeding

Clay-colored Sparrows expanded their breeding range eastward into western New York and southern New England beginning about 1970 (Veit 2000; Grant and Knapton 2020). They now occur regularly in fall, especially early October.

Fall. The first Rhode Island record was of 1 at Sakonnet 8 September 1955 (RBo). There were 29 records during the period 1959–1979, nearly all of them from Block Island, mostly in late September and October. Only 4 records are from the mainland (Kingston, Little Compton, and Green Hill, twice).

Fall maxima. There have been 3 records of 3 individuals at a time at Block Island, all in October. There are 3 Block Island specimens in the MAS collection taken by JBa

7–14 October 1960. Twenty-four individuals were reported during the fall of 2012, and there is a total of 100 records since 2005.

Spring. The few spring occurrences include 1 at Moonstone 19 May 1956 (DLK, FPe, KKi, EWe); 1 banded, Block Island, 27 May 2005 (ELa, KGa); and additional May reports from 2000 (2), 2003 (1), 2010 (2), and 2011 (1).

Winter. One, Portsmouth 14 December 1991 (RLF); 1, Matunuck 25–26 December 1981 (DLK); 1, ASRI feeder in Smithfield 26 November 1988–30 April 1989 (AFr et al.; photo); 1, South Kingstown, 27 February–15 March 1991 (STs, MJM, DFi, JO et al.); 1, South Kingstown CBC 22 December 2012; 1, URI campus 23 December 2017 (SSM; photos). There are also at least 3 January and 1 February reports 2005–2018. The latest is 1 at Block Island, 17 February 2014 (SSM; photos).

Field Sparrow (*Spizella pusilla*)

Status

Fairly common summer resident, common transient and winter resident.

Breeding

Field Sparrows prefer scrub-oak and pitch pine forest and open areas with small red cedars. Plantations of small Christmas trees, powerline corridors, and natural gas easements have provided suitable habitat in recent decades. Singing males have been reported on territory in the interior from 13 March onward; general arrival occurs in late April and early May. Field Sparrows breed in the protected eastern sections of the 3 major Narragansett Bay islands and are fairly common in the eastern Narragansett Bay region, as well as in Tiverton and Little Compton. At Block Island, despite a wealth of brushy habitat, they were historically unreported in the breeding season until the discovery of a singing male near Dickens Point, 30 June–1 July 1982 (REn).

Summer maxima. 40, Foster to Coventry, 16–24 June 1986; 46, Richmond to Hopkinton, 15–21 June 1987 (all RLF). Field Sparrows, along with Prairie Warblers and Brown Thrashers, have gradually declined since 1900 as pastures have succeeded to secondary growth and forest.

Nonbreeding

Field Sparrows are fairly common migrants in coastal areas in fall, particularly at Block Island. A peak in transient numbers occurs in early October.

Fall maxima. 125, Block Island, 13 October 1958 (ASRI); 49, Lonsdale, 10 November 1988 (REn).

Winter. In late fall and early winter Field Sparrows are highly gregarious and somewhat nomadic. They are more common at this season within a few miles of the coast, especially in the Narragansett Basin north along the Blackstone to Woonsocket. While probably more frequent inland than many other "half-hardy" species, the relatively few inland winter records are probably more reflective of a paucity of field work in these districts during the colder months than of a lack of birds.

Winter maxima. 122, Newport–Westport CBC, 22 December 1957; 50, East Greenwich, 21 January 1956 (RWa); 33, Lincoln, 11 November 1981 (REn). All counts of more than 50 birds since 2005 are from South Kingstown CBC; e.g., 89, South Kingstown CBC, 22 December 2007 (STs et al.). Wintering flocks are regularly encountered in milder coastal districts through February. They are relatively scarce on Block Island during winter.

Spring. Late migrants were found at Middletown to 24 and 25 May (JBa). One was banded 4 June 1976 at Block Island (ELa).

Fox Sparrow (*Passerella iliaca*)

Status

Uncommon to fairly common transient and winter resident.

Nonbreeding

Fox Sparrows migrate early in spring and late in fall.

Spring. Considerable irregularity in the spring migration results from weather conditions and perhaps food supply. Large numbers have accumulated at feeders after early spring snowstorms. The peak movement occurs in late March. There are several records for late April, and 1 at Hopkinton 2 May 1981 (DWa). One was killed at Johnston on the remarkable date of 21 May 1891 (specimen lost).

Spring maxima. 100, Kimball Sanctuary, 17 March 1939 (MSo); 100, Kingston, 17–18 March 1967 (DLK).

Fall. Fox Sparrows are probably at least as common in late fall as they are in early spring, but counts are usually smaller due to a more protracted migration and a lack of snowstorms to concentrate the birds. The earliest transient appeared at Warren 18 September 1942 (RBo), and 3 were seen at Block Island 1 October 1933 (EDi). They are usually not common until late October, and peak numbers pass through in November.

Fall maxima. "Hundreds," Block Island, 20 November 1933 (EDi); 150, Kenyon, 5 November 1965 (ESa); 63, Swan Point Cemetery, 17 November 1964 (BRi).

Winter maxima. 59, Napatree CBC, 18 December 2005 (SSM, GW et al.); 54, Napatree CBC, 22 December 2013 (SSM, GW et al.); 29, Napatree CBC, 18 December 2011; 26, Block Island CBC, in both 1963 and 1968, a combined 77 on the 3 CBCs in the winter of 1969–1970, and 91 in 1963–1964. These counts were vastly more than are normally reported and roughly correspond to the year of highest abundance (1966) in Massachusetts (Veit and Petersen 1993).

American Tree Sparrow (*Spizelloides arborea*)

Status

Common transient and winter resident; substantial decline in abundance since the 1970s.

Nonbreeding

Tree Sparrows are winter residents in Rhode Island. They are found evenly distributed in open habitats throughout the state. Most arrive during the last few days of October and the first week of November.

Fall maxima. 100, Newport County, 11 November 1957 (SSD).

Winter. In late fall and winter, Tree Sparrows aggregate in small flocks of up to 50 birds. They are, nevertheless, one of the most common winter sparrows along the coast and near Narragansett Bay. In winter, Tree Sparrows are less numerous at Block Island than on the mainland. On the mainland, large winter counts are occasionally the result of concentration at feeders after heavy snowfall.

Winter maxima. 1500, Kimball Sanctuary, 31 January 1931 (MSo); 716, Newport–Westport CBC, 23 December 1956 (DLE et al.); 287, South Kingstown CBC, 22 December 2001; 200, Narragansett, 28 December 1939 (WBD). They have declined since about 1970, likely the result of milder winter temperatures.

Extreme dates. 7 October 1906, Providence (1; HSH), to 10 May 1941, Providence (WBD), and 10 May 1956, Middletown (1; JBa).

Dark-eyed Junco (*Junco hyemalis*)

Status

Abundant migrant and winter resident. Rare breeder.

Breeding

Dark-eyed Juncos breed in small numbers in the northwestern portions of Rhode Island, especially near groves of hemlocks and white pines. The first breeding birds were found on territory in the hemlock groves east of Bowdish Reservoir, in Burrillville and Glocester, in 1970 (SSD). Up to 3 singing males were present, and a pair with an empty nest was found 13 June 1974 (RLF). Another was found carrying food north of Wakefield Pond, Burrillville, 20 June 1974 (RLC). Two pairs were feeding young in the same area 7–8 August 1982 (RLF). They are also in the Hemlock Ledges area of Durfee Hill Management Area (3 agitated birds, 9 June 1982 [REn]; 2 on 28 June 1990 [REn]). Farther south, 2 pairs were in the hemlock swamp near Foster Center 6–24 June 1981 (REn). In the Arcadia area, a pair with nest and eggs was found among the few hemlocks and many white pines on Frosty Hollow Road, Exeter, 25 June 1977 (FPe, ESa, CWo, DLK). They may also breed in the Carolina Management Area, where only white pines are present and where birds have been seen on the suggestive dates of 25 August 1971 and 19 May 1972 (DWa).

Nonbreeding

Dark-eyed Juncos declined significantly as fall migrants at both Kingston and Block Island banding stations 1969–2001 (Osenkowski et al. 2012), and this is consistent with their decline throughout the Northeast 1966–2015 (BBS).

Spring. Dark-eyed Juncos begin moving about in late March and April. Most wintering birds disappear by mid-April, but there are many records of stragglers into May. One was banded at Block Island 24 May 1971 (ELa).

Fall. Migration of juncos can be spectacular in Rhode Island, especially at Block Island and elsewhere along the immediate coastline. The largest numbers of migrants are found in late October and November; the earliest fall record for Napatree was 14 September, and the earliest fall record since 2005 was 21 September. A wandering bird was at Kingston, 9 August 1969 (KKi). Otherwise, the earliest fall records away from possible breeding areas are for 1 at Block Island on 4 September 1919 (EDi), 3 at Green Hill on 6 September 1976 (DWa), and 1 banded at Block Island on 11 September 1975 (ELa).

Fall maxima. 10,000, Block Island, 13 October 1958, when 1800 flew northwestward from Sandy Point toward the mainland (JBa, Ian Nisbet, RLF et al.). Similarly, more than 2000 left Grove Point 9 October 1995 (RLF et al.); 1050 left Sandy Point 13 October 1962 (MSl); 500, Block Island, 6 November 1960 (JBa). Compared with Block Island counts, fall numbers along the mainland coast are smaller (to 200 per day per locality) and still smaller in the interior (to 100 per day per locality).

Winter. In winter, juncos are common residents in coastal districts. They are moderately common in inland areas at that season.

Winter maxima. 1146, South Kingstown CBC, 26 December 2009 (STs et al.); 1062, South Kingstown CBC, 22 December 2012; 220, Tiverton, 1 February 1942 (RBo).

Subspecies

Juncos with convex blackish bibs that contrast with bright rufous flanks, formerly classified as *J. h. cismontanus* of the interior of western Canada, have been reported frequently in fall, collected at Block Island 24 September 1960 (JBa) and banded there 11 October 1957 (JBa). One wintered at DLK's feeder in Kingston, 1 January to 19 April 1990; 1 was seen in Warren 15 November 1988 (RBo); 1 was in Woonsocket 2 December 2007 (MLy). There are approximately 36 other records of birds reported as "Oregon Juncos," and they are almost certainly not "pure" "Slate-colored" Juncos (*J. h. hyemalis*). Virtually all such birds have appeared at feeders with other juncos during the winter, and dates have ranged from 20 October to 16 April.

White-crowned Sparrow (*Zonotrichia leucophrys*)

Status

Uncommon spring and fairly common fall transient; uncommon but increasing in winter.

Nonbreeding

Spring. Numbers of White-crowned Sparrows appearing in both spring and fall vary considerably from year to year. Peak numbers occur during mid-May in spring, but some arrive much earlier, in mid-April. The largest numbers appear in coastal areas. Some of the latter are likely individuals that wintered nearby. Earlier migrants include 1, Napatree, 31 March 2006 (CRa); 1, Green Hill, 20 April 1975 (DWa); 1, Quicksand Pond, 20 April 1990 (RBo et al.). Late-spring records include 1 at Jamestown 1–2 June 1971 (MDa, KKi) and 1 at Block Island 12 June 1928 (EDi).

Spring maxima. 14 banded, Middletown, 13 May 1956 (JBa); 9, Block Island, 7 May 2002 (John Magill)

Fall. White-crowned Sparrows are fairly common to common transients, far more numerous along the coast than inland. Peak numbers appear in mid-October, and a few early migrants appear in early September. They are particularly numerous at Block Island.

Fall maxima. 200, Block Island, 13–14 October 1974 (CWo et al.); 175, Block Island, 4–6 October 1974 (ASRI); 150, Block Island, 14 October 1956 (ASRI). Numbers on the mainland are lower, never more than 100 birds per day per locality.

Winter. The first winter White-crowned Sparrow in Rhode Island was at the Kimball feeder, Charlestown, 22 December 1940 to 6 January 1941 (M. Southwick), and the second was at Jamestown 12 January–5 May 1957 (MDa). Between 1957 and 1969 there were 7 additional records of individuals, including several in February and March and 1 that survived at a feeder in Westerly to 1 May 1961 (ESa). During the

1970s, White-crowned Sparrows increased and were found in small flocks. Twelve were at Block Island 26 December 1971 (v.o.), and 12 were also there on the CBC in 1977; 15 were on Aquidneck Island 15 December 1974 (RLF, DWa). Thirty birds were in Portsmouth and Middletown 22 December 1985 (RLF, REn, CRa), and 18 successfully overwintered in Little Compton to 29 February 1996 (RBo). The best winter location on Aquidneck Island was the farm at the Portsmouth Abbey, where rank weeds in recently cultivated garden plots, abundant multiflora-rose hedgerows, and uncovered bunker silage provided excellent habitat. This site was destroyed for the construction of a golf course in 2000.

Winter maxima. 144, Newport–Westport CBC, 18 December 2010 (REm et al.); 60, South Kingstown CBC, 2 January 2011 (STs et al.).

Subspecies

The common subspecies in our area is *Z. l. leucophrys*. An adult of the well-marked taiga race *Z. l. gambelii* was collected on 13 May 1956 at Middletown (JBa; specimen confirmed by Sutton, in MCZ). Sight records of adult *gambelii* include 1 at a feeder in Cowesett 16 November 1975 through April 1976 (KKi et al.); adults with *pink* (not orange) bills and *white* lores at Matunuck 28 November 1997 (PAB, D. Sonneborn), and Portsmouth 20 December 1975 (RLF), appeared to have been intermediate between *gambelii* and *leucophrys*.

Golden-crowned Sparrow (*Zonotrichia atricapilla*)

Status

Vagrant. One record.

Nonbreeding

One was seen briefly by HWi, P. Capobianco, and others in Middletown at the southwest corner of St. Mary's Pond 22 November 1998. There are 9 records for Massachusetts.

Harris's Sparrow (*Zonotrichia querula*)

Status

Vagrant. Two records.

Nonbreeding

There are 2 Rhode Island records, the first of an adult seen and then banded at Block Island 17–18 November 1965 by MSl. Another, an immature, was carefully observed at Charlestown 1 January 1968 by CWo and FPe and seen the next day by SSD. There is a CBC record not mentioned by Veit and Petersen (1993) for nearby Acoaxet, Massachusetts, on 17 December 1978 (DLE, REm).

White-throated Sparrow (*Zonotrichia albicollis*)

Status

Rare summer resident, abundant migrant, and common winter resident.

Historical

Prior to 1900, White-throated Sparrows appeared only as migrants in Rhode Island. Howe and Sturtevant (1899) knew of only one Rhode Island winter record, and neither of the Mearnses (1902) ever saw one in winter in 3 years of living at Newport. Hathaway (1913) stated that they wintered occasionally and cited two additional records; later, he noted another in Cranston in 1916. Sturtevant, who spent his whole life in Middletown, knew of several that wintered in Warren in 1908–1909. His first contact with wintering birds was apparently one he collected in Middletown in March 1917, but he then discovered 15 there on 8 February 1919. Farther inland, Arnold noted one that lingered at Woonsocket to 28 December 1918, but otherwise never saw them in winter between 1896 and 1938. Hathaway (1943) called them an "occasional winter resident" but gave few wintering records prior to the 1930s.

After the long-term grazing of most of Rhode Island's coastal area in the nineteenth century, succession to brushy conditions began in the 1920s and 1930s. Among the outcomes of this coastal plant succession has been a dramatic change in winter numbers of this species in relation to other sparrows, and White-throated Sparrow is now one of the most abundant passerines during winter, outnumbering Song and American Tree Sparrows, as well as Dark-eyed Juncos.

Breeding

White-throated Sparrows occasionally nest in Rhode Island. Birds were seen carrying nesting material on DLK's farm in Kingston in May 1963, and a juvenile was noted 3–25 July 1963 (DLK). They have been heard singing in mid-June along the Pawcatuck River in South Kingstown, in Foster (pair in hemlock swamp 3 July 1981; CRa), Scituate, West Greenwich, Arcadia, Burrillville, Glocester, and irregularly at Greene. A single bird with a nest and 2 eggs was found in an Atlantic white cedar swamp near Escoheag Hill 7 June 1962 (RLC; Champlin 1962). Individuals remaining into summer near feeders (e.g., to 24 June 1982 at Warwick; through July 1958 at Ruecker's in Tiverton; into July 1972 at Jamestown and Swan Point Cemetery; through July several times at Kingston [DLK]) could well have nested.

Nonbreeding

White-throated Sparrows declined significantly as fall migrants at both Kingston and Block Island banding stations 1969–2001 (Osenkowski et al. 2012), and this is consistent with a significant decline through the entire Northeast 1966–2015 (BBS).

Spring. New migrants appeared at Dickens Point, Block Island, in early April during the years 1916–1933 (EDi); they appeared 15 April at Providence (HSH). They are now most numerous during late April and the first week of May, with a peak about

1 May. Most birds have passed through by 20 May, but there are numerous records of lingerers or late transients; e.g., 1, Block Island, 10 June 1958 (EDi).

Spring maxima. 500, Block Island, 1 May 1975 (ELa); 450, Block Island, 1 May 1972 (CWo); 60, Swan Point Cemetery, 22 April 2001 (MLy).

Fall. The earliest transients arrive in mid-September; e.g., 1 banded, Block Island, 15 September 1995 (ELa). Peak migration occurs in October and early November.

Fall maxima. 577, Block Island VDC, 13 November 2000 (SSM et al.); 250, Block Island, 1–3 October 1982 (RFa et al.); 140 along the South Shore, 3 October 1954 (DLK).

Winter maxima. 1096, Newport–Westport CBC, 14 December 2013 (REm et al.).

Vesper Sparrow (*Pooecetes gramineus*)

Status

Uncommon migrant; formerly an uncommon breeder, now extirpated.

Historical

Howe and Sturtevant (1899) considered Vesper Sparrows to be common summer residents and collected breeding specimens from Middletown and Bristol. Doe collected eggs in Bristol in 1884 and 1887. Vesper Sparrows bred regularly in that era near the coast; HSH and others found them breeding in 1900 at Sakonnet and along Ocean Drive, Newport. Hathaway also considered them to be abundant breeders in the 1880s and 1890s in the then-undeveloped southern sections of Cranston and Providence. Four sets of eggs are extant from Cranston, 1883 to 1908 (CED). Vesper Sparrows bred regularly around HSH's home in the Auburn area of Cranston in the 1880s and persisted in this area (including one site near present-day Roger Williams Park) well into the 1920s. Doe found nests on Warwick Neck in 1882 and 1897, and some were still present in Warwick in 1931. Doe also took a clutch of four eggs in 1913 from Glocester. The decline of the species was lamented by Madison (in litt.), who said that "thirty years ago we could count on seeing the Vesper Sparrows whenever we went afield" (meaning mainly near Providence, especially Cranston). The species was of legendary abundance up through the 1930s on the moorlands at the southwestern corner of Block Island (EDi). They began to decline during the Depression, however, and she later wrote in her diaries that "there are very few on Block Island where once they were numerous." By 1945, they had virtually disappeared. A pair still present in 1947 at the south end of the island (RCC) was the last sign of their presence there in summer.

Vesper Sparrows also largely disappeared from many other regular locations in the 1940s, with the last known occurrences during the breeding season for many areas as follows: Warren and Scituate (1943); Charlestown (1944); Burrillville, Glocester, and Prudence Island (1946); and Conanicut Island (1948). In the mid-1950s, Vespers were still locally present at Kingston in 1953 (DLK) and in several Newport County locations including central Aquidneck Island, in southern Little Compton, and in the Seapowet area of Tiverton. They persisted very intermittently at Matunuck at least to 1956. In the 1960s, the only known locations were near extensive potato fields on sand plains at West Kingston, at Usquepaug, and especially north of Wood River Junction (DLK et al.). By the 1970s, they could be found only at Wood River Junction, and territorial birds found in summer were noted there through the late 1970s. Two singing males were present there 22 April 1982 (CRa). These fields were greatly enlarged and converted to turf farming in the mid-1980s. Thus, the discovery of a territorial pair 2–7 July 1981 in a nonagricultural location at Arcadia (REn, CRa) was of considerable interest. The habitat, shared with nearby Nashville Warblers, included 20-ft (6 m) scattered pitch pines with a grassy, savanna-like understory.

The decline of Vesper Sparrows in Rhode Island parallels similar reductions in Cliff Swallows, Bobolinks, Eastern Meadowlarks, Upland Sandpipers, Grasshopper Sparrows, and Savannah Sparrows: all grassland species.

Nonbreeding

Spring. General arrival of Vesper Sparrows in the 1880s was during early to mid-April; they were found twice at Woonsocket on 31 March at the turn of the twentieth century (C. M. Arnold), and 10 were seen at Neutaconkanut Hill 31 March 1905 (Mason).

Fall. Most migrants are found from late September through November. A pair of wanderers was present at Napatree Point 25 August 1984 (CRa).

Winter. A cumulative total of 188 was recorded on the Newport–Westport CBC, 1949–1969, with a maximum of 69 in 1953, but only 20 were found on this census in the period 1970–1987 and only 18 in the period 1986–2016, with a maximum of 3 each in 2009 and 2010.

Winter maxima. 20, Tiverton, 26 December 1948 (RCC et al.); 20, Charlestown Beach, 17 February 1963 (DLK); 12, Little Compton, 27 February 1949 (DLE). The largest wintering flock of recent years was 6 or 7 birds present near Cards Pond, 28 December 1985–13 January 1986 (CBC) and 5 at Ninigret, 12 November 2000 (STs); the largest count since 2005 was 5, Napatree CBC, 23 December 2007 (SSM, GW et al.). There have been 66 reports of Vesper Sparrows in Rhode Island since 2005, all but 1 of these during fall and early winter.

LeConte's Sparrow (*Ammospiza leconteii*)

Status

Vagrant. Five records, perhaps some recent increase.

Nonbreeding

The first Rhode Island record was of 1 found 21 January 1993 in a field of common timothy off Mitchell's Lane in Middletown (DFi). This accommodating individual was seen by dozens of observers, and even inadvertently touched (DFi, DLK, RBo et al.), to 18 March 1993 (DFi). Many excellent photographs, some from as close as 2 ft (0.6 m), are on file (*AB* 47:318; see Photo 26). The records since then are 1, Napatree, 26 November to at least 5 December 2003 (CRa); 1, Avondale, 4–5 November 2006 (CPe et al.); 1, Newport–Westport CBC, 19–23 December 2009 (REm et al.); 1, Newport–Westport CBC, 16 December 2012 (RFa, REm et al.).

Seaside Sparrow (*Ammospiza maritima*)

Status

Fairly common summer resident and transient; rare in winter.

Historical

Howe and Sturtevant (1899) indicated that Seaside Sparrows bred only in the Point Judith marshes, but Sturtevant (1897) mentioned specimens taken in spring and summer on the formerly extensive Sachuest marshes in Middletown and concluded that they bred there as well. Hathaway (1913) called them "an uncommon summer resident in nearly all of the salt marshes of our southern coast" (perhaps too generous an assessment based on modern information), and mentioned a 1907 nest from Quonochontaug. Hathaway noted that they markedly decreased at Quonochontaug and perhaps elsewhere with mosquito ditching of the salt marshes in the 1930s, but that numbers recovered significantly after 1938 when some of the ditches were filled with hurricane-deposited sediments.

Breeding

Seaside Sparrows are more particular in their choice of nesting habitat than are sharp-tailed sparrows and less likely to breed in ditched marshes. They currently breed at the following locations:

Winnapaug Pond. Eight birds in June 1982 (CRa).

Quonochontaug Marsh. Five to 12 birds 1974–1988 (DLK; DWa).

Ninigret Pond. Twenty birds in 1982 (CRa).

Potter Pond. Five pairs in 1942, now 1–2 pairs (DLK).

Jerusalem. Two to 3 pairs regularly (DLK).

Galilee. Mainly south of the present Escape Road. Formerly numerous, perhaps 10–15 pairs, but later extirpated by human activities; see below.

Fox Hill Meadows, Conanicut Island. One or more pairs bred in 1936–1940 (A. M. Clark); none recently.

Sachuest marshes. Formerly common; still present on remnant marsh in 1937–1938 but unreported thereafter (v.o.).

Ocean Drive, Newport. A perhaps wandering male 25–30 June 1983 (RTL).

Briggs Marsh. One pair irregularly in 1940s (DLE).

Quicksand Pond. One to 2 pairs regularly (DLE); 4 males 3 June 1984 (RBo); 7, 7 July 1971 (DLE), but no breeding here since about the mid-1980s, possibly because of higher water levels and *Phragmites* encroachment (REm).

They appear to be more numerous in upper Narragansett Bay: a nest with 5 eggs from Barrington is in the Doe collection, collected by N. R. Hall (FlMNH #54217). Upper Narragansett Bay sites include 10–18 pairs on The Tongue, Hundred Acre Cove; 13–16 pairs in the northeast marsh, Hundred Acre Cove; and 3–5 pairs in the upper Warren River marsh, all in 1981 (SRi). Birds were reported in the 1980s from Seapowet (RFa), an estuary off the Sakonnet Passage in Tiverton several miles north of the outer coast. Seaside Sparrows do not nest at Block Island, and the only records there involve transients; e.g., 6, 2–3 October 1971 (ASRI). Stoll and Golet (1983) show the importance to Seaside Sparrows of medium-sized salt-water cordgrass in unditched salt marshes containing *Spartina*-lined salt pannes and pools. These features are typical of the Barrington and Warren River marshes. Seaside Sparrows sometimes use ditched marshes with abundant hightide-bush, particularly along the outer coast, but unditched marshes with *Spartina*-lined salt pannes provide the best habitat. A total of 27 nests were located throughout Rhode Island in 1981–1983 by W. DeRagon, and these contained eggs during the period 23 May–21 August (DeRagon 1988).

Berry et al. (2015) resurveyed the entire state in 2007–2008 and found a total of 68 pairs of Seaside Sparrows nesting; this represented a 31% decline from 1982.

Nonbreeding

Spring. Most appear during early May. A migrant appeared at Napatree Point 27 April 1985 (CRa). There are 4 records for 29 April. Migrants and wanderers are still moving in late May; e.g., 1 at Napatree Point 27 May 1986 (CRa).

Fall. The breeding population may begin to disperse in late summer (e.g., 1 at Napatree Point 13 August 1992; CRa) (Hill 1965). There are 5 records from late December.

Extreme dates. 19 April 1959, Jerusalem (DLK), and 20 April 2006, Napatree (CRa), to 1 January 1951, Briggs Marsh (DLK, RBo).

Nelson's Sparrow (*Ammospiza nelsoni*)

Status

Fairly common migrant; rare in winter.

Nelson's Sparrow was separated from Saltmarsh Sparrow by the AOU in 1998. The newly separate species includes geographically and morphologically distinct subspecies: *A. n. nelsoni* of freshwater marshes and meadows of western and central Canada, *A. n. subvirgatus* of the salt marshes of northern New England and Atlantic Canada, and *A. n. altera* of the James Bay region at the south end of Hudson Bay (Greenlaw 1993). *Ammospiza n. subvirgatus* is a fairly common migrant through Rhode Island in both spring and fall, whereas *A. n. nelsoni* and *A. n. altera* occur annually in small numbers. At Napatree, *subvirgatus* is uncommon and seen as often in spring as in fall. *Nelsoni* and *altera* dominate the fall flight but are unrecorded in spring (CRa).

Nonbreeding

Spring. *A. n. subvirgatus* migrates through Rhode Island in late May to early June. Dates of sight records range from 23 May to 8 June, the latest 8 June 1940 at Quonochontaug (DLK); a probable *subvirgatus* in the URI fields, Kingston, 1 June 1973 (RLF, DLK); 1, Peace Dale, 23 May 1984 (RLF); 1, Spar Island, Mount Hope Bay, 5 June 1984 (RLF).

Fall. *A. n. subvirgatus* is more common in fall than spring and occurs mainly in October and early November in salt marshes. Those found in shrubby pastures or in wet meadows are likely *A. n. nelsoni* or *A. n. altera*.

Fall maxima. Eight, Napatree, 9 October 2010 (CRa); 4, Block Island, 2–4 October 1989 (RBo, DLK); 1, Lonsdale, 2 October 1948 (RCC, WBD). Later records include 1, Warren, 25 November 1956 (RBo); 1, Warren, 21 December 1958 (v.o.). There are also 3 winter sight records of very buffy "sharp-tailed sparrows" that were likely *A. n. nelsoni or A. n. altera*.

Saltmarsh Sparrow (*Ammospiza caudacuta*)

Status

Common summer resident and transient; rare winter resident.

Breeding

Saltmarsh Sparrows breed in salt marshes, both near the outer coast and along the shores of Narragansett Bay as far north as Warwick and Barrington. In a 3-year study of this species in Rhode Island salt marshes, W. DeRagon, SRi, and others located 162 nests and found eggs between 26 May and 30 August (Reinert et al. 1981; DeRagon 1988; DeRagon and Golet 1989). Combining this count with other estimates of Rhode Island marshes suggests a total of perhaps 350 nesting pairs. As for other salt marsh birds, Saltmarsh Sparrows are negatively affected by ditching of marshes for mosquito "control" (Clarke et al 1984).

Summer maxima. 61, Prudence Island, 8 June 1990 (CRa, REn); 50, Seapowet Marsh, Tiverton, 28 May 1950 (RBo); 40, Seapowet Marsh, 4 July 2013 (REm). DiQuinzio et al. (2001) banded 446 Saltmarsh Sparrows, 271 of which were juveniles, in the Galilee marshes 1993–1997.

Nonbreeding

Spring. Migrants usually appear in mid to late April; 1, Seapowet, 19 April 1951 (RBo); 2, Potter Pond, 19 April 1959 (DLK); 1, Quonochontaug, 21 April 1986 (DLK, DFi); 3, Hundred Acre Cove, 22 April 1981 (SRi).

Fall maxima. 51, Quonochontaug, 3 October 1975 (DWa); 19, South Shore, 2–7 November 1982 (CRa, REn). It is not clear what proportion of these birds were Nelson's Sparrows; recent observations suggest that northeastern *A. nelsoni subvirgatus* migrate through in October, after the majority of *A. caudacuta* have departed, but this pattern requires confirmation.

Winter. There were no records later than 1 January until recent years; e.g., 1, Block Island PDC, 20 February 2012 (REm, DFi, STa). There is still only 1 midwinter record for the South Shore: 2, Galilee 14 January 1979 (RAC). This is surprising in light of the numbers wintering in Massachusetts. An interesting group of December observations materialized from a study of upper Narragansett Bay salt marshes (Reinert et al. 1981), particularly from the upper Warren River and Seapowet marshes; e.g., 1 or 2 birds seen repeatedly at Seapowet, 18 January–19 March 1981 (SRi). Another individual seen at Cards Pond 2 April 1957 (DLK) was probably also a surviving winterer. Again, what proportion of these late-fall migrants are Saltmarsh rather than Nelson's Sparrows requires confirmation.

Henslow's Sparrow (*Centronyx henslowii*)

Status

Formerly uncommon and local breeder; now a rare migrant. There is 1 winter record.

Historical

Howe and Sturtevant (1899) listed only a single record of Henslow's Sparrow, one collected in late April 1874 at Cranston by F. T. Jencks. Hathaway later recorded another collected 6 October 1898 near the Pawtuxet River in Warwick. They were first found on territory near Brightman's Pond, Westerly, where two singing males were noted 6–8 June 1901 in adjoining wet meadows (Hathaway). After additional field work, HSH heard seven singing in a field of some 20 acres (8 ha) in extent in Charlestown (Green Hill, fide DLK; Hathaway 1913). He collected two males in this field 5 June 1910 and found two nests there 28 May 1911 by dragging a rope (eggs extant).

A nest and five eggs were taken on the high, then-open south end of Prudence Island 4 June 1910 by CED (FlMNH #54213, #54214). In some of the earliest extensive exploration of the Tiverton–Little Compton uplands, a singing male was found with Grasshopper Sparrows on a hill overlooking Peckham's Marsh in the summer of 1936 (RCC, AEE), and one to two males were present at this location 1936–1940 (v.o.). Another one to two males were present at another location off Taylors Lane, Little Compton, in 1940 (RCC, DLE). In 1937, WHD found a territorial male in Middletown, but otherwise there were no records from the Narragansett Bay islands other than Prudence. The species may have persisted a few more years in Newport County; e.g., two seen somewhere in Little Compton 21 August 1944 (Williston), but territorial birds could not be found at the two principal Little Compton locations after 1940 (DLE).

Breeding

In Rhode Island, Henslow's Sparrows preferred formerly cultivated meadows that had passed into a stage of succession where goldenrods and grasses prevailed but where small bayberry bushes and other shrubs were sometimes present in small quantities (cover painting; Herkert 2007). In 1942, traveling by bicycle, DLK found 20 pairs in a fairly complete but not exhaustive search between Bonnet Shores and Quonochontaug. Kraus's records show that peak abundance in South County came about 1938–1942. On 1 May 1940, in a single field south of the intersection of Matunuck Beach Road and Cards Pond Road, DLK found 50 birds, many of these singing, but the majority were apparently migrants and did not remain. Conservation efforts in Illinois (Herkert 2007) suggest Henslow's Sparrows might also be restored to Rhode Island.

During the period 1940–1949, Henslow's Sparrows could be found at about 5 locations from Bonnet Shores to Charlestown but were irregular at some of these and numbers were sharply down between 1946 and 1948. In 1949, however, a final rally occurred and DLK found 25 in early May at Green Hill; 12 were still present in June, and 2 additional birds were at Bonnet Shores. Only 1 was present at Green Hill in 1950, and from then on, they remained in very low numbers. After 1950, they seem to have been confined to 1–2 places at Green Hill and at "Blackberry Hill" in Matunuck until about 1955, when the habitat was destroyed by succession and a housing development. At Green Hill, up to 4 birds were found in the late 1950s, but the last record was 15 May 1960 (DLK).

Nonbreeding Henslow's Sparrows are even more secretive and difficult to find as migrants than Grasshopper Sparrows.

Spring. The earliest known spring arrival date is 9 April 1921 at Quonochontaug (HSH); another appeared at Kimball Sanctuary 19 April 1932 (MSo). At nesting locations, birds sang until mid-August and were rarely seen after that time.

Fall. Even when they were fairly common, Henslow's Sparrows were rarely detected in fall. On 14 October 1932, HSH found 5 killed under wires on a foggy night on Charlestown Post Road, suggesting that that is the peak migration time for Henslow's Sparrows. One was at Block Island 8 October 1965 (v.o.), and another was mist-netted, banded, and photographed at Block Island 9 October 1968 (ELa). The latest fall record is for 1 found in the marsh at East Matunuck 4 November 1939 (DLK). There is an exceptional midwinter record from a field in Charlestown, where 1 was found by HDa 30 January 1963 and collected the following day by Dr. George West (in URI collection).

Savannah Sparrow (*Passerculus sandwichensis*)

Status

Uncommon and local summer resident; common spring and common to very common fall transient. Fairly common in winter.

Historical

Howe and Sturtevant (1899) considered this sparrow of field and pasture to be a "not uncommon summer resident, especially near salt water," in large measure based on their experience in the Middletown area. There is a series of 10 breeding-season birds collected in the (former) Middletown marshes in June 1900 by Mearns (in USNM), as well as others collected there by ESt about the same time. Most of these once-extensive marshes were flooded a few years later to create Gardiners Pond. Hathaway called this species "the most abundant sparrow in the fields of the Sakonnet Point area" in June 1900. This area was then virtually treeless, and he found several nests and flying young. Mearns (1900b, 1902) called it an abundant summer resident in the former fields about Fort Adams, Newport, in 1899–1902. Farther north nearer the head of Narragansett Bay, HSH found them to be common on 29 May 1889 on the "tongue" (a salt-marsh peninsula in Hundred Acre Cove, Barrington), at or near which place they were likely breeding. He found a nest and eggs on the salt marshes along the Warren River, Warren, on 24 May 1889, and also considered them to be common there at the time. They bred regularly in fields in Bristol (juvenile collected, 7 July 1897; RHH, BMS). According to HSH, before the turn of the twentieth century Savannah Sparrows were "common at most places along the shore" in the breeding season, but since then a vast decline has occurred in their breeding numbers. Formerly rare in winter (Howe and Sturtevant 1899); Shattuck shot one on 24 January 1900 at Third Beach, Middletown, and claimed it as the second New England winter record. Howe and Sturtevant found a small flock there on 22 December 1900, however, and other records suggest that they were probably overlooked in winter.

Breeding

During the early twentieth century, many of the salt marshes and barrier beaches were bordered by hayfields and grazing land and crisscrossed by stone walls. The marshes themselves were often grazed or mowed. These mown hayfields provided ideal nesting habitat for Savannah Sparrows. As much of this habitat has been replaced by human development, Savannah Sparrows have substantially declined. They are now most numerous on the remaining moorlands at Block Island, but also nest in pastures and shortgrass meadows on the mainland.

Nonbreeding

Spring. Some arrivals are occasionally detected in mid to late March: 20 March 1971, Barrington (LBe); 22 March 1963, Kingston (DLK); and 22 March 1988, Wallum Lake (SMi). Maximum numbers are found in May, with totals of up to 25 per day per locality.

Fall. Savannah Sparrows may be found in numbers along the coast in late August, but peak numbers occur in late September and early October.

Fall maxima. 579 left Sandy Point, Block Island, headed for the mainland 13 October 1958 (Ian Nisbet, RLF); 300, South County, 18 September 1975 (RBo).

Winter. Some survive the winter, especially in weed fields near the coast.

Winter maxima. 228, Newport–Westport CBC, 15 December 2012 (REm et al.); to 70 per day per locality in early January; 20, Galilee, 3 February 1957 (DLK).

"Ipswich Sparrow" (*Passerculus (sandwichensis) princeps*)

Status

Uncommon winter resident and migrant.

Nonbreeding

"Ipswich" Sparrows inhabit coastal dunes, salt marshes, and adjacent fields. They are most common at Quicksand Pond, Briggs Beach, beaches in the Sachuest area (at least formerly), dunes north of Narragansett Pier, and dunes formed on dredge spoil near Jerusalem, as well as at many other South Shore locations such as Trustom Pond, Charlestown Beach, Quonochontaug, and Napatree Point. They are also regular in the marsh edge adjacent to Beane's Point (Andy's Way) on Block Island in November and December, occasionally to February (REm). They are rare inland; 1 was reported seen at Fields Point, Providence, 31 January 1905 (E. H. Mason). One was seen in the Kickemuit River area of Warren 17–18 November 1983 (RBo). Formerly more numerous, "Ipswich" Sparrows were once considered to be "a common migrant and not uncommon winter resident" by Howe and Sturtevant (1899). Frederick T. Jencks shot a number among the now-developed dunes west of Point

Judith 27 November 1884. On 6 November 1898, HSH shot 7 and saw several more in these same dunes.

Maxima. 11, Newport–Westport CBC, 21 December 1958 (DLE et al.); 8, Block Island, 8 November 2008 (SSM et al.); 8, Napatree, 31 March 2010 (CRa); 5, Matunuck, 4 November 1939 (DLK); 5, Napatree Point, 24 November 1995 (CRa); 5, Napatree Point, 4 December 1958 (DLK, CW).

Extreme dates. 26 September 1946, Jerusalem (DLK), and 4 October 2005, Fogland Beach (RFa), to 15 May 2005, Napatree (CRa).

Song Sparrow (*Melospiza melodia*)

Status

Abundant summer resident and transient, common and widespread in winter.

Breeding

Song Sparrows prefer "edge" or early second-growth habitats. They are more common in the eastern and southern parts of the state, nearer civilization and the coast. They are, however, also common summer residents in the wooded uplands of the interior, where they breed in recently disturbed habitats as well as in shrub swamps.

Summer maxima. 325, Cumberland to Lincoln, 16–22 June 1984 (RLF); 219, Foster to Coventry, 16–24 June 1986 (RLF). Song Sparrows are especially abundant in the maritime scrub of the coastal zone and breed on most of the islands in Narragansett Bay; e.g., 638, Block Island, 18–22 June 1990 (CRa, REn); 62, Prudence Island, 8 June 1990 (CRa, REn).

Nonbreeding

Song Sparrows declined significantly as fall migrants at both Kingston and Block Island 1969–2001 (Osenkowski et al. 2012). This probably reflects the gradual reforestation of the area over that period.

Spring. Migrants are most numerous in March and April. Spring counts may exceed 100 per day per locality.

Fall. They are abundant in fall with a peak in mid-October to mid-November when hundreds are seen per day per locality.

Winter. Numbers on the Newport–Westport CBC totaled 380 in the mild season of 1952–1953. The extent of winter emigration and replacement of the breeding population with northern birds is unknown. Upstate areas lacking them in winter see a return of territorial birds in mid to late March or earlier and a gradual departure in September and October.

Winter maxima. 886, Newport–Westport CBC, 15 December 2007; 859, Newport–Westport CBC, 14 December 2013.

Lincoln's Sparrow (*Melospiza lincolnii*)

Status

Rare spring and uncommon fall transient.

Nonbreeding

Spring. Lincoln's Sparrows are secretive and often go undetected unless they sing; most spring maxima are from mist nets. Most migrants pass through in late May or even the first few days of June. There are some early records: 1, Carolina 4 May 1972 (DWa); Swan Point Cemetery 8 May 1990 (DFi, RFa); and Kingston 9–10 May 1991 (DLK). One was singing at Napatree Point 11 May 1996 (CRa).

Spring maxima. Eight banded, Block Island, 16–26 May 1974 (ELa); 7 banded, Block Island, 19 May 1996 (ELa); 3, Block Island 18 May 1957 (WBD, DLE).

Fall. The earliest fall record is for 1 at Napatree Point 31 August 1982 (JGe, RGe). Another remained aboard a boat at Cox Ledge all day, 2 September 1978 (RAC, CWo). Peak numbers pass through in late September and October. Lincoln's Sparrows decline in abundance through November; the latest record is of 1 at Block Island 24 November 1960 (S. Keith).

Fall maxima. Ten, Block Island, 3–4 October 1981 (RFa); 9, Block Island, 11–13 October 1958 (ASRI); 5, Swan Point Cemetery, 11 October 2001 (RFa et al.). There are 129 records since 2005, all but about 5 in fall between 10 September and 29 December. Recent maxima do not exceed 4 birds per day per locality.

Winter. There are December reports from Tiverton and Block Island, and another one wintered at Kimball Sanctuary in Charlestown in 1937–1938 (ESo et al.). One was on the South Kingstown CBC 27 December 1998 (DLK et al.), and 2 were on the Newport–Westport CBC 19 December 2009.

Swamp Sparrow (*Melospiza georgiana*)

Status

Fairly common summer resident, common to abundant migrant, and uncommon winter resident.

Breeding

Swamp Sparrows breed in open wooded swamps, near major watercourses, and along the upper margins of cattail marshes (Reinert 1979; Ellis 1980). They are found in the upper Pawcatuck estuary in Westerly; commonly at Fresh Pond, Quonochontaug; in the Moonstone Brook area near Trustom Pond; at least formerly near Point Judith; and at several points (usually in freshwater habitats) along outer Newport

County (Ocean Drive, Sachuest marshes, Long Pond, Briggs Marsh). They are much less common on the Narragansett Bay islands. One was seen carrying food at the edge of a small salt marsh on Patience Island 4 June 1985 (REn, CRa, RLF). They were not present at Block Island until 1 was found singing in Franklin Swamp on the west side, 16 June 1981 (RBo).

Summer maxima. 25 pairs, Diamond Bog, Carolina, summer 1977 and 1978 (Frank Golet, SRi); 30 males in 1 mi (1.6 km), Buckeye Brook, Charlestown, 29 April 1983 (CRa); 37, Chipuxet and upper Pawcatuck Rivers, 2–3 July 1984 (CRa, REn); 40, Great Swamp, 20 April 1992 (JMu).

Nonbreeding

Spring. Most have arrived on territory by late April, and they are scarce at migrant traps such as Swan Point Cemetery after mid-May.

Fall. First arrivals at Block Island have been in mid-September and peak numbers occur in late October or early November.

Fall maxima. 110, Avondale, 21 October 2009 (DFi, STa); 91, Tiverton, 28 October 2005 (RFa); 75, Warren, 19 October 1980 (RBo); 50, Warren's Point after a cold front, 17 October 1982 (RBo).

Winter. Moderate numbers winter within a few miles of the coast and Narragansett Bay around brooks and small marshes and regularly survive; e.g., 18 probable winterers along the South Shore 19 March 1949 (DLK). Inland, Swamp Sparrows have remained regularly into winter north to the East Providence Reservoir and the Lonsdale marshes and have sometimes survived at both locations to late February.

Winter maxima. 168, Newport–Westport CBC, 1952–1953; 95 on 2 Newport–Westport CBCs since 2005.

Green-tailed Towhee (*Pipilo chlorurus*)

Status

Vagrant. Three records.

Nonbreeding

One appeared at a feeder in Perryville 4 December 1995–23 February 1996 (BCG et al.). A second was on the Newport–Westport CBC in Middletown 18 December 2010 to 30 January 2011 (PLE, RTL, m.obs.; see Photo 27). A third individual was found by Sue Collin at Sachuest NWR 19 January 2023, and subsequently seen and photographed by numerous observers until at least 20 April 2023.

Eastern Towhee (*Pipilo erythrophthalmus*)

Status

Abundant summer resident, common transient, and fairly common winter resident.

Breeding

Eastern Towhees are one of the most numerous and widespread breeding birds in Rhode Island. They are most numerous in summer in the northern and northwestern parts of the state and most numerous in winter at the immediate coast. They seem to have reached peak abundance in the 1950s and 1960s and have somewhat declined since then (Hagen 1993); these changes undoubtedly follow the overall pattern of reforestation (see below).

Summer maxima. 630, Richmond to Hopkinton, 15–21 June 1987; 569, Foster to Coventry, 16–24 June 1986; 243, Conanicut Island, 22–25 June 1985 (all RLF); 50 pairs, Block Island, 8–20 June 1981 (RBo).

Nonbreeding

Eastern Towhees declined as fall migrants at Kingston but not Block Island during 1969–2001 (Osenkowski et al. 2012).

Spring. General arrival of migrants occurs during mid-April; a singing bird in North Smithfield 24 March 1984 (HCa) and 1 at Napatree Point 31 March 1984 (CRa) may have been new arrivals or overwinterers.

Fall. Migrants peak during October.

Fall maxima. 255, Swan Point Cemetery, 6 October 1997 (RFa, DVa); numbers drop through November.

Winter. Howe and Sturtevant (1899) and Allen (1909) knew of only 1 winter record: Newport, 14 January 1896. Towhees have greatly increased since then, probably because of both the increase of secondary growth and a warming climate.

Winter maxima. 96, Conanicut Island, 27 December 1956 (JBa); 35, Conanicut Island, 28 February 1975 (KKi, FPe). Highest winter count in last 10 years is 44, South Kingstown CBC, 26 December 2009.

YELLOW-BREASTED CHAT (ICTERIIDAE)

Yellow-breasted Chat (*Icteria virens*)

Status

Rare summer resident, formerly more numerous; uncommon spring and fairly common fall transient. Rare in winter.

Historical

Howe and Sturtevant (1899) noted a 20-year increase. This increase was perhaps related to regeneration of thickets that followed the decline of sheep farming in coastal areas. There are many indications that Yellow-breasted Chat were common at the turn of the twentieth century and that they were found farther north than at any time since; e.g., collected at Providence 4 July 1882 (HSH), Pawtucket 9 June 1893 (Monroe; CED), Cranston 1 June 1897 and 26 May 1901 (CED), Warwick in 1900 and 1902 (HSH), "Kent County" in 1907 (Remington), and Bristol in 1885 and 1887 (CED). There are several pre-1900 egg sets from Manton in the Flanagan collection.

Their status during 1910–1930 is unclear, but suggestions of a decline are echoed by similar statements about Massachusetts in Griscom and Snyder (1955). By 1931, however, they were again numerous, with four pairs at Kimball Sanctuary, Charlestown (MSo), and numerous records from other coastal areas through the 1930s. In 1938, Russell (in litt. to RCC) noted that he was "surprised at the numbers on Jamestown." During the period 1936–1940, A. M. Clark estimated one pair per 2 acres (0.8 ha) of brushy habitat on Conanicut Island but noted a 50% drop in 1939 compared with the previous 3 years. Clement and AEE noted one on territory in Lincoln Woods 21 June 1936, the northernmost for Rhode Island.

Kraus found 12 pairs in 1942 between Bonnet Shores and Quonochontaug, and WBD noted eight pairs in 1946 on Prudence Island. As late as 1953, there were four pairs at Jamestown (MDa). Since the mid-1950s, they have declined and are now rare.

Breeding

Three territorial male chats were found "close together" near Sakonnet in July 1972, and 3 males were there again at the same location in 1974 (DLE, REm). In addition, a few June observations from Jamestown, East Greenwich, Narragansett, Green Hill, Charlestown, and Westerly were made in the 1970s. Nevertheless, the overall abundance in Rhode Island during the 1970s was much lower than 20 years previously and that low abundance is even more pronounced now.

The abandonment of agriculture during the early 1900s was followed by substantial regrowth of coastal brush. This growth was beneficial to chats up to about 1950, but since then the thickets and brush have grown to forests that are not suitable for chats. Encapsulating the successional process and its impact on chats is the closing of the Charlestown Naval Air Station north of Ninigret Pond. Closure of this base in the 1960s allowed vegetative regrowth that provided a considerable area of suitable habitat. Other than 2 males on territory on Prudence Island, 21–22 June 2004 (REn), this area had the only territorial chats in Rhode Island since the 1970s. One was found there 20–21 July 1981 (DLK, DRu, DVT), 3 were there in 1982–1983 (v.o.), and 8 on 18 May 1984 (RAC). One was present across Ninigret Pond on the well-vegetated barrier beach 16–26 June 1982 (CRa). During the 1990s, most of the vegetation on the old air base had grown up to the point that the habitat was no longer suitable, and no chats have been seen there since.

Nonbreeding

Spring. Yellow-breasted Chats are difficult to detect because of their secretive nature. That, and the recent increase of wintering birds, makes migratory dates difficult to determine. A first bird of the spring was detected at Kimball Sanctuary 27 April 1930, when they were common breeders there. Most spring arrivals seem to be in the first week of May, with known transients at Swan Point Cemetery to 23 May and numerous records from coastal areas that lack breeders into the first week of June. One was banded at Block Island 13 June 1967 (ELa).

Fall. Chats have an extended fall migration, from mid-August into December, with a slight peak in mid-September. Usually, only single birds are seen.

Fall maxima. Four, Sakonnet, 12 October 1957 (DLE, RBo); 2, Cox Ledge, 15 September 1974 (v.o.).

Winter. Chats now linger annually into early winter, nearly always very close to the coast and particularly where multiflora-rose is present. By 1992, there were 68 reports for the CBCs, even more than for the Common Yellowthroat. This increase in winter seems likely to be connected to the regrowth of vegetation in coastal areas, where little winter habitat or adequate food sources were available before about 1940. The presence of Yellow-breasted Chats in early winter remains paradoxical because their recent establishment as winter residents has coincided with their decline as breeders, perhaps suggesting that winterers derive from midwestern populations.

Winter maxima. Maxima for anywhere in North America are often on CBCs in southern New England, including Rhode Island; e.g., 7, Newport–Westport CBC, 1 January 2009; 6 on same CBC in December 2011 and 2012.

There are fewer reports later in winter. For example, since 2005, there are 27 records for January, 4 for February, and none for March. One was found dead in East Greenwich 10 January 1958 (RLF), its crop crammed with rose hips. There are occasional reports for late January. One was found dead in Saunderstown 18 January 2000 (PAB) and sent to AMNH (CRa). Two were found dead on Sachuest Point after a cold snap, 19 January 1992 (AMNH #831242 and #831243; CRa et al.). One survived in Westerly into February 1950 but was later found dead (ESa). One was seen on Sachuest Point 7 February 1993 (CPe), and another was there 18 February 1995 (JSJ). One survived to 13 February 1962 at North Kingstown (Smith), and another was seen on Block Island 15 February 1999 (CRa et al.). One remained at Westerly to 19 February 1958 (K Ga), and another was at Sachuest on 27 February 1996 (fide N. Logothets). There is a report of 1 at Jamestown 31 March 1991, which implies winter survival. Yellow-breasted Chats regularly survive the winter in Massachusetts (Veit and Petersen 1993). It seems that Yellow-breasted Chats have been exploiting winter habitats made newly suitable by warming winter climate and planting of fruiting plants but are still suffering high winter mortality during severe winters.

BLACKBIRDS (ICTERIDAE)

Yellow-headed Blackbird (*Xanthocephalus xanthocephalus*)

Status
Rare migrant, mainly in fall.

Nonbreeding
The first 2 Rhode Island records were at Block Island 1 December 1919 and 23 October 1920 (EDi). Yellow-headed Blackbirds have extended their breeding range eastward into Ontario, and their frequency of occurrence has increased on the East Coast, especially since the 1960s. Seasons with high reproduction in the eastern part of their range are followed by high numbers of vagrants on the East Coast (Veit 1997, 2000; Zawadzki et al. 2019). Most East Coast vagrants occur in late August and early September, but this is a hardy bird, and there are records for almost every month. During the years 1919–1993, 29 birds were recorded in Rhode Island between August and December, 10 between March and May, and 2 in January–February. There are 17 records (all single birds) since November 2005; these span several months—all but February, June, and July. Only 3 are from Block Island. Early fall birds are usually at the immediate coast (and sometimes at sea) while those late in fall and winter appear at feeders and in mixed flocks with cowbirds in pastures.

Fall. Following early Block Island records, singles were at Jamestown 14–22 September 1936 (H. Russell Jr.) and Hope Valley 2–21 February 1957 (Segar; photographed by RBo). Beginning in the 1960s, Yellow-headed Blackbirds have appeared almost annually, with an annual total of 4 birds in 1976 (1 banded at Block Island 6 April [ELa, RAC]; 3 in Barrington 26 September–5 October [HWi]). Earliest fall record is 1, Napatree Point, 6 August 2015 (CRa).

Winter. One, Warwick, 21 January 2012 (B. Walsh); 1, 3 February–21 March 2020 in Little Compton (REm).

Bobolink (*Dolichonyx oryzivorus*)

Status
Fairly common, but declining, breeder in grasslands. Fairly common migrant.

Historical
Sir Charles Blagden, suggesting they were common on Aquidneck Island during the Revolution, wrote concerning the "Quongqueedle" (Bobolink) at Newport, 1776–1780, "[T]he males appear some time before the females, which then appear all of a sudden and pair immediately" (Howe 1905). Howe and Sturtevant (1899) described Bobolinks as abundant summer residents. At the turn of the twentieth

century, Bobolinks were plentiful in many parts of Warren, Bristol, coastal portions of Tiverton and Little Compton, on Conanicut and Aquidneck Islands, and even bred "abundantly" in fields at Price's Neck, Ocean Drive, Newport (HSH), over much of the Point Judith peninsula, and throughout the South County coastal plain. They still bred in many now-urbanized upstate locations (e.g., Cranston, Warwick, East Providence) into the 1920s and 1930s and lingered to 1943 in Barrington (HNG). They were breeding near Chepachet and in Burrillville at least to 1946, at Howard to 1955 or later, and near the Wenscott Reservoir in Lincoln at least to 1971 (CWo). Bobolinks were formerly abundant breeders throughout the state but have decreased with the disappearance of agriculture. They require hay and alfalfa fields, and these fields must be mowed regularly to avoid rapid succession. The disappearance of horse-drawn transportation after the 1920s, and dairy farms after that, have reduced the need for hay. Production of nursery stock, potatoes, corn, turf farming, and the spread of housing developments have supplanted much land formerly used for growing hay.

Breeding

Ferren surveyed the state for nesting Bobolinks beginning in 1972 and found that they were most numerous in the remaining pastures and farmlands on the Narragansett Bay islands. At Block Island, Bobolinks bred regularly at Dickens Point and probably elsewhere on the island in the 1880s, but the only mention in EDi's diaries of summering birds after 1912 is of 1 seen in the Corn Neck area 21 June 1921. They have also bred in Little Compton at several sites since at least the 1970s: 24 males, 11 females, and 5 fledglings were in a traditional breeding field south of the Commons on 16 June 2016 (REm). In northern Cumberland, 44 birds were present at one site, 14 June 2003 (MLy).

Nonbreeding

Spring. As transients in spring, Bobolinks were formerly common but have declined, and it is now unusual to see more than 50 per day per locality. The earliest record is for 1 at Jerusalem 21 April 1946 (Coman). General arrival of breeders is during the last few days of April and early May.

Spring maxima. 200, Block Island, 18 May 1957 (WBD); 60, Conanicut Island, 20 May 1990 (ChP et al.).

Fall. A few birds begin to drift into nonbreeding areas in early July, and migrants are sometimes noted moving south by late July; e.g., 40 over HSH's shorebird blind at Quonochontaug 23 July 1911. Southbound transients reach a peak in the first week of September. Planted millet and cornfields attract large numbers of Bobolinks, which often roost overnight there during fall. Woodruff and Mahoney banded 580 at NBS in Middletown 27 August–11 September 1963 and estimated that 2000 were feeding in a 2-acre (0.8-ha) field of millet during that time.

Fall maxima. 3000 at a millet field, Trustom Pond NWR, 6 September 1983 (DVT); 2000 going to roost, Middletown, 7 September 1959 (JBa); 1650, Watch Hill, 6 September 1980 (RLF); 1000, Sakonnet, 8 September 1955 (RBo); 700+, Great Swamp, 1 September 1971 (DWa). Bobolinks are uncommon after September, but a few flocks of migrants have been seen through October.

Extreme dates. 21 April (above) to 19 November 1944, Warren (1 collected; RBo).

Eastern Meadowlark (*Sturnella magna*)

Status

Uncommon summer resident in grasslands. Fairly common migrant and winter resident.

Breeding

Eastern Meadowlarks have been declining for more than a century. They were formerly very common as breeders at Block Island, e.g., near Dickens Point, but had substantially declined by 1945 (EDi) and have been absent there since the late 1960s (MSl, RLF, RBo, REn, CRa). During the 1960s, they were most numerous in the Little Compton, Middletown–Portsmouth, and Matunuck areas of southern Rhode Island. Meadowlarks prefer hayfields and abandoned pastures but move on when these succeed to secondary growth. Thus, they have left most former farmlands that have either been developed or grown up to forest. Meadowlarks have been absent from Little Compton during the breeding season since about 2001. A pair was present in a Seapowet field on 6 June 2009 but apparently departed when the field was subsequently mowed in mid-June (REm).

Nonbreeding

Spring. Migrants appear from mid-March through April.

Fall. Although not very common at Napatree, Eastern Meadowlarks appear in a pronounced peak of migrants during November. Birds usually appear along the beach after strong cold fronts (CRa). Meadowlarks tend to gather in cultivated fields near salt water in late fall; they feed on grain in the fields and roost at night in the *Spartina* grass of the salt marsh. Wintering flocks are now found principally in fields along the shores of the Sakonnet Passage of Tiverton–Little Compton; on the 2 larger Narragansett Bay islands and elsewhere around the shores of Narragansett Bay, occasionally north to Gaspee Point; and along the South County coastal plain. Wintering flocks exceeding 100 birds were once commonplace at the turn of the twentieth century despite year-round hunting of meadowlarks as "marsh quail," but winter flocks over 50 in recent years have been rare. The decline in winter is very evident from CBC data; e.g., a high of 501 on the mainland CBCs in 1952 with an average of 378 during the period 1950–1960, compared with a high of 115 and an average of only 50 during the period 1980–1990. They were formerly common at Block Island in winter (303 on the 1932 CBC), but an average of less than 8 (maximum 40) was the rule in the period 1962–1994.

Winter maxima. 54, Seapowet, 29 January 2001 (RFa, DVa). Since 2005, highest CBC tallies were 44 on the Newport–Westport CBC in 2 consecutive years: 17 December 2005 and 16 December 2006. But as of 2016, the 10-year average on this CBC had dropped to 16.3, and the last report was of a single bird in 2015 (REm).

Western Meadowlark (*Sturnella neglecta*)

Status

Vagrant. Four records.

Nonbreeding

One was present at Dickens Point, Block Island 5–16 April 1921, and another appeared at the same location 26–29 June 1934. Elizabeth Dickens observed both birds and heard them singing from nearby stone walls. One was seen and heard at a golf course north of Sakonnet Point 17–25 May 1975 (REm, DLE); another appeared, and its song was recorded at almost the same location 29–30 April 1988 (RBo, HMP, RKl; DLE, DLK, RAC et al.). Western Meadowlarks have been slowly extending their breeding range eastward through the Great Lakes area, and a *magna–neglecta* nesting has been documented in the lower Hudson Valley (Lanyon 1966).

Orchard Oriole (*Icterus spurius*)

Status

Uncommon breeder and migrant; increasing.

Historical

Orchard Orioles may have been most numerous around 1900, and then again in recent decades, with a long period of reduced numbers in between. Howe and Sturtevant (1899) called them "rare summer resident[s]" but stated that they "had possibly been increasing in numbers" and cited nests in Barrington, Newport, and Warwick Neck. Doe collected nine sets of eggs in the Warwick area between 1885 and 1901. Louis Mearns (1901) recorded nests at five places near Fort Adams, Newport, between 1899 and 1901. LeRoy King found two pairs near Brenton Point in 1900, and HSH found a nest at Westerly in 1903. Hathaway (1943) regarded Orchard Orioles as "very rare, irregular, local summer resident[s], formerly found in West Barrington, Bristol, Westerly, and Charlestown but rarely seen in recent years." Unmentioned were nestings at Lonsdale in 1932 and thereafter (AEE) and near Seapowet in 1936 (Ruecker). There seems to have been a gradual increase thereafter, perhaps due to maturation of ornamental trees (including evergreens) and northern red cedars in coastal areas and an increase in suburban development. Breeding was reported at Wakefield (1946 on), several places in Tiverton (1952 on), Jamestown (1958 on), and Cross Mills (1960 on).

Breeding

Orchard Orioles currently breed in South County, mainly within 15 mi (24 km) of the ocean along the southern and eastern fringes, in the small interior towns north of the South County moraine (Carolina, Shannock, Hope Valley, Kingston, Wakefield), as well as in the western Narragansett Bay area (Narragansett to Warwick). They also breed on the 2 larger Narragansett Bay islands as well as in Tiverton and Little Compton. They occasionally breed in the Providence metropolitan area and north in the Blackstone Valley to Lincoln and Cumberland. They have been reported in June west to western Coventry (Greene) and western Burrillville, with a pair breeding annually at Wallum Lake 1981–1983 (SMi). Migrants and wanderers occur at Block Island in late spring, and a pair nested at least once in a planted spruce there 28 June 2003 (CNu, m.obs.).

Summer maxima. Ten, Narragansett to South Kingstown, 23–28 May 1984 (RLF); 8, Richmond to Hopkinton, 15–21 June 1987 (RLF).

Nonbreeding

Spring. Peak arrival is in mid-May, with transient wanderers into June; e.g., 2, Block Island, 8 June 1972 (RLF). Early migrants often appear after periods of southerly airflow in April; e.g., 1, Jamestown, 16 April 1961 (MDa); 1, Newport, 21 April 1980 (RSt); 1, Westerly, 23 April 1967 (ESa).

Spring maxima. Seven, Tiverton, 9 May 1971 (SSD et al.); 4, Swan Point Cemetery, 5 May 2011 (RFa et al.).

Fall. Orchard Orioles depart shortly after they finish breeding in July. Up to 3 per day have been seen at Napatree Point during August. Some records of fall migrants are 1, Jamestown, 16–17 August 1963 (MDa); 1, Napatree Point, 24 August 1992 (CRa); 1, Wallum Lake, 25 August 1981 (SMi); 1, Middletown, 4 September 1989 (POn).

Fall maxima. A flock of 9 (with several Baltimores) in a brushy field edge near Briggs Marsh, 8 August 2005 (REm). A small greenish oriole, thought to be this species, was well seen at Kingston on the exceptional date of 8 November 1952 (DLK), and another was at Warren 17–18 November 1997 (RBo).

Bullock's Oriole (*Icterus bullockii*)

Status

Vagrant. About six records.

Nonbreeding

An immature bird, photographed in color by RBo, was seen regularly at a feeder in Tiverton in December 1956 by numerous observers; the photographs were submitted to G. Sutton, who considered the bird to be *bullockii*. Other well-seen individuals include 2 birds at feeders in Warwick 20 December 1965 to 3 February 1966 (KKi, DLK et al.), and 1 gray-backed bird with a white belly at Block Island 25 September 1971 (RBo, WBD); a similar observation is of an immature male at Block Island 2 October 1972 (KPA). Another was present at Swan Point Cemetery 9 December 1985–17 January 1986 (CRa, REn; RBo).

Baltimore Oriole (*Icterus galbula*)

Status

Common summer resident and migrant; rare but increasing in early winter.

Breeding

Baltimore Orioles nest in oak and mixed deciduous forests, commonly in residential areas but also including unbroken tracts of woods. They also nest in more protected parts of the larger Narragansett Bay islands and outer coast. At Block Island, individuals and occasional pairs have been present into June.

Summer maxima. 338, Foster to Coventry, 16–24 June 1986; 290, Cumberland to Lincoln, 16–22 June 1984; 28, Conanicut Island, 22–25 June 1985 (all RLF). Four were on Prudence Island 8 June 1990 (CRa, REn).

Nonbreeding

Spring. Baltimore Orioles arrive in late April and early May; e.g., 22 April 1973 at Block Island (ELa); 22 April 1980 at Matunuck (v.o.).

Spring maxima. 35, Block Island, 21 May 1966 (WBD); 25, Block Island, 10 May 1959 (MSl); 25, Block Island, 14 May 1932 (EDi). June wanderers have been evident at Block Island as late as 21 June. By mid-July, males stop singing, and migration of adults peaks in late August and early September.

Fall. Baltimore Orioles were rare after September until the 1950s and 1960s, when they increased. Late records such as 1, Woonsocket, 18 November 1914 (Arnold), and another anonymously reported from Warren 18 December 1926, were highly unusual.

Fall maxima. 50 adult males, Middletown, 20 August 1959 (JBa); 50 migrating, Block Island, 30 August 2012 (CRa).

Winter. An immature male remained into January 1938 in Tiverton (Ruecker), and an adult male was seen in Warren on the remarkable date of 27 March 1938 (RBo). In more recent years, the number of individuals that have attempted to winter at feeders has substantially increased.

Winter maxima. A flock of 11 visited a feeder at Jamestown 22 December 1978 (MDa). There are several records suggesting successful overwintering: 1, Bristol, to 6 February 1984 (C. House); 1, Wakefield, 6 March 1975 (fide CWo). One remained at a feeder in Fiskeville to 2 April 1957 (JGe, RGe).

Red-winged Blackbird (*Agelaius phoeniceus*)

Status

Abundant summer resident and migrant; fairly common winter resident.

Historical

Red-wings have increased dramatically continent-wide, owing to both a warming climate and increased availability of corn and other grains in agricultural fields. Arrival dates of red-wings in spring have changed considerably since the late 1800s. Blagden (1903) said the males "appeared [in Newport] the 10th or 12th of March [in 1776–1780]; in a fortnight or 3 weeks came the females." The earliest date known to Howe and Sturtevant (1899) was 7 March. Hathaway recorded the arrival of red-wings and grackles in Cranston between 2–16 March during the period 1903–1910 and as far as is known never encountered them in February in Rhode Island between 1890–1910. In contrast, red-wings have generally arrived 2 weeks earlier than this in recent years. The first February appearance was a flock of eight at Middletown 21 February 1909 (ESt). Others appeared at Dickens Point, Block Island, on 13 February 1939 and 14 February 1935 (EDi).

Breeding

Red-winged Blackbirds are ubiquitous summer residents. They nest in cattail marshes, the edges of salt marshes, wet meadows, brushy swamps, and hayfields.

Summer maxima. 434, western Aquidneck Island, 15–16 June 1990; 310, Cumberland to Lincoln, 16–22 June 1984; 253, Narragansett to South Kingstown, 23–28 May 1984 (all RLF). An average of 50 pairs occupied the Blackstone River marshes at Lonsdale during the 1980s (REn, CRa).

Nonbreeding

Spring. By early to mid-March (the period of earliest arrival 80 years ago), red-wings are now frequently abundant.

Spring maxima. 1000, West Kingston, 2 March 1984 (FPe, KKi); 1000, Matunuck, 10 March 1973 (DLK).

Fall. Because of their early arrival and breeding season, red-wings begin moving away from nesting areas in early July. Migrants are often seen at coastal vantage points and at sea from boats at that time. Peak counts are later in fall, in September and October. Red-wings form roosts with other blackbirds in stands of *Phragmites* and under bridges.

Fall maxima. 10,000, Island Park, 21 October 1986 (RBo); 7500, Warwick, 21 September 2008 (CPe, DFW); 6000, Tiverton, 23 October 1983 (HCa); 2000, Little Compton, 30 October 1993 (DLE); 2000 migrating past Watch Hill, 15 November 2015 (CRa); 1700, Newport, 27 October 1940 (DLK).

Winter maxima. 1146, South Kingstown CBC, 21 December 2013 (STs et al.); 1000, Little Compton, 14 December 1991 (RAC et al.); 300 winterers, Tiverton, 31 January 1992 (DLE); 250, Warren, 17 January 1993 (RBo); 100, Lonsdale Marsh, 8 December 1980 (REn).

Brown-headed Cowbird (*Molothrus ater*)

Status

Common to abundant breeder and migrant; fairly common in winter.

Breeding

Brown-headed Cowbirds are obligate brood parasites that cause serious conservation problems for some species whose nests they exploit, especially Kirtland's and Cerulean Warblers (Lowther 2020). The clearing of forests throughout North America exposed many woodland birds to parasitism by cowbirds; formerly, cowbirds (because they preferred prairie habitats) could not access these birds. Brown-headed Cowbirds nest throughout Rhode Island, and, although there are no endangered species that could be vulnerable here, there are locally rare species that could be affected.

Nonbreeding

Spring. The earliest record of sizable numbers is for 500 at West Kingston on 8 March 1994 (DLK). They are common in spring in farmland and suburbs from mid-March through mid-April (flocks rarely to 500), when they are particularly noticeable at feeders. Birds spread out during the breeding season in May and June and remain in small groups through July.

Fall. They are abundant in late summer at mixed roosts of blackbirds and starlings; e.g., 1500+, Barrington, 13 September 1978 (Barbara Spicer). Numbers gradually decline through the fall.

Fall maxima. 3000, Barrington, 12 November 1951 (HNG); 2500, Warwick, 21 September 2008 (DFW, CPe); 1100+ going to roost in Portsmouth, 18 December 1982 (RLF et al.). Numbers of Brown-headed Cowbirds have increased with suburban development and an increase in feeders.

Winter. Mixed flocks of starlings and Brown-headed Cowbirds persist through the winter; e.g., 957, Napatree CBC, 18 December 2011 (GW et al.).

Rusty Blackbird (*Euphagus carolinus*)

Status

Uncommon migrant, uncommon but regular in winter. Substantial population decline since 1960 (Greenberg and Droege 1999).

Nonbreeding

Rusty Blackbirds are at times common, but they are usually quiet and inconspicuous migrants, typically remaining in wooded swamps but sometimes visiting weedy gardens and agricultural fields. The North American population has declined by 90% since about 1960 (Greenberg and Droege 1999).

Spring. Howe and Sturtevant (1899) noticed that Rusty Blackbirds were less common during spring in Rhode Island than farther inland in Massachusetts, a pattern that is still as true today for Rusty Blackbirds as it is for other Canadian-zone breeders. The first migrants appear in late February and March.

Spring maxima. 55, Kingston, 24 March 1969 (DLK); 45, Diamond Hill, 27 April 1981 (fide RLF); 40, Green Hill, 19 March 1977 (DWa). Seven were seen at Napatree Point 5 May 1984 (CRa) and 8 at Cumberland 8 May 1993 (JO). There are records of lingerers or perhaps prospecting birds at George Washington State Forest 11 June 1991 (JSJ, BCG) and Carolina 18 June 1989 (DWa).

Fall. Most are seen during late September to early November, but 1 was seen 12 September 1941 at Kingston (DLK). Most fall counts do not exceed 50 birds.

Fall maxima. 1000, Kingston, 9 October 1938 (DLK); 200+, Block Island, 8–11 October 1965 (ASRI); 120, North Kingstown, 18 October 1977 (DWa). More recently, 60+, Slater Memorial Park, 30 December 2007 (MLy); 66, South Kingstown CBC, 22 December 2007.

Winter. Rusty Blackbirds were first recorded in winter at Apponaug 4 February 1912 (specimen, HSH), and they are now regular in southern Rhode Island.

Winter maxima. 182, South Kingstown CBC, 27 December 1998 (DLK et al.); 80, Westerly, 8 February 2012 (CRa); 75, Kingston, 17 February 1985 (DLK); 65, Lincoln Woods, 15 February 2002 (SDz); 52, Lonsdale Marsh, 18 January 1986 (REn); 30, Cumberland 27 January 1993 (BCG).

Brewer's Blackbird (*Euphagus cyanocephalus*)

Status

Vagrant. About six records.

Nonbreeding

Three Brewer's Blackbirds were found at Portsmouth 22 December 1968 and another at the same location 17 December 1972 (both by CBl, previously familiar with the species). A female was observed at close range at a Barrington feeder 24 February 1983 (BSp). Two males were seen at Sakonnet 13 November 1983 (RBo), and an adult male with bright purple splotches on the neck was seen at 20 ft (6 m) and very carefully described from DLK's feeder in Kingston, 14 February 1985. Another was found in Swan Point Cemetery 13 October 1999 (RFa, DFi). Like Clay-colored Sparrows and Western Meadowlarks, Brewer's Blackbirds have been expanding their range eastward in the northern Midwest (Walkinshaw and Zimmerman 1961) and now occur with some regularity in winter in the southeastern states. Brewer's Blackbirds are now of nearly annual occurrence in southern New England, where they frequent farms and barnyards, and sometimes survive the winter (Veit and Petersen 1993), although there are few recent confirmed records from Rhode Island.

Common Grackle (*Quiscalus quiscula*)

Status

Abundant resident and migrant; fairly common in winter.

Breeding

Common Grackles have increased dramatically since the late 1800s, when they were uncommon in Rhode Island. Clement knew of only 5 colonies in Tiverton and Little Compton about 1938, and the species was entirely absent on Prudence Island in 1946 (WBD). Now greatly increased, they are abundant and ubiquitous in these locations, including Block Island and all coastal areas. Grackles nest in trees, especially conifers, so their scarcity 100 years ago may have been due to a lack of suitable nesting sites. They have benefited from the planting of conifers in suburban areas and from the ecological succession of abandoned farmlands. Common Grackles nest singly or semicolonially in conifers, shrubbery, and vines, often near water, and they also use vines on the sides of buildings and even lower portions of Osprey nests. Grackles are least numerous in the forested western sections, but even there, modest "colonies" are found in swamps within the most remote woodlands.

Nonbreeding

Spring. Since grackles often spend the winter in or near Rhode Island, there are no "first arrivals" in spring. The date of 12 February 1937 long stood as the earliest historic appearance of definite migrants at Woonsocket, brought on by a record period of mild weather (Arnold). There is still an obvious increase in movements as the weather warms in February or March. In very cold winters, increased activity does not occur until later, often mid-March. David Emerson considered the arrival of 30 grackles in North Kingstown 27 January 1989 (v.o.) during mild weather, and 700 grackles and 50 Red-winged Blackbirds at Little Compton 27 January 1990, after a long period of record mild temperatures, to be probable early arrivals from farther south. New arrivals are now found almost annually in mid to late February. They are usually common by early March in recent decades; e.g., 100 in Barrington 27 February 1990 (JCB), 500 in Tiverton on 2 March 1986 (RBo), and 1000 in South County 9 March 1974 (RBo, HWi). On the other hand, first arrivals have been slowed occasionally by severe weather; as late as 16 March in the bleak spring of 1978. There is usually a peak movement in late March or early April and migration continues actively at least to the end of April, with some continued migration or wandering movements into mid-May; e.g., 1, Cox Ledge, 14 May 1977 and 28 May 1984 (CWo; DLE); 20 departed southeastward from Point Judith 24 May 1980 (RLF).

Fall. Roosts of apparent nonbreeders in pine woods or *Phragmites* occur in the middle of the breeding season (e.g., 1350 going to roost at Galilee 26 May 1980 [RLF]; 6000 roosting at Island Park 16 June 1990 [RLF]), and such nocturnal aggregations increase to thousands in late summer and early fall. Grackles become increasingly gregarious both in their feeding and roosting habits as fall progresses.

Fall maxima. 25,000 roosting in *Phragmites* at Island Park, 21 October 1986 (RBo et al.); 14,000 going to roost at Sakonnet, 29 October 1972 (DLE, REm). A flock took 20 minutes to pass west across the Sakonnet Passage at Seapowet and was estimated to contain more than 150,000 birds (RBo). This total is currently 1 of the 2 highest historic day-counts among bird species in Rhode Island (see also Tree Swallow account).

Winter. Large flocks in late fall and winter are particularly attracted by standing cornfields, but the species has probably derived most of its great increase in winter from the proliferation of backyard feeders.

Winter maxima. 10,000, Wickford, 8 January 1999 (STs); 6148, South Kingstown CBC, 2 January 2011 (STs et al.); 5000, Charlestown, 23 December 1976 (DWa); 3351, South Kingstown CBC, 21 December 2013 (STs et al.).

Subspecies

Both pure "Bronzed" Grackle (*Q. q. versicolor*) and "Ridgway's" Grackle (formerly *Q. q. ridgwayi*) breed commonly in Rhode Island. All Rhode Island references to "Purple" Grackle (*Q. q. stonei*) pertain to "Ridgway's" Grackle, an intergrade population between northern *versicolor* and true *stonei*, the latter breeding from northern New Jersey south. "Ridgway's" Grackle is the dominant form in southern Rhode

Island, including Block Island, but is also found regularly with *versicolor* in northeastern and eastern sections. Hathaway noted an apparent difference in nesting habits, the "Bronzed" tending to nest colonially in evergreens and "Ridgway's" breeding in the vicinity of swamps. Wintering birds and transients are largely *versicolor*. A midwinter series of specimens taken by the author from East Providence were all of this race, but there are several sight records of "Ridgway's" for winter.

Boat-tailed Grackle (*Quiscalus major*)

Status
Vagrant. Slowly increasing.

Nonbreeding
Boat-tailed Grackles breed near salt marshes and barrier beaches along the middle Atlantic coastal states north to Long Island and central Connecticut. Prior to 1985, there were a handful of records of large *Quiscalus* grackles north to Nova Scotia. A flock of 5 female Boat-tailed Grackles appeared at Quonochontaug 18 October 1985 and 3 remained through 21 April 1986 (DFi, DLK et al.; photos by DFi, RBo). Possibly connected to this incursion were the first birds known in Massachusetts, 2 females in Essex County on 24 April 1986 (Forster; in Veit and Petersen 1993). Three (additional?) females were at the original Quonochontaug location 11–16 April 1987 (SHa, DLK, DFi et al.; DLK). After a 20-year gap, another female was at the Charlestown Breachway 2 June 2006 (RRV, CSM, LM-W). Another Boat-tailed Grackle was seen by many at Charlestown Breachway 11–12 May 2013 (T. Auer), and, at the same time, a male and female were seen at Block Island 12 May–2 June 2013 (STs et al.). One was at Jerusalem 12–14 April 2015 (CPe, m.obs.).

NEW WORLD WARBLERS (PARULIDAE)

Ovenbird (*Seiurus aurocapilla*)

Status
Very common summer resident; common spring and fairly common fall transient.

Breeding
Ovenbirds are found in lavish abundance in the wooded country of western Rhode Island; CRa ranked them as the most abundant species at 116 sample sites in the dry Arcadia Management Area in 1989–1992, ahead of Eastern Towhee and Veery. William R. Eddleman, sampling maple swamps in South County, similarly ranked Ovenbirds first in abundance, ahead of Veery and Eastern Towhee. They ranked second after Eastern Towhee among more than 30 species inhabiting pure stands of pitch pine in western Coventry during the 1980s.

Summer maxima. 502, Foster to Coventry, 16–24 June 1986; 432, Richmond to Hopkinton, 15–21 June 1987 (RLF); 152, Arcadia, 13–15 June 1992 (CRa); 152, Burlingame Management Area, 29 May–6 June 2006 (CRa); 9, Conanicut Island, 22–25 June 1985 (RLF). They do not breed at Block Island.

Nonbreeding

Ovenbirds in spring are conspicuous because of their unmistakable song; they are ubiquitous as migrants or newly arrived summer residents. Their arrival dates are shifting earlier, probably in response to warming climate, and there are many records for the third week of April; general arrival is now late April. Ovenbirds declined significantly as fall migrants at both Block Island and Kingston from 1969–2001 (Osenkowski et al. 2012).

Spring maxima. 51, Nooseneck area, 31 May 1989 (REn); 24 banded, Block Island, 17 May 1977 (ELa).

Fall. Fall migrants have appeared in early August, but peak migration is late September to mid-October. Ovenbirds are fairly common (though broadly dispersed and quiet) during migration in late August and September. They rank as 10th most abundant among warblers mist-netted in fall at Block Island (ELa).

Extreme dates. 11 April 1995, Fisherville Brook (MJM), and 21 April 1892, Johnston (1 shot; in RWPM), to 15 December 1957, Wilbour Woods (RLF, SSD), and 20 December 1975, Portsmouth (RLF, T. E. O'Connell). There is 1 midwinter record: 8–13 February 2016, Little Compton (GDe).

Worm-eating Warbler (*Helmitheros vermivorum*)

Status

Uncommon and local summer resident and rare transient; greatly increased in recent years.

Historical

This Appalachian species was not mentioned in Howe and Sturtevant (1899), but old notebooks of HSH suggest that it may have been recorded before 1899. Some of the first reports included birds at Block Island 18 May 1926 and 10 May 1927 (EDi); at Kimball Sanctuary 22 May 1930, 12 May 1935, and 8 May 1940 (MSo); and at Warren 6 September 1940 (RBo). Hathaway searched all suitable habitat in earlier years, and DLK similarly investigated numerous potential locations in South County in the 1940s, without success. Thus, there is no evidence that Worm-eating Warblers bred in Rhode Island woodlands before the 1970s.

Breeding

Beginning in the early 1970s, Worm-eating Warblers were found breeding at 4, possibly more, sites in Rhode Island. The available evidence suggests this was a colonization rather than birds being overlooked in the past. Worm-eating Warblers are more widespread and numerous now compared with 30 years ago. Sixty point counts in the Burlingame Management Area detected 5 birds 6–13 June 2006 (CRa), and 7 were detected on a walking survey of Weetamoo Woods in Tiverton 19 June 2006 (REm).

Newport County Sin and Flesh Brook, Tiverton. Luxuriant stands of mountain-laurel and maturing oaks along the rocky sides of the stream valley provide excellent habitat. There were no Worm-eating Warblers there in the 1930s and early 1940s, judging from the extensive breeding-season work of RCC and RBo, but from 1947–1954, RBo located singing birds in this area during May. He suspected breeding but was unable to obtain further evidence. Similarly, on 11–12 June 1972, RLF found a displaying pair on territory but no nest (RLF). On 11 June 1974, 3 pairs were behaving in a territorial fashion (RLF). On 13 June 1976, the first Rhode Island nest and 4 young were photographed in the same area by RSp. In the 1980s, a broader distribution of birds in Tiverton was evident, as birds were seen 15 May 1983 off Highland Road (HCa), and 6 were in Tiverton 30 June 1988; this included 2 males south of Lafayette.Road (REn). The Weetamoo Woods in Tiverton produced 11 birds (9 pairs) 23–24 June 1991 (RLF).

Kent County. A singing bird was present along the stream at Parker Woodland 11 May–4 June 1975 (JGe, RGe), a pair with 2 young was located there 1 July 1976 (JGe, RGe), and an adult with a single chick was seen at the same location 19–27 June 1977 (RAC, DLK). North of Watchaug Pond, Charlestown, singing birds have been seen regularly since 1983 on the old Clawson Farm, Burlingame State Park, in a rocky area of second growth bordering a wooded swamp (CRa, DRo et al.). The numbers of transients banded or seen on the Kraus farm, a mile east of Kingston, suggest the presence of a small breeding population in the vicinity. In West Greenwich, 3 were in the Wickaboxet area, 14 July 1990 (DMK).

Washington County. There are extensive areas of suitable habitat in Hopkinton, particularly west and south of Rockville on rocky slopes densely covered with mountain-laurel and rhododendron. One was found near Long Pond in July 1982 (v.o.), and another west of Ell Pond 20 June 1987 (RLF). South of Ashville Pond, both members of an agitated pair were carrying food on a steep and rocky laurel-covered slope 22 June 1987 (RLF). Beginning in the late 1970s, the species was reported during June and July on a dry hillside in the Carolina Management Area (CRa, RAC, DRo et al.). A large amount of rocky, oak-covered habitat also exists in the vicinity of Beach Pond, western Exeter, where 2 singing males were seen southeast of the pond on 22 May 1985 (CRa, REn), and a male was on territory through June 1989 (CRa, REn).

Providence and the Northeast. An adult was seen carrying food in 1979 off Round Top Road (RLC). A pair was seen in Foster 19 May 1982. In Burrillville, the species has been regularly reported near Wallum Lake since 1982. At nearby Wakefield Pond, in western Burrillville, 1 was singing 4 June 1980 (JGe, RGe). Worm-eating Warblers have also invaded northeastern Rhode Island, with singing males located at Limerock in 1989 and in the Diamond Hill area and several other sites in Cumberland since 1989 (DMK).

Nonbreeding

Spring. The earliest records in spring are for individuals banded at Block Island 11 April 1971 (ELa) and others seen at Arcadia 19 April 1986 (CRa), Swan Point Cemetery and Smithfield 21–22 April 1993 (RBo; BCG), Middletown 24 April 1958 (JBa), and Block Island 24 April 1987 (ELa) and 25 April 1991 (REn). Other records of nonterritorial birds are generally distributed from 1 May through 1 June. Like a number of other southern or Appalachian species, Worm-eating Warblers are somewhat more frequent in spring than in fall; through 1975, 42 (or 63%) of 67 transient reports were for spring.

Fall. In fall, Worm-eating Warblers migrate early compared with other locally breeding warblers. A migrant or wanderer was discovered in a maple swamp in Matunuck 25 July 1971 (SSD). Others have been seen or banded at DLK's farm in Kingston 11 August–13 September. Elsewhere, the majority of sight reports have occurred during late August and the first 2 weeks of September. One was banded at Kingston 3 October 1998 (DLK et al.) and another much later at Block Island 2 November 1999 (ELa et al.).

Extreme dates. 11 April 1971, Block Island (1 banded; ELa), to 2 November 1999, Block Island (1 banded; ELa).

Louisiana Waterthrush (*Parkesia motacilla*)

Status

Uncommon and local summer resident, mainly along streams in western and northern Rhode Island; rare transient.

Historical

Louisiana Waterthrushes were found on territory in the Snake Den area of Johnston in 1877 and in West Greenwich in 1879 (both by F. T. Jencks). Another was shot in West Kingston 1 May 1879 (collection of W. Brewster), and the first nest was found in Kent County in 1900.

Breeding

Since 1900, Louisiana Waterthrushes have gradually colonized many or most streams in the western parts of the state, and they are now found south and east to

Charlestown (north of Watchaug Pond), Kingston, Wakefield, and Warwick. They do not breed on the Narragansett Bay islands, but they may have bred at one time in the Tiverton highlands (in the Sin and Flesh Brook area), where up to 3 singing males have been noted at 2 locations; none have been in this area in breeding season since about 2000 (REm). They may also have bred in Little Compton; e.g., 1 at Adamsville 20 May 1944 (Williston), but none have been seen here during breeding season since at least 1970 (REm).

Their breeding habitat usually includes a running stream or brook, unlike Northern Waterthrushes, which are often found in stagnant bogs and swamps. In 1907, however, a nest was found within a short distance of a Northern Waterthrush nest in the Great Swamp, Richmond, where little or no running water was present (HSH, Bent, Flanagan). Both species breed within earshot of each other at Parker Woodland in Coventry.

Summer maxima. Seven, Stepstone Falls, 12 June 1989 (CRa); 5, Scituate, 8 May 1993 (CRa); 4, Scituate to Glocester, 30 May 1981 (RBu, HWi).

Nonbreeding

Spring. Louisiana Waterthrushes are uncommon in spring away from breeding sites. After Pine Warblers, they are the earliest of the warblers to arrive, the earliest appearances on territory including birds 2 April 1986 at Parker Woodland, Coventry (JGe, RGe); 3 April 1983 in Hopkinton (DWa); 4 April 2000 at the Gilbert Stuart birthplace (REm); and 5 April 2010 in Foster (JMu). There are 4 reports from Block Island during the 7–11 April range of dates. The latest certain migrant is 1 that was banded at Block Island, 17 May 1971 (ELa).

Fall. Louisiana Waterthrushes are one of the earliest migrants in fall. Most leave during July: singles were banded 22 July and 6 August 1960 at Middletown (JBa) and 1 banded at Block Island 6 August 1971 (ELa). The latter was the only Louisiana Waterthrush among more than 50,000 fall landbirds that have been captured at Block Island over the course of 30 years by all banders combined.

Extreme dates. 2 April 1986, Parker Woodland (JGe, RGe), to 6 August 1971, Block Island, and 9 September 2011, Napatree (CRa); the latter was well seen, the only Napatree record.

Northern Waterthrush (*Parkesia noveboracensis*)

Status

Fairly common summer resident; uncommon spring, fairly common fall, transient.

Historical

None of the early oologists such as CED ever encountered Northern Waterthrushes in their ramblings of Rhode Island's northwestern towns, where the species is now common. Hathaway found the first breeding birds in 1906 (including three pairs and a nest) in the Great Swamp, Richmond.

Breeding

Highest densities of Northern Waterthrushes are found at the edges of swamps or river floodplains, such as 13 males detected in 9 point counts along Poquiant Brook, Charlestown, 6 June 2007 (CRa). Densities may be comparable in central South County, that is, in the general vicinity of the Great Swamp, South Kingstown (Miller 1999). They do not breed on the larger Narragansett Bay islands. In Tiverton and Little Compton, RCC found territorial birds in the late 1930s, and 2 pairs were located in this area in 1991. The breeding habitat consists of sphagnum moss, usually in standing water influenced by acidic leachate from glacial till. Most of the red maple–sphagnum swamps in Rhode Island that are inhabited by Northern Waterthrushes are characterized by an understory of the relatively southern-distributed sweet pepperbush, along with other plants of the southern coastal plain such as Atlantic white cedar and occasional great laurel.

Summer maxima. 20, Richmond to Hopkinton, 15–21 June 1987; 18, Glocester to Burrillville, 17–23 June 1982; 13, Poquiant Brook, 6 June 2007 (CRa).

Nonbreeding

Northern Waterthrushes declined significantly as fall migrants at Block Island, but not Kingston, 1969–2001 (Osenkowski et al. 2012).

Spring. Arrival of resident breeders is later than for Louisiana Waterthrushes. Local breeders appear in late April, a week or more earlier than migrants at Swan Point Cemetery and other migrant "traps." General arrival of migrants is in early May.

Spring maxima. 15 banded, Block Island, 19 May 1996 (MHC, SMC). Occasional individuals after 20 May and lingerers into June. One was banded at Block Island on 16 June 1973 (ELa).

Fall. Peak migration is in late July and August, but the migration is extended, with individuals commonly seen through October, and there is an increasing number of later records: 1, Block Island, 12 December 1924 (EDi); 1 photographed at Galilee, 23 December 1989 (DFi, JHo); another seen along upper Moonstone Beach Road 26 December 1994 (STs). One (not mentioned by Veit and Petersen 1993) was present in nearby Acoaxet, Massachusetts, 18–29 December 1982 (DLE, REm), and 1 was on the South Kingston CBC, 26 December 2016 (GLB).

Extreme dates. 13 April 1973, West Greenwich (FPe), 19 April 2012, Great Swamp (JSJ, BCG), and 21 April 1941 and 1948, Kingston (DLK), to 12 January 1974, Perryville (ESa).

Golden-winged Warbler (*Vermivora chrysoptera*)

Status

Rare transient.

Historical

The population of Golden-winged Warblers increased from about 1874 through the 1930s in eastern Massachusetts (mainly Essex, Middlesex, and Norfolk Counties, away from the coastal plain) and reached a peak in the early 1940s. Since then, they have steadily declined and by the 1990s had essentially disappeared (Veit and Petersen 1993). Never as common in Rhode Island, Golden-winged Warblers were confined to northeastern parts of the state during their short tenure (~1930–1956) as a rare breeder. Their Rhode Island history corresponds well to their rise and fall in Massachusetts and Connecticut.

Howe and Sturtevant (1899) mentioned a Golden-winged Warbler seen by F. T. Jencks in Providence in May 1880. Two others were shot by F. T. Stainton 3 June 1892 in Cranston. The only other known occurrence before the modern period is a migrant seen at Block Island 12 May 1914 (EDi). Records increased after 1931, and breeding was first confirmed at the Wenscott Reservoir, North Providence, where territorial birds were present 1931–1943 (WBD, CWo, DLK et al.) and, less definitely, at Diamond Hill, where a male was on territory in a small locust grove 18 June 1941 (DLE). In 1936, RCC and AEE found "several singing there in a cut-over clearing" in Lincoln Woods. They were found regularly at Lincoln Woods through the 1940s, for example, five on territory in June 1948 (WBD). In June 1956, GRa and RLF found at least five pairs there, one pair feeding a small fledgling that obligingly perched on the author's shoulder during observations in dense brush. After the late 1950s, Golden-winged Warblers disappeared from this location and all of Rhode Island. The three known or suspected locations were wholly within the towns of Cumberland, Lincoln, and neighboring North Providence, a region characterized by many rocky outcroppings and a generally richer flora than that of more granitic western parts of northern Rhode Island. Small black locusts were present at each of these localities.

A more recent effort to find territorial Golden-wingeds in Cumberland and Lincoln in June 1984 (RLF) revealed 65 Blue-wingeds and a Brewster's, but no Golden-wingeds. Gill (1980) indicates that in many parts of its range the Golden-winged has disappeared within 50 years of the appearance of significant numbers of Blue-wingeds (cf. Confer and Knapp 1981). Since 2000, there have been only 14 reports, 10 in fall and 4 in spring, and no suspected breeding.

Nonbreeding

Spring. Golden-winged Warblers were only occasionally encountered during May warbler waves, even in 1930–1960, when they bred regularly in Rhode Island and eastern Massachusetts. There were 31 records of spring migrants in 20 years 1958–1977. Most of these occurred during the first 2 weeks of May.

Fall. A measure of their rarity as coastal transients is the total of only 5 banded at Block Island 1959–1977 (ELa), whereas DLK has banded them at Kingston on 8 occasions in a fraction of the net-hours over the same period. The earliest fall transients at the latter location were individuals 1 August 1957, 11 August 1965, and 14 August 1969 (DLK). There is a peak in transient reports in late August. The latest records include individuals banded at Block Island 29 September 1971 and 1 October 1993 (ELa); another was banded at Kingston on the exceptional date of 12 October 1976 (DLK). Two were reported in August 2012.

Extreme dates. 1 May 1983, Providence (DFi), to 12 October 1976, Kingston (DLK).

Blue-winged Warbler (*Vermivora cyanoptera*)

Status

Common summer resident and fairly common transient.

Historical

Blue-winged Warblers extended their breeding range northeastward during the late-nineteenth and early twentieth centuries; before 1900, their breeding range did not extend east of the Appalachians, and they were rare in southern New England even as transients or wanderers. Their population growth and spread were likely driven by the succession of open fields to secondary forests that followed the decline of agriculture throughout eastern North America during the twentieth century (Rosenberg et al. 2016).

Other than a nesting in Glocester 3 May 1890 (nest and four eggs are extant in the Doe collection; FlMNH #53434, #55601), Blue-winged Warblers were unrecorded by Howe and Sturtevant (1899), who omitted mention of a bird collected in Pawtucket 3 May 1883, the specimen formerly at RWPM. Hathaway (1901d) mentioned another he collected at Hopkinton 7 June 1901. Doe found occasional summer resident Blue-winged Warblers during the early 1900s and collected a nest in Cranston 2 June 1911, another nest and four eggs in Glocester 8 June 1911, and four eggs at West Glocester in 1915.

Blue-winged Warblers appeared more frequently during the 1920s, especially in South County (e.g., seven singing between Bradford and Quonochontaug 15 May 1921; HSH), and they spread to the east and north as well. They bred regularly in South County during the 1940s and appeared in Tiverton and Little Compton about the same time. While probably breeding at the Wenscott Reservoir from 1931 on, they were scarce in the northwestern part of the state during the 1940s, for example, none in an all-day trip to Burrillville, Glocester, and North Smithfield 8 June 1946 (RCC, WBD, DLK). A nest was found in Lincoln Woods in 1948 (WBD).

They increased within South County during the 1950s and 1960s and in western and northern Rhode Island during the 1970s.

Breeding

Counts in interior districts by RLF include 70, Glocester to Burrillville 17–23 June 1982; 167, Foster to Coventry 16–24 June 1986; 44, Exeter to West Greenwich 29 May–5 June 1983; 85, Richmond to Hopkinton 15–21 June 1987; 56, Narragansett to South Kingstown 23–28 May 1984; 65, Cumberland to Lincoln 16–22 June 1984.

They are found throughout the state but are less numerous near the coast. They are nevertheless present on the larger Narragansett Bay islands; 26, Conanicut, 22–25 June 1985 (RLF); 22, Conanicut, 22 May 1988 (ChP). They do not breed on Block Island.

Nonbreeding

Spring. General arrival is in late April and early May. Most migrants have passed through by mid-May, but individuals were at Block Island 9 June 1981 and 11 June 1984 (RBo).

Fall. Individuals have appeared 4 July 1955 at Middletown, where they do not (or did not then) breed (JBa). Peak movement is in mid to late August, earlier than for many warblers.

Fall maxima. 15, White Brook, 31 August 1992 (RS). Blue-winged Warblers are rare after September; 2 were at Block Island 3–4 October 1981 (RFa et al.), 1 was banded at Kingston 7 October 1987 (DLK), and 1 was seen there 8 October 1993 (DLK). The latest are 1 banded at Kingston 17 October 1992 (DLK), 1 seen at Little Compton, 17 October 2005 (GDe), and 1 banded, Block Island, 21 October 1997 (ELa, KGa).

As with Golden-winged Warblers, Blue-wingeds are less common at Block Island in fall than on the nearby mainland. While Elise Lapham has rarely caught Blue-wingeds at Block Island (only 15 banded 1959–1977), Doug Kraus banded 674 on his farm at Kingston during the period 1956–1990 (including 12 banded 8 August 1986; DLK) using a single mist net. A seasonal total of 69 was banded at this location in 1970.

Extreme dates. 21 April 1932, Bradford (MSo), 24 April 1999, Exeter (SMH), and 24 April 1976, Charlestown (ESa), to 21 October 1997, Block Island (1 banded; ELa, KGa).

Hybrids. Through 2005, there are 44 Rhode Island records of "Brewster's Warbler" and 44 records of "Lawrence's Warbler," the 2 most common hybrid types between Blue-winged and Golden-winged Warblers. Since 2005, the ratio has been 2 Brewster's to 5 Lawrence's.

Black-and-white Warbler (*Mniotilta varia*)

Status

Common summer resident and transient.

Breeding

Black-and-white Warblers are one of the most common warblers breeding in Rhode Island. They are most numerous in mixed coniferous–deciduous woodlands in northwestern and western Rhode Island. They nest on the 3 larger Narragansett Bay islands and bred in Providence in 1906 (CED).

Summer maxima. 175, Glocester to Burrillville, 17–23 June 1982 (RLF); 123, Foster to Coventry, 16–24 June 1986 (RLF); 12 pairs, Prudence Island, June 1946 (WBD); 12 pairs, Conanicut Island, 22–25 June 1985 (RLF). Black-and-white Warblers are less numerous at coastal locations than farther inland but are fairly common on the Narragansett Bay islands. They do not nest at Block Island, however.

Nonbreeding

Black-and-white Warblers declined significantly as fall migrants at Block Island, but not Kingston, 1969–2001 (Osenkowski et al. 2012).

Spring. General arrival is during mid to late April. Black-and-white Warblers are one of the earlier warbler migrants through Rhode Island.

Spring maxima. 90, Block Island, 12 May 2002 (DJA, STs); 52 banded, Block Island, 1 May 1966 (ELa); 50, Lincoln, 5 May 1985 (SDz; RBu et al.); 30, Kingston, 25 April 1980 (GLB); 19 banded, Block Island on 14 May 1974 (ELa). Migrants occur in diminishing numbers through late May, and the latest transients are 1 at Block Island 1 June 1971 (ELa) and another there 5 June 1972 (RLF).

Fall. Fall migration begins in early August; e.g., 1 seen 30 July 1955 and another collected 6 August 1955 at NBS (JBa), 1 seen 10 August 1960 and another banded 13 August 1971 at Block Island (ELa). They are common transients in late August and September but rarely are as numerous as in spring.

Fall maxima. 30, Block Island, 3–4 October 1981 (RFa et al.); 20, Block Island, 12 September 1959 (SSD). They occur regularly through October and there are several November records; e.g., 21 November 1953, Tiverton (WBD, MDa); 24 November 1993, Tiverton (DFi); and 28 November 1952, Wakefield (DLK).

Extreme dates. 11 April 1994 and 12 April 1986, both Wallum Lake (SMi), to 16–29 December 2015, Providence (DFi, STa).

Prothonotary Warbler (*Protonotaria citrea*)

Status

Rare spring and very rare fall vagrant. Rare breeder.

Breeding

Two male Prothonotary Warblers were found in the Great Swamp area of the Pawcatuck River, 1 on 6 May 1988 (HWi) and another 4 May 1989 (HWi et al.). These records sparked the interest of RWa and CRa, whose repeated investigations led to the discovery of 2 and probably 3 males plus a female along a mile or so of this same stretch of river, 4 May 1992. On 5 June of that year, a coordinated search revealed a nest in a Downy Woodpecker hole about 7 ft (2 m) up a red maple stub and overlooking a pool of open water about 100 yd (91 m) from the Pawcatuck River (REn, CRa, RWa). Young birds, present in the nest 13 June (RWa), had apparently fledged by 14 June, when an infertile egg was collected (CRa, RLF; photos of the adults by RWa, CRa, RLF; photo of adult and young bird by RWa). This part of the river is between South Kingstown and Charlestown, in the Great Swamp immediately west of the impoundment. This area harbors what is perhaps Rhode Island's densest population of American Redstarts. The forest contains green ash, red maple, pin-oak, and black tupelo in a lush swamp with a distinctly Carolinian flavor. Two warblers were present on the Pawcatuck River 3–8 June 2010 (CRa), but breeding was not confirmed. Another pair nested at Tillinghast Pond in West Greenwich 12 to at least 17 June 2012 (m.obs.).

A presumptive Prothonotary × Yellow Warbler hybrid was carefully studied at Simmons Pond, Little Compton, 8 June 1992 (DLE, RBo).

Nonbreeding

For birds not associated with potential breeding habitat, there are 11 reports since 2005, 4 of which are in fall.

Spring. Prothonotary Warblers have appeared annually in spring since the 1970s, often with as many as 5 birds per year. The records span the dates 29 March to 9 June. Most have occurred near the coast, often following southwesterly airflow and rising temperatures and some following coastal storms. About one-third of the spring records are from inland locations.

Fall. There are 13 records from fall, 6 of these since 1975, spanning the dates 15 August to 17 October.

Extreme dates. 29 March 2005, Block Island (KGa), 1 April 2000, Gilbert Stuart birthplace (JSJ, RFa, DFi), 31 March to 1 April 2005, Newport (JMa, Linda Gardrel et al.), and 3 April 1920, Point Judith (1 dead; HSH), to 17 October 1956, Block Island (1 banded; JBa).

Swainson's Warbler (*Limnothlypis swainsonii*)

Status
Vagrant. One record.

Nonbreeding
A Swainson's Warbler was mist-netted and banded at Block Island 22 May 2011 (KGa). There is a pattern of increasingly frequent occurrence in the Northeast, likely related to their concurrent population increase in the noncoastal portions of their southeastern range (BBS).

Tennessee Warbler (*Leiothlypis peregrina*)

Status
Uncommon to fairly common transient in spring and fall.

Historical
Formerly very rare; only two records were known to Howe and Sturtevant (1899) for 1886 and 1898, and another was collected in 1901. They remained little known in New England until 1916 (Peters 1916), when two were recorded in Massachusetts. They have since become regular but somewhat erratic spring migrants in Rhode Island, sometimes common in May warbler waves but occasionally nearly absent. Along with Cape May and Bay-breasted Warblers, its history of fluctuations is correlated with infestations of Spruce Budworm (*Choristoneura fumiferana*). They are more regular but still only fairly common in fall. A general increase seems to have occurred in the twentieth century, and some further increase was apparent through the 1970s (see Hill and Hagan 1991).

Nonbreeding
Spring. This is one of the later migrants in spring, and peak migration is approximately 20–30 May. There are a few records for the first few days of June, the latest being 1 at Block Island 5 June 1972 (RLF). In many spring seasons few or none are seen.

Spring maxima. 25+, Lincoln, 22 May 1983 (SDz); 22, Swan Point Cemetery, 25 May 1974 (DLE, REm).

Fall. The earliest arrivals include individuals at Kingston 13 August 1982, 13 August 1988, and 18 August 1977 (DLK). Tennessee Warbler abundance fluctuates with outbreaks of Spruce Budworm; in budworm years they have been common as migrants in Rhode Island (Hill and Hagan 1991). In other years, they can be scarce, and only a few individuals are seen.

Fall maxima. 25+, Block Island, 3–4 October 1981 (RFa et al.); 10, Middletown, 3 September 1955 (JBa); 8, Block Island, 11–13 October 1958 (ASRI).

Extreme dates. 23 April 1970, Block Island (ELa, HLa), and 25 April 1983, Napatree Point (CRa), to 11 November 1969, Charlestown (DLK).

Orange-crowned Warbler (*Leiothlypis celata*)

Status

Uncommon late-fall transient; rare winterer. Rare spring migrant.

Historical

Two specimens are mentioned in Howe and Sturtevant (1899), and another was collected in Cranston 3 December 1874 (Purdie 1877). A male was collected at East Providence 9 May 1891, acquired by HSH, and donated to RWPM, but the specimen appears not to be extant. They increased thereafter (Wright 1917) but were still considered rare by Hathaway (1943).

Nonbreeding

Fall. Orange-crowned Warblers are fairly regular in small numbers along the outer coast and at Block Island, but total numbers seen vary considerably from year to year.

Fall maxima. Six, Middletown, 1 October 1956 (JBa).

Winter. Orange-crowned Warblers linger into early winter, and there are now at least 15 CBC reports since 2005.

Winter maxima. Five, Providence, 16 December 2014 (DFi, STa); 3, Portsmouth, 23 December 1956 (RLF). Late-winter occurrences include 1 at Touisset Marsh 25–27 January 2002 (MT; B. Forbes); 1, Sachuest Point 28 January 2001 (RCC et al.); 1, Block Island 15 February 1999 (CRa et al.); 1, Napatree 15 February 1999 (SSM); 1, Block Island 18 February 2002 (CNu, DFi); 1, Block Island 20 February 2012 (SSM et al.); 1, Wilbour Woods, Little Compton, 24 March 1959 (RLF); and another that remained at a feeder in West Kingston 9 February–6 April 1992 (GTa, ATa).

Spring. One was photographed at Providence 1 May 1983 (DFi). There are at least 6 records from Swan Point Cemetery 5–22 May. One was at Seapowet 14 May 1977 (HWi), and another was collected at Block Island 21 May 1929 and placed in the Block Island Collection, but the specimen has apparently disappeared. There are 4 spring records since 2005.

Nashville Warbler (*Leiothlypis ruficapilla*)

Status

Uncommon summer resident and fairly common spring and fall transient; has increased as a migrant since 1990 but declined as a breeder.

Historical

They are said to have been quite common at the height of agricultural development in central Massachusetts in the era 1842–1890 (Griscom and Snyder 1955) but reportedly declined sharply after 1900 with regrowth of the forest. On 31 May 1890, William A. Sprague and CED found a pair and five eggs in Burrillville, the nest "on the ground under a tuft of running briers" and in an open space surrounded by small birches (female, nest, and eggs collected; FlMNH #90066–90068). Another nest and five eggs were discovered by Angell and Remington at Greene 22 May 1906. A third instance involved a nest and four eggs collected by CED among wintergreen in a "seven or eight-year cut-over area" in Glocester 4 June 1916 (FlMNH #52852, #54510).

A pair was found near the Wenscott Reservoir, North Providence, 24–30 July 1931, and was thought to have been breeding (WBD). During the period 1940–1956, DLK found nine singing males in the Rockville to Wincheck area. Another was noted in Scituate in June 1957 (KKi). Kraus found a male singing 4 June 1949 at the junction of Route 165 and Wood River in Exeter, and territorial birds were seen in the same area during the late spring or summer of 1967 (MDa, KKi), 1971 (SSD, LBe), 1974 (KKi, MDa), and 1979 (JGe, RGe). A male was seen carrying food into an acidic patch of young birches in an oak clearing in southwest Glocester 21 June 1982 (RLF).

Breeding

It was CRa who first became convinced that Nashvilles were also nesting in sandplain habitats in Rhode Island through his discovery of territorial males in the Kingston pine barrens in 1979 and in the Arcadia and northeast West Greenwich pine barrens in 1981. A survey of 5 sandplain areas in southwestern Rhode Island in June 1983 located a total of 47 birds (RLF). During the 1980s, Nashville Warblers were most numerous in the Arcadia Management Area (at least 27 territories in 1983 where 8000 acres [3237 ha] were burned in 1951), with smaller numbers in South Kingstown, Charlestown, western Coventry, and northeastern West Greenwich. A territorial male was in eastern Little Compton 19 June 1985 (RBo). Nesting Nashville Warblers on the sand plains of Rhode Island have been in or near young pitch pines on dry, acidic, and relatively flat soils, often with an admixture of scrub-oak, other small oaks, and occasional small red maples, together with an understory of lowbush blueberries, huckleberries, sheep-laurel, and sometimes bearberry or trailing arbutus. A few pairs have been in purely deciduous vegetation, without pines. Forest fires have played a major role in creating much of the best habitat.

That being said, Nashville Warblers are now rare or absent as breeders in Rhode Island. The most recent summer record of a potential breeder was in 1995 from a former nesting location in the Great Swamp (SSM).

Nonbreeding

Spring. Nashville Warblers appear in moderate numbers beginning in late April and early May. They appeared in larger numbers and earlier in spring concurrent with their increase as a breeder.

Spring maxima. Ten, Swan Point Cemetery, 13 May 2011 (RFa).

Fall. Dispersing residents are found in late July and August; peak numbers of other migrants occur in late September and early October. Late birds include singles at Block Island 11 November 2001 (CRa, SSM et al.) and 13 November 1997 (CRa et al.), and 18 November, Quonochontaug (CPe, STs). There are several recent CBC records including South Kingstown (2002), Napatree (2002), and Newport–Westport (2016).

Fall maxima. 15, Block Island, 4 October 2008 (DFi); 6, Block Island, 11–13 October 1958 (ASRI).

Extreme dates. 25 April (3 times) to 16 December 2000, Middletown (REn et al.), and 25 December 1973, Little Compton (JAG).

Virginia's Warbler (*Leiothlypis virginiae*)

Status

Vagrant. One record.

Nonbreeding

A Virginia's Warbler was seen and photographed at Walker Farm, Barrington, 8 October 2006 (STa, DFi; m.obs.). Single records from Maine, New Jersey, New Brunswick, and Labrador are the only others for the Atlantic coast.

Connecticut Warbler (*Oporornis agilis*)

Status

Uncommon to rare fall transient; one spring record.

Historical

Connecticut Warblers were more numerous before the turn of the twentieth century. Eight specimens were collected during the period 1894–1898 in Warwick and Cranston, a number that would be difficult to encounter there today. Veit and Petersen (1993) discuss the numbers seen and collected (to 60+ per season) by Brewster in Cambridge, Massachusetts, during the 1870s and 1880s. The reasons for their decline are unknown, but possible habitat changes on the wintering grounds may have affected this south-central Canadian breeder, which takes a direct overwater route from eastern North America to northern South America in fall. Veit and Petersen (1993) discuss the tendency of Connecticut Warblers to use maple swamps in the interior in preference to drier, more exposed coastal habitats. Since birders focus their efforts at the coast and not at inland swamps during fall, it is possible Connecticut Warblers are overlooked.

Nonbreeding

The earliest date is for 1 at Middletown 24 August 1960 (JBa). Another was collected 3 September 1962 at Middletown (RLF). The bulk of reports are from mid-September through the second week of October (to 4 per day). One was carefully observed at Cowesett 7 November 1966 (KKi), another at Swan Point Cemetery 9 November 1994 (RFa), and another was shot at Warwick 12 November 1898 (fide HSH).

Spring. Connecticut Warblers are rare anywhere on the East Coast during spring. There is 1 Rhode Island record: a male carefully observed at Westerly 18 May 1975 (ESa).

MacGillivray's Warbler (*Geothlypis tolmiei*)

Status

Vagrant. Two records.

Nonbreeding

One immature male was collected after being caught in a mist net at Block Island, 28 September 1972 (KPA). It is #808837 at AMNH; identification confirmed on the basis of measurements and plumage details (RRV). Another was seen and photographed at Blackstone Park, Providence, by numerous observers 7 November 2022 (DFi, STa).

Mourning Warbler (*Geothlypis philadelphia*)

Status

Uncommon transient.

Nonbreeding

Spring. Mourning Warblers are one of the latest spring migrants through Rhode Island; most are found after 20 May. An exceptionally early record is for 1 at Swan Point Cemetery 8 May 1993 (JCB). Most are reported 20 May–6 June. Singles were banded at Block Island 11 June 1968 and 13 June 1972 (ELa). The latest records were a singing male at Swan Point Cemetery 18 June 1983 (DFi) and 1 at Glocester/Durfee Hill 28 June 1997 (BCG, JMu).

Spring maxima. Six banded, Block Island, 24–25 May 1974 (ELa).

Summer. There is 1 Rhode Island record of a bird in midsummer: an adult female at Kingston 28 July 1962 (DLK); an adult male was in nearby Seekonk, Massachusetts, 21 July 1990 (JR). These may have been early migrants or perhaps prospecting birds; Mourning Warblers breed in northwestern Massachusetts (Veit and Petersen 1993; Petersen and Meservey 2003).

Fall. Fall migrants occur from mid-August through late September, with occasional records to mid-October. The latest was 1 banded at Kingston 22 October 1977 (DLK).

At Block Island, Mourning Warblers outnumber Connecticut Warblers in fall. Between 1959 and 1987, 83 Mourning and 49 Connecticut Warblers were captured in mist nets and banded (ELa et al.). For comparison, at Kingston between 1956 and 1990 the proportions were more nearly equal: 36 Mournings and 35 Connecticuts (DLK). This supports the notion that Mourning Warblers are more coastal, and Connecticuts more inland, in their distributions during fall.

Kentucky Warbler (*Geothlypis formosa*)

Status

Vagrant both in spring and in fall; increasing in frequency. One breeding record.

Historical

Kentucky Warblers bred north to the lower Hudson Valley of New York until 1942, disappeared (Bull 1964), then expanded northward again in the 1970s and 1980s. There is an 1892 breeding record from Connecticut (and an 1890 record of a vagrant from Rhode Island), as well as other more recent breeding records from the Mystic, Connecticut, area (Bevier and Askins 1994). Andrle and Carroll (1988) provide an account of their historical fluctuation in New York and describe the most common breeding habitat as maple–beech woodland, usually near water and often on a slope. All the Rhode Island records, except for the 1890 one referred to above, are since 1959.

Breeding

Three adult Kentucky Warblers were found by Evan Lipton in Little Compton 26 June 2017; on 28 June, Liam Waters returned and saw the 3 adults plus 4 newly fledged young. This is the northernmost nesting record for the East Coast.

Nonbreeding

There are 7 reports since 2005, 1 of which is in fall.

Spring. Kentucky Warblers have appeared in spring annually since the 1970s. Up to 10 individuals per season have been found, spanning the dates 5 May to 9 June, with a peak around 25 May. The majority of Kentucky Warblers are found on the basis of the very loud and distinctive song of the male; this fact may account for their relative scarcity in fall.

Fall. There are 9 records from fall (5 in mist nets) spanning the dates 11 August–9 October.

Extreme dates. 27 April 2012, Swan Point Cemetery (J. Sweeney), to 24 November 1960, Block Island (JBa; 1 collected, specimen at MCZ). There is 1 summer record: 1, Little Compton, 15 June 2007 (GDe), suggesting prospecting or even breeding.

Common Yellowthroat (*Geothlypis trichas*)

Status

Abundant summer resident and transient. Rare in winter.

Breeding

Common Yellowthroats are one of Rhode Island's most common breeding warblers. Overall, the species is remarkably common everywhere in the state, even in western sections. They are most numerous in open and secondary growth, but also in forest understory, especially in forest gaps. Because of their preference for open habitats, their abundance and distribution has fluctuated during the reforestation of the state.

Summer maxima. 300+ territorial males, Block Island, June 1972 (RLF); 310, Foster to Coventry, 16–24 June 1986 (RLF); 285, Narragansett to South Kingstown, 23–28 May 1984 (RLF); 50 along the Blackstone River at Lonsdale, 10 July 1983 (REn, CRa).

Nonbreeding

Common Yellowthroats declined significantly as fall migrants at Kingston but not Block Island 1969–2001 (Osenkowski et al. 2012).

Spring. Birds have appeared 21 April 1976 at Barrington (JCB) and 21 April 1987 at Napatree Point (CRa, Jay Manning). General arrival is during the first week of May.

Spring maxima. Counts of 50–100 per day per locality have often been made during the last 2 weeks of May; 37 banded, Block Island, 25 May 1971 (ELa). Migrants occur into June; e.g., 1 at Cox Ledge 8 June 1983 (RBo).

Fall. Migrants begin appearing in late July; e.g., 1, Cox Ledge, 29 July 1978 (v.o.). Banding suggests that most of Block Island's breeding population leaves before the end of August (ELa, HLa).

Fall maxima. 290 left Napatree Point in early morning 9 September 2017 (CRa).

Winter. After the mild fall of 1953, 10 were seen on the Newport–Westport CBC 27 December 1953. Common Yellowthroats have been found nearly annually since the 1960s on CBCs, including 3, Block Island, 17 December 1992, and 3, Fogland, 15 December 1979 (FAP). They are rare after December. One was seen in Little Compton 27 January 1980 (DLE), and another remained at Perryville to 8 February 1979 (v.o.). One at Carolina 30 March 1973 (DWa) may have survived the winter. The same applies to a winter-plumaged bird seen on Patience Island 6 April 1983 (CRa, REn).

Hooded Warbler (*Setophaga citrina*)

Status

Uncommon and local summer resident and uncommon to rare transient.

Historical

Only one transient record was known to Howe and Sturtevant (1899), and the first nest was discovered in Hopkinton in 1934 by HSH, who in subsequent years noted an eastward movement in South County from previously well-established breeding areas in Connecticut.

Breeding

Prior to the breeding bird atlas in the 1980s, Hooded Warblers were known as summer residents from the Beach Pond and Arcadia areas of Exeter south through the Rockville area of Hopkinton (8 males 9 June 1940; DLK) to Westerly (rarely, Woody Hill); Charlestown, particularly near Watchaug Pond; and South Kingstown, particularly along the Ministerial Highway north of the South County moraine. Hopkinton now holds the largest numbers, especially in the area from Rockville and Canonchet west to the Connecticut border. This small part of Rhode Island, relatively unique geologically, is a heavily forested region of varied topography that contains dense mountain-laurel and rhododendron, apparently the most favorable habitat for both Hooded and Worm-eating Warblers. Among 40 Hooded Warblers found in Hopkinton 19–21 June 1987 (RLF), the majority was in this limited area. This area is still the epicenter of the breeding range in Rhode Island. Hooded Warblers are also fairly common in the Burlingame Management Area, where 90 point counts in 2006 yielded 17 males (CRa).

Hooded Warblers are rare north of Exeter. Two were found in western Coventry 7 June 1987 (CRa) and in Scituate during July 1956 (RLF). Some of the best habitat in northern Rhode Island is in Glocester, and birds were noted in laurel near Wallum Lake in the early 1940s (DLK). There were no further observations in the 2 northwestern towns until 3 June 1993, when RWa found a pair sharing the same habitat with Black-throated Blue Warblers in the "Sprague Hill" area of Glocester.

Hooded Warblers have bred for years in Tiverton. Two pairs were along the laurel-rich Sin and Flesh Brook in 1941 (RCC), and 5 singing males were present there on 4 June 1942 (RCC). The species has been continually present since; the high count was 10 males in Tiverton, 7 of these along Sin and Flesh Brook, 24 June 1972 (RLF). Since 2003, Robert Emerson has generally found no more than 1–2 pairs per breeding season during his annual June surveys of the Tiverton highlands (Weetamoo Woods/Fort Barton), suggesting a decline in this area since the 1970s and 1980s.

Nonbreeding

Spring. Early spring records include a male at Swan Point Cemetery 8 April 1985 (RBo); others appeared at Block Island 17 April 1961 (MSl) and Middletown 18 April 1957 (JBa); general arrival is late April and early May. Spring movement has been noted to 3 June.

Fall. There are records of fall transients in nonbreeding areas from early August to mid-October.

Extreme dates. 8 April 1985, Swan Point Cemetery (RBo), to 1 November 1953, Little Compton (1 male; RBo) and Matunuck, 28 November–2 December 2002 (1; CNu, STs, RFa et al.).

American Redstart (*Setophaga ruticilla*)

Status

Common summer resident; common to abundant transient.

Breeding

American Redstarts are most numerous in mature deciduous forests of the inland parts of the state, and they are also common in red maple swamps on the coastal plain. Areas dominated by pine–oak forests contain relatively low densities. They breed in the wind-blown brush within 200 yd (183 m) of the Beavertail lighthouse.

Summer maxima. 100, Foster, 21–24 June 1986 (RLF); 33 in riverine swampland of red maple, green ash, and pin-oak at Wordens Pond, 3 July 1984 (REn, CRa); 19, Prudence Island, 18 June 2003 (REn). They were absent from Block Island as breeders until 1979 and 1981, when the first pairs appeared at 2 separate locations. (ELa; RBo).

Nonbreeding

American Redstarts declined significantly as fall migrants at both Block Island and Kingston 1969–2001 (Osenkowski et al. 2012).

Spring. American Redstarts are rather late migrants; first arrivals are often in early to late May, and peak numbers occur during late May.

Spring maxima. 40, Prudence Island, 14 May 1946 (WBD); 31 banded, Block Island, 27 May 1985 (ELa). Late transients regularly occur until 10 June.

Fall. The first fall migrants appear in mid to late July (19 July 2006, Napatree; CRa), and peak numbers are in late August and early September.

Fall maxima. 40, Block Island, 25 September 1982 (RLF; STs); 30, NBS, Middletown, 24 August 1955 (JBa). A few are regularly noted through mid-October and occasionally to late October. There are several early November records. An adult male was seen at Little Compton 29 November 1953 (CWo).

Extreme dates. 27 April (3 times, 1959 to 1996) to 15 December 2001, Portsmouth (RLF).

Cape May Warbler (*Setophaga tigrina*)

Status

Uncommon spring and fairly common fall transient. Abundance fluctuates with Spruce Budworm (*Choristoneura fumiferana*) infestations in boreal forests.

Historical

Cape May Warblers were listed as rare by Howe and Sturtevant (1899), and probably legitimately so, as they were also considered rare in the New York City area (Griscom 1923). According to Griscom, they increased after about 1909. They were especially abundant in Rhode Island and the rest of the Northeast during a major Spruce Budworm infestation in the late 1970s and early 1980s (Morse 1978).

Nonbreeding

The abundance of Cape May Warblers, like that of Bay-breasted and Tennessee Warblers, is positively correlated with "outbreaks" of Spruce Budworm caterpillars that have been occurring with increasing frequency in boreal forests (Baltz and Latta 2020). Maximum counts of Cape May Warblers in Rhode Island during the 1970s and 1980s almost certainly resulted from exceptionally high reproductive success in boreal forests due to the massive food supply. Cape May Warblers were uncommon from about 1990 to 2010, when numbers began to increase once again.

Spring. Cape May Warblers are uncommon transients through Rhode Island; it is unusual to ever find more than 2–3 per day per locality. In some years they are scarce: only 1 was found in Rhode Island during spring 1971 and again in 1982; there were 2 in spring 2000. The earliest dates are for birds at Sakonnet 28 April 1979 (RBo, HMP, RKl) and Westerly 30 April 1957 (Sullivan and West). Other dates run from 2 and 4 May on, and peak migration is 15–25 May, almost as late as Blackpoll Warblers. Early June records are rare, the latest being 5 June 1972 at Block Island (RLF).

Fall. Cape May Warblers fluctuate considerably in abundance but are always much more numerous in fall, especially at the coast. At Block Island during the 1970s and 1980s, they were often abundant and occasionally outnumbered all warblers except Yellow-rumped and Common Yellowthroat. The earliest fall arrival was 9 August 1955 at Middletown (JBa). They are regular at coastal locations in late August and most common through September into the first week of October.

Fall maxima. 300, Block Island, 3–4 October 1981 (RFa et al.); 200, Block Island, 13 September 1958 (SSD). The largest mainland count is 75, Charlestown, 15 October 1981 (GLB).

Extreme dates. 28 April 1979, Sakonnet (RBo, HMP, RKl), to 15 December 2001, Block Island, and 17 December 2015, Block Island CBC (SSM). One wintering record is of 1, 15 January to 3 April 2016, at a feeder in Wakefield (J. Restivo).

Cerulean Warbler (*Setophaga cerulea*)

Status

Rare migrant, more frequent during spring. One breeding record.

Historical

Cerulean Warblers breed in large bottomland and lakeside deciduous trees in the eastern Mississippi drainage, east to western and central New York, and locally in the lower Hudson Valley. They have been slowly increasing and extending their range in New York (Bull 1974) and have been found breeding in Vermont, New Hampshire, Connecticut, and central Massachusetts. There are about 95 records of Cerulean Warblers in Rhode Island, about 73 of these since 1975. Their occurrence in Rhode Island can be divided into 5 periods:

1. 1879–1909: 4 May records of vagrants: Cumberland, 22 May 1878 (C. M. Carpenter); Pawtucket, 22 May 1879 (F. B. Webster); Lonsdale, 14 May 1893 (W. Deardon); Johnston, May 1909 (H. E. Walter)
2. 1910–1951: no records at all
3. 1952 to mid-1970s: resumption and gradual increase of spring vagrants
4. 1976 through late 1980s: considerable further increase in vagrants and the discovery of regularly summering birds
5. 1988 on: confirmed as a breeder but subsequently (after 1990) decreased again

Prospecting and Breeding

Beginning in the 1960s and 1970s, Cerulean Warblers appeared in seemingly appropriate nesting habitat during the summer. The first such observation was in Kingston 26 June 1962 (G. West). The next was a singing male found 13 June 1976 (RLF) in tall bottomland hardwoods along the brook at Parker Woodland at the same location where the first Rhode Island nest of Acadian Flycatcher would be found a few weeks later. One or 2 male Ceruleans were present at Parker Woodland 2–25 June 1977, and a single bird was noted there 22 May 1979 and 24 May–26 June 1980. At Austin Farm Road in Exeter, a male remained on territory 28 May–23 June 1977. One or 2 males were present in Exeter 6 June–7 July 1978, 25 May–27 July 1979, 28 May–26 June 1980, and 14 May–29 July 1982. Other individuals have been seen during the breeding season in West Greenwich and Burrillville. The first breeding occurred in Scituate in 1988, when adults and fledged young were found 31 July (JGe, RGe). None have been found since then, but there was a bird that may have been on territory in Burrillville 26 May–9 June 2016 (D. Pedro, m.obs.).

Nonbreeding

There are a total of 4 reports for 2005–2020, 3 of these in spring.

Spring. Cerulean Warblers occurred annually in spring from the 1970s to the 1990s. Reports span the dates 2 May through June. Rarely are more than 1 per day per locality recorded. One was at Block Island 17 May 2005 (STs).

Fall. Cerulean Warblers are still rare in fall. The records include an immature female banded at Kingston 17 August 1978; 1 seen, Kingston, 20 August 1979; another banded there 18 August 1991 (all DLK); and 1 seen at Warren's Point 16 September 1995 (REm).

Extreme dates. 1 May 2004, Swan Point (1; RFa, DVa et al.), and 2 May 1962, Block Island (1 banded; MSl), to 16 September 1995, Warren's Point (REm).

Northern Parula (*Setophaga americana*)

Status

Rare summer resident and fairly common to common transient.

Historical

The disappearance of this species as a breeding bird from Rhode Island and most of the central Northeast coast is thought to be attributable to the disappearance of the lichen *Usnea trichodea* ("old man's beard"), which was formerly common in maple and cedar swamps within a few miles of the coast and Narragansett Bay. Howe and Sturtevant (1899) mentioned breeding at Kingston (probably in the Great Swamp), in Tiverton, and at Mount Hope in Bristol. Hathaway's egg collection contains seven clutches collected in Westerly and Quonochontaug between 1896 and 1901. His notes state, "Charlie Doe took one set of eggs [in Westerly]. Breeds commonly at Kingston swamp and other places in the southern part of the state." Hathaway (1900b) also noted the finding of two nests in *Usnea* on the ungrazed, wooded islands in Briggs Marsh, Sakonnet, in early June 1900 and stated that four to five pairs inhabited these islands, which were then covered by an unusual (for the turn of the twentieth century) growth of tall oaks and a few cedars. John H. Flanagan and Charles H. Remington (in Remington and Commissioners 1908) reported finding a nest in Kent County in 1905. Hathaway found several males singing near Watchaug Pond 22 June 1919. Hathaway wrote RCC in 1937 that "the swamp at Watchaug Pond and the cedars at Quonochontaug Pond were the places where the Parula bred . . . [these were] . . . denuded of *Usnea* 10 years or so ago." According to Hathaway (1943), *Usnea* had practically disappeared in Rhode Island by 1933, and he dated the disappearance of the parulas to approximately that time. For many years, the last record of breeding was of an adult with young found by MSo at Watchaug Pond to the west of Kimball Sanctuary 28 June 1933, where the habitat included a brook through a cedar swamp with "plenty" of *Usnea*. Elsewhere in southern New England and nearby areas, the species also essentially disappeared, persisting at Westport, Massachusetts, at least until 1937 (RCC, WHD). On Long Island, they also ceased breeding about 1934, with only two later records to 1951 (Bull 1964).

Breeding

The breeding bird atlas revealed a small number of territorial birds in the interior, the most intriguing of which was a female carrying nesting material into a tree in Glocester on 22 May 1981 (CRa). Raithel returned later to find the bird gone. Another series of observations involved a male that remained on territory at Ponagansett, Scituate, through June and July 1982, 1984, and 1986 without any nesting evidence materializing (JGe, RGe et al.). Territorial birds arrived late April 1994 in a part of the Great Swamp just north of Wordens Pond. A nest and incubating female were observed in early June (CRa, REn, RWa). Another male was singing in the maple swamp southwest of Watchaug Pond 16 June 1994 (CRa), and they continue to breed at these locations. Raithel has canvassed most swamps in Rhode Island looking for *Usnea trichodea*, which is the preferred nesting material for parulas in this area. The only 2 places that contain both *Usnea* and parulas are the 2 sites mentioned above, the Great Swamp, and the swamp west of Watchaug Pond. *Usnea trichodea* is much more widely distributed than supposed and may be recovering from air quality issues that reduced it during the nineteenth and twentieth centuries, but most sites have only a little of it and may therefore not be able to support parulas. It is not unusual to hear male parulas advertising as though they are territorial well into June, but these may just be floaters that are not in appropriate nesting habitat.

Nonbreeding

Spring. Northern Parulas are one of the earlier warbler migrants; they commonly arrive in Rhode Island during mid to late April and peak counts are usually in early May.

Spring maxima. 100, Swan Point Cemetery, 8 May 1934 (DLK); 50, Neutaconkanut Hill, 3 May 1905 (Mason), 12 May 1962 (CWo), and 14 May 1934 (DLK); 30,

Lincoln Woods, 30 May 1939 (DLK); 14 banded, Block Island, 1 May 1966 (E Di). Counts have seldom exceeded 10 per day per locality since 1970.

Recent maxima. 45, Block Island, 12 May 2002 (DJA, STs); 45, Block Island, 15 May 2005 (STs et al.); 35, Swan Point Cemetery, 10 May 2013 (RFa); 20, Lincoln, 5 May 1985 (RBu). Migrants continue through early June; e.g., 4, Foster, 4 June 1982 (RBu); 1, Block Island, 11 June 1973 (RLF); 1, Block Island, 21 June 1990 (CRa).

Fall. Early migrants appear in early August; e.g., Green Hill 4 August 1946 (DLK) and Kingston 12 August 1958 (DLK). Parulas seem to have steadily increased from the 1950s to 2015 or so, a pattern supported by BBS data (BBS; accessed 24 January 2023), perhaps a result of warming climate. Rare after mid-October; 1 was observed at Fiskeville 16 November 1979 (JGe, RGe).

Fall maxima. 100, Block Island, 2 October 2010 (DFi, STs); 50, Block Island, 29–30 September 2007 (ASRI); 35, Block Island, 23 September 1983 (FAP, STs, JDi).

Extreme dates. 14 April 1962, Middletown (REW), 15 April 2000, Jamestown (fide MT), and 19 April 1976, Swan Point Cemetery (5; REm, CWo), to 1 December 1957, NBS (1 found dead; fide JBa), 26 December 2011, South Kingstown CBC (STs et al.), and 26 December 2012, Napatree (CRa).

Magnolia Warbler (*Setophaga magnolia*)

Status

Common transient; possibly a rare breeder.

Breeding

During June 1955 and again in 1956, RCC observed Magnolia Warblers singing in spruce and pine groves along Tunk Hill Road, Scituate, but did not find a nest or young. In June 1976, RLF surveyed the entire watershed using taped songs without success. A singing male there 6 June 1987 (v.o.) suggests the possibility of future breeding. In western Massachusetts, Magnolia Warblers breed fairly commonly in mixed forests similar in composition to some forests in western Rhode Island. There is a record of a wandering male in a mixed hemlock–deciduous habitat in Pulaski State Park, Burrillville, 30 June 1985 (REn).

Nonbreeding

Magnolia Warblers declined significantly as fall migrants at Block Island, but not at Kingston, from 1969–2001 (Osenkowski et al. 2012).

Spring. This common Canadian-zone species generally arrives in early to mid-May, and peak numbers occur in the last 2 weeks of May. Occasional transients appear during the first few days of June. The latest of these were 1 at Barden Reservoir 9 June 1986 (RBu et al.), and 3 different birds at Block Island 8, 10, and 11 June 1972 (RLF, ELa).

Spring maxima. 78 banded, Block Island, 19 May 1996 (MHC, SMC); 75, Block Island, 12 May 2002 (D. Abbott, STs); 44 banded, Block Island, 18 May 1992 (ELa); 37, Block Island, 21 May 2010 (DFi, STa); 35, Block Island, 18 May 1957 (WBD); 30, Napatree Point, 16 May 2013 (CRa).

Fall. The earliest arrivals are individuals 12 August 1986 at Kingston (DLK, DFi), 19 August 1962 at Providence (RRi), and 20 August 1986 at Kingston (DLK, DFi). Peak counts occur in early to mid-September.

Extreme dates. 29 April 1932, Providence (fide HSH), to 16 November 1980, Westerly (CRa).

Bay-breasted Warbler (*Setophaga castanea*)

Status

Uncommon to fairly common spring, and fairly common fall, transient.

Nonbreeding

Spring. Bay-breasted Warblers are late-spring transients, with peak numbers appearing about the third week of May. The latest is 1 at Breakheart Pond 9 June 1992 (JGe, RGe).

Spring maxima. 15, Block Island, 21 May 1961 (WBD, DLE); 12, Block Island, 12–14 May 1999 (CRa, REn); 12, Swan Point Cemetery, 17 May 1977 (v.o.); 10+, Lincoln, 22 May 1983 (SDz); 10, Block Island, 17 May 2007 (STs et al.).

Fall. Peak numbers appear in late August and early September. The earliest fall transient was 3 August 1955 at Middletown (JBa). One was seen at Block Island 17 October 1981 (DLE, REm); another was banded at Kingston 18 October 1977 (DLK).

Fall maxima. Ten, Block Island, 3–4 October 1981 (RFa et al.).

Like Cape May and Tennessee Warblers, Bay-breasted Warblers fluctuate in abundance with outbreaks of Spruce Budworm (*Choristoneura fumiferana*) in Canadian-zone spruce and fir forests (Venier et al. 2020). Numerous authors have tied unusual numbers of fall migrants to this fluctuating food supply's effect on the production of young (Morse 1978). Citing field and banding data, Veit and Petersen (1993) describe a boom in Massachusetts transient numbers 1975–1979, during a major budworm outbreak.

Extreme dates. 3 May 1952, Swan Point Cemetery (CWo), to 13 November 1974, Providence (1 freshly dead; CWo).

Blackburnian Warbler (*Setophaga fusca*)

Status

Uncommon to rare breeder and fairly common transient.

Historical

Howe and Sturtevant (1899) were unaware of breeding records of Blackburnian Warbler; however, Walter Angell saw a female carrying nesting material into a white pine in Greene in 1903 but could not find the nest. Similarly, HSH found one to two singing males at Wallum Lake in May and June 1911–1912. Charles Doe found a pair, nest, and three young "ready to fly" in a birch associated with a large white pine grove on "Dandelion Hill," Glocester, 23 June 1912. He collected an adult along with two of the young (an extant fledgling is FlMNH #2922). Doe found another nest and four eggs in Glocester 10 June 1923.

Blackburnian Warblers were regularly found on territory and confirmed as breeders in hemlock groves at George Washington State Park/Pulaski Management Area east of the Bowdish Reservoir during the period 1941–1953 (RCC, DLE et al.), and four were seen there 26 June 2004 (MLy). More recently, the species has been noted on territory in hemlocks or white pines at several other locations in Glocester and Burrillville, as well as near Foster Center. In sum, it seems likely that small numbers of Blackburnian Warblers have been breeding in northwestern Rhode Island throughout the historical period.

Breeding

Blackburnian Warblers increased as breeders with gradual reforestation of the state but have more recently (after 1990 or so) declined. They are more common farther south in the Arcadia Management Area in western Exeter and in parts of West Greenwich and Richmond than they were in the 1950s. Most birds in these areas have been found in the abundant white pines. Up to 3 males were found near hemlocks in the Breakheart Brook area during the 1970s (m.obs.), and 4 males were found 16 June 1971 east of "Lewis City," in semi-mature white pines on a dry slope lacking an understory (RLF). During 1989–1995, however, an extensive June census of the Arcadia area found only a single territorial male (CRa). In summer, Blackburnian Warblers have been present annually in planted spruces in the Scituate Reservoir watershed (up to 12 males in 1973; RLF, FPe). Territorial birds were present in a 20-acre (8-ha) stand of white pine in Fiskeville during the 1980s (JGe, RGe).

Nonbreeding

Spring. Blackburnian Warblers are fairly common migrants in spring, with first arrivals during early May and a peak in late May. Transients have been found at Block Island to 9 June.

Spring maxima. 27, Block Island, 13 May 2002 (DJA, STs); 19, Swan Point Cemetery, 9 May 2016 (RFa); 15, Sakonnet Point, 3 June 1984 (RBo).

Fall. Blackburnian Warblers are relatively uncommon as fall migrants; it is unusual to see more than 5 birds per day per locality. They are rare after early October. A wandering immature remained at NBS 30 July–16 August 1955 (JBa).

Fall maxima. Five, Block Island, 27 September 2011 (CNu).

Extreme dates. 29 April 1990, Swan Point Cemetery (DFi et al.), to 18–19 November 2006, Central Beach (PLE, STs).

Yellow Warbler (*Setophaga petechia*)

Status

Common to abundant summer resident and very common transient.

Breeding

Yellow Warblers achieve their greatest abundance in the fog-drenched thickets of arrowwood, multiflora-rose, and willow at the outer coast and in maple swamps and brush surrounding ponds and streams in the interior. In the northeastern part of the state, sweeter soils provide greater floral diversity, and territorial birds are occasionally present in mesic upland brush with little or no wetland vegetation.

Summer maxima. 244, Aquidneck Island, 15–16 June 1990; 212, Conanicut Island including 110 on Beavertail, 22–25 June 1985; 187, Narragansett to South Kingstown, 23–28 May 1984 (all RLF).

Nonbreeding

Spring. This tropical-wintering species appears early in spring, sometimes 2 weeks earlier than many Canadian-zone breeding warblers. The earliest arrivals (see also "Extreme dates") include 1 at Block Island 21 April 1960 (fide MSl) and 1 in the Great Swamp 21 April 1993 (SSM). Shaibal Mitra (City University of New York–Staten Island, pers. comm.) has noted a "striking disparity in arrival dates and buildups of local breeding populations between sheltered spots such as the Great Swamp and the ocean-cooled maritime thickets of the outer coast. Both habitats host dense breeding populations and experience sequential arrival and buildup of territorial breeders," with Great Swamp birds arriving earlier. General arrival is in late April and early May. Migrants continue to arrive through May, and the latest date for a transient or wanderer is for 1 at Cox Ledge 10 June 1978 (MBO staff).

Spring maxima. 100+, Block Island, 20 May 1973 (WBD); 97, Conanicut Island, 20 May 1990 (CaP et al.); 25, Lonsdale, 9 May 1990 (REn).

Fall. There is a peak in numbers during early to mid-August, when diurnal migrants can regularly be seen short-hopping westward over the barrier beaches. Yellow Warblers have a protracted fall migration season that extends from mid-July to October, the variation likely due to arrival of migrants from differing breeding areas. Two were banded at Block Island 3 October 1996 (MHC, SMC), and others seen at Napatree, 4 October 1997 (MJM); Sakonnet, 5 October 1957 (SSD, RLF); Charlestown Beach, 6 October 1990 (RLF, PAB); Block Island, 12 October 1964 (RTL); Charlestown Beach, 6 October 1990 (RLF, PAB); and Block Island, 12 October 1964 (RTL). The latest date at Napatree Point is 9 October (CRa).

Fall maxima. 27 migrants, Napatree, 15 August 2013 (CRa).

Extreme dates. 8 April 1962, Middletown (REW, RLF), to 25 October 1981, Sachuest Point (RBo), and 7 November 2001, Trustom NWR (CNu, DFi).

Chestnut-sided Warbler (*Setophaga pensylvanica*)

Status

Fairly common, but declining, summer resident and transient.

Breeding

Chestnut-sided Warblers are fairly common in wet woods and acid swamps but are local and less common in Rhode Island than in neighboring states to the north or west. There are fewer in western Rhode Island, yet sizable numbers inhabit some of the wooded swamps of South County (17 in the Great Swamp, 25–26 May 1984; RLF). They are present on northern Conanicut Island (4 there 9 June 1981; RLF) and Prudence Island (11 in southern sections, 6–8 June 1990; REn, CRa). They appear to be absent from Aquidneck Island as well as from Block Island. They were fairly common in eastern Tiverton, Little Compton, and Wilbour Woods in the 1950s and 1960s (DLE), but have since declined and have not bred since the 1990s. Sample counts in western Rhode Island towns include 50, Glocester to Burrillville, 17–23 June 1982; and 50, Foster to Coventry, 16–24 June 1986 (both RLF); but they are now rare or absent from much of the state. They are abundant only in western Glocester, near Sprague Hill, where 19 males were recorded during a point-count transect conducted 11 June 2012 (CRa).

Nonbreeding

Spring. The earliest spring appearances were at Kimball Sanctuary 21 April 1932 (MSo) and Kingston 24 April 1954 (DLK). General arrival is in early May. Maximum counts in spring through the 1990s were on the order of 20 per day per locality, but they are now much reduced; the largest spring maxima since 2005 are 4, Swan Point Cemetery, 19 May 2014 (RFa), and 4, Swan Point Cemetery, 10 May 2009 (RFa). Late migrants include birds at Green Hill 5 June 1977 (CWo, ESa) and Block Island 10 June 1981 (RBo).

Fall. Breeding birds depart by early August, when some migrants appear; e.g., 1 at Napatree Point, 3 August 1991 (CRa). Most migrants pass through by mid-September. The latest birds have all been individuals at Block Island: 8 October 1959 (RLF, JBa), 14 October 1962 (ASRI), and 1 banded on 20 October 1981 (ELa).

Fall maxima. 30, Warren, 18 August 1941 (RBo); 10, Block Island, 23 September 1983 (FAP, STs, JDi). Recent daily maxima for fall do not exceed 4 birds per day, mostly at Block Island.

Extreme dates. 21 April 1932, Kimball Sanctuary (MSo), to 20 October 1981, Block Island (1 banded; ELa), 21 October 2013, Block Island (1 seen; CNu), and Block Island, 22 October 1996 (1 banded; ELa).

Blackpoll Warbler (*Setophaga striata*)

Status

Common spring and common to abundant fall transient.

Nonbreeding

Blackpoll Warblers declined significantly as fall migrants at Block Island, but not at Kingston, 1969–2001 (Osenkowski et al. 2012).

Spring. Until recently, Blackpoll Warblers were one of the latest migrants to appear in spring; since about 2000, they have appeared earlier and are now regularly reported in the first few days of May. Migrants continue to be seen into mid-June: 17 June 1900 at Wakefield (HSH) and 17 June 1969 at Providence (BRi).

Spring maxima. 200, Little Compton, 26 May 1951 (RBo); 68, Block Island, 21 May 2010 (DFi, STa); 50, Swan Point Cemetery, 14 May 1934 (DLK).

Fall. Early fall transients have appeared 9 August 1959 at the East Providence Reservoir (RLF) and 10 August 1957 at Middletown (JBa); peak abundance is in late September to mid-October. There are several records for late November; e.g., Warren 25 November 1948 (RBo) and Middletown 27 November 1955 (JBa).

Fall maxima. 151, Pulaski State Park, 14 September 2003 (MLy).

Unlike Cape May and Bay-breasted Warblers, Blackpolls do not seem to fluctuate in abundance dramatically with Spruce Budworm (*Choristoneura fumiferana*) outbreaks.

Extreme dates. 1 May 1904, Cranston (HSH), and 1 May 1979, Swan Point Cemetery (RBu), to 3 December 1983, Carolina (CRa, REn), and 22–23 December 2004, Perryville (STs).

Black-throated Blue Warbler (*Setophaga caerulescens*)

Status

Uncommon summer resident and common transient.

Historical

Black-throated Blue Warblers are limited to habitats with substantial stands of mountain-laurel. The first nest was discovered in Burrillville on 30 May 1911 by HSH. He had found a male on territory there in 1906, and three males in 1910. Hathaway's initial discovery was closely followed by a nest found in Glocester 16 June 1911 by CED. Subsequently, 13 more clutches were found in Glocester from 1912–1924 (CED, mostly in FlMNH). Since then, territorial birds were found in Glocester, Burrillville, Foster, Coventry (feeding young at Greene, 1953; FPe), and West Greenwich. They have been found at several places in Hopkinton, in West Kingston, several sites in Exeter, and Liberty. Even farther south, a pair was present sporadically for many years near the Indian Cedar Swamp, Charlestown (1920s through 1940s; HSH, DLK), and there are two mid-June and early July records for South Kingstown. One was seen in laurel thickets in Tiverton from 1–16 June 1996 (DLE) and another in similar habitat at Weetamoo Woods on 1 July 1999 (REm); otherwise, there is no evidence of nesting in eastern Rhode Island.

Breeding

Most surprisingly, given the above history, not a single Black-throated Blue Warbler was found during breeding bird atlas work in the 1980s, despite considerable investigation of appropriate oak–laurel woodlands. Following this failure, a number of territorial pairs have again been located by REn in the Sprague Hill area of Glocester because of special efforts since 1988. He located 13 singing males and nesting activity at 5 sites there during 1993. They also nest in Buck Hill Wildlife Management Area in Burrillville: 6 males along 1 point-count transect, 11 June 2012 (CRa).

Nonbreeding

Black-throated Blue Warblers increased significantly as fall migrants at Kingston, but not Block Island, 1969–2001 (Osenkowski et al. 2001).

Spring. Black-throated Blue Warblers are fairly common migrants in spring, with occasional counts to 5 per day. Peak numbers occur in mid-May. The latest straggler was 1 at Block Island 8 June 1919 (EDi).

Spring maxima. 15, Swan Point Cemetery, 10 May 1989 (DFi et al.); 12, Swan Point Cemetery, 14 May 1983 (DLE).

Fall. They are somewhat more numerous than in spring, especially coastally. The earliest appearance in nonbreeding areas was 1 at Napatree Point 19 August 1986 (CRa). They are most common in September and early October. The bulk of coastal records are for late September and early October. There are numerous records through late October; 5 banded at Block Island 17 October 1973 (ELa); 5 seen, Block Island, 17 October 1976 (RLF).

Fall maxima. 55, Block Island, 9 October 1995 (RLF et al.); 50, Block Island, 3–4 October 1984 (RFa et al.).

Extreme dates. 22 April 1972, South County (RMa), to 18 November 1976, Jamestown (RLC).

Palm Warbler (*Setophaga palmarum*)

Status

Common spring and abundant fall transient; uncommon in winter.

Historical

Howe and Sturtevant (1899) listed only one record of Palm Warbler. Hathaway (1943) listed one other record prior to 1939, a spring bird shot in West Greenwich during the week of 17–24 May 1902.

Nonbreeding

There are 2 subspecies that occur in Rhode Island: *Setophaga p. palmarum* ("Western" Palm Warbler) and *S. p. hypochrysea* ("Yellow" Palm Warbler). "Western" Palm Warblers, which breed east to eastern Ontario, migrate north in spring west of the Appalachians and are rare on the East Coast at that season. "Yellow" Palms are the breeding form of the Northeast and winter on the Gulf Coast. They are the common form here in spring but uncommon in fall. "Yellow" Palms migrate north in spring slightly earlier, and south in fall slightly later, than "Western" Palms.

Spring. Palm Warblers are one of the earliest warblers to pass through, arriving regularly by early April: 3 April 2005 at Neutaconkanut Hill (Mason), 2 at the Great Swamp 4 April 1999 (SSM), and 2 on 5 April 1997 at Lonsdale (DLE). Peak migration is during the last week of April. Migrants are still encountered to about 8 May, with occasional lingerers remaining to mid-May. The latest record is for Swan Point Cemetery 23 May 1958 (CWo).

Spring maxima. 100, Wenscott Reservoir, 24 April 1933 (DLK); 100, East Providence Reservoir, 27 April 1953 (RLF). These are mainly "Yellow" Palm Warblers (*S. p. hypochrysea*).

"Western" Palm Warblers (*S. p. palmarum*) are rare in spring in Rhode Island. Two were banded at Block Island 2 May 1970 (ELa), and individuals were seen at Middletown 11 May 1956 and 18 May 1957 (both JBa). Other spring "Westerns" may be successful winterers from the previous fall.

Fall. "Western" Palms arrive earlier than do the smaller numbers of "Yellow" Palms, with records for 12 and 22 August. The earliest records for "Yellow" Palms are 2, Napatree Point 28 August 1980 (RLF); 1, Little Compton 7 September 1949 (DLE).

Fall maxima: These are mainly "Western" Palms: 600, Block Island, 30 September–2 October 1994 (RFa et al.); 200, Block Island, 8–10 October 1965 (ASRI); 200, Block Island, 1–2 October 2005 (ASRI). In the interior they are almost always found in smaller numbers; e.g., 50, Arnold Mills Reservoir, 9 November 1991 (DMK); 60, Cumberland, 10 October 1993 (DMK). They frequent exposed hedgerows in agricultural country or the edges of salt marshes and the protected edges of agricultural fields, often feeding on the ground. Abundance varies with the severity of the early winter season.

Winter maxima: Mainly "Western" Palms. 42, Newport–Westport CBC, 23 December 1950; 31, Middletown to Portsmouth, 21 December 1991 (REn, RLF et al.); 18, Nonquit Pond, 22 December 1985 (SDz et al.). Records of "Western" Palms suggesting winter survival are 1, Trustom Pond NWR, 2 March 1991 (RS, JMu); 1, Point Judith, 4 March 1973 (FAP); 1, Middletown, 11 March 1963 (REW); and 1, Bonnet Shores, 12 March 1983 (GLB). One was found dead 6 March 1949 at Block Island (EDi).

Late records of "Yellow" Palms include 2, Portsmouth, 15 December 1974 (RLF, A. Sanborn); 2, Middletown, 1 January 1953 (RBo); individuals at Point Judith, 2 January 1966 (DLK); and Charlestown, 4 January 1980 (DLK). One remained in Portsmouth through January 1953, a record mild season (RBo).

Pine Warbler (*Setophaga pinus*)

Status

Common breeder and migrant. Uncommon to rare in winter.

Historical

There are conflicting accounts of the status of Pine Warblers in Rhode Island before about 1930. Since they have fluctuated in abundance more recently (since about 1960), it seems possible this occurred in the past as well. Howe and Sturtevant (1899) stated that they occurred "wherever there is Pitch Pine growth." However, the species was given only "hypothetical" breeding status by Remington and Commissioners (1908). Probably most indicative of their rarity in the period 1882–1924 is that their eggs were unknown to CED or to other local oologists. However, Pine Warblers occurred regularly in the breeding season at Woonsocket 1903–1921 (Arnold), and there are old breeding-season specimens from Johnston and from the Ten Mile River in East Providence. Nests remained undiscovered until 1910, when Hathaway (1913) found them breeding at Washington. As fires set by Native Americans promoted the growth of pitch pine (Bromley 1935), it seems likely that Pine Warblers may have been even more numerous prior to 1800 than they are now.

Breeding

With the gradual regrowth of Rhode Island's forests in 1930–1970, there appears to have been a substantial increase in the Rhode Island nesting population. The extent of the increase was clarified during the early 1980s by the breeding bird atlas. Probably 80% of Pine Warbler territories in Rhode Island are in white pines. Planted groves of red pine have also been readily accepted, more so when near water.

Summer maxima. 61, Exeter to West Greenwich, 29 May–5 June 1983 (RLF); 59, Richmond to Hopkinton, 2–8 June 1984 (RLF); 48, Carolina Management Area, 6–11 June 2005 (CRa); 15 singing males, Arcadia, 17 June 1971 (RLF).

Nonbreeding

Spring. Pine is the earliest warbler to appear in spring, usually preceding Louisiana Waterthrushes and Yellow-rumped, Palm, and Black-and-white Warblers. The earliest spring date is for a singing male with other early migrants at Matunuck 16 March 1946 (DLK). Another appeared at Jamestown, where they do not breed, 19 March 1991 (ChP), and 1 was at Exeter 18 March 2012 (SSM). Peak numbers of migrants occur in mid-April. Late transients occur very sporadically into May; e.g., 2 banded, Block Island 7–8 May 1966 (E. Dickerson).

Spring maxima. 32, Great Swamp, 19 April 2003 (SSM et al.); 20, Carolina, 31 March 1986 (SSM).

Fall. Numbers of migrants seen in fall are rather small (5 per day per locality) but have increased since about 1990.

Fall maxima. 50, George Washington Management Area, 11 September 2005 (MLy); 30, Block Island, 2 October 2010 (STa, DFi); 20, Wood River Junction, 15 September 1946 (DLK).

Winter. Several Pine Warblers are found every December, and there is an increasing number of winterers. Later records include 2 seen in Westerly 23 January 1953 (ESa) and another that remained there, 25 January–19 February 1967 (ESa). One remained from January to mid-March 1986 at a feeder in Charlestown (T. Smith, DLK). Two birds attended a Wakefield suet feeder through February and March 1990 (MJM).

In 1991–1992, 1 attended the same feeder through 22 January (MJM). Two were in West Kingston 18 February 2005 (JSJ) and 1 in Little Compton 3 March 2005 (GDe).

Winter maxima. Nine, Middletown, 17 December 1994 (CRa, REn).

Yellow-rumped Warbler (*Setophaga coronata*)

Status

Abundant transient and uncommon summer resident in interior. Common winter resident at the coast.

Historical

As a summer resident in Rhode Island, this northern warbler has enjoyed a mystique only exceeded by the Nashville Warbler. They were unsuspected of breeding by HSH and the early oologists, despite an enhanced breeding status in Massachusetts during the broad era of the "field pine" at the turn of the twentieth century (Forbush 1929). It may be only a random circumstance that CED and other oologists never found "Myrtles" breeding in western sections, but there were, in fact, very few groves of white pine in CED's era (1882–1930) that would have supported breeding birds. The few extensive local pine stands that survived the heavy lumbering early in the twentieth century were prostrated or ripped apart by the 1938 hurricane, and groves of adequate size and density only began to reappear in any numbers after about 1965. For decades, the only known nesting location was in Greene, near the home of FPe, where large white pines adjoined the house near extensive woodland. Here, territorial pairs were present almost annually between 1948 and the 1980s, with a nest found on one occasion in the 1950s and parents noted feeding young several times. Remarkable now in light of current knowledge is the fact that, prior to the 1980s and the inception of the breeding bird atlas, the only suggestion of breeding elsewhere was a male singing in pine–oak–hemlock woods in Glocester on 24 June 1951 (RCC).

While Rhode Island "missed out" on breeding Yellow-rumps during the bygone era of the field pines, the post–1938 reappearance of similarly pure stands of sizable white pine has lately created the best Rhode Island habitat in 200 years. Even so, extensive stands with sizable trees are still scarce, and this warbler remains a relatively local summer resident. It is exceptionally vulnerable to lumbering because the habitat becomes acceptable for only about a decade before the trees reach adequate size for cutting. Even so, widespread maturation of such pine stands (usually on dry, acidic sites, and best near water) has caused a great proliferation of known breeding stations.

Breeding

Much of the best habitat for Yellow-rumped Warblers (stands of white pine) is within a few miles of Connecticut in the towns between Burrillville and Exeter. Atlas workers found territorial birds at approximately 15 sites, including a male accompanied by 2 begging young at Durfee Hill 24 June 1982 (RLF) and another male carrying food near Wickaboxet Pond in West Greenwich 24 June 1984 (RLF). Breeding locations were found eastward to north-central Coventry and in the Mishnock Pond area. Nearby in the proposed "Big River Watershed "area, a remarkable 22 pairs were located in 1989 (REn). Eighteen territorial pairs were in the Arcadia Management Area between 1989 and 1992 (CRa).

Nonbreeding

"Myrtle" Warblers declined significantly as fall migrants at both Block Island and Kingston banding stations 1969–2001 (Osenkowski et al. 2012). This is consistent with their general decline in New England 1966–2015 (BBS), even though they increased during the same time farther west.

Spring. Chris Raithel has seen "Myrtle" Warblers on several occasions in mid-February and early March near his house in Hope Valley. Since Yellow-rumped Warblers probably do not ordinarily winter in this area, such reports suggest that some northward movement can occur by then. In upstate areas, a few individuals have appeared during early April. Spring habitat preferences include the tops of maples and oaks draped with insect-laden catkins or the wooded margins of small inland ponds, far from the low brush inhabited in winter. Peak migration is from 25 April through mid-May (usually to 50 per day per locality). The latest transients were at Swan Point Cemetery 5 June 1971 (BRi) and Matunuck 12 June 1983 (v.o.). One landed on a boat about 4 mi (6 km) south of Sakonnet 23 July 1955 (Greenway).

Spring maxima. 3500, Block Island, 12 May 2002 (D Abbott, STs); 100, Swan Point Cemetery, 8 May 1992 (DFi, DLE, RBo et al.); 100, Lincoln, 5 May 1985 (SDz).

Fall. Yellow-rumped Warblers arrive in the second or third week of September (earliest: 2, 31 August 2012, Sachuest NWR; RWa). Peak abundance is near 15 October. The species' lavish abundance at Block Island in October is a legendary feature of the migration there, and island-wide totals of 10,000–20,000 have often been estimated. More than 2500 passed in a steady stream through a single clump of bushes en route to the north tip of the island (Sandy Point) after a cold front in the early morning of 17 October 1982 (RLF). They are far less numerous at inland locations.

Winter. In early winter, numbers in coastal areas often remain high, particularly in mild years if bayberry fruits are abundant. Inland, however, Yellow-rumpeds are now very uncommon in winter. In a long-past era, when inland regions were generally in an earlier successional stage, the "Myrtle" was often fairly common in winter; e.g., to 15 per day on Glocester CBCs 1900–1908. They were also formerly more common in the upper Narragansett Bay area in winter; e.g., 100 at Fields Point 8 January 1905 (Mason), but they have since declined.

Winter maxima. 1500, Little Compton, 24 December 1964 (B. Harrington); 1479, Block Island CBC, 22 December 1997 (CRa et al.); 1318, Block Island PDC, 15 February 1999 (SSM et al.); 1000 going to roost toward NBS from Sachuest Point, 9 February 1985 (A. Weimer). Most winter totals on the mainland coast are now well under 300, and in much of the coastal area, only a few dozen can now be located on a winter day. The widespread succession of bayberry to other vegetation such as shadbushes and the proliferation of housing developments along the mainland coast have reduced wintering habitat and food supply.

Subspecies

There are a few carefully documented sight records of "Audubon's" Warblers (*S. c. auduboni*) in Rhode Island, 3 of these from Block Island. One was seen 8 October 1977, Block Island (David Roach). It was described as similar to nearby "Myrtle" Warblers but possessed a "strikingly yellow throat patch." Others were seen 22 October 2001, Block Island (JSJ, STs); 17 December 2015, Block Island CBC (SSM); and 11 February 2006, Misquamicut (DFi). "Audubon's" Warblers have been collected twice in Massachusetts and reported there on several other occasions (Veit and Petersen 1993).

Yellow-throated Warbler (*Setophaga dominica*)

Status

Vagrant. Now occurring annually, mainly in spring.

Nonbreeding

Spring. Yellow-throated Warblers have been found with increasing frequency, and there are now multiple individuals reported each year. The spring records span the dates 16 April to 27 May, with a slight concentration in late April and early May. Yellow-throated Warblers are appearing with increasing frequency, probably at least in part owing to changing climate, as is true of a number of southeastern passerines. Since 2005, there have been about 12 records in spring, including 5 May 2006, Block Island (JSJ, STs); 19–21 April 2008, Miantonomi Park (m.obs.); 5–16 May 2010, Miantonomi Park (WM); 28 April 2012, Miantonomi Park (Drew Wheelan); 13 May 2013, Miantonomi Park (Wayne Munns); 3 May 2015, Rome Point (Wayne Munns); and 15 May 2015, Miantonomi Park (v.o.).

Spring maxima. No more than 1 per locality per day.

Summer. One or 2 territorial males were present in a grove of white pines south of Breakheart Pond, western Exeter, in 1982 and 1983. Two males were there

23–28 June 1982, and 1 of them persisted until 1 July (A. Moorhouse, CRa et al.). In 1983, single males were seen 11 and 31 May (CRa, RLF). The habitat at Breakheart Pond seems similar to that found where Yellow-throated Warblers nest in New Jersey and Delaware.

Fall. There are 13 records from fall, and 6 of the 13 are since 2005, as follows: 19 August 1970, Kingston (DLK, FPe); 3–5 October 1980, Block Island (ASRI, RFa et al.); 17 October 1992, Warren's Point (REm); 12 October 1998, Warren's Point (REm); 13 November 2000, Block Island (SSM); 23 September 2001, Block Island (SSM); 6 October 2003 (dead), Narragansett (MJM); 18–19 October 2005, Block Island (Pete McCalmont); 8 October 2008, Little Compton (GDe); 28 August 2009, Block Island (fide RFa); 21 October 2011, Ninigret (Erin King); 22 September 2012, Block Island (DFW); 1–11 November 2012, Scarborough (Jen Scranton).

Extreme dates. 14 April to 11 November.

Prairie Warbler (*Setophaga discolor*)

Status

Common summer resident and fairly common transient. Very rare in early winter.

Historical

Charles E. Doe collected at least 79 egg sets in Rhode Island and surrounding areas 1889–1915, including 40 sets from Cranston alone.

Breeding

Prairie Warblers are common and broadly distributed in a variety of shrubby and second-growth habitats. Along with Field Sparrows and Brown Thrashers, they have been very slowly decreasing with the succession of Rhode Island woodlands and more effective fire suppression. A single-day count that would be difficult to duplicate today was of 18+ in Burrillville, Glocester, and North Smithfield, 8 June 1946 (RCC, WBD, DLK). Prairie Warblers are fairly common breeders on the Bay islands, and a pair or 2 have nested since 1973 on Block Island (RLF, RBo).

Summer maxima. 47, Exeter to West Greenwich, 29 May to 5 June 1983; 41, Richmond to Hopkinton, 15–21 June 1987; 33, Foster to Coventry, 16–24 June 1986 (all RLF).

Nonbreeding

Prairie Warblers are scarce at regular migrant traps such as Swan Point Cemetery and Block Island, both spring and fall. Daily maxima for obvious transients rarely exceed 3 birds.

Spring. Prairie Warblers are regular spring transients throughout, the earliest dates including individuals at Swan Point Cemetery 22 April 1974 (SSD) and Green Hill 23 April 1976 (DWa). Usual arrival is in late April with peak counts in early May.

Spring maxima. 18, Conanicut Island, 20 May 2008 (CaP et al.).

Fall. Early fall transients or wanderers have been banded at Block Island 4 July 1975 and 30 July 1970 (ELa), and CRa has recorded them twice at Napatree on 5 August. Most migrants appear between mid-August and mid-September. The fall migration of Prairie Warblers is rather extended; they are regular through October and there are a number of November records.

Fall maxima. 40, Warren, 25 August 1941 (RBo).

Winter. There are 3 early winter records: 1 collected somewhere in Rhode Island, 4 December 1884 (Jencks 1885); 1 seen at 5 ft (1.5 m) in Portsmouth, 14 December 1991 (RLF, E. Neumuth); 1 photographed at Block Island, 18 December 1990 (DFi, RBo et al.); another seen there, 22 December 1997 (DLE, REm et al.); 1, NBS, 28–29 December 2006 (RTL).

Black-throated Gray Warbler (*Setophaga nigrescens*)

Status

Vagrant. Six records, all in spring and summer.

Nonbreeding

The spring and summer records of this western North American warbler, and lack of fall ones, is similar to the pattern of records in western Massachusetts (Bagg and Elliott 1937) and likely involve 1-year-old birds. The first Rhode Island record was of 1 singing male at Block Island 5–21 April 1921 (EDi), followed by a second one, also a singing male, there 26–29 June 1934 (EDi). A third was watched at leisure by EDi in her yard at Dickens Point, Block Island, on 14 May 1943 during a particularly heavy wave of migrants. A female-plumaged bird appeared 20 April 1974 at Swan Point Cemetery in Providence (REm, SSD) and was seen again the next day (DLE, FAP et al.; photos by P. Duffy examined). On 11 May 1975, another female-plumaged bird appeared at the same location and was studied for 2 hours at close range (REm, DLE, SSD et al.). Most unusual was a singing male that remained "on territory" near Breakheart Pond, Exeter, 6–29 July 1979 (DRu, DLK et al.).

Townsend's Warbler (*Setophaga townsendi*)

Status

Vagrant. Three records.

Nonbreeding
An adult male was found freshly dead in the dunes at Quicksand Pond 11 May 1990 by DLE and REm. The specimen, the first for New England, is MCZ #332835. Another was banded at Block Island 24 October 1994 (JPa). A third was at Trustom Pond NWR 25–26 November 2017 (STs, PLE; photos; see Photo 28).

Black-throated Green Warbler (*Setophaga virens*)

Status
Fairly common summer resident and common transient.

Historical
Black-throated Green Warblers were formerly more common and widespread, breeding throughout western Rhode Island south to the South County moraine, in Warren, in the Tiverton Highlands south to northern Little Compton, and even on Prudence Island (1946; WBD). They formerly bred not only in white pine and eastern hemlock but in several other habitats. Douglas L. Kraus found them regularly in pitch pines at Carolina in the 1940s; RCC located a nest in northern red cedars in the Sin and Flesh Brook area of Tiverton in 1942 and reported them to be fairly common there.

Breeding
Breeding in white cedar swamps is an apparently routine but relatively local occurrence; e.g., pair singing near Chapmans Pond 27 June 1984 (CRa, REn), and 2 males at the Great Swamp fens 14 June 1992 (CRa, RLF). The species is now rare away from large stands of those evergreens. Atlas observers found them at several Tiverton locations, primarily in hemlocks in Weetamoo Woods south of Lafayette Road. Most of the hemlocks have now succumbed to the Woolly Adelgid (*Adelges tsugae*), and Black-throated Greens are no longer regular here. Robert Emerson has found them only 6 times in the breeding season from 2003–2013, and none since then.

Currently, they breed south to Westerly and Charlestown and east to Fiskeville and eastern Exeter. They are now most common in the western halves of the towns from Burrillville to Exeter, where they nest in planted evergreens at the Scituate Reservoir watershed, at Carolina, and in the Watchaug Pond area. They also nest in groves of pitch pine and white pine. Sample counts by the author include 20, Glocester to Burrillville, 17–23 June 1982; 18 in Foster to Coventry, 16–24 June 1986; 18 in Exeter to West Greenwich, 29 May–5 June 1983; 15 in Richmond to Hopkinton, 15–21 June 1987. One-day counts include 23 along Frosty Hollow Road, Exeter, 20 June 1989 (REn); 10 in Glocester on 28 June 1979 (REn), and 14+ there on 12 July 1980 (REn); 48 in Arcadia, 1–25 June 1990 (CRa); and 17, George Washington Management Area, 26 June 2004 (MLy).

Nonbreeding

Spring. Black-throated Green Warblers are common transients in spring. Local breeders arrive by mid-April, and migrants appear a bit later, by late April or early May. Peak numbers (to 10 per day per locality) occur in mid-May.

Spring maxima. 35, Block Island, 12 May 2002 (DJA, STs). The latest were individuals at Sakonnet 7 June 1975 (RBu) and Block Island 8 June 1972 (RLF).

Fall. Early migrants appear in mid-August; e.g., 2 August 1946 at Green Hill (DLK) and 16 August 1960 at Middletown (JBa). Peak numbers (to 15 per day per locality) occur in late September and October. There are at least 13 records for November, the latest at West Kingston 26 November 1990 (LAM) and Green Hill 29 November 1973 (DWa).

Extreme dates. 9 April 2002, Miantonomi Park (v.o.), and 12 April 1977, Green Hill (DWa), to 28 December 1974, Green Hill (DWa).

Canada Warbler (*Cardellina canadensis*)

Status

Fairly common summer resident and transient; perhaps declining.

Historical

This northern warbler was rare as a summer resident at the turn of the twentieth century and was listed as "breeding" at Westerly and Johnston by Howe and Sturtevant (1899). It took CED more than 30 years of tramping western Rhode Island to find the first nest and eggs, in Glocester 2 June 1913 (FlMNH #55368, #53367) The second was found near Watchaug Pond in 1915 (HSH). They were still considered rare summer residents until the 1930s, when records markedly increased.

Breeding

Canada Warblers are widespread summer residents, although variable in abundance from year to year. Like Northern Waterthrushes, they are usually (but not always) found in and near sphagnum bogs (Miller 1999). They are now common breeders in northwestern towns in cool swamps and the wetter portions of hemlock and white pine woods. They are fairly common in the western parts of most of the other towns south to Westerly and east in the interior of South County to within 2 mi (3 km) of the ocean at Watchaug Pond, in the swamp along Matunuck School House Road, at Narragansett, and near Wickford.

Summer maxima. 14, Richmond to Hopkinton, 15–21 June 1987; 13, Glocester to Burrillville, 17–23 June 1982; 11, Narragansett to South Kingstown, 23–28 May 1984 (all RLF).

Canada Warblers breed on Conanicut but not on the other Narragansett Bay islands, and were fairly common in Tiverton; e.g., 12 pairs in June 1942 (RCC, RBo). They have since declined or disappeared there.

Nonbreeding
Canada Warblers declined significantly as fall migrants at both Block Island and Kingston 1969–2001 (Osenkowski et al. 2012).

Spring. Canada Warblers are late-spring migrants, and peak numbers occur in mid to late May. The earliest records are for 29 April 1962 at Kingston (DLK et al.) and 30 April 1976 at Block Island (banded; ELa). Transients are commonly encountered in nonbreeding coastal districts and elsewhere in the first few days of June and at Swan Point Cemetery to 6 June; the latest migrant was seen at Block Island on 8 June 1972 (RLF).

Spring maxima. 22 banded, Block Island, 24 May 1974 (ELa).

Fall. Canada Warblers are regular transients from mid-August to mid-September but are rare after that.

Extreme dates. 29 April 1962, Kingston (DLK), to 27 October 1995, Swan Point (1 male; BCG), and 29 October 1974, Little Compton (1; JAG).

Wilson's Warbler (*Cardellina pusilla*)

Status
Fairly common spring and uncommon fall transient.

Nonbreeding
Spring. Wilson's Warblers are among the later spring migrants, with peak numbers during late May. The earliest record in spring is for 1 banded at Block Island on the exceptional date of 24 April 1979 (ELa). Individuals appeared at Fiskeville 30 April 1999 and 1 May 1984 (JGe, RGe). The majority of transients pass through from mid to late May. The latest was 1 at Rockville 3 June 1979 (RAC).

Spring maxima. Seven banded, Block Island, 19 May 1996 (MHC, SMC); 5, Swan Point, 16 May 2002 (RFa et al.); 4 males, Swan Point Cemetery, 19 May 2001 (RFa et al.).

Fall. The earliest fall migrant appeared at Kingston 2 August 1962 (DLK). They are uncommon in fall and occur from early September through mid-October.

Fall maxima: 4 banded, Block Island, 11 September 2003 (ELa, KGa).

There are a surprising number of late-fall records of Wilson's Warblers for Rhode Island, despite their comparative scarcity here as fall transients. There are at least 9 records for November and at least 8 for December: 1, Green Hill, 2 December

1976 (DWa); a male in Cumberland, 3 December 1983 (H. Barton, K. Barton); an adult male collected near Sandy Point, Portsmouth, 19 December 1971 (RLF; MCZ #329717); 1 female or juvenile, Portsmouth, 20 December 1986 (RLF); 1 male, Portsmouth, 18 December 1999 (RLF); Little Compton, 1 December 2005 (GDe); Block Island CBC, 14 December 2009 (CRa et al.); Newport–Westport CBC, 15 December 2012 (REm et al.).

Extreme dates. 24 April 1979, Block Island (1 banded; ELa), to 18 December 1999, 19 December 1971, and 20 December 1986, all Portsmouth (RLF).

CARDINALINE FINCHES (CARDINALIDAE)

Summer Tanager (*Piranga rubra*)

Status

Vagrant. Large increase since the 1950s. More frequent in spring than fall.

Nonbreeding

Spring. Through 1977, there were about 52 records of Summer Tanagers in spring, spanning the dates 7 April to 8 June. Since 2005, there have been 59 reports, 4 of these in fall. Like most southeastern breeding species, Summer Tanagers migrate rather early in spring and most Rhode Island records are from April. Early dates include males seen at Moonstone 5 April 1959 (DLK) and Matunuck 6 April 1984 (JZy), and singles collected at Block Island 7 April 1901 (Capt. Sisson) and Newport 8 April 1958 (JBa). One was seen 1 June 1956 at Middletown (JBa), and another in Saunderstown 2 June 2005 (PAB). During a remarkable flight of other late transients, 1 was seen at Bonnet Shores 7 June 1968 and 2 the next day in Exeter (both DWa). During the period 8–19 May 2000, 6 individuals were scattered from Block Island to Swan Point Cemetery.

Spring maxima. Three, Block Island, 16 May 2011 (JSJ); 3, Block Island, 12 April 1961 (MSl); 3, Block Island, 21 May 1966 (WBD).

Fall. There are 18 records for fall through 1977. The earliest was 1 banded at Middletown 18 August 1957 (JBa). Other records extend from 31 August and early September to 13 October; more than half are in early October. There is 1 record of a bird at Wakefield 25 October–6 November 1954 (DLK et al.). Block Island has produced a disproportionately large number of records compared with the mainland. All fall records are coastal.

Extreme dates. 5 April 1959, Moonstone (DLK), to 11–14 December 2012, Block Island (D. Mawhinney).

Scarlet Tanager (*Piranga olivacea*)

Status

Common summer resident and transient.

Breeding

Scarlet Tanagers are most numerous in northern and western sections of the state in oaks with scattered white pines but are also found in pitch pines and stunted oaks on the coastal plain and more mesic woodlands dominated by white ash. They are less numerous toward the coast and scarce on the Narragansett Bay islands and Bristol peninsula.

Summer maxima. 194, Foster to Coventry, 16–24 June 1986; 108, Richmond to Hopkinton, 15–21 June 1987; 40, Cumberland to Lincoln, 16–22 June 1984; 6 males, Conanicut Island, 22–25 June 1985 (all RLF).

Nonbreeding

Spring. Most Scarlet Tanagers pass through Rhode Island during early to mid-May, but there are quite a few records of single birds appearing much earlier in spring, usually after spells of unusually warm weather with southwesterly airflow. The earliest record is 4 April 1891 (Southwick and Critchley 1891). Another was found dead at Block Island 13 April 1932 (EDi). There is a specimen in the Dwight collection at AMNH, taken at Centerdale 18 April 1898 (H. Kellogg, discovered by J. Bull). Two males were photographed at Sakonnet 18 April 1984 (RBo), and another that subsequently died was picked up at Jamestown 18 April 1983 (CaP). Others have been seen at Cranston 19 April 1976 (JGe, RGe), Coventry 20 April 1964 (Fortin), and Middletown 23 April 1969 (JGe, RGe).

Spring maxima. 38, Phenix to Cranston, 27 May 1968 (DWa).

Fall. Most migrants appear during September and early October. One appeared at Napatree Point 27 October 1981 (CRa). Another was seen 11 November 2001 at Block Island (SCo). A very late bird was seen in Tiverton 21 November 1953 (DLE et al.).

Fall maxima. 40, Block Island, 4 October 1969 (WRP et al.); 15, Block Island, 1–3 October 1982 (RFa et al.).

Extreme dates. 4 April 1891 to 21 November 1953.

Western Tanager (*Piranga ludoviciana*)

Status

Vagrant. About 10 records.

Nonbreeding

The occurrence of Western Tanagers on the U.S. East Coast has fluctuated, and these fluctuations are related to rates of population growth within the core range (Veit 1990, 2000; Zawadzki et al. 2019). Peak numbers appeared in New England during the 1950s, and there appears to be an upswing of records during 2005–2013. These fluctuations seem most likely to be real, but see Davis (1972) for issues of identification. A Western Tanager visited the Ingraham feeder on Taylor's Lane, Little Compton, 11–20 November 1953, verified by RBo. Another appeared at the same feeder in "late November and early December" 1955. On 29 November 1960, JBa saw 1 at Block Island that had been reportedly present for several weeks. Another appeared at Ruecker Sanctuary, Tiverton, 6–10 December 1982 (HCa, J. Calhoun). More recent records include 1, Ninigret Park, 26 December 2002 to 18 January 2003 (CPe et al.); 1, Block Island, 11 November 2002 (REm, DFi, STa et al.); another at Block Island 1 October 2009 (DFi, STa); another at Block Island 10 January 2013 (M. Komosinski); 1, Napatree CBC (in Watch Hill), 27 December 2015 (GW et al.); and 1 on the South Kingstown CBC 28 December 2016 (GLB, DFW). Records were clustered in 2002–2003 and 2013–2016.

Northern Cardinal (*Cardinalis cardinalis*)

Status

Very common resident.

Historical

Northern Cardinals were first noted in the state when a female appeared at Warren 23–30 January 1917 (Leach; see Madison 1917). Other early records of vagrants include individuals at Middletown 16 April 1932 (fide MLy), Block Island 15 May 1933 (EDi), and Jamestown 13 July 1938 (A. M. Clark). One was seen 8 October 1938 in Warren (RBo), and another appeared there 30 November 1942, remaining through the entire winter (RBo). Vagrants appeared with increasing frequency during the 1950s, and 12 were reported between 1950 and 1955. Northern Cardinals were continuously present at Westerly beginning in 1954, and adults were seen feeding young there 8 July 1957 (ESa, RCC). There was a sharp increase in reports during October and November 1957. Many spent the winter of 1957–1958 at feeders, and there were widespread records of breeding in 1958. There were 10 breeding pairs found in South County in 1960 (DLK), and in 1960–1963 they were "generally distributed" in the

Providence area. During the winter of 1966–1967, 683 were counted in a statewide feeder survey (Beddall 1963).

Breeding

Cardinals are most common in southern and eastern parts of the state. They do not seem to be entirely dependent on bird feeders for winter survival but congregate at them wherever available.

Summer maxima. 177, Conanicut Island, 22–25 June 1985; 149, Foster to Coventry, 16–24 June 1986; 148, Cumberland to Lincoln, 16–22 June 1984 (all RLF).

Nonbreeding

Since they are resident throughout the state, detection of migrant cardinals is difficult. Nevertheless, cardinals certainly do migrate (as confirmed by banding), and occasionally influxes are noted; e.g., a loose flock of 10 moving northward with a vast horde of other migrants along the east shore of the Pawcatuck estuary at Watch Hill 4 November 1979 (RLF). At Napatree, where they do not breed, Chris Raithel's records show a weak peak in late April and a strong peak in early November, with a maximum of 10, 29 October 2010.

Winter maxima. 497, Newport–Westport CBC, 18 December 2010.

Rose-breasted Grosbeak (*Pheucticus ludovicianus*)

Status

Fairly common summer resident and transient.

Breeding

Rose-breasted Grosbeaks prefer extensive deciduous forests and are uncommon in the oak–pine forests typical of much of Rhode Island. They are almost always outnumbered by Scarlet Tanagers in these habitats. In southeastern Massachusetts, Rose-breasted Grosbeaks are similarly scarce or absent from pine–oak barrens on Cape Cod and the outer islands (Veit and Petersen 1993).

Summer maxima. 72, Foster to Coventry, 16–24 June 1986; 60, Cumberland to Lincoln, 16–22 June 1984; 26, Narragansett to South Kingstown, 23–28 May 1984; 15, Conanicut Island, 22–25 June 1985 (all RLF).

Nonbreeding

Spring. Migrants arrive in numbers during early May and are common through the second and third weeks of May. Some arrive much earlier, usually during or after unusually warm weather with southerly airflow, including southerly storms. One was at DLK's farm in Kingston 8 April 1971 and attended the feeder for several days. Others were seen 13 April 1993 at Tiverton (WBo) and 14 April 1943 at Kimball Sanctuary, Charlestown (E. Southwick). There are 16 additional records for 17–22 April. Late transients occur at Block Island through early June, the latest being 1 banded there 14 June 1972 (ELa, HLa) and another seen 17 June 1982 (CRa).

Fall. Migrants occur from early August to mid-October. They are rare after October, but there are a number of November records, often at feeders.

Fall maxima. 20, Block Island, 23 September 1983 (FAP, STs).

Winter. There is 1 record for early winter, a male with bright red underwing coverts seen at Middletown on 15 December 1979 (REn).

Black-headed Grosbeak (*Pheucticus melanocephalus*)

Status

Vagrant. Five records.

Nonbreeding

The first Black-headed Grosbeak for Rhode Island appeared at the Gladding feeder in Bristol in mid-February 1953 and was confirmed by RBo. Another was seen at a feeder in Tiverton 15–22 November 1953 by WBD, DLE, RBo, and 50 other observers. Another attended a feeder in Wickford January through March 1960 and was examined by DLK, CWo, SSD, JBa, RBo, and many others. There is a record of a nearly adult male, very buffy orange below, seen in Tiverton 14 October 1987 (RBo, HMP). More recently, a young male appeared in Charlestown 7–19 December 2008 (M. St. Sauveur). Veit (2000) showed fluctuations in abundance of Black-headed Grosbeaks in Massachusetts, with peak numbers appearing during the 1950s.

Blue Grosbeak (*Passerina caerulea*)

Status

Rare spring and uncommon fall vagrant. Most appear at the coast.

Historical

The first Blue Grosbeak in Rhode Island was at Drownville 12 October 1899 (F. T. Jencks), but they then went unreported until 1950, when expansion in the south began to yield vagrants to New England (Griscom and Snyder 1955; Bull 1964).

Nonbreeding

Spring. Like Indigo Buntings, most spring Blue Grosbeaks occur during early May, but there are also many earlier records, usually following warm weather with southerly airflow. Two were at Portsmouth 1–8 April 1958 (JBa), 1 was found dead at Block Island after a storm 18 April 1929 (EDi), a male was with Indigo Buntings at

Matunuck 18–19 April 1984 (v.o.), male was at Burdickville 18–26 April 1984 (ESa), and 1 was at Wickford 23–25 April 1975 (TMC). There are a number of records from late April and early May.

Spring maxima. Nine, Block Island, 29 April 2012 (DFi, STa).

Fall. The earliest fall date is 9 September 1956 at Block Island (DLK). Rare after October; 1 was at Sunset Farm 4 November 2012 (v.o.). Though less common on the mainland, there are fall records north to Greene, Warwick, and Swan Point Cemetery.

Fall maxima. Six, Block Island, 3–4 October 1981 (RFa et al.); 5, Block Island, 4–6 October 1985 (RFa et al.); 4, Block Island, 29 September 1957 (RLF, DLK); 3, Kingston, 20 September 1951 (DLK).

Winter. There is a unique record of 1 that remained at a feeder in Adamsville from 1 January 1970 through the first week of April 1970 (Wayne Petersen, m.obs.).

Indigo Bunting (*Passerina cyanea*)

Status

Fairly common summer resident and transient.

Historical

Howe and Sturtevant (1899) and Forbush (1929) both noted that Indigo Buntings were most common and regular as breeders in the northern and western portions of the state and that they were rare or absent in the southeastern section. A set of eggs was collected in Edgewood 2 June 1908 (CED).

Breeding

Indigo Buntings frequently colonize habitats soon after the habitats reach suitability because of ecological succession and then later abandon these habitats as they turn into closed forest. In June 1991, CRa found at least 13 males in Arcadia, where none had been noted during intensive field work during the previous 10 years. A similar influx of about a dozen singing (but apparently nonbreeding) males occurred at Arcadia in mid-July 1992; none had been present there the previous month. They were apparently absent from Aquidneck Island prior to 1957, when a June incursion brought several singing males to NBS in Middletown and others were seen elsewhere on the island (JBa). They bred there at least through the mid-1960s (REW, RLF). Three singing males were present on tiny Dutch Island 21 June 1981 (RLF), the birds apparently attracted by the burning of most of the island's vegetation in the previous year. They bred for many years at Swan Point Cemetery, close to the shores of the Seekonk River (7, 13–18 May 1983; v.o.) but have since disappeared.

Summer maxima. 26, Cumberland to Lincoln, 16–22 June 1984 (RLF); 11, Glocester to Burrillville, 17–23 June 1982 (RLF); 3, Carolina, 7 July 1984 (CRa).

Nonbreeding

Spring. Indigo Buntings arrive on territories during early to mid-May. However (seemingly more often than other Neotropical migrants), they tend to appear much earlier than this (especially at the immediate coast) following bouts of unseasonably mild weather accompanied by southerly airflow, sometimes but not always associated with southerly storms. The earliest such appearances were 23 March 1954 at Block Island (Lee) and 27 March 1969 at Matunuck (KOp). There are a great many records during mid to late April, under the conditions described above. Maximum counts are usually on the order of 10 per day per locality.

Spring maxima. Seven, Little Compton, 28 April 2005 (GDe); 6, Block Island, 27 April 2012 (DFi, STa).

Fall maxima. 15, Block Island, 3–4 October 1981 (RFa et al.); 9, Block Island, 29 September–1 October 1989 (RFa et al.); 9, Narragansett, 1 October 2011 (m.obs.); 7, Block Island, 6 October 1984 (DLE, REm). One was banded at Block Island 8 November 1964 (MSl), and others were seen there on 11 November 2006 and 10 November 2007 (REm, DFi, STa).

Winter. One wintered at the Kimball Sanctuary feeder in Charlestown, 5 January–22 March 1975 (CWo et al.), the only winter record for the state.

Painted Bunting (*Passerina ciris*)

Status

Vagrant. Increasing in frequency in the Northeast.

Nonbreeding

Painted Buntings occurred in the Northeast during the late 1800s and early 1900s, and then again beginning in about 1960, after a lengthy hiatus. Early records were often dismissed as being of escaped cage birds (Griscom and Snyder 1955), a conclusion that seems unlikely in light of the recent pattern of vagrancy. In Rhode Island, Howe and Sturtevant (1899) listed a record of 1 that was shot in Scituate by Daniel Seamans in the summer of 1882. After a break of 50 years, a male was seen at close range in Hope Valley 25 September 1953 (G. Segar); a male was examined at leisure in Jamestown 14 May 1972 (MDa et al.), as was another male in Jamestown 3 May 1974 (fide MDa). A male was at a feeder in Middletown 24 May 1980 (Donegan), and a male was at Westerly 19 May 1982 (S. Moorhouse, ESa); a male was in Swan Point Cemetery 22 May 1983 (E. Marks). A male was found 21 November 1993 at Ruecker Sanctuary in Tiverton (W. Nichols). A female or immature was closely observed at Block Island 12 October 1974 (T. M. and P. Curry). One was in Lincoln, 26 May 2000 (fide MTu). A female or young male was along the roadside in Matunuck 12–18 December 2009 (L. Gardrel, m.obs.), 1 was at Newport 21 February to 1 April 2014 (Bobbi Smith, fide RFa), and a male visited a feeder in Middletown 11–19 February 2015 (J. Clarkson).

Dickcissel (*Spiza americana*)

Status

Uncommon fall and rare spring transient.

Historical

These open-country breeders of the eastern prairies have varied in abundance and distribution considerably during the past 150 years. Until about 1880, Dickcissels bred throughout much of eastern North America, north to New England and south to the Carolinas. At that time, they also occurred during migration through New England (Forbush 1929; Griscom and Snyder 1955; Gross 1956). As agriculture declined during the late-nineteenth century, Dickcissels disappeared from the eastern portion of their breeding range. Beginning in the 1950s, Dickcissels increased as migrants to the East Coast, mainly in fall, and were regularly reported in Rhode Island.

Gross stated that "during the nineteenth century [Dickcissels] nested and was [*sic*] a common species in a wide range extending from the Carolinas and Pennsylvania, New Jersey, and New York to Rhode Island and Massachusetts." The evidence for their having nested in Rhode Island is vague, but such nesting seems highly likely given their overall distribution at that time. The first migrant was shot in Newport 5 September 1888 (W. Robinson). The next record was of one seen 2–5 June 1922 at Block Island (EDi). Another appeared 22 December 1922 at Pawtuxet, was caught and banded by Henry E. Childs Sr., and brought alive to Forbush for verification. They were then unreported until one was seen in Tiverton 23 December 1939 (WHD, DLK). Others were sporadically seen thereafter through the mid-1940s. In 1952–1953 there were eight fall reports, with six lingering on CBCs and a remarkable ten at feeders in March. Six were reported on CBCs in December 1953. Five were seen in one small tree at Charlestown 23 August 1955 (RLF), and another was noted at Jerusalem (RLF et al.). This increase in occurrence after World War II may reflect increased farming of grain in the Midwest at that time.

Nonbreeding

Dickcissels are most often found at the coast in September and October, where they associate with sparrows and blackbirds in weed-choked gardens and fields. Late fall and winter birds are mainly at feeders, where they often associate with House Sparrows and frequently remain through the winter into spring. They were most common from the 1950s through the early 1980s, after which they declined, but as of 2017, they seem to be increasing again. These fluctuations in abundance likely correspond to population growth or shrinkage within the core range, as described by Veit (2000) and Zawadzki et al. (2019) for other species of western vagrants.

Spring. There are many March records and at least 5 for April. One was seen 10 May 1950 at Tiverton (Ruecker) and others 15 May 1992 at Smithfield (BCG), 17 May 1953 at Wickford (fide RLF), 17 May 1969 at Jamestown (MDa), 17 May 1986 at Westerly (RMa), and 28 May 2005 at Napatree (CRa). There is 1 Block Island report for June.

Fall. The earliest fall date is 8 August 1954 at Tiverton (DLE). There have been about 72 reports since 2005, all from 19 August to 5 January.

Fall maxima. Six in flock with Bobolinks and House Finches, Trustom Pond NWR, 18 August 1988 (Jim Twining); 4, Avondale Farm Preserve, 15 October 2004 (RFa). One was reported on a pelagic trip to Block Canyon 19–20 August 2011 (CPe et al.). In recent years, totals of 20 or more have been reported during a fall season.

APPENDIX A: *Plant Species Mentioned in the Text*

Plant names follow Gould et al. 1998

Mosses, Liverworts, Lichens, Green Algae			
Clubmosses	*Lycopodium* spp.	Sea lettuce	*Ulva* spp.
Sphagnum moss	*Sphagnum* spp.	Old man's beard lichen	*Usnea trichodea*

Ferns			
Cinnamon fern	*Osmunda cinnamomea*	Bracken fern	*Pteridium aquilinum*
Rock polypody ferns	*Polypody* spp.	Christmas fern	*Polystichum acrostichoides*
Hay-scented fern	*Dennstaedtia punctilobula*		

Trees			
Balsam fir	*Abies balsamea*	Witch-hazel	*Hamamelis virginiana*
Norway spruce	*Picea abies*		
Black spruce	*Picea mariana*	American elm	*Ulmus americana*
Eastern hemlock	*Tsuga canadensis*	American beech	*Fagus grandifolia*
Eastern white pine	*Pinus strobus*	American chestnut	*Castanea dentata*
Red pine	*Pinus resinosa*	White oak	*Quercus alba*
Pitch pine	*Pinus rigida*	Chestnut-oak	*Quercus prinus*
Japanese black pine	*Pinus thunbergiana*	Scrub-oak	*Quercus ilicifolia*
Atlantic white cedar	*Chamaecyparis thyoides*	Black oak	*Quercus velutina*
		Northern red oak	*Quercus rubra*
Northern red cedar	*Juniperus virginiana*	Pin-oak	*Quercus palustris*
		Scarlet oak	*Quercus coccinea*
Tulip-tree	*Liriodendron tulipifera*	Yellow birch	*Betula alleghaniensis*
Sycamore	*Platanus occidentalis*	Black birch	*Betula lenta*
		White birch	*Betula papyrifera*

Gray birch	*Betula papulifolia*	American holly	*Ilex opaca*
White poplar	*Populus alba*	Sugar maple	*Acer saccharum*
Aspen	*Populus* spp.	Red maple	*Acer rubrum*
Beaked willow	*Salix bebbiana*	Ash-leaved maple	*Acer negundo*
Willow	*Salix* spp.	Staghorn-sumac	*Rhus typhina*
Wild black cherry	*Prunus serotina*	Shining sumac	*Rhus copallinum*
Common pear	*Pyrus communis*	White ash	*Fraxinus americana*
Apple	*Pyrus malus*		
Shadbush	*Amelanchier* spp.	Green ash, red ash	*Fraxinus pennsylvanica*
Black locust	*Robinia pseudoacacia*	Arrowwood	*Viburnum dentatum*
Black tupelo	*Nyssa sylvatica*		

Shrubs, Wildflowers

Sweet fern	*Comptonia peregrina*	Black huckleberry	*Gaylussacia baccata*
		Spotted wintergreen	*Chimaphila maculata*
Sweet gale	*Myrica gale*		
Northern bayberry	*Myrica pennsylvanica*	Whorled loosestrife	*Lysimachia quadrifolia*
Northern pitcher-plant	*Sarracenia purpurea*	Raspberries	*Rubus* spp.
		Multiflora-rose	*Rosa multiflora*
Sundews	*Drosera* spp.	Black chokeberry	*Aronia melanocarpa*
Sweet pepperbush	*Clethra alnifolia*		
Great laurel	*Rhododendron maximum*	Beach-pea	*Lathyrus maritimus*
Rhododendron	*Rhododendron* spp.	Purple loosestrife	*Lythrum salicaria*
Sheep-laurel	*Kalmia angustifolia*	Winterberry	*Ilex verticillata*
Mountain-laurel	*Kalmia latifolia*	Wild sarsaparilla	*Aralia nudicaulis*
Leatherleaf	*Chamaedaphne calyculata*	Buttonbush	*Cephalanthus occidentalis*
Wintergreen, teaberry	*Gaultheria procumbens*	Hightide-bush	*Iva frutescens*
		Goldenrods	*Solidago* spp.
Trailing arbutus	*Epigaea repens*	Pickerelweed	*Pontederia cordata*
Bearberry	*Arctostaphylos uva-ursi*	Canada mayflower	*Maianthemum canadense*
Lowbush blueberries	*Vaccinium* spp.	Bullbrier, catbrier	*Smilax rotundifolia*
Highbush blueberry	*Vaccinium corymbosum*	Pink lady-slipper	*Cypripedium acaule*

Grasses

Cow-wheat	*Melampyrum lineare*	Salt-water cordgrass	*Spartina alterniflora*
Wild celery	*Vallisneria americana*	Switchgrass	*Panicum virgatum*
		Little bluestem	*Schizachyrium scoparium*
Common eelgrass	*Zostera marina*		
Beachgrass	*Ammophila breviligulata*	Common cattail	*Typha latifolia*
		Narrow-leaved cattail	*Typha angustifolia*
Common timothy	*Phleum pratense*		
Common reed	*Phragmites australis*		

APPENDIX B: *List of Observers*

List is alphabetized by last name of observer.

DJA	Dennis J. Abbott
KPA	Kenneth P. Able
TAd	Thaire Adamson
CCA	Charles Allin
DAt	David Atwater
TAu	Tom Auer
CAv	Charles Avenengo
DB	Donna Baer
JBa	James Baird
LBe	Larry Benjamin
GBi	Grace Bissell
CBl	Charles Blake
DBo	Dickson Boenning
WBo	Walter Bosse
RBo	Richard Bowen
CBr	Charles Brown
JCB	Janet C. Bryan
FBu	Francine Buckley
PAB	P. A. Buckley
RBu	Robert Bushnell
HCa	Hobson Calhoun
PCa	Peter Capobianco
MHC	Mark H. Carr
SRC	Sarah R. Carr
SMC	Susan M. Carr
RLC	Richard L. Champlin
HCJ	Henry E. Childs Jr.
HEC	Henry E. Childs Sr.
NCl	Nancy Clayton
RCC	Roland C. Clement
JC	Jim Cody
SCo	Scott Comings
RAC	Robert A. Conway
SCr	Sidney Crawford
SSD	Severyn S. Dana
MDa	Maybelle Davenport
JDa	Jane Davis
THD	Thomas H. Davis
HDa	Harold Day
WBD	William B. Dean
GDe	Geoff Dennis
RDe	Robert Dewire
EDi	Elizabeth Dickens
ElDi	Elise Dickerson
JDi	Joe DiConstanzo
CED	Charles E. Doe
BD	Beverly Downing
MD	Mason Downing
SRD	Susan Roney Drennan
WHD	William H. Drury Jr.
SDz	Stephen Dziadosz
WRE	William R. Eddleman
DLE	David L. Emerson
REm	Robert Emerson
REn	Richard Enser
AEE	Alfred E. Eynon

RFa Rachel Farrell
RLF Richard L. Ferren
DWF Davis W. Finch
DFi Daniel Finizia
RFl Ron Flores
NF Nancy Fullerton
AFr Adam Fry
KGa Kim Gaffett
TSG T. Scott Gamwell
CGa Charles Gardiner
DGa Donald Gardner
GGa George Gardiner
JGa Joan Gardiner
LOG Lee O. Gardner
BCG Barbara Gearhart
WG William Gearhart
JGe Juanita Gee
RGe Richard Gee
HNG Harold N. Gibbs
BG Bryan Glemboski
RG Richard Graefe
JAG John A. Gwynne
LAH Lisa A. Hanson
CHa Charles Harten
HSH Harry S. Hathaway
ALH Alfred L. Hawkes
CHa Charles Haydock
SHa Stephen Haydock
SMH Sheila M. Hempstead
NPH Norman Hill
JHo Jan Hoftun
KHo Karen Holmes
MHo Millie House
RHH Reginald Heber Howe Jr.
SKe Seth Kellogg
LKi LeRoy King
KKi Katherine (Kay) Kinsey
RKl Richard Kleber
DMK Daniel M. Kowal
DLK Douglas L. Kraus
ELa Elise Lapham
HLa Helen Lapham
RTL Reynold T. Larsen
GLB Geoff LeBaron
PLE Paul L'Etoile
ILe Irving Leven
PJL Patricia J. Lindsay
MLy Mark Lynch
AMc Andrew MacLachlan
RMc Robert MacLachlan
LAM Lisa McCurdy
JMa John Magill
RMa R. Madison
DMe David Merchant
JMe John Merchant
JEM Jim Meyers
BMi Bruce Mielke
SMi Sigrid Miller
SSM Shaibal Mitra
ASM Arnold Moorhouse
CSM Carolyn S. Mostello
LM-W Laila Mostello-Wetherbee
JMu Jim Murphy
MJM Mary Jo Murray
JEM James E. Myers
WM Wayne Munns
CNu Christian Nunes
POn Patty O'Neill
KOp Kay Oppersdorff
JO Jay Osenkowski
MPa Moe Parenteau
HMP Henry M. Parker
JPa Jeffrey Parrish
PP Peter Paton
JP Jean Patric
CPe Carlos Pedro
DPe Dylan Pedro
FPe Frances Perry
MPe Marion Perry
WRP Wayne R. Petersen
APo Arthur Poutry
NPo Nancy Poutry
CaP Candy Powell
ChP Chris Powell

FAP	Fred A. (Pat) Pratt
LP	Lisa Pritchard
GRa	George Rae
CRa	Christopher Raithel
SRi	Stephen Reinert
SRs	Scott Restivo
BRi	Beverly Ridgely
RRi	Robert Ridgely
WCR	William C. Russell
DRi	David Rider
HR	Hope Rider
DRo	David Roach
DRn	David Rodman
SRo	Steven Rodrigues
RRo	Ray Rogers
DRu	Dante Russillo
ESa	Eloise Saunders
ASi	Andrew Silverman
MSl	Merrill Slate
RSm	Robert Smart
KS	Ken Smith
CS	Charlotte Smith
ESo	Everett Southwick
MSo	Mary Southwick
DSp	Daryl Speicher
BSp	Barbara Spicer
RSp	Richard Spicer
RSt	Richard Stackpole
GS	George Sprague
FSt	Frank Sterrett
JSJ	Janice St. Jean
AES	Alan E. Strauss
ESt	Edward Sturtevant
JSu	John Sullivan
RS	Ruth Sunn
LS	Lori Suprock
ATa	Ann Taft
GTa	Gardner Taft
LT	Larry Taft
STa	Sue Talbot
DVT	Donald V. Tiller
TTr	Timothy Traver
STs	Scott Tsagarakis
MT	Mike Tucker
GV	George Vallere
DVa	Diane Vary
RRV	Richard R. Veit
RWa	Robert Wadman
DWa	Dallas Wait
EWe	Elizabeth Weeks
JHW	Julianne H. Whitcomb
CW	Charles Wilde
EW	Elizabeth Wilde
MWh	Maureen Whitman
GWi	Glenn Williams
HWi	Hugh Willoughby
DFW	Doug Wilson
CWo	Charles Wood
REW	Robert E. Woodruff
CGW	Carlos G. Wright
JZe	Joe Zeranski
JZy	Joseph Zbyrowski

References

Able, K.P., and J.R. Belthoff. 1998. Rapid 'evolution' of migratory behaviour in the introduced House Finch of eastern North America. Proceedings Biological Sciences 265:2063–2071.

Acosta, M.A., L.L. Manne, and R.R. Veit. 2021. Does population size predict transatlantic vagrancy in gulls of the North Atlantic? Frontiers in Ecology and Evolution 10:1–13

Adriaens, P., and C. Gibbins. 2016. Identification of the *Larus canus* complex. Dutch Birding 38:1–64.

Aldrich, J.W. 1946. The U.S. races of Bobwhite. Auk 63:493–508.

Allen, G.M. 1901. The European Woodcock (*Scolopax rustica*) in R.I. Notes on Rhode Island Ornithology 2:6.

Allen, G.M. 1909. Fauna of New England II: List of the *Aves*. Boston: Boston Society of Natural History.

Allin, C., G.G. Chasko, and T.P. Husband. 1987. Mute Swans in the Atlantic flyway: A review of the history, population growth and management needs. Transactions Northeastern Section of the Wildlife Society 44:32–47.

Ames, P.L., and G.S. Mersereau. 1964. Some factors in decline of the Osprey in Connecticut. Auk 81:173–184.

Andrews, R. 1990. Coastal waterbird colonies: Maine to Virginia 1984–85, an update of an atlas based on 1977 data, showing colony locations, species and nesting pairs at both time periods. Part 1. Maine to Connecticut. Newton Corner, MA: U.S. Fish and Wildlife Service.

Andrle, R.F., and J.R. Carroll, eds. 1988. The atlas of breeding birds of New York State. Ithaca, NY: Cornell University Press.

AOU (American Ornithologists' Union) Check-list Committee. 1983. Check-list of North American Birds, 6th ed. Lawrence, KS: American Ornithologists' Union.

Arnold, J.M., S.A. Oswald, I.C.T. Nisbet, P. Pyle, and M.A. Patten. 2020. Common Tern (*Sterna hirundo*), version 1.0. In: Birds of the World, ed. S.M. Billerman. Ithaca, NY: Cornell Lab of Ornithology. https://doi.org/10.2173/bow.comter.01.

Arvisais, K., and R.J. Craig. 2018. Forest birds of Connecticut and Rhode Island: Maps of distributions and population densities. Bird Conservation Research 24:1–126.

Ashley, G.H. 1915. Rhode Island coal. Washington, DC: U.S. Geological Survey Bulletin 615.

Aubry, Y. 1984. First nests of the Common Black-headed Gull in North America. American Birds 38:366–367.

Badyaev, A.V., V. Belloni, and G.E. Hill. 2020. House Finch (*Haemorhous mexicanus*), version 1.0. In: Birds of the World, ed. A.F. Poole. Ithaca, NY: Cornell Lab of Ornithology. https://doi.org/10.2173/bow.houfin.01.

Bagg, A.C., and S.A. Eliot. 1937. The birds of the Connecticut River valley. Northampton, MA: Hampshire Bookshop.

Bagg, A.M., and H.M. Parker. 1951. The Turkey Vulture in New England and eastern Canada. Auk 68:315–333.

Baird, J. 1959a. The first breeding record of the Brown Creeper in R.I. Narragansett Naturalist 2:91.

Baird, J. 1959b. Magnificent Frigatebird seen in Middletown. Narragansett Naturalist 2:90.

Baird, J. 1960. Lost at sea: An account of a Prothonotary Warbler in R.I. Narragansett Naturalist 3:36–39.

Baird, J. 1961. A winter record of the Forster's Tern for Rhode Island. Wilson Bulletin 73:89.

Baird, J. 1962. Ash-throated Flycatcher in Rhode Island. Auk 79:272.

Baird, J., A.M. Bagg, I.C.T. Nisbet, and C.S. Robbins. 1959. Operation Recovery: Report on mist-netting along the Atlantic Coast in 1958. Bird-Banding 30:143–171.

Baird, J., A.M. Bagg, C.S. Robbins, and J.V. Dennis. 1958. Operation Recovery: the Atlantic Coast netting project. Bird-Banding 29:137–168.

Baird, J., and I.C.T. Nisbet. 1959. Observations of diurnal migration in the Narragansett Bay area of Rhode Island, in fall 1958. Bird-Banding 30:171–181.

Baird, J., and I.C.T. Nisbet. 1960. Northward fall migration on the Atlantic Coast and its relation to offshore drift. Auk 77:119–149.

Baird, S.F. 1887. Occurrence of Cory's Shearwater (*Puffinus borealis*) and several species of jaegers in large numbers in the vicinity of Gay Head, Massachusetts, during the autumn of 1886. Auk 4:71–72.

Baird, S.F., T.M. Brewer, and R. Ridgway. 1874. A history of North American birds. Vol. 1: Land birds. Boston: Little, Brown.

Baird, S.F., T. Brewer, and R. Ridgeway. 1884. The water birds of North America. Vol. 1. Memoirs of the Museum of Comparative Zoology, Harvard, 12. Boston: Little, Brown.

Baird, S.F., J. Cassin, and G.N. Lawrence. 1860. The birds of North America: The descriptions of species based chiefly on the collections in the Museum of the Smithsonian Institution. Philadelphia: J.B. Lippincott.

Baltz, M.E., and S.C. Latta. 2020. Cape May Warbler (*Setophaga tigrina*), version 1.0. In: Birds of the World, ed. P.G. Rodewald. Ithaca, NY: Cornell Lab of Ornithology. https://doi.org/10.2173/bow.camwar.01.

Banks, R.C., C. Cicero, J.L. Dunn, A.W. Kratter, P.C. Rasmussen, J.V. Remsen, J.D. Rising, and D.F. Stotz. 2003. Forty-fifth supplement to the American Ornithologists' Union Check-list of North American Birds. Auk 121:985–995.

Barrows, W.B. 1889. The English Sparrow (*Passer domesticus*) in North America: Especially in its relations to agriculture. Washington, DC: Department of Economic Ornithology and Mammalogy.

Bechard, M.J., T.R. Swem, J. Orta, F.D. Boesman, E.F.J. Garcia, and J.S. Marks. 2020. Rough-legged Hawk (*Buteo lagopus*), version 1.0. In: Birds of the World, ed. S.M. Billerman. Ithaca, NY: Cornell Lab of Ornithology. https://doi.org/10.2173/bow.rolhaw.01.

Beddall, B.G. 1963. Range expansion of the Cardinal and other birds in the northeastern states. Wilson Bulletin 75:140–156.

Bednarz, J.C., D. Klem Jr., L.J. Goodrich, and S.E. Senner. 1990. Migratory counts of raptors at Hawk Mountain, Pennsylvania, as indicators of population trends. Auk 107:96–109.

Bellrose, F.C. 1976. Ducks, geese and swans of North America. Harrisburg, PA: Stackpole Books.

Bildstein, K.L., J.P. Smith, E.R. Inzuna and R.R. Veit, eds. 2008. State of North America's birds of prey. Series in Ornithology 3. Cambridge, MA: Nuttall Ornithological Club; Washington, DC: American Ornithologists' Union.

Bendire, C. 1895. Life histories of North American birds: With special reference to their breeding habits and eggs. Smithsonian Institution Special Bulletin. Washington, DC: Government Printing Office.

Benkman, C.W. 1989. On the evolution and ecology of island populations of crossbills. Evolution 43:1324–1330.

Benkman, C.W. 1993. The evolution, ecology, and decline of the Red Crossbill of Newfoundland. American Birds 47:225–229.

Benkman, C.W., and M.A. Young. 2020. Red Crossbill (*Loxia curvirostra*), version 1.0. In: Birds of the World, ed. S.M. Billerman, B.K. Keeney, P.G. Rodewald, and T.S. Schulenberg. Ithaca, NY: Cornell Lab of Ornithology. https://doi.org/10.2173/bow.redcro.01.

Bent, A.C. 1919–1958. Life histories of North American Birds. U.S. National Museum Bulletin.

Berry, W.J., S.E. Reinert, M.E. Gallagher, S.M. Lussier, and E. Walsh. 2015. Population status of the Seaside Sparrow in Rhode Island: A 25-year assessment. Northeastern Naturalist 22(4): 658–671.

Beuth, J.M., S.R. McWilliams, P.W. Paton, and J.E. Osenkowski. 2017. Habitat use and movements of Common Eiders wintering in southern New England. Journal of Wildlife Management 81(7): 1276–1286.

Bevier, L.R., and R. Askins, eds. 1994. The atlas of breeding birds of Connecticut. Hartford: Department of Environmental Protection, State Geological and Natural History Survey of Connecticut.

Bierregaard, R.O., Jr., A. Ben David II, T.D. Baird, and R.E. Woodruff. 1975. First northwestern Atlantic breeding record of the Manx Shearwater. Auk 92:145–147.

Blagden, C. 1903. The letters of Sir Charles Blagden to Sir Joseph Banks. Bulletin of the New York Public Library 7(11): 407–446.

Blokpoel, H., and G.D. Tessier. 1986. The Ring-billed Gull in Ontario: A review of a new problem species. Canadian Wildlife Service Occasional Paper 57.

Blomberg, E.J., B.C. Tefft, J.M. Reed, and S.R. McWilliams. 2012. Evaluating spatially explicit viability of a declining Ruffed Grouse population. Journal of Wildlife Management 76(3): 503–513.

Bock, C.E., and L.W. Lepthien. 1972. Winter eruptions of Red-breasted Nuthatches in North America, 1950–1970. American Birds 26:558–561.

Bowen, R. 1967. A flock of Black-headed Gulls in Rhode Island. Records of New England Birds 24:5.

Bowen, R. 1979. Townsend's Solitaire (*Myadestes townsendi*) occurrence in Rhode Island. American Birds 33:142.

Boyle, W.J., Jr. 2011. The birds of New Jersey. Princeton, NJ: Princeton University Press.

Brewer, R., G.A. McPeek, and R.J. Adams. 1991. The atlas of breeding birds of Michigan. East Lansing: Michigan State University Press.

Brewster, W. 1881. Early arrival in New England of the Least Bittern. Bulletin of the Nuttall Ornithological Club 6:186.

Brewster, W. 1890. The Little Brown Crane in Rhode Island. Auk 7:89.

Brewster, W. 1895. A remarkable flight of Pine Grosbeaks (*Pinicola enucleator*). Auk 12:245.

Brewster, W. 1896. Occurrence of the Wood Ibis in Bristol County, Massachusetts. Auk 13:341.

Brewster, W. 1909. The European Widgeon in Rhode Island. Auk 26:186–188.

Brinkley, E.S., P.A. Buckley, L.S. Bevier, and A.M. Byrne. 2011. Photo essay: Redpolls from Nunavut and Greenland visit Ontario. North American Birds 65:206–233.

Bromley, S.W. 1935. The original forest types of southern New England. Ecological Monographs 5:61–89.

Browning, M.R. 2022. Reassessment of the taxonomic status of Thayer's Gull. Dutch Birding 44:137–144.

Browning, M.R., and J.A. Kushlan. 2022. The Great White Heron is a species. Journal of Heron Conservation 7:1–13

Bryant, W.L. 1929. Lewis' Woodpecker in Rhode Island. Auk 46:113–114.

Buckley, P.A. 1968. The status of the Eared Grebe on the eastern Great Lakes and the Atlantic Coast. Audubon Field Notes 22:4.

Buckley, P.A., and F.G. Buckley. 1984. Expanding Double-crested Cormorant and Laughing Gull populations on Long Island. Kingbird 34:146–155.

Buckley, P.A., E.B. Massiah, M.B. Hutt, F.G. Buckley, and H.F. Hutt. 2009. The birds of Barbados. British Ornithologists' Union Checklist Series 24.

Buffum, B., S.R. McWilliams, and P.V. August. 2011. A spatial analysis of forest management and its contribution to maintaining the extent of shrubland habitat in southern New England, United States. Forest Ecology and Management 262(9): 1775–1785.

Bull, J. 1964. Birds of the New York area. New York: Harper and Row.

Bull, J. 1974. Birds of New York State. Ithaca, NY: Cornell University Press.

Bull, J. 1976. Supplement to the Birds of New York State. Ithaca, NY: Special Publication of the Federation of New York State Bird Clubs.

Cabe, P.R. 2020. European Starling (*Sturnus vulgaris*), version 1.0. In: Birds of the World, ed. S.M. Billerman. Ithaca, NY: Cornell Lab of Ornithology. https://doi.org/10.2173/bow.eursta.01.

Caduc, E.E. 1914. Starlings (*Sturnus vulgaris*) at Providence, Rhode Island. Auk 31:249.

Cairns, W.E., and I.A. McLaren. 1980. Status of the Piping Plover on the east coast of North America. American Birds 34:206–208.

Calder, I.M., ed. 1935. Colonial captivities, marches and journeys. New York: Macmillan.

Caron, C.M, and P.W. Paton. 2007. Population trends and habitat use of Harlequin Ducks in Rhode Island. Journal of Field Ornithology 78:254–262.

Carroll, D., and B.L. Swift. 2000. Status of the Trumpeter Swan in New York State. Kingbird 50:232–236.

Chadbourne, A.P. 1889. An unusual flight of Killdeer Plover (*Aegialitis vocifera*) along the New England coast. Auk 6:255.

Chadwick, J. 1982 Wild Turkeys have returned. Audubon Society of Rhode Island Report 16:25, 28–29.

Champlin, R.L. 1962. New nesting record. Narragansett Naturalist 5:81–82.

Chapman, F.M. 1895, 1912, 1932, 1939. Handbook of birds of eastern North America. New York: D. Appleton.

Chapman, F.M. 1908. Camps and cruises of an ornithologist. New York: D. Appleton.

Chesser, R.T., S.M. Billerman, K.J. Burns, C. Cicero, J.L. Dunn, A.W. Kratter, I.J. Lovette, N.A. Mason, P.C. Rasmussen, J.V. Remsen Jr., D.F. Stotz, and K. Winker. 2020. Sixty-first supplement to the American Ornithological Society's Check-list of North American birds. Auk 137:1–24.

Chesser, R.T., K.J. Burns, C. Cicero, J.L. Dunn, A.W. Kratter, I.J. Lovette, P.C. Rasmussen, J.V. Remsen Jr., J.D. Rising, D.F. Stotz, and K. Winker. 2017. Fifty-eighth supplement to the American Ornithologists' Union Check-list of North American birds. Auk 134:751–773.

Chesser, R.T., K.J. Burns, C. Cicero, J.L. Dunn, A.W. Kratter, I.J. Lovette, P.C. Rasmussen, J.V. Remsen, Jr., D.F. Stotz, and K. Winker. 2019. Sixtieth supplement to the American Ornithological Society's Check-list of North American birds. Auk. 136:1–23.

Clark, A.M. 1939. Breeding of Acadian Flycatcher at Jamestown. Bulletin of New England Bird Life 3:9.

Clarke, J.A., B.A. Harrington, T. Hruby, and F.E Wasserman. 1984.The effect of ditching for mosquito control on salt marsh use by birds in Rowley, Massachusetts. Journal of Field Ornithology 55:160–180.

Clement, R.C. 1946. Obituary: Harry S. Hathaway. Auk 63:633–634.

Clement, R.C., and R.E. Woodruff. 1962. A history of the nesting gulls and terns of Rhode Island, 1899–1961. Narragansett Naturalist 5:68–73, 108–111, 125–128.

Confer, J.L., and K. Knapp. 1981. Golden-winged Warblers and Blue-winged Warblers: The relative success of a habitat specialist and generalist. Auk 98:108–114.

Conway, R.A. 1979. Field checklist of Rhode Island Birds. Rhode Island Ornithological Club Bulletin 1.

Conway, R.A. 1992. Field checklist of Rhode Island Birds. 2nd ed. Rhode Island Ornithological Club Bulletin 1.

Cottrell, G.W., Jr. 1949. The southern heron flight of 1948. Bulletin of the Massachusetts Audubon Society 33:101–104, 155–160.

Craig, R.J. 2017. Forest birds of Connecticut and Rhode Island. Pomfret, CT: Arts and Academic Publishing.

Curtis, J.T. 2007. First record of Lesser Sand Plover (*Charadrius mongolus*) from Florida. Florida Field Naturalist 35:26–29.

Custer, T.W., C.M. Bunck and C.J. Stafford. 1985. Organochlorine concentrations in prefledging Common Terns at three Rhode Island colonies. Colonial Waterbirds 8:150–154.

Custer, T.W., and W.E. Davis Jr. 1982. Nesting by one-year-old Black-crowned Night-Herons on Hope Island, Rhode Island. Auk 100:784–786.

Custer, T.W., J.C. Franson, J.F. Moore, and J.E. Myers. 1986. Reproductive success and heavy metal contamination in Rhode Island Common Terns. Environmental Pollution, Series A 41:33–52.

Custer, T.W., G.L. Hensler, and T.E. Kaiser. 1983. Clutch size, reproductive success, and organochlorine contaminants in Atlantic coast Black-crowned Night Herons. Auk 100:699–710.

Davis, D.E. 1960. The spread of the Cattle Egret in the United States. Auk 77:421–424.

Davis, P. 2014. Albatrosses in the western North Atlantic: A compendium of reports. North American Birds 68:4–39.

Davis, T.H. 1972. How often does the Western Tanager occur in the east? Eastern Bird Banding Association News 26: 713–714.

Davis, W.E., Jr. 1997. The Nantucket Oldsquaw flight: New England's greatest bird show? Bird Observer 25:16–22.

Deane, R. 1902. Unusual abundance of the Snowy Owl (*Nyctea nyctea*) in New England and Canada. Auk 19:271–283.

Deane, R. 1905. Another large flight of Snowy Owls. Auk 22:101.

Deane, R. 1906. Unusual abundance of the Snowy Owl (*Nyctea nyctea*). Auk 23:283–298.

Deane, R. 1907a. American Hawk Owl (*Surnia ulula caparoch*) in Rhode Island. Auk 24:219.

Deane, R. 1907b. Great Gray Owl (*Scotiaptex nebulosa*) in Rhode Island. Auk 24:215.

Deane, R. 1907c. The Snowy Owl (*Nyctea nyctea*) not generally abundant in the winter of 1906–1907. Auk 24:217–219.

Deane, R. 1907d. Unusual abundance of the American Goshawk (*Accipiter atricapillus*). Auk 24:182–186.

DeLuca, W.V., B.K. Woodworth, C.C. Rimmer, P.M. Marra, P.D. Taylor, K.P. McFarland, S.A. MacKenzie and R.D. Norris. 2015. Transoceanic migration by a 12 g songbird. Biology Letters 11:20141045.

DeRagon, W.R. 1988. Breeding ecology of Seaside and Sharp-tailed Sparrows in Rhode Island salt marshes. MS thesis, University of Rhode Island, Kingston.

DeRagon, W.R., and F.C. Golet. 1989. Effects of spring tides on nest success of Sharp-tailed Sparrows in Rhode Island salt marshes. American Ornithologists' Union 107th Meeting, Pittsburgh, PA.

Devillers, P. 1977. The skuas of the North American Pacific coast. Auk 94:417–429.

Dexter, S.N. 1887. Golden Eagle in Rhode Island. Forest and Stream 28:106.

Dexter, S.N. 1900. Remarks on the autumn water bird migration at Sakonnet. Notes on Rhode Island Ornithology 1:2–3.

Dickens, E. 1934. Rare birds on Block Island, Rhode Island. Auk 51:247.

Dickinson, E.C., and L. Christidis, eds. 2014. The Howard and Moore complete checklist of birds of the world. Vol. 2: Passerines, 4th ed. Eastbourne, UK: Aves Press.

Dickinson, E.C., and J.V. Remsen, eds. 2013. The Howard and Moore complete checklist of birds of the world. Vol. 1: Non-passerines, 4th ed. Eastbourne, UK: Aves Press.

DiQuinzio, D.A., P.W. Paton, W.R. Eddleman, and J. Brawn. 2001. Site fidelity, philopatry, and survival of promiscuous Saltmarsh Sharp-tailed Sparrows in Rhode Island. Auk 118:888–899.

Dorr, B.S., J.J. Hatch, and D.V. Weseloh. 2020. Double-crested Cormorant (*Phalacrocorax auritus*), version 1.0. In: Birds of the World, ed. A.F. Poole. Ithaca, NY: Cornell Lab of Ornithology. https://doi.org/10.2173/bow.doccor.01.

Drilling, N., R.D. Titman, and J.S. Marks. 2020. Mallard (*Anas platyrhynchos*), version 1.0. In: Birds of the World, ed. S.M. Billerman. Ithaca, NY: Cornell Lab of Ornithology. https://doi.org/10.2173/bow.mallar3.01.

Drury, W.H., Jr. 1973, 1974. Population changes in New England seabirds. Bird-Banding 44:267–313; 45:1–15.

Dunwiddie, P.W. 1990. Postglacial vegetation history of coastal islands in southeastern New England. National Geographic Research 2:178–195.

Eaton, G.F. 1898. The prehistoric fauna of Block Island, as indicated by its ancient shell heaps. American Journal of Science 6:136–159.

Eaton, R.J. 1931. Great Black-Backed Gull (*Larus marinus*) breeding in Essex County, Massachusetts. Auk 48:588–589.

Elliot, J.J., and R.S. Arbib. 1953. Origin and status of the House Finch in the eastern United States. Auk 70:31–37.

Ellis, H.K. 1980. Ecology and breeding biology of the Swamp Sparrow (*Melospiza georgiana*) in a southern Rhode Island peatland. MS thesis, University of Rhode Island, Kingston.

Enser, R.W. 1992. The atlas of breeding birds in Rhode Island (1982–1987). Providence: Rhode Island Department of Environmental Management.

Erskine, A.J. 1988. The changing patterns of Brant migration in eastern North America. Journal of Field Ornithology 59:110–119.

Erwin, R.M., and C.E. Korschgen.1979. Coastal waterbird colonies: Maine to Virginia, 1977. U.S. Fish and Wildlife Service FWS/OBS-79/08.

Eynon, A.E. 1935. Ring-necked Duck (*Nyroca collaris*) at Lonsdale, Rhode Island. Auk 53:181.

Eynon, A.E. 1936. Rough-winged Swallow breeding in Rhode Island. Auk 53:83.

Ferren, R.L., and J.E. Myers. 1990. Cormorants invading heronry on Little Gould Island: A natural disaster in the offing. Audubon Society of Rhode Island. Report 25:3.

Ferren, R.L., and J.E. Myers. 1998. The maritime nesting birds of Rhode Island: A history and survey of the state's beach and island breeders: Cormorants, gulls, terns, herons, plovers, oystercatchers, and more. West Kingston: Rhode Island Division of Fish and Wildlife.

Finnegan, S. 2001. First Yellow-nosed Albatross (*Thalassarche chlororhynchos*) for New Jersey. Records of New Jersey Birds 27:58–62.

Forbush, E.H. 1916. A history of the game birds, wild-fowl and shore-birds of Massachusetts and adjacent states, 2nd ed. Boston: Massachusetts State Board of Agriculture.

Forbush, E.H. 1925, 1927, 1929. Birds of Massachusetts and other New England states, 3 vols. Norwood, MA: Massachusetts Department of Agriculture.

Forster, R. 1994. The great Bohemian Waxwing flight of 1994. Bird Observer 22:137–140.

Fry, A. 1992. Bird walks in Rhode Island. Woodstock, VT: Backcountry Publications.

Gill, F.B. 1980. Historical aspects of hybridization between Blue-winged and Golden-winged Warblers. Auk 97:1–18.

Gillihan, S.W., and B.E. Byers. 2020. Evening Grosbeak (*Coccothraustes vespertinus*), version 1.0. In: Birds of the World, ed. A.F. Poole and F.B. Gill. Ithaca, NY: Cornell Lab of Ornithology. https://doi.org/10.2173/bow.evegro.01.

Giraud, J.P., Jr. 1844. The birds of Long Island. New York: Wiley and Putnam.

Gleason, S.C. 1988. Elizabeth Dickens and the birds of Block Island. Audubon Society of Rhode Island Report 23:2, 7.

Gordon, M. 1955. A western North Atlantic record for the Frigate [White-faced] Petrel (*Pelagodroma marina hypoleuca*). Auk 72:81–82.

Gordon, M.S. 1955. Summer ecology of oceanic birds of southern New England. Auk 72:138–147.

Gould, L.L., R.W. Enser, R.E. Champlin, and I.H. Stuckey. 1998. Vascular flora of Rhode Island: A list of native and naturalized plants. Kingston, RI: Rhode Island Natural History Survey.

Grant, A.M. 1899. Birds. In: 14th Annual Report of the State Board of Agriculture, pp. 50–82. Providence.

Grant, T.A., and R.W. Knapton. 2020. Clay-colored Sparrow (*Spizella pallida*), version 1.0. In: Birds of the World, ed. A.F. Poole. Ithaca, NY: Cornell Lab of Ornithology. https://doi.org/10.2173/bow.clcspa.01.

Greenberg, R., and S. Droege. 1999. On the decline of the Rusty Blackbird and the use of ornithological literature to document population changes. Conservation Biology 13:553–559.

Greenlaw, J.S. 1993. Behavioral and morphological diversification in Sharp-tailed Sparrows (*Ammodramus caudacutus*) of the Atlantic coast. Auk 110:286–303.

Griscom, L. 1914a. The Acadian Chickadee (*Penthestes hudsonicus littoralis*) at Watch Hill, Rhode Island. Auk 31:254.

Griscom, L. 1914b. The Arkansas Kingbird (*Tyrannus verticalis*) in Rhode Island. Auk 31:248.

Griscom, L. 1915. Prairie Horned Lark in Rhode Island in summer. Auk 32:229.

Griscom, L. 1923. Birds of the New York City region. New York: American Museum of Natural History.

Griscom, L. 1939. The Ring-necked Duck as a transient in the northeastern states. Auk 56:134–137.

Griscom, L. 1949. The Birds of Concord. Cambridge, MA: Harvard University Press.

Griscom, L. 1955. The Yellow-billed Cuckoo flight of 1954. Bulletin of the Massachusetts Audubon Society 39:151–156.

Griscom, L., and D.E. Snyder. 1955. Birds of Massachusetts. Salem, MA: Peabody Museum of Salem.

Gross, A.O. 1927. The Snowy Owl migration of 1926–27. Auk 44:479–493.

Gross, A.O. 1956. The recent reappearance of the Dickcissel in eastern North America. Auk 73:66–70.

Groth, J.G. 1993. Evolutionary differentiation in morphology, vocalizations, and allozymes among nomadic sibling species in the North American Red Crossbill (*Loxia curvirostra*) complex. University of California Publications in Zoology 127:1–143.

Hagen, J.M., III. 1993. Decline of the Rufous-sided Towhee in the eastern United States. Auk 110:863–874.

Hamilton, T.R. 1992. House Finch winter range expansion as documented by Christmas Bird Counts, 1950–1990. Indiana Audubon Quarterly 70:147–153.

Harrison, P., M. Perrow, and H. Larsson. 2021. Seabirds: The new identification guide. Cerdanyola del Vallès, Spain: Lynx Edicions.

Harrop, A.H. J., J.M. Collinson, S.P. Dudley, C. Kehoe, and the British Ornithologists Union Records Committee. 2014. The British list: A checklist of the birds of Britain, 8th ed. Ibis 155:635–676.

Hasbrouck, E.M. 1944. The status of Barrow's Golden-eye in the eastern United States. Auk 61:544–554.

Hatch, J.J. 1984. Rapid increase of Double-crested Cormorants nesting in southern New England. American Birds 38:984–988.

Hathaway, H.S. 1900a. The Acadian Sharp-tailed Sparrow in Rhode Island. Notes on Rhode Island Ornithology 1:20–21.

Hathaway, H.S. 1900b. Birds observed at Sakonnet Point and vicinity. Notes on Rhode Island Ornithology 1:15.

Hathaway, H.S. 1900c. The European Ruff (*Pavoncella pugnax*) in Rhode Island. Notes on Rhode Island Ornithology 1:20.

Hathaway, H.S. 1900d. First record of the Blue Goose in Rhode Island, with correction of three erroneous records, and other notes. Notes on Rhode Island Ornithology 1:19–20.

Hathaway, H.S. 1900e. First record of the Greater Redpoll (*Acanthus linaria rostrata*) in R.I. Notes on Rhode Island Ornithology 1:11–13.

Hathaway, H.S. 1900f. A home-loving Osprey. Bird-Lore 2:89.

Hathaway, H.S. 1900g. A new bird for Rhode Island. Notes on Rhode Island Ornithology 1:4.

Hathaway, H.S. 1900h. Odd notes. Notes on Rhode Island Ornithology 1:4–5.

Hathaway, H.S. 1900i. Spring migration at South Auburn. Notes on Rhode Island Ornithology 1:11–13.

Hathaway, H.S. 1900j. Wintering robins and bluebirds. Notes on Rhode Island Ornithology 1:9–10.

Hathaway, H.S. 1901a. Capture of two Bald Eagles (*Haliaeetus leucocephalus* (Linn). Notes on Rhode Island Ornithology 2:24.

Hathaway, H.S. 1901b. Second record of the Henslow's Sparrow for Rhode Island. Notes on Rhode Island Ornithology 2:11.

Hathaway, H.S. 1901c. Unusual abundance of the Coot and other notes. Notes on Rhode Island Ornithology 1:20–21.

Hathaway, H.S. 1901d. Various notes. Notes on Rhode Island Ornithology 2:15–16.

Hathaway, H.S. 1901e. Winter notes. Notes on Rhode Island Ornithology 2:11–12.

Hathaway, H.S. 1902. Birds observed at Point Judith, Rhode Island, November 5, 1901. Notes on Rhode Island Ornithology 3:2–3.

Hathaway, H.S. 1905. The American Osprey. Apteryx 1:81–88.

Hathaway, H.S. 1906. The water-thrush nesting in Rhode Island. Auk 23:463–464.

Hathaway, H.S. 1913. Notes on the occurrence and nesting of certain birds in Rhode Island. Auk 30:545–558.

Hathaway, H.S. 1943. Bird-notes. Unpublished manuscript.

Hathaway, H.S., and L.d. Mearns. 1901. Arrival and departure notes, 1900. Notes on Rhode Island Ornithology 2:8.

Hazard, R.G., II. 1907. The Great Carolina Wren in southern Rhode Island. Auk 24: 446–447.

Hazard, R.G., II. 1908. Breeding of the Carolina Wren in Rhode Island. Auk 25:480.

Heath, J.A., P.C. Frederick, J.A. Kushlan, and K.L. Bildstein. 2020. White Ibis (*Eudocimus albus*), version 1.0. In: Birds of the World, ed. A.F. Poole. Ithaca, NY: Cornell Lab of Ornithology. https://doi.org/10.2173/bow.whiibi.01.

Henny, C.J., F.C. Schmid, E.M. Martin, and L.L. Hood. 1973. Territorial behavior, pesticides, and population ecology of Red-shouldered Hawks in central Maryland, 1943–1971. Ecology 54:545–554.

Herkert, J.R. 2007. Evidence for a recent Henslow's sparrow population increase in Illinois. Journal of Wildlife Management 71:1229-1233.

Hill, N.P. 1945. The 1944 hurricane in New England. Auk 62:410–413.

Hill, N.P. 1965. The Birds of Cape Cod, Massachusetts. New York: William Morrow.

Hill, N.P., and J.M. Hagan III. 1991. Population trends of some northeastern North American landbirds: A half century of data. Wilson Bulletin 103:165–182.

Hoffmann, R. 1904. Guide to the birds of New England and eastern New York. Boston: Houghton Mifflin.

Holt, D.W., J.P. Lortie, B.J. Nikula, and R.C. Humphrey. 1986. First record of Common Black-headed Gulls breeding in the United States. American Birds 40:204–205.

Howe, R.H., Jr. 1899. On the birds' highway. Boston: Small, Maynard.

Howe, R.H., Jr., ed. 1900a. Another Cape May Warbler record. Notes on Rhode Island Ornithology 1:15.

Howe, R.H., Jr. 1900b. Another 1886 R.I. record for *Puffinus borealis*. Notes on Rhode Island Ornithology 1:5.

Howe, R.H., Jr. 1900c. An error. Notes on Rhode Island Ornithology 1:4.

Howe, R.H., Jr., ed. 1900f. A rare capture. Notes on Rhode Island Ornithology 1:22.

Howe, R.H., Jr. 1900g. Review of "The Journal of William Jefferay, Gentleman." Notes on Rhode Island Ornithology 1:1–2.

Howe, R.H., Jr., ed. 1900h. Spring arrival notes. Notes on Rhode Island Ornithology 1:10.

Howe, R.H., Jr. 1901a. Additional bibliography to Howe and Sturtevant's "Birds of Rhode Island." Notes on Rhode Island Ornithology 2:12.

Howe, R.H., Jr. 1901b. A tern colony. Notes on Rhode Island Ornithology 2:5.

Howe, R.H., Jr. 1901c. A trip to Cormorant Rock. Notes on Rhode Island Ornithology 2:7.

Howe, R.H., Jr. 1901d. *Uria troile* in Rhode Island. Notes on Rhode Island Ornithology 2:11.

Howe, R.H., Jr. 1901e. Waterbird shooting about Newport from 1867 to 1874. Notes on Rhode Island Ornithology 2:9–11, 13–15, 20–23.

Howe, R.H., Jr. 1902. The Labrador Savannah Sparrow. Auk 19:85–86.

Howe, R.H., Jr. 1903. A supplement to "The Birds of Rhode Island." Privately printed.

Howe, R.H., Jr. 1904. Killdeers at Allens Harbor, Rhode Island. Auk 21:485.

Howe, R.H., Jr. 1905. Sir Charles Blagden, earliest of Rhode Island ornithologists. American Naturalist 39:397–404.

Howe, R.H., Jr. 1919. Carolina Wren (*Thryothorus l. ludovicianus*) nesting in Rhode Island. Auk 27:79.

Howe, R.H., Jr., and G.M. Allen. 1901. The birds of Massachusetts. Cambridge, MA: privately printed.

Howe, R.H., Jr., and E. Sturtevant. 1899. The birds of Rhode Island. Privately printed.

Howell, S.N.G., I. Lewington, and W.C. Russell. 2014. Rare birds of North America. Princeton, NJ: Princeton University Press.

Hughes, J.M. 2020. Yellow-billed Cuckoo (*Coccyzus americanus*), version 1.0. In: Birds of the World, ed. P.G. Rodewald. Ithaca, NY: Cornell Lab of Ornithology. https://doi.org/10.2173/bow.yebcuc.01.

Hume, E.E. 1942. Ornithologists of the United States Army Medical Corps. Baltimore: John Hopkins Press.

Jencks, F.T. 1880a. Capture of the Carolina Wren and other rare birds in Rhode Island. Bulletin of the Nuttall Ornithological Club 5:237.
Jencks, F.T. 1880b. Capture of the Hudsonian Titmouse in Rhode Island. Bulletin of the Nuttall Ornithological Club 6:54.
Jencks, F.T. 1880c. Least Bittern. Ornithologist and Oologist 5:78.
Jencks, F.T. 1881. Least Bittern. Ornithologist and Oologist 6:1.
Jencks, F.T. 1884a. Another Gerfalcon in Rhode Island. Auk 1:94.
Jencks, F.T. 1884b. Bird notes. Random Notes on Natural History 1:9.
Jencks, F.T. 1884c. Brunnich's Guillemot in Rhode Island. Random Notes on Natural History 1:6.
Jencks, F.T. 1884d. A Short-eared Owl's nest. Random Notes on Natural History 1:8.
Jencks, F.T. 1884e. A trip to a heronry. Ornithologist and Oologist 9:103.
Jencks, F.T. 1885. Prairie Warbler killed in Rhode Island. Random Notes on Natural History 2:15.
Jencks, F.T. 1886a. Common [Great] Cormorants in Rhode Island. Random Notes on Natural History 5:34.
Jencks, F.T. 1886b. The Ipswich Sparrow in Rhode Island. Random Notes on Natural History 3:17–18.
Johnsgard, P.A., and D. Hagemeyer. 1969. The Masked Duck in the United States. Auk 86:691–695.
Kadlec, J.A., and W.H. Drury Jr. 1968. Structure of the New England Herring Gull population. Ecology 49:644–676.
Kamm, M., J. Walsh, J. Galluzo, and W. Petersen. 2013. The Massachusetts breeding bird atlas 2. Lincoln: Massachusetts Audubon Society.
Kenneally, J. 1966. A second record of the Say's Phoebe in Rhode Island. Records of New England Birds 22:2.
Kenney, R.D. 1995. Where, oh where have all the whales gone? Audubon Society of Rhode Island Report 29:1–6.
Kessel, B., D.A. Rocque, and J.S. Barclay. 2020. Greater Scaup (*Aythya marila*), version 1.0. In: Birds of the World, ed. S.M. Billerman. Ithaca, NY: Cornell Lab of Ornithology. https://doi.org/10.2173/bow.gresca.01.
King, L. 1900. Odd notes of interest. Notes on Rhode Island Ornithology 1:7.
King, L. 1901a. Autumn notes from Rhode Island. Notes on Rhode Island Ornithology 2:7.
King, L. 1901b. A trip to Cormorant Rock. Notes on Rhode Island Ornithology 2:6.
King, L. 1902a. Notes from Newport. Notes on Rhode Island Ornithology 3:24.
King, L. 1902b. Notes on the water and shore birds about Middletown, autumn of 1901. Notes on Rhode Island Ornithology 3:4.
King, L. 1902c. Spring notes from Rhode Island on the migration of water birds. Notes on Rhode Island Ornithology 3:14–16.
King, L. 1902d. The tern colony at Gull Rock. Notes on Rhode Island Ornithology 3:16.
King, L. 1902e. A trip to Sakonnet. Notes on Rhode Island Ornithology 2:23.
King, L. 1903. Early record of the Piping Plover in Rhode Island. Auk 20:305.
King, L. 1904. Occurrence of the Ruff and other birds in Rhode Island. Auk 21:85.
Koch, S.L., and P.W. Paton. 2014. Assessing anthropogenic disturbances to develop buffer zones for shorebirds using a stopover site. Journal of Wildlife Management 78(1): 58–67.
Kolbeinsson, Y. 2017. Birding Iceland. https://notendur.hi.is/yannk/status_meralb.html (accessed September 2020).
Kress, S.W., E.H. Weinstein, and I.C.T. Nisbet. 1983. The status of tern populations in northeastern United States and adjacent Canada. Colonial Waterbirds 6:84–106.
Krohn, W.B., R.B. Allen, J.R. Moring, and A.E. Hutchinson. 1995. Double-crested Cor-

morants in New England: Population and management histories. Colonial Waterbirds 18:99–109.

Kus, B., S.L. Hopp, R.R. Johnson, and B.T. Brown. 2020. Bell's Vireo (*Vireo bellii*), version 1.0. In: Birds of the World, ed. A.F. Poole. Ithaca, NY: Cornell Lab of Ornithology. https://doi.org/10.2173/bow.belvir.01.

Kushlan, J.A., and J.W. Prosper. 2009. Little Egret (*Egretta garzetta*) nesting on Antigua: A second nesting site in the Western Hemisphere. Journal of Caribbean Ornithology 22(2): 108–111.

Labiskey, R.F., J.A. Harper, and F. Greeley. 1964. The influence of land use, calcium, and weather on the distribution and abundance of pheasants in Illinois. Illinois Natural History Survey Biological Notes 51.

Lanyon, W.E. 1966. Hybridization in meadowlarks. Bulletin of the American Museum of Natural History 134: article 1.

Lapham, E. 1971. First Rhode Island record of Bewick's Wren. Auk 88:168.

Lawton, C.H. 1887–1888. The water-birds of Newport, R.I. Proceedings Newport Natural History Society, pp. 16–17.

LeBaron, G. 1982. Best hawk migration yet in Rhode Island. Audubon Society of Rhode Island Report 16:18.

Lee, D.S. 1984. Petrels and storm-petrels in North Carolina's offshore waters: Including species previously unrecorded for North America. American Birds 38:151–163.

Lee, D.S. 1986. Seasonal distribution of marine birds in North Carolina waters, 1975–1986. American Birds 40:409–412.

Lee, D.S. 1989. Jaegers and skuas in the western North Atlantic: Some historical misconceptions. American Birds 43:18–20.

Lee, D.S., and J. Booth Jr. 1979. Seasonal distribution of offshore and pelagic birds in North Carolina waters. American Birds 33:715–721.

Lee, D.S., D.B. Wingate, and H.W. Kale II. 1981. Records of tropicbirds in the North Atlantic and Upper Gulf of Mexico, with comments on field identification. American Birds 35:887–890.

Levine, E., ed. 1998. Bull's birds of New York State. Ithaca, NY: Comstock Publishing Associates.

Livermore, S.T. 1877. History of Block Island. Reprinted in 1961. Block Island, RI: Block Island Committee for Republication for the Tercentenary.

Longcore, J.R., D.G. McAuley, G.R. Hepp, and J.M. Rhymer. 2020. American Black Duck (*Anas rubripes*), version 1.0. In: Birds of the World, ed. A.F. Poole and F.B. Gill. Ithaca, NY: Cornell Lab of Ornithology. https://doi.org/10.2173/bow.ambduc.01.

Loring, P.H., P.W. Paton, J.E. Osenkowski, S.G. Gilliland, J.P.L. Savard, and S.R. McWilliams. 2014. Habitat use and selection of Black Scoters in southern New England and siting of offshore wind energy facilities. Journal of Wildlife Management 78(4): 645–656.

Lowther, P.E. 2020. Brown-headed Cowbird (*Molothrus ater*), version 1.0. In: Birds of the World, ed. A.F. Poole and F.B. Gill. Ithaca, NY: Cornell Lab of Ornithology. https://doi.org/10.2173/bow.bnhcow.01.

Ludlum, D. 1976. New England weather book. Boston: Houghton Mifflin.

Lynch, J. J. 1935. European Wigeon at Newport, Rhode Island. Auk 52:78.

Lynch, P.J. 2019. A field guide to Cape Cod. New Haven, CT: Yale University Press.

MacConnell, W.P. 1974. Remote sensing land use and vegetative cover in Rhode Island. Kingston, RI: Cooperative Extension Service, University of Rhode Island.

Mackay, G.H. 1891. The scoters in New England. Auk 8:279.

Mackay, G.H. 1894. Habits of the Double-crested Cormorant in Rhode Island. Auk 11:18.

Mackay, G.H. 1911. Old records for Massachusetts and Rhode Island. Auk 28:119–120.

Mackay, G.H. 1929. Shooting journal (1865–1922). New York: Cosmos Press.
Madison, H.L. 1917. Cardinal in Rhode Island. Bird-Lore 19:2.
Mayfield, H. 1977. Brown-headed Cowbird: Agent of extinction. American Birds 31:107–113.
Maynard, C.J. 1896. The birds of eastern North America with original descriptions [. . .]. Newtonville, MA: C.J. Maynard.
Mayo, T.W., P.W. Paton, and P.V. August. 2015. Responses of birds to humans at a coastal barrier beach: Napatree Point, Rhode Island. Northeastern Naturalist 22(3): 501–512.
McDaniel, J.W. 1973. Vagrant albatrosses in the western North Atlantic and Gulf of Mexico. American Birds 27:563–565.
McKinney, R.A., and P.W. Paton. 2009. Breeding birds associated with seasonal pools in the northeastern United States. Journal of Field Ornithology 80:380–386.
Mearns, E.A. 1901. Winter birds of Lake Worden, Washington County, R.I., November 21 to December 24, 1900. Notes on Rhode Island Ornithology 2:1–5.
Mearns, L.d. 1900a. Birds observed at Chepachet, R.I. Notes on Rhode Island Ornithology 1:21–22.
Mearns, L.d. 1900b. Notes from Newport. Notes on Rhode Island Ornithology 1:22; 3:13–15.
Mearns, L.d. 1900c. Spring arrival and departure notes 1900. Notes on Rhode Island Ornithology 1:18, 22.
Mearns, L.d. 1901. Birds observed on Prudence Island, Narragansett Bay, Rhode Island. Notes on Rhode Island Ornithology 2:18–19.
Mearns, L.d. 1902. A list of birds observed on the island of Rhode Island and adjacent waters. Notes on Rhode Island Ornithology 3:6–14, 17–23.
Mendall, H.L. 1958. The Ring-necked Duck in the Northeast. Orono: University of Maine Press.
Mendall, H.L., and C.M. Aldous. 1943. The ecology and management of the American Woodcock. Orono: Cooperative Wildlife Research Unit, University of Maine.
Merk, F. 1985. History of the westward movement. New York: Alfred A. Knopf.
Merriam, C.H. 1877. A review of the birds of Connecticut. New Haven, CT: Tuttle, Morehouse, and Taylor.
Merrow, J.S. 1990. The influence of area and habitat on the avian community in red maple swamps in southern Rhode Island. MS thesis, University of Rhode Island, Kingston.
Migliaccio, F. 1993. Miss Dickens' collection finds a new home. Audubon Society of Rhode Island Report 28:1.
Miller, N.A. 1999. Landscape and habitat predictors of Canada Warbler (*Wilsonia canadensis*) and Northern Waterthrush (*Seiurus noveboracensis*) occurrence in Rhode Island swamps. MS thesis, University of Rhode Island, Kingston. https://digitalcommons.uri.edu/theses/601
Mitchell, C., and R.D. Hearn. 2004. Pink-footed Goose *Anser brachyrhynchus* (Greenland/Iceland population) in Britain and Ireland 1960/61–1999/2000. Waterbird Review Series. Slimbridge, UK: The Wildfowl and Wetlands Trust/Joint Nature Conservation Committee.
Mitra, S.S. 1996. Sexual dimorphism and plumage elaboration in socially-monogamous birds. PhD dissertation, University of Chicago, Chicago.
Mitra, S.S. 1999. Ecology and behavior of Yellow Warblers breeding in Rhode Island's Great Swamp. Northeastern Naturalist 6(3): 249–262.
Mitra, S.S. 2002. Unexpected centers of winter landbird density during the 101st CBC. Bird Observer 30:20–27.
Mitra, S.S., and C.J. Raithel. 2001. Seasonal trends in species richness and density among landbirds wintering on Block Island, Rhode Island. North American Birds 55:7–12.
Mitra, S.S., D. Wilson, R. Emerson, R. Farrell, R. Ferren, P. L'Etoile, P. Paton, C. Raithel, and S. Tsagarakis. 2021. Second report of the Rhode Island Avian Records Committee. Bird Observer 49:29–53.

Monroe, B.L., Jr., and A. Barron. 1980. The Fork-tailed Flycatcher in North America. American Birds 34:842–845.

Montevecchi, W.A., D.K. Cairns, A.E. Burger, R.D. Elliot, and J. Wells. 1987. The status of Common Black-headed Gull in Newfoundland and Labrador. American Birds 41:197–203.

Morse, D.H. 1978. Populations of Bay-breasted Warblers and Cape May Warblers during an outbreak of the Spruce Budworm. Wilson Bulletin 90:404–413.

Mowbray, T.B., F. Cooke, and B. Ganter. 2020. Snow Goose (*Anser caerulescens*), version 1.0. In: Birds of the World, ed. P.G. Rodewald. Ithaca, NY: Cornell Lab of Ornithology. https://doi.org/10.2173/bow.snogoo.01.

Murphy, R.C. 1911. An unusual occurrence of the Pine Grosbeak in Rhode Island. Auk 28:369–370.

Murphy, R.C., and W. Vogt. 1933. The Dovekie influx of 1932. Auk 50:325–349.

Murphy, W.L. 1982. The Ash-throated Flycatcher in the east: An overview. American Birds 36:241–245.

Murray, B.G., Jr. 1989. A critical review of the transoceanic migration of the Blackpoll Warbler. Auk 106: 8–17.

Myers, J.E. 1976. Hunting and the hunter in Rhode Island. Providence: Rhode Island Division of Fish and Wildlife.

Myers, J.E. 1980. The Mourning Dove in R.I. Providence: Rhode Island Division of Fish and Wildlife.

Myers, J.E. 1982. Glossy Ibis and "Big" Gould. Audubon Society of Rhode Island Report 16:2.

Neidermyer, W.J., and J.J. Hickey. 1977. The Monk Parakeet in the United States, 1970–75. American Birds 31:273–279.

Newton, I. 2008. The migration ecology of birds. London: Academic Press.

Nichols, J.D. 1927. Brown Pelican and summering Bonaparte's Gulls at Newport, R.I. Auk 44:555–556.

Nichols, L.N. 1933. Rhode Island bird records 1781–1804. Compiled from Tom Hazard's Diary. Proceeding Linnaean Society of New York 33–34.

Nichols, L.N. 1934. Rhode Island bird records 1781–1804. Compiled from Tom Hazard's Diary. Proceeding Linnaean Society of New York 34–36.

Nichols, R.L., and A.F. Marston. 1939. Shoreline changes in Rhode Island produced by a hurricane of September 21, 1938. Bulletin of the Geological Society of America 50:1357–1370.

Nisbet, I.C.T. 1971. The Laughing Gull in the Northeast. American Birds 25:677–683.

Nisbet, I.C.T. 1973. Terns in Massachusetts: Present numbers and historical changes. Bird-Banding 44:27–55.

Nisbet, I.C.T. 2002. Common Tern (*Sterna hirundo*), version 2.0. In: Birds of the World, ed. A.F. Poole and F.B. Gill. Ithaca, NY: Cornell Lab of Ornithology. https://doi.org/10.2173/bna.618.

Nisbet, I.C.T., and W.H. Drury. 1967. Orientation of spring migrants studied by radar. Bird-Banding 38:173–186.

Nisbet, I.C.T., R.R. Veit, S.A. Auer, and T.P. White. 2013. Marine birds of the eastern United States and the Bay of Fundy. Nuttall Ornithological Monographs 29.

Nuttall, T. 1832. A manual of the ornithology of the United States and Canada. Cambridge, MA: Hilliard and Brown.

Olsen, S., D.D. Robadue Jr., and V. Lee. 1980. An interpretive atlas of Narragansett Bay. Narragansett: Coastal Resources Center, University of Rhode Island, Marine Bulletin 40.

O'Neill, P.A. 1995. Black-tailed Gull sighting. Bird Observer 23:284–287.

Orwig, D.A., and D.R. Foster. 1998. Forest response to the introduced Hemlock Woolly Adelgid in southern New England, USA. Journal of the Torrey Botanical Society 125:60–73.

Osenkowski, J.E., P.W. Paton, and D. Kraus. 2012. Using long-term constant-effort banding data to monitor population trends of migratory birds: A 33-year assessment of adjacent coastal stations. Condor 114(3): 470–481.

Oswald, J.A., M.G. Harvey, R.C. Remsen, D.U. Foxworth, S.W. Cardiff, D.L. Dittman, L.C. Megna, M.D. Carling, and R.D. Brumfield. 2016. Willet be one species or two? A genomic view of the evolutionary history of *Tringa semipalmatus*. Auk 133:593–614.

Palmer, R.S. 1949. Maine birds. Bulletin of the Museum of Comparative Zoology, Harvard, 102.

Parchman, T.L., C.W. Benkman, and S.C. Britch. 2006. Patterns of genetic variation in the adaptive radiation of New World crossbills (*Aves: Loxia*). Molecular Ecology 15(7): 1873–1887.

Parker, J.W., and J.C. Ogden. 1979. The recent status of the Mississippi Kite. American Birds 33:119–129.

Parrish, J.D. 1997. Patterns of frugivory and energetic condition in Nearctic landbirds during autumn migration. Condor 99(3): 681–697.

Parrish, J.D. 2000. Behavioral, energetic, and conservation implications of foraging plasticity during migration. Studies in Avian Biology 20:53–70.

Paton, P. 1998. Study shows Saw-whet Owls common fall migrants in R.I. Audubon Society of Rhode Island Report 32:3.

Paton, P., K. Winiarski, C. Trocki, and S. McWilliams. 2010. Spatial distribution, abundance and flight ecology of birds in nearshore and offshore waters of Rhode Island. Interim technical report for the Rhode Island Ocean Special Area Management Plan 2010. Kingston: Department of Natural Resources Sciences, University of Rhode Island, Technical Report 11.

Pearce-Higgins, J.W., and R.E. Green. 2014. Birds and climate change: Impacts and conservation responses. Cambridge: Cambridge University Press.

Peters, J.I. 1916. Additional autumn records for the Tennessee Warbler (*Vermivora peregrina*) in Massachusetts. Auk 33:78.

Petersen, W.R., and W.R. Meservey, eds. 2003. Massachusetts breeding bird atlas. Lincoln: Massachusetts Audubon Society.

Phillips, A.R. 1975. Semipalmated Sandpiper: Identification, migrations, summer and winter ranges. American Birds 29:799–806.

Phillips, J.C. 1915. The old New England Bobwhite. Auk 32:204–207.

Phillips, J.C. 1929. Shooting journal of George H. Mackay. Cambridge, MA: privately printed.

Poole, A., and P. Spitzer. 1983. An Osprey revival. Oceanus 26:49–54.

Poor, H.H. 1946. The chickadee flight of 1941–42. Proceeding Linnaean Society of New York 54–57:16–27.

Post, P.W. 1979. An irruption of Tufted Titmice in the Northeast. American Birds 33:249–250.

Powel, Col. J.H. 1883–1884. A list of birds shot near Newport. Proceedings Newport Natural History Society, pp. 42–43.

Powers, K.D. 1983. Pelagic distributions of marine birds off the northeastern United States. Woods Hole, MA: National Oceanic and Atmospheric Administration Technical Memorandum NMFS-F/NEC-27. Powers, K.D., and J. Cherry. 1983. Loon migrations off the coast of the northeastern United States. Wilson Bulletin 95:125–132.

Purdie, H.A. 1877. Notice of a few birds of rare or accidental occurrence in New England. Bulletin of the Nuttall Ornithological Club 2:20–22.

Purdum, C.C. 1902. The catbird again in Rhode Island in winter. Auk 19:292–293.

Pyle, P. 1997. Identification guide to North American birds. Part 1: Columbidae to Ploceidae. Bolinas, CA: Slate Creek Press.

Quinn, A.W. 1976. Rhode Island geology for the non-geologist. Kingston: Rhode Island Natural History Survey.

Read, H.B. 1887. Our local raptors, Bristol County, Mass. Ornithologist and Oologist 12:117–120.

Reed, C. 1910. Bird guide. 2 vols. New York: Doubleday, Page.

Reinert, S.E. 1979. Breeding ecology of the Swamp Sparrow (*Melospiza georgiana*) in a southern Rhode Island peatland. MS thesis, University of Rhode Island, Kingston.

Reinert, S.E., F.C. Golet, and W.R. Deragon. 1981. Avian use of ditched and unditched salt marshes in southeastern New England. Proceedings Northeastern Mosquito Control Association 27:1–23.

Remington, C.H., and the Commissioners of Birds. 1908. A checklist of Rhode Island nesting birds. n.p.: Commissioners of Birds, C. H. Remington, chairman.

Rising, J. D., and J. Avise. 1993. An application of genealogical concordance principles to the taxonomy and evolutionary history of the Sharp-tailed Sparrow (*Ammodramus caudacutus*). Auk 110:844–856.

Rives, W.C. 1883–1884. The birds of Newport. Proceedings Newport Natural History Society, pp. 28–41.

Rives, W.C. 1887a. Cory's Shearwater at Newport, R.I. Auk 5:108.

Rives, W.C. 1887b. Wilson's Phalarope in Rhode Island. Auk 4:73.

Robbins, C.S., D. Bystrak, and P.H. Geissler. 1986. The Breeding Bird Survey: Its first fifteen years, 1965–1979. Washington, DC: U.S. Fish and Wildlife Service Resource Publication 157.

Roberts, P.H. 1996. Major Broad-winged Hawk flights reported by the Eastern Massachusetts Hawk Watch. Bird Observer 24:211–212.

Romagosa, C.M. 2020. Eurasian Collared-Dove (*Streptopelia decaocto*), version 1.0. In: Birds of the World, S.M. Billerman. Ithaca, NY: Cornell Lab of Ornithology. https://doi.org/10.2173/bow.eucdov.01.

Root, T. 1988. Atlas of wintering North American Birds. Chicago: University of Chicago Press.

Rosenberg, K.V., T. Will, D.A. Buehler, S.B. Swarthout, W.E. Thogmartin, R.E. Bennett, and R.B. Chandler. 2016. Dynamic distributions and population declines of Golden-winged Warblers. In: Golden-winged Warbler ecology, conservation, and habitat management, ed. H.M. Streby, D.E. Andersen, and D.A. Buehler, pp. 3–28. Studies in Avian Biology 49. Boca Raton, FL: CRC Press.

Sauer, J.R., and S. Droege. 1990. Recent population trends of the Eastern Bluebird. Wilson Bulletin 102:239–251.

Schiff, S., and A. Wollin. 1999. Region 10: Marine. Kingbird 49:89–95.

Schorger, A.W. 1955. The Passenger Pigeon. Madison: University of Wisconsin Press.

Sealy, S.C., and H.R. Carter. 2012. Body condition and survival of vagrant Long-billed Murrelets, *Brachyramphus perdix*, in North America. Canadian Field-Naturalist 126:6–14.

Shields, M. 2020. Brown Pelican (*Pelecanus occidentalis*), version 1.0. In: Birds of the World, ed. A.F. Poole. Ithaca, NY: Cornell Lab of Ornithology. https://doi.org/10.2173/bow.brnpel.01.

Sibley, D. 1993. The birds of Cape May. Franklin Lakes, NJ: New Jersey Audubon Society.

Sirkin, L. 1995a. Block Island geology. Watch Hill, RI: Book and Tackle Shop.

Sirkin, L. 1995b. On the geology of Great Salt Pond [Block Island]. Audubon Society of Rhode Island Report 30:1.

Slack, R.S., and C.B. Slack. 1981. Fall migration of Peregrine Falcons along the Rhode Island coast. Journal of Field Ornithology 52:61–62.

Smith, P.W. 1985. Jackdaws reach the New World. American Birds 39:255–258.

Snyder, D. 1961. First record of the Lesser Frigatebird (*Fregata ariel*) in North America. Auk. 78:265.

Southwick, J.M. 1887–1888. Our birds of Rhode Island. Proceedings Newport Natural History Society, pp. 3–15.
Southwick, J.M. 1889. Capture of the Little Brown Crane. Ornithologist and Oologist 14:159.
Southwick, J.M. 1902. The Leach's Petrel taken at Providence. Notes on Rhode Island Ornithology 3:3–4.
Southwick, J.M., and J.M. Critchley. 1891. Early arrival of Scarlet Tanager at Apponaug, R.I. Ornithologist and Oologist 16:78.
Southwick, J.M., and F.T. Jencks. 1883. Capture of Sparrow Hawk and Snowy Owls. Ornithologist and Oologist 8:24.
Southwick, J.M., and F.T. Jencks. 1884a. Common Cormorant in Rhode Island. Random Notes on Natural History 2:34.
Southwick, J.M., and F.T. Jencks. 1884b. Curious death of a Saw Whet Owl. Random Notes on Natural History 1:4.
Southwick, J.M., and F.T. Jencks. 1884c. Early woodcock. Random Notes on Natural History 1:8.
Southwick, J.M., and F.T. Jencks. 1884d. An English Corn Crake in Rhode Island. Random Notes on Natural History 1:3.
Southwick, J.M., and F.T. Jencks. 1884e. Great Gray Owl, spectral owl. Random Notes on Natural History 1:3.
Southwick, J.M., and F.T. Jencks. 1884f. Late crossbills. Random Notes on Natural History 1:6.
Southwick, J.M., and F.T. Jencks. 1884g. Night Heron in winter. Random Notes on Natural History 1:9.
Southwick, J.M., and F.T. Jencks. 1884h. A Nonpareil in Rhode Island. Random Notes on Natural History 1:8.
Southwick, J.M., and F.T. Jencks. 1884i. Note on Connecticut Warbler, Scarlet Tanager, and Maryland Yellowthroat. Random Notes on Natural History 1:6.
Southwick, J.M., and F.T. Jencks. 1884j. Prothonotary Warbler. Random Notes on Natural History 1:3.
Southwick, J.M., and F.T. Jencks. 1884k. [Two Brunnich's Murres and Night Herons, winter record]. Random Notes on Natural History 1:8.
Southwick, J.M., and F.T. Jencks. 1884l. A very rare bird in Rhode Island. Random Notes on Natural History 1:6.
Southwick, J.M., and F.T. Jencks. 1885a. The Ipswich Sparrow in Rhode Island. Random Notes on Natural History 2:17.
Southwick, J.M., and F.T. Jencks. 1885b. A King Eider. Random Notes on Natural History 1:9.
Southwick, J.M., and F.T. Jencks. 1885c. Snowy Owl. Ornithologist and Oologist 10:48.
Southwick, J.M., and F.T. Jencks. 1885d. Snowy Owl at Newport. Random Notes on Natural History 2:23.
Southwick, J.M., and F.T. Jencks. 1886a. Capture of Barn Owl and Razor-billed Auk. Random Notes on Natural History 3:91.
Southwick, J.M., and F.T. Jencks. 1886b. Capture of Yellow-crowned Night Heron at Tiverton. Random Notes on Natural History 3:49.
Southwick, J.M., and F.T. Jencks. 1886c. Note on a scarcity of Snowy Owls. Random Notes on Natural History 3:9.
Southwick, J.M., and F.T. Jencks. 1886d. Purple Gallinule. Random Notes on Natural History 3:79.
Southwick, J.M., and F.T. Jencks. 1886e. Tennessee Warbler in Rhode Island. Random Notes on Natural History 3:79.

Spitzer, P., A. Poole, and M. Scheibel. 1983. Initial population recovery of breeding Ospreys in the region between N.Y. City and Boston. In: Proceedings of the 1st International Osprey/Bald Eagle Conference, ed. D. Bird, pp. 231–241. Montreal: McGill University Press.

Spitzer, P., R.W. Risebrough, W. Walker II, R. Herandez, A. Poole, and D. Puleston. 1978. Productivity of Ospreys in Connecticut–Long Island increases as DDE residues decline. Science 202:333–335.

Stein, R.C. 1963. Isolating mechanisms between populations of Traill's Flycatchers. Proceedings American Philosophical Society. 107:21–50.

Stiles, E.W. 1984. Expansions of mockingbird and multiflora rose in the northeastern United States and Canada. American Birds 36:358–364.

Stoll, M.J., and F.C. Golet. 1983. Status of the Seaside Sparrow in Rhode Island. Audubon Society of Rhode Island Report 17:57–61.

Sturtevant, E. 1897. The Seaside Sparrow at Middletown, R.I. Auk 14:322.

Sturtevant, E. 1899. The Carolina Wren breeding in Rhode Island. Auk 17:284.

Suprock, L. 1997. Another successful Osprey season. Briefings [publication by Rhode Island Division of Fish and Wildlife], January 1997.

Taber, W. 1951. The northern element in the summer bird life of south-central New England. Wilson Bulletin 63:69–74.

Talbot, H.A. 1884a. Brunnich's Guillemot in Rhode Island. Random Notes on Natural History 1:6.

Talbot, H.A. 1884b. Notes from Warwick Neck. Ornithologist and Oologist 9:58.

Taylor, A.O. 1924. Snowy Egret (*Egretta candidissima*) at Newport, Rhode Island. Auk 41: 473.

Taylor, A.O. 1934. Yellow-crowned Night-Heron in Rhode Island. Auk 51:369.

Telfair, R.C., II. 2020. Cattle Egret (*Bubulcus ibis*), version 1.0. In: Birds of the World, ed. S.M. Billerman. Ithaca, NY: Cornell Lab of Ornithology. https://doi.org/10.2173/bow.categr.01.

Terborgh, J. 1989. Where have all the birds gone? Princeton, NJ: Princeton University Press.

Thomson, B.F. 1958. The changing face of New England. New York: Macmillan.

Tierney, N., R.A. Tierney, Þ. Björnsson, Y. Kolbeinsson, S.E. Townsend, and A. Petersen. 2020. Status and distribution of the Common Shelduck *Tadorna tadorna* in Iceland, with special reference to the core area in Borgarfjörður. Wildfowl 70:127–147.

Townsend, C.W. 1909. A Carolina Wren invasion of New England. Auk 26:263–269.

Trocki, C.L., and P.W. Paton. 2006. Comparison of two foraging habitats used by Glossy Ibis during the breeding season in Rhode Island. Northeastern Naturalist 13(1): 93–102.

Underwood, L.A., and T.J. Stowe. 1984. Massive wreck of seabirds in eastern Britain, 1983. Bird Study 31:79–88.

USGS (U.S. Geological Survey) Eastern Ecological Science Center. n.d. North American Breeding Bird Survey Results and Analysis 1966–2015. https://www.mbr-pwrc.usgs.gov/bbs/specl15.html (accessed 20 September 2019).

Van Tyne, J. 1926. An unusual flight of Arctic Three-toed Woodpeckers. Auk 43:469–474.

Veit, R.R. 1978. Some observations of South Polar Skuas (*Catharacta maccormicki*) on Georges Bank. American Birds 32:300–302.

Veit, R.R. 1985. Long-tailed Jaegers wintering along the Falkland Current. American Birds 39:873–877.

Veit, R.R. 1990. Do vagrant birds in Massachusetts reflect population growth and dispersal rather than weather patterns? Bird Observer 18:86–91.

Veit, R.R. 1997. Long-distance dispersal and population growth of Yellow-headed Blackbird (*Xanthocephalus xanthocephalus*). Ardea 85:135–143.

Veit, R.R. 2000. Vagrants as the expanding fringe of a growing population. Auk 117:242–246.

Veit, R.R. 2005. Yellow-nosed Albatross at Tuckernuck Island, Massachusetts. Bird Observer 33:284–288.

Veit, R.R. 2008. Why expect clusters of vagrants? Bird Observer 36:213–217.

Veit, R.R., and P.A. Guris. 2009. Recent increases in alcid abundance in the New York Bight and New England waters. New Jersey Birds 34:83–87.

Veit, R.R., and M.A. Lewis. 1996. Dispersal, population growth, and the Allee effect: Dynamics of the House Finch invasion of eastern North America. American Naturalist 148:255–274.

Veit, R.R., and L.L. Manne. 2015. Climate and changing winter distribution of alcids in the northwest Atlantic. Frontiers in Ecology and Evolution 3:38–47.

Veit, R.R., and W.R. Petersen. 1993. Birds of Massachusetts. Lincoln: Massachusetts Audubon Society.

Veit, R.R., E. Velarde, M. Horn, and L.L. Manne. 2021. Long distance vagrancy leads to colonization in Europe by Elegant Terns *Thalasseus elegans*. Frontiers in Ecology and Evolution 9:824.

Venier, L., S. Holmes, and J.M. Williams. 2020. Bay-breasted Warbler (*Setophaga castanea*), version 1.0. In: Birds of the World, ed. A.F. Poole. Ithaca, NY: Cornell Lab of Ornithology. https://doi.org/10.2173/bow.babwar.01.

Vennesland, R.G., and R.W. Butler. 2020. Great Blue Heron (*Ardea herodias*), version 1.0. In: Birds of the World, ed. A.F. Poole. Ithaca, NY: Cornell Lab of Ornithology. https://doi.org/10.2173/bow.grbher3.01.

Vickery, P.D., C.D. Duncan, W.J. Sheehan, and J.V. Wells. 2020. Birds of Maine. Princeton, NJ: Princeton University Press/Nuttall Ornithological Club.

Vickery, P.D., and R.P. Yunick. 1979. The 1978–1979 Great Gray Owl incursion across northeastern North America. American Birds 33:242–244.

Vierling, K.T., V.A. Saab, and B.W. Tobalske. 2020. Lewis's Woodpecker (*Melanerpes lewis*), version 1.0. In: Birds of the World, ed. A.F. Poole. Ithaca, NY: Cornell Lab of Ornithology. https://doi.org/10.2173/bow.lewwoo.01.

Vinicombe, K.E., and A.H.J. Harrop. 1999. Ruddy Shelducks in Britain and Ireland 1986–94. British Birds 92:225–255.

Walkinshaw, L.H., and D.A. Zimmerman. 1961. Range expansion of Brewer Blackbird into eastern North America. Condor 63:162–177.

Wallace, R.D., and R. Wigh. 2007. Pelagic birds of the southern South Atlantic Bight. North American Birds 61:198–207.

Walsh, K.J.E., J.L. McBride, P.J. Klotzbach, S. Balachandran, S.J. Camargo, G. Holland, T.R. Knutson et al. 2016. Tropical cyclones and climate change. Wiley Interdisciplinary Reviews: Climate Change 7:65–89.

Walton, R.K. 1984. Birds of the Sudbury River Valley: An historical perspective. Lincoln: Massachusetts Audubon Society.

Watson, G.E., D.S. Lee, and E.S. Backus. 1986. Status and subspecific identity of White-faced Storm-Petrels in the western North Atlantic Ocean. American Birds 40:401–408.

Weaver, R.J. 1961. A European Ruff in Newport. Narragansett Naturalist 4:30.

Weaver, R.L. 1940. The Purple Finch invasion of northeastern United States and the Maritime Provinces in 1939. Bird-Banding 11:79–105.

Weber, K. 1997. A century of dedication: The first hundred years of the Audubon Society of Rhode Island. n.p.: Audubon Society of Rhode Island.

Weir, D.N., A.C. Kitchener, and R.Y. McGowan. 2000. Hybridization and changes in the distribution of Iceland Gulls *Larus glaucoides/kumlieni/thayeri*. Journal of Zoology London 252:517–530.

West, D.J., and J.M. Speirs. 1959. The 1956–57 invasion of Three-toed Woodpeckers. Wilson Bulletin 71:348–363.

White, T.P., R.R. Veit, and M.C. Perry. 2009. Feeding ecology of Long-tailed Ducks *Clangula hyemalis* wintering on the Nantucket Shoals. Waterbirds 32:293–299.

Whitman, H.S. 1982. Elizabeth Dickens: The bird lady of Block Island. Cornwall, CT: Still Pond Press.

Williams, T.C., and J.M. Williams. 1979. An oceanic mass migration of land birds. Scientific American Birds 239:166–176.

Williams, T.C., J.M. Williams, L.C. Ireland, and J.M. Teal. 1977. Autumnal bird migration over the western North Atlantic Ocean. American Birds 31:251–267.

Williams, T.C., J.M. Williams, L.C. Ireland, and J.M. Teal. 1978. Estimated flight time of transatlantic autumnal migrants. American Birds 32:275–280.

Wilson, W.H. 1999. The northward expansion of the Tufted Titmouse in Maine. Northeastern Naturalist 6:231–237.

Winiarski, K.J., M.L. Burt, E. Rexstad, D.L. Miller, C.L. Trocki, P.W.C. Paton, and S.R. McWilliams. 2014. Integrating aerial and ship surveys of marine birds into a combined density surface model: A case study of wintering Common Loons. Condor 116:149–161.

Winiarski, K.J., D.L. Miller, P.W. Paton, and S.R. McWilliams. 2013. Spatially explicit model of wintering Common Loons: Conservation implications. Marine Ecology Progress Series 492:273–283.

Wormington, A., and J.H. Leach. 1992. Concentrations of migrant diving ducks at Point Pelee National Park, Ontario, in response to invasion of zebra mussels, *Dreissena polymorpha*. Canadian Field-Naturalist 106(3): 376–380.

Wright, A.H. 1910. Some early records of the Passenger Pigeon. Auk 27:428–443.

Wright, H.W. 1914. Acadian Chickadees in Boston and vicinity in the fall of 1913. Auk 31:236–242.

Wright, H.W. 1917. The Orange-crowned Warbler as a fall and winter visitant in the region of Boston, Massachusetts. Auk 34:11–27.

Wright, H.W. 1918. The Labrador Chickadee (*Penthestes hudsonicus nigricans*) in its return flight from the fall migration of 1916. Auk 35:37–40.

Wright, H.W. 1921a. Bohemian Waxwing (*Bombycilla garrula*) in New England. Auk 38:59.

Wright, H.W. 1921b. The mockingbird in the Boston region and in New England and Canada. Auk 38:417.

Wright, T.A. 1941. Study of the fall food supply of the Ring-necked Pheasant and Bobwhite Quail in Washington County, R.I. Journal of Wildlife Management 5:279–296.

Zawadzki, L.C., G.T. Halgrimsson, R.R. Veit, N. Gillies, and T. Guilford. 2021. Vagrancy and colonization of Lesser Black-backed Gulls (*Larus fuscus*) in North America is linked with population trends in Greenland. Frontiers in Ecology and Evolution 9: 879.

Zawadzki, L.C., R.R. Veit, and L.L. Manne. 2019. The influence of population growth and wind on vagrancy in a North American passerine. Ardea 107(2): 131–147.

Zeranski, J.D., and T.R. Baptist. 1990. Connecticut birds. Hanover, NH: University Press of New England.

Index of Common and Scientific Names

Note: Page numbers in **boldface** refer to main species accounts.

Subject Index